T0237621

Fiber Optic Communications

Gerd Keiser

Fiber Optic Communications

 Springer

Gerd Keiser
Newton Center, MA, USA

ISBN 978-981-33-4667-3   ISBN 978-981-33-4665-9   (eBook)
https://doi.org/10.1007/978-981-33-4665-9

This Springer imprint is published by the registered company Springer Nature Singapore Pte Ltd.
The registered company address is: 152 Beach Road, #21-01/04 Gateway East, Singapore 189721,
Singapore

To

Ching-Yun, Nishla, Keith, Kai, and Neyla
for their loving patience and encouragement

Preface

The discipline of optical fiber communications has undergone a fascinating journey in the past several decades and is still growing rapidly worldwide. Especially exciting during this time was the 2009 Nobel Prize in Physics received by Sir Charles Kuen Kao for his pioneering insight, in the 1960s, into using glass fibers as a data transmission medium and for his enthusiastic international promotions in the further development of low-loss fibers. As a result of the promotions by Kao, the first ultrapure fiber was fabricated in 1970, only four years after his prediction. Modern sophisticated telecom networks based on optical fiber technology now have become an integral and indispensable part of society. Applications include services such as database queries, home shopping, interactive video, remote education, telemedicine and e-health, high-resolution editing of home videos, blogging, and large-scale high-capacity e-science and grid computing. Due to the importance of these networks to modern life, the communication services are expected to always be available and to function properly. Such stringent demands require careful engineering in all technological aspects ranging from component development through system design and installation to network operation and maintenance.

To master the skills needed to work in the optical fiber communications discipline, this book presents the fundamental principles for understanding and applying a wide range of optical fiber technologies to modern communication networks. The sequence of topics takes the reader systematically from the underlying principles of photonic components and optical fibers through descriptions of the architectures and performance characteristics of complex optical networks to essential measurement and test procedures required during network installation, operation, and maintenance. By mastering these fundamental topics, the reader will be prepared not only to contribute to disciplines such as current device, communication link, or equipment designs, but also to understand quickly any further technology developments for future enhanced networks.

To accomplish these objectives, the sequence of chapters addresses the following topics:

- Chapter 1 gives the motivations and advantages for using optical fibers, the spectral bands of interest, methods used to boost the transmission capacity of an optical fiber, and what standards are being applied.
- Despite its apparent simplicity, an optical fiber is one of the most important elements in a fiber link. Chapter 2 gives details on the physical structures, constituent materials, and lightwave propagation mechanisms of optical fibers.
- Chapter 3 gives details on the attenuation behavior and signal dispersion characteristics of the wide variety of common optical fibers. In addition, the chapter discusses international standards for manufacturing optical fibers.
- Chapter 4 addresses the structures, light-emitting principles, operating characteristics of light sources, and optical signal modulation techniques.
- How to couple the light source to a fiber is described in Chap. 5, as well as how to join two fibers in order to ensure a low optical power loss at the joints.
- Chapter 6 covers the structures and performances of photodetectors. Because an optical signal generally is weakened and distorted at the end of link, the photodetector must possess a high sensitivity, have a fast response time, and add minimum noise effects to the system. In addition, its size must be compatible with that of the fiber output.
- The lightwave receiver detects an arriving optical signal and converts it into an electrical signal for information processing. Chapter 7 describes receiver principles and functions, which include signal detection statistics and eye diagram error measurements schemes.
- Chapter 8 discusses digital link design methods including power budgets and bandwidth limitations. In addition, the topics include power penalties, basic coherent detection schemes, and details of error control methods for digital signals.
- Chapter 9 examines the concepts of analog links for sending radio frequency (RF) signals at microwave frequencies over optical fibers. An expanding application of these techniques is for broadband radio-over-fiber networks.
- Chapter 10 addresses wavelength division multiplexing (WDM), examines the functions of a generic WDM link, and discusses international standards for different WDM schemes. It also includes descriptions of passive WDM devices, such as fiber Bragg gratings, thin-film filters, and various types of gratings.
- Chapter 11 describes the concepts for creating optical amplification and the applications of these devices. Among the topics are semiconductor optical amplifiers, doped fiber amplifiers, and Raman amplification schemes.
- Chapter 12 is devoted to the origins and effects of nonlinear processes in optical fibers. Some of these nonlinear effects degrade system performance and need to be controlled, whereas others can have beneficial uses.
- Chapter 13 is devoted to optical networking concepts for long distance, metro, data center, and access networks. Among the topics are optical add/drop multiplexing and optical cross-connects, wavelength routing, optical packet switching, optical burst switching, and passive optical networks.

- Chapter 14 discusses performance measurement and monitoring. The topics include measurement standards, test instruments for fiber link characterization, evaluation of link performance through eye pattern measurements, error monitoring, network maintenance, and fault management.

Use of This Book

This book provides the basic material for a senior-level or postgraduate course in the theory and application of optical fiber communication technology. It also will serve well as a working reference for practicing engineers dealing with the design and development of components, transmission equipment, test instruments, and cable plants for optical fiber communication systems. The background required to study the book is that of typical senior-level engineering students. This includes introductory electromagnetic theory, calculus and elementary differential equations, and basic concepts of optics as presented in a basic physics course. Concise reviews of several background topics, such as optics concepts, electromagnetic theory, and basic semiconductor physics, are included in the main body of the text.

To assist readers in learning the material and applying it to practical designs, 147 examples and 75 drill problems are given throughout the book. A collection of 187 homework problems is included to help test the reader's comprehension of the material covered and to extend and elucidate the text.

Numerous references are provided at the end of each chapter as a start for delving deeper into any given topic. Because optical fiber communications bring together research and development efforts from many different scientific and engineering disciplines, there are hundreds of articles in the literature relating to the material covered in each chapter. Even though not all these articles can be cited in the references, the selections represent some of the major contributions to the fiber optics field and can be considered as a good introduction to the literature. Supplementary material and references for up-to-date developments can be found in specialized textbooks and various conference proceedings.

To help the reader understand and use the material in the book, Appendix A gives an overview of the international system of units and a list of various standard physical constants. Appendix B presents a brief review of the concept of decibels. Appendices C through E provide listings of acronyms, Roman symbols, and Greek symbols, respectively, that are used in the book.

Newton Center, USA   Gerd Keiser

The original version of the book was revised: Incorrect sentences in Chaps. 1, 3, 5, 8, 10 and 13 has been corrected. The correction to this book is available at https://doi.org/10.1007/978-981-33-4665-9_15.

Acknowledgements

For preparing this textbook on *Fiber Optic Communications*, I am extremely grateful to the numerous people worldwide with whom I had countless discussions over the years and who helped me in many different ways. The list of these people is too long to include here, but I truly appreciate their help. In addition, I would like to thank Loyola D'Silva of Springer for his assistance and expert guidance, together with the other editorial and production team members of Springer, to produce the book. As a final personal note, I am grateful to my wife Ching-Yun and my family members Nishla, Keith, Kai, and Neyla for their patience and encouragement during the time I devoted to writing this book.

Contents

About the Author

Dr. Gerd Keiser is a research professor at Boston University and a consultant for the telecom and biophotonics industries at PhotonicsComm Solutions. Previously, he was involved with developing and implementing telecom technologies at Honeywell, GTE, and General Dynamics. His technical achievements at GTE earned him the prestigious Leslie Warner Award. In addition, he has served as an adjunct professor of Electrical Engineering at Northeastern University, Boston University, and Tufts University, and was an industrial advisor to the Wentworth Institute of Technology. Formerly, he was a visiting chair professor in the Electronics Engineering Department at the National Taiwan University of Science and Technology. He also was a visiting researcher at the Agency for Science, Technology, and Research (A*STAR) in Singapore and at the University of Melbourne, Australia. He is an IEEE life fellow, an OSA fellow, and a SPIE fellow. In addition, he has served as an associate editor and reviewer of several technical journals and is the author of five postgraduate-level books. He received his B.A. and M.S. degrees in mathematics and physics from the University of Wisconsin and a Ph.D. in Physics from Northeastern University. His professional experience and research interests are in the general areas of optical networking and biophotonics.

Chapter 1
Perspectives on Lightwave Communications

Abstract The concept of using optical fibers for communications proposed by Kao and Hockman in 1966 spawned an entire new industry based on photonics technology. This chapter describes the motivations and the progressive successes behind optical fiber communications. The discussion notes that many innovative optical fiber and photonic component developments were created to achieve high-speed links. In addition, a great deal of effort was expended in devising installation procedures, creating network test and monitoring equipment, and formulating a variety of international standards.

People have considered using optical methods for communicating over long distances ever since ancient times. In the era around 1000 BC the Greeks and Romans used optical transmission links employing methods such as smoke signals and beacon fires for sending alarms, calls for help, or announcements of certain events. Improvements of these optical transmission systems were not pursued very actively due to technological limitations at the time. For example, the speed of sending information over the communication link was limited because the transmission rate depended on how fast the senders could move their hands, the optical signal receiver was the error-prone human eye, line-of-sight transmission paths were required, and atmospheric effects such as fog and rain made the transmission path unreliable. Thus it turned out to be faster, more efficient, and more dependable to send messages by a courier over the road network [1].

Subsequently, no significant advances for optical communications appeared until the invention of the laser in the early 1960s. With the potential of high-speed laser-based transmission capacities in mind, experiments using atmospheric optical channels then were carried out. However, the high cost of developing and implementing such systems, together with the limitations imposed on the atmospheric optical channels by rain, fog, snow, and dust, make such extremely high-speed links economically unattractive.

The original version of this chapter was revised: All the incorrect sentences have been corrected. The correction to this chapter can be found at https://doi.org/10.1007/978-981-33-4665-9_15

G. Keiser, *Fiber Optic Communications*,
https://doi.org/10.1007/978-981-33-4665-9_1

At the same time it was recognized that an optical fiber could provide a more reliable transmission channel because it is not subject to adverse environmental conditions [2, 3]. Initially, the extremely large losses of more than 1000 dB/km made optical fibers appear impractical. This changed in 1966 when Kao and Hockman [4] speculated that the high losses were a result of impurities in the fiber material, and that the losses potentially could be reduced significantly in order to make optical fibers a viable transmission medium. In 2009 Charles K. C. Kao was awarded the Nobel Prize in Physics for his pioneering insight and his enthusiastic international follow-ups in promoting the further development of low-loss optical fibers. These efforts led to the first ultrapure fiber being fabricated in 1970, only four years after Kao and Hockman's prediction [5]. This breakthrough led to a series of technology developments related to optical fibers. These events finally allowed practical optical fiber based *lightwave communication systems* to start being fielded worldwide in 1978.

The goal of this book is to describe the various technologies, implementation methodologies, and performance measurement techniques that make fiber optic telecom systems possible. The reader can find additional information on the theory of light propagation in fibers, the design of links and networks, and the evolution of optical fibers, photonic devices, and optical fiber communication systems in a variety of reference books [6], tutorial papers [7-12], textbooks [13-20], and conference proceedings [21-23].

This chapter is organized as follows:

- Section 1.1 gives the motivations behind the development of optical fiber transmission systems.
- Section 1.2 defines the different spectral bands that describe various operational wavelength regions used in optical communications.
- Section 1.3 reviews decibel notation for expressing optical power levels.
- Section 1.4 illustrates the basic hierarchy for electrically multiplexing digitized information streams used on optical links.
- Section 1.5 describes basic optical multiplexing methods for greatly increasing the information-handling capacity of optical links.
- Section 1.6 introduces the functions and implementation considerations of the key elements used in optical fiber links.
- Section 1.7 describes the evolution and advances in fiber optic telecom networks that have resulted from the progressive introduction of emerging technologies.
- Section 1.8 lists the main classes of standards related to optical communication components, system operations, and installation procedures.

Next, Chaps. 2–12 describe the purpose and performance characteristics of the major elements in an optical link. These elements include optical fibers, light sources, photodetectors, passive optical devices, optical amplifiers, and active optoelectronic devices used in multiple-wavelength networks. Chapters 13 and 14 show how the elements are put together to form links and networks and explain measurement methodologies used to evaluate the performance of lightwave components and links.

1.1 Reasons for Fiber Optic Communications

1.1.1 The Road to Optical Networks

Figure 1.1 shows the schematic of a generic optical fiber structure. Most other fiber constructions are based on material, size, and layering variations of this fundamental configuration. A standard fiber consists of a solid glass cylinder called a *core*. The core is the region in which light propagates along the fiber. This is surrounded by a dielectric *cladding*, which has a different material property from that of the core in order to achieve light guiding in the fiber. A standard cladding diameter is 125 μm for most types of fibers. A polymer buffer coating with a nominal 250 μm diameter surrounds these two layers to protect the fiber from mechanical stresses and environmental effects. Finally an outer protective polymer jacket with a nominal 900 μm diameter encapsulates the fiber. Chaps. 2 and 3 give details on the structural and performance characteristics of common optical fibers.

The first generation optical fiber had a 50 μm core diameter and a 125 μm cladding diameter. Six such buffered fibers were enclosed in a single optical cable, which was used in the first installed optical fiber links in the late 1970s. These commercial links were used for transmitting telephony signals at about 6 Mb/s over distances of around 10 km. As research and development progressed, the sophistication and capabilities of these systems increased rapidly during the 1980s to create links carrying aggregate data rates beyond terabits per second over distances of hundreds of kilometers. These achievements were based on new technology developments using single-mode optical fiber with nominally 9 μm core diameters.

Starting in the 1990s there was a burgeoning demand on communication network assets for bandwidth-hungry services such as database queries, home shopping, high-definition interactive video, remote education, telemedicine and e-health, high-resolution editing of home videos, blogging, and large-scale high-capacity e-science and Grid computing. This demand was fueled by the rapid proliferation of personal computers (PCs) and sophisticated smart phones coupled with a phenomenal increase in their storage capacity and processing capabilities. Furthermore, the widespread

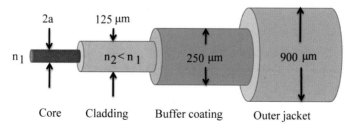

Fig. 1.1 Schematic of a generic optical fiber structure; nominal core diameters are $2a = 9$ μm for single-mode and 50 μm for multimode fibers

availability and continuous expansion of the Internet, and an extensive choice of remotely accessible application programs and information databases, resulted in a dramatic rise in PC and mobile device usage. To handle the ever-increasing demand for high-bandwidth services in locations ranging from homes and mobile devices to large businesses and research organizations, telecom companies worldwide greatly enhanced the capacity of fiber lines. This was accomplished by adding more independent signal-carrying wavelengths on individual fibers, increasing the transmission speed of information being carried by each wavelength, and utilizing more advanced signal modulation techniques with improved spectral efficiency.

1.1.2   Benefits of Using Optical Fibers

The advantages of optical fibers compared to copper wires include the following:

Long Distance Transmission Optical fibers have lower transmission losses compared to copper wires. Consequently data can be sent over longer distances, thereby reducing the number of intermediate repeaters needed to boost and restore signals in long transmission spans. This reduction in equipment and components decreases system cost and complexity.

Large Information Capacity Optical fibers have wider bandwidths than copper wires, so that more information can be sent over a single physical line. This property decreases the number of physical lines needed for sending a given amount of information.

Small Size and Low Weight The low weight and the small dimensions of fibers offer a distinct advantage over heavy, bulky wire cables in crowded underground city ducts or in ceiling-mounted cable trays. This feature also is of importance in aircraft, satellites, and ships where small, low-weight cables are advantageous.

Immunity to Electrical Interference An especially important feature of an optical fiber relates to the fact that it is a dielectric material, which means it does not conduct electricity. This makes optical fibers immune to the electromagnetic interference effects seen in copper wires, such as inductive pickup from other adjacent signal-carrying wires or coupling of electrical noise into the line from any type of nearby electronic equipment.

Enhanced Safety Optical fibers offer a high degree of operational safety because they do not have the problems of ground loops, sparks, and potentially high voltages inherent in copper lines. However, precautions with respect to possible high-intensity laser light emissions need to be observed to prevent eye damage.

Increased Signal Security An optical fiber offers a high degree of data security because the optical signal is well confined within the fiber and an opaque coating around the fiber absorbs any light signal emissions. This feature is in contrast to copper wires where electrical signals potentially could be tapped off easily. Thus optical fibers are attractive in applications where information security is important, such as financial, legal, government, and military systems.

1.2 Optical Wavelength Bands

1.2.1 Electromagnetic Energy Spectrum

All communication systems use some form of electromagnetic energy to transmit signals. The *spectrum* of electromagnetic (EM) radiation is shown in Fig. 1.2. *Electromagnetic energy* is a combination of electrical and magnetic fields and includes power, radio waves, microwaves, infrared light, visible light, ultraviolet light, X rays, and gamma rays. Each discipline takes up a portion (or band) of the electromagnetic spectrum. The fundamental nature of all radiation within this spectrum is that it can be viewed as electromagnetic waves that travel at the speed of light, which is about $c = 3 \times 10^8$ m/s in a vacuum. Note that the speed of light s in a material is smaller by the refractive-index factor n than the speed c in a vacuum, as described in Chap. 2. For example, $n \approx 1.45$ for silica glass, so that the speed of light in this material is about $s = c/n = 2 \times 10^8$ m/s.

The physical properties of the waves in different parts of the spectrum can be measured in several interrelated ways. These are the length of one period of the wave, the energy contained in the wave, or the oscillating frequency of the wave.

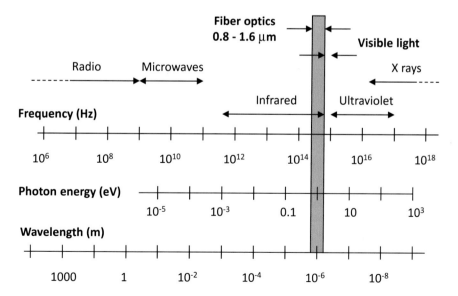

Fig. 1.2 The spectrum of electromagnetic radiation

Whereas electrical signal transmission tends to use frequency to designate the signal operating bands, optical communication generally uses *wavelength* to designate the spectral operating region and *photon energy* or *optical power* when discussing topics such as signal strength or electro-optical component performance. However, note that in some cases the units of optical frequency are used, for example, when dealing with nonlinear effects in fibers.

As can be seen from Fig. 1.2, there are three different ways to measure the physical properties of a wave in various regions in the EM spectrum. These measurement units are related by some simple equations. First of all, in a vacuum the speed of light c is equal to the wavelength λ (Greek letter lambda) times the frequency v (Greek letter nu), so that

$$c = \lambda v \tag{1.1}$$

where the frequency v is measured in cycles per second or *hertz* (Hz).

Example 1.1 Two commonly used wavelength regions in optical communications fall in spectral bands centered around 1310 and 1550 nm. What are the frequencies of these two wavelengths?

Solution Using $c = 2.99793 \times 10^8$ m/s, then from Eq. (1.1) the corresponding frequencies are $v(1310$ nm$) = 228.85$ THz and $v(1550$ nm$) = 193.41$ THz.

An important concept in optical communications is the relationship between the width of a narrow wavelength band $\Delta\lambda$ centered around λ and its corresponding frequency band Δv. This can be found by differentiating the rearranged Eq. (1.1) given by $v = c/\lambda$, which yields $\Delta v = c \, \Delta\lambda/\lambda^2$. More details on this relationship and its applications are given in Chap. 10.

The relationship between the energy E of a photon and its frequency (or wavelength) is determined by the equation known as *Planck's Law*

$$E = hv = hc/\lambda \tag{1.2}$$

where the parameter

$$h = 6.63 \times 10^{-34} \text{J-S} = 4.14 \times 10^{-15} \text{eV-s}$$

is *Planck's constant*. The unit J means *joules* and the unit eV stands for *electron volts*, which is equal to 1.60218×10^{-19} J. In terms of wavelength (measured in units of μm), the energy in electron volts is given by

$$E(eV) = \frac{1.2406}{\lambda(\mu m)} \tag{1.3}$$

Example 1.2 Show that photon energies decrease with increasing wavelength. Use wavelengths at 850, 1310, and 1550 nm.

Solution Using Eq. (1.3) yields $E(850 \text{ nm}) = 1.46$ eV, $E(1310 \text{ nm}) = 0.95$ eV, and $E(1550 \text{ nm}) = 0.80$ eV.

Figure 1.2 shows the optical spectrum ranges from about 5 nm in the ultraviolet region to 1 mm for far-infrared radiation. In between these limits is the 400-to-700 nm *visible band*. Optical fiber communication uses the *near-infrared spectral band* ranging from nominally 770–1675 nm.

The Telecomunications Sector of The International Telecommunications Union (ITU-T) has designated six spectral bands for use in optical fiber communications within the 1260-to-1675 nm region [24]. These *long-wavelength band* designations arose from the attenuation characteristics of optical fibers and the performance behavior of an erbium-doped fiber amplifier (EDFA), as described in Chaps. 3 and 10, respectively. Figure 1.3 shows and Table 1.1 defines the regions and the origins of their designations, which are known by the letters O, E, S, C, L, and U.

Traditionally fiber optic telecommunications organizations expressed interest in transmitting high-capacity information over long distances. Thus the emphasis was

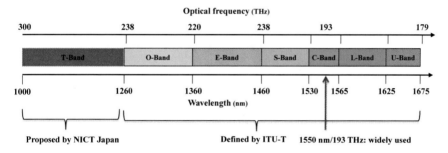

Fig. 1.3 Designations of spectral bands used for optical fiber communications

Table 1.1 Spectral band designations used in optical fiber communications

Name	Designation	Spectrum (nm)	Origin of name
Thousand band	T-band	1000–1260	Thousands of potential additional transmission channels
Original band	O-band	1260–1360	Original region used for single-mode fiber links
Extended band	E-band	1360–1460	Link use can extend into the region for fibers with low water content
Short band	S-band	1460–1530	Wavelengths are shorter than the C-band but higher than the E-band
Conventional band	C-band	1530–1565	Wavelength region used by EDFAs
Long band	L-band	1565–1625	Gain of an EDFA decreases steadily to 1 at 1625 nm in this band
Ultra-long band	U-band	1625–1675	Region beyond the response capability of an EDFA

on using the 1260–1675 nm spectral region, because optical fibers exhibit low optical power losses in this range. Then in 2012 the National Institute of Information and Communications Technology (NICT) of Japan proposed using the 1000–1260 nm spectral band, which is designated as the T-band. Here the symbol "T" stands for "thousand." The motivation for examining this spectral band was for addressing the rapidly increasing need for moderate distance links within data centers, the developments of Internet of Things (IoT), and the support of fifth-generation (5G) wireless systems [25]. Although the optical signal attenuation in standard telecom fibers is higher in the T-band compared to the ITU-T bands, the losses are tolerable for the relatively short transmission distances of up to several kilometers used for applications in the T-band. However, other fiber types with potentially lower losses in the T-band are being considered.

The 770-to-910 nm band is used for shorter-wavelength multimode fiber systems. Thus this region is designated as the *short-wavelength* or *multimode fiber band*. Later chapters describe the operational performance characteristics and applications of optical fibers, electro-optic components, and other passive optical devices for use in the short- and long-wavelength bands.

1.2.2   Optical Windows and Spectral Bands

Figure 1.4 shows the operating range of optical fiber systems and the characteristics of the four key components of a link: the optical fiber, light sources, photodetectors, and optical amplifiers. Here the dashed vertical lines indicate the centers of the three main legacy operating wavelength bands of optical fiber systems, which are the short-wavelength region, the O-band, and the C-band. One of the principal characteristics of an optical fiber is its attenuation as a function of wavelength, as shown at the top in Fig. 1.4. Early applications in the late 1970s made exclusive use of the 770-to-910 nm wavelength band where there was a low-loss window and GaAlAs optical sources and silicon photodetectors operating at these wavelengths were available. Originally this region was referred to as the *first window* because around 1000 nm there was a large attenuation spike due to absorption by water molecules. As a result of this spike, early fibers exhibited a local minimum in the attenuation curve around 850 nm.

By reducing the concentration of hydroxyl ions (OH–) and metallic impurities in the fiber material, in the 1980s manufacturers could fabricate optical fibers with very low losses in the 1260-to-1675 nm region. This spectral band is called the *long-wavelength region*. Because the glass still contained some water molecules, initially a third-order absorption spike remained around 1400 nm. This spike defined two low-loss windows, these being the *second window* centered at 1310 nm and the *third window* centered at 1550 nm. These two windows now are called the O-band and C-band, respectively.

The desire to use the low-loss long-wavelength regions prompted the development of InGaAsP-based light sources, InGaAs photodetectors, and InGaAsP optical

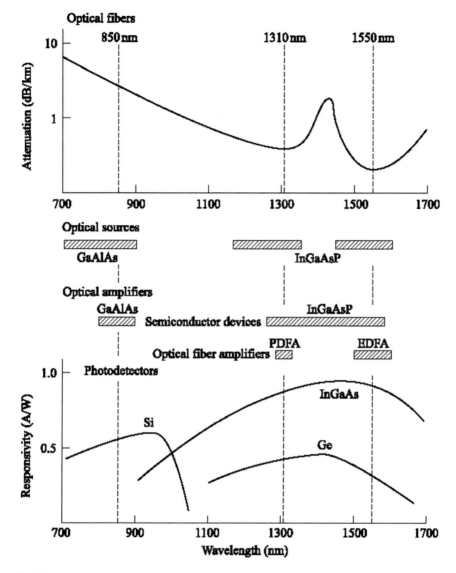

Fig. 1.4 Characteristics and operating ranges of the four key optical fiber link components

amplifiers that can operate in the 1310 and 1550 nm regions. In addition, doping optical fibers with rare-earth elements such as Pr, Th, and Er creates optical fiber amplifiers (called PDFA, TDFA, and EDFA devices, respectively). These devices and the use of Raman amplification gave a further capacity boost to high-capacity long-wavelength systems.

Special material-purification processes can eliminate almost all water molecules from the glass fiber material, thereby dramatically reducing the water-attenuation peak around 1400 nm. This process opens the E-band (1360-to-1460 nm) transmission region to provide around 100 nm more spectral bandwidth in these specially fabricated fibers than in conventional single-mode fibers.

Systems operating at 1550 nm provide the lowest attenuation, but the signal dispersion as a function of distance in a standard silica fiber is larger at 1550 nm than at 1310 nm. Manufacturers overcame this limitation first by creating dispersion-shifted fibers for single-wavelength operation and then by devising non-zero dispersion-shifted fiber (NZDSF) for use with multiple-wavelength implementations. The latter fiber type has led to the widespread use of multiple-wavelength S-band and C-band systems for high-capacity, long-span terrestrial and undersea transmission links. These links routinely carry traffic at 10 Gb/s over nominally 90 km distances between amplifiers or repeaters. By 2010 links operating at 100 Gb/s were being installed and in 2017 the IEEE P802.3bs Task Force ratified the 400GbE (Gigabit Ethernet) standard. This standard established the foundation for industrial deployment of 400GbE in the global network [26–28].

1.3 Decibel Notation

As the following chapters of this book describe, a critical consideration when designing and implementing an optical fiber link is to establish, measure, and/or interrelate the optical signal levels at each of the elements of a transmission link. Thus it is necessary to know parameter values such as the optical output power from a light source, the power level needed at the receiver to properly detect a signal, and the amount of optical power lost at each of the constituent elements of the transmission link.

Reduction or attenuation of signal strength arises from various loss mechanisms in a transmission medium. For example, electric power is lost through heat generation as an electric signal flows along a wire, and optical power is attenuated through scattering and absorption processes in a glass or plastic fiber or in an atmospheric channel. To compensate for these energy losses, amplifiers are used periodically along a channel path to boost the signal level, as shown in Fig. 1.5.

A standard and convenient method for measuring attenuation through a link or a device is to reference the output signal level to the input level. For guided media such as an optical fiber, the signal strength normally decays exponentially. Thus for convenience one can designate signal attenuation or amplification in terms of a logarithmic power ratio measured in *decibels* (dB). The dB unit is defined by

$$\text{Power ratio in dB} = 10 \log \frac{P_2}{P_1} \tag{1.4}$$

where P_1 and P_2 are the electrical or optical power levels of a signal at points 1 and 2 in Fig. 1.6, and *log* is the base-10 logarithm. The logarithmic nature of the decibel allows a large ratio to be expressed in a fairly simple manner. Power levels differing

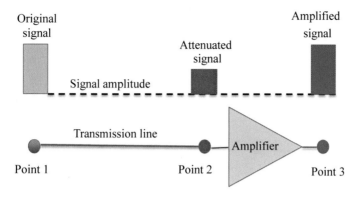

Fig. 1.5 Periodically placed amplifiers compensate for energy losses along a link

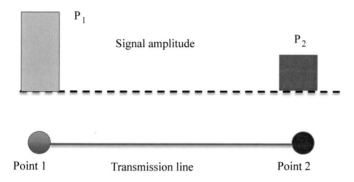

Fig. 1.6 Example of pulse attenuation in a link. P_1 and P_2 are the power levels of a signal at points 1 and 2

by many orders of magnitude can be compared easily when they are in decibel form. For example, a power reduction by a factor of 1000 is a -30 dB loss, an attenuation of 50% is a -3 dB loss, and a 10-fold amplification of the power is a $+10$ dB gain. Another attractive feature of the decibel is that to measure the changes in the strength of a signal, one merely adds or subtracts the decibel loss or gain numbers in a series of connected optical link elements (e.g., optical fibers, couplers, power splitters, or amplifiers) between two different points.

Example 1.3 Assume that after traveling a certain distance in some transmission medium, the power of a signal is reduced to half, that is, $P_2 = 0.5\,P_1$ in Fig. 1.6. At this point, using Eq. (1.4) the attenuation or loss of power is

$$10\log\frac{P_2}{P_1} = 10\log\frac{0.5P_1}{P_1} = 10\log 0.5 = 10(-0.3) = -3 \text{ dB}$$

Thus, -3 dB (or a 3 dB attenuation or loss) means that the signal has lost half its power. If an amplifier is inserted into the link at this point to boost the signal back to its original level, then that amplifier has a 3 dB gain. If the amplifier has a 6 dB gain, then it boosts the signal power level to twice the original value.

Example 1.4 Consider the transmission path from point 1 to point 4 shown in Fig. 1.7. Here the signal is attenuated by 9 dB between points 1 and 2. After getting a 14 dB boost from an amplifier at point 3, it is again attenuated by 3 dB between points 3 and 4. Relative to point 1, the signal level in dB at point 4 is

$$\text{dB level at point 4} = (\text{loss in line 1}) + (\text{amplifier gain}) + (\text{loss in line 2})$$
$$= (-9\text{dB}) + (14\text{dB}) + (-3\text{dB}) = +2\text{dB}$$

Thus the signal has a 2 dB (a factor of $10^{0.2} = 1.58$) gain in power in going from point 1 to point 4.

Table 1.2 shows some sample values of power loss given in decibels and the percent of power remaining after this loss. These types of numbers are important when considering factors such as the effects of tapping off a small part of an optical signal for monitoring purposes, for examining the power loss through some optical element, or when calculating the signal attenuation in a specific length of optical fiber.

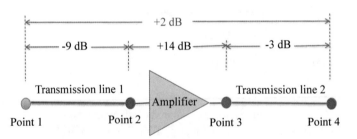

Fig. 1.7 Example of signal attenuation and amplification in a transmission path

Table 1.2 Representative values of decibel power loss and the remaining percentages	Power loss (in dB)	Percent of power left
	0.1	98
	0.5	89
	1	79
	2	63
	3	50
	6	25
	10	10
	20	1

Because the decibel is used to refer to ratios or relative units, it gives no indication of the absolute power level. However, a derived unit can be used for this purpose. Such a unit that is particularly common in optical fiber communications is the dBm (simply pronounced *dee bee em*). This unit expresses the power level P as a logarithmic ratio of P referred to 1 mW. In this case, the power in dBm is an absolute value defined by

$$\text{Power level (in dBm)} = 10 \log \frac{P(\text{in mW})}{1 \text{ mW}} \qquad (1.5)$$

An important rule-of-thumb relationship to remember for optical fiber communications is 0 dBm = 1 mW. Therefore, positive values of dBm are greater than 1 mW and negative values are less than 1 mW.

Example 1.5 Consider three different light sources having the following optical output powers: 50, 1, and 50 mW. What are the power levels in dBm units?

Solution Using Eq. (1.5) to express the light levels in dBm units shows that the output powers of these sources are -13 dBm, 0 dBm, and $+17$ dBm, respectively.

Example 1.6 Consider a product data sheet for a photodetector that states that an optical power level of -32 dBm is needed at the photodetector to satisfy a specific performance requirement. What is the power level in nW (nanowatt) units?

Solution Equation (1.5) shows that -32 dBm corresponds to a power in nW of

$$P = 10^{-32/10} \text{ mW} = 0.631 \text{ } \mu\text{W} = 631 \text{ nW}$$

Table 1.3 lists some examples of optical power levels and their dBm equivalents.

Table 1.3 Examples of optical power levels and their dBm equivalents

Power	dBm equivalent
200 mW	23
100 mW	20
10 mW	10
1 mW	0
100 μW	-10
10 μW	-20
1 μW	-30
100 nW	-40
10 nW	-50
1 nW	-60
100 pW	-70
10 pW	-80
1 pW	-90

1.4   Digital Multiplexing Techniques

To handle the continuously rising demand for high-bandwidth services from users ranging from individuals to large businesses and research organizations, telecom companies worldwide are implementing increasingly sophisticated digital multiplexing techniques that allow a larger number of independent information streams to share the same physical transmission channel simultaneously. This section describes some common electrical digital signal multiplexing techniques [29–32].

1.4.1   Basic Telecom Signal Multiplexing

Table 1.4 gives examples of information rates for some typical telecom services. To send these services from one user to another, network providers combine the signals from many different users and send the aggregate signal over a single transmission line. This scheme is known as *time-division-multiplexing* (TDM) wherein N independent information streams, each running at a data rate of R b/s, are interleaved electrically into a single information stream operating at a higher rate of $N \times R$ b/s. To get a detailed perspective of this methodology, this section looks at the multiplexing schemes used in telecommunications.

Early applications of fiber optic transmission links were mainly for large capacity telephone lines. These digital links consisted of time-division-multiplexed 64-kb/s voice channels. The multiplexing was developed in the 1960s and is based on what is known as the *plesiochronous digital hierarchy* (PDH). Figure 1.8 shows the digital transmission hierarchies used in the North American and the European-based telephone networks.

The fundamental building block in the North American network is a 1.544 Mb/s transmission rate known as a *DS*1 rate, where *DS* stands for *digital system*. It is formed by time-division-multiplexing twenty-four voice channels, each digitized at a 64 kb/s rate (which is referred to as *DS*0). *Framing bits*, which indicate where an information unit starts and ends, are added along with these voice channels to yield the 1.544 Mb/s bit stream. Framing and other control bits that may get added to an

	Type of service	Data rate
Table 1.4 Examples of information rates for some typical services	Video on demand/interactive video	1.5–6 Mb/s
	Video games	1–2 Mb/s
	Remote education	1.5–3 Mb/s
	Electronic shopping	1.5–6 Mb/s
	Data transfer or telecommuting	1–3 Mb/s
	Video conferencing	0.384–2 Mb/s
	Voice (single phone channel)	33.6–56 kb/s

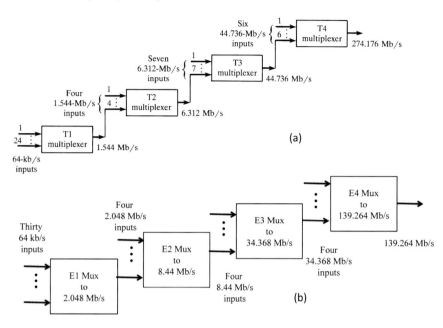

Fig. 1.8 Digital transmission hierarchies used in the **a** North American and **b** European-based telephone networks

information unit in a digital stream are called *overhead* bits. At any multiplexing level a signal at the designated input rate is combined with other input signals at the same rate.

In describing North American telephone network data rates, one sees terms such as $T1$, $T3$, and so on. Often the terms Tx and DSx (e.g., $T1$ and $DS1$ or $T3$ and $DS3$) are used interchangeably. However, there is a subtle difference in their meaning. Designations such as $DS1$, $DS2$, and $DS3$ refer to a *service type*; for example, a user who wants to send information at a 1.544 Mb/s rate would subscribe to a $DS1$ service. Abbreviations such as $T1$, $T2$, and $T3$ refer to the data rate the transmission line technology uses to deliver that service over a physical link. For example, the $DS1$ service is transported over a physical wire or optical fiber using electrical or optical pulses sent at a $T1 = 1.544$ Mb/s rate.

Telephone networks in other countries use either European- or Japanese-based multiplexing hierarchies. Similar to the North American hierarchies, basic 64 kb/s channels are combined but at different multiplexed bit-rate levels as shown in Table 1.5. Most countries outside of North America and Japan (such as in Europe, South and Central America, Africa, Australia, and most of Asia) use the European-based multiplexing hierarchy labeled by $E1$, $E2$, $E3$, and so on. Figure 1.8b shows the number of channels multiplexed at each bit-rate level up to $E4$. For example, multiplexing thirty 64 kb/s channels and adding required overhead bits results in a 2.048 Mb/s $E1$ rate.

Table 1.5 Digital multiplexing levels used in North America, Europe, and Japan

Multiplexing level	Number of 64 kb/s channels	Bit rate (Mb/s)		
		North America	Europe	Japan
DS0	1	0.064	0.064	0.064
DS1	24	1.544		1.544
E1	30		2.048	
	48	3.152		3.152
DS2	96	6.312		6.312
E2	120		8.448	
E3	480		34.368	32.064
DS3	672	44.736		
	1344	91.053		
	1440			97.728
E4	1920		139.264	
DS4	4032	274.176		
	5760			397.200

The TDM scheme is not restricted to multiplexing voice signals. For example, at the $DS1$ or $E1$ level, any 64 kb/s digital signal of the appropriate format could be transmitted as one of the 24 or 30 input channels shown in Fig. 1.8. As noted there and in Table 1.5, the main multiplexed rates for North American applications are designated as $DS1$ (1.544 Mb/s), $DS2$ (6.312 Mb/s), and $DS3$ (44.736 Mb/s).

Example 1.7 As can be seen from Fig. 1.8, at each multiplexing level some overhead bits are added for synchronization purposes. What is the overhead for $T1$?

Solution At $T1$ the overhead is 1544 kb/s $- 24 \times 64$ kb/s $= 8$ kb/s.

1.4.2 Multiplexing Hierarchy in SONET/SDH

With the advent of high-capacity fiber optic transmission lines in the 1980s, service providers established a standard signal format called *synchronous optical network* (SONET) in North America and *synchronous digital hierarchy* (SDH) in other parts of the world [33–35]. These standards define a synchronous frame structure for sending multiplexed digital traffic over optical fiber trunk lines. The basic building block and first level of the SONET signal hierarchy is called the *Synchronous Transport Signal—Level 1* (STS-1), which has a bit rate of 51.84 Mb/s. Higher-rate SONET signals are obtained by byte-interleaving N of these STS-1 frames (where one byte is a group of 8 bits), which then are scrambled and converted to an *Optical Carrier— Level N* (OC-N) signal. Thus the OC-N signal will have a line rate exactly N times

Table 1.6 Common SDH and SONET line rates and their popular numerical name

SONET level	Electrical level	SDH level	Line rate (Mb/s)	Popular rate name
OC-1	STS-1	–	51.84	–
OC-3	STS-3	STM-1	155.52	155 Mb/s
OC-12	STS-12	STM-4	622.08	622 Mb/s
OC-48	STS-48	STM-16	2488.32	2.5 Gb/s
OC-192	STS-192	STM-64	9953.28	10 Gb/s
OC-768	STS-768	STM-256	39,813.12	40 Gb/s

that of an OC-1 signal. For SDH systems the fundamental building block is the 155.52 Mb/s *Synchronous Transport Module—Level 1* (STM-1). Again, higher-rate information streams are generated by synchronously multiplexing N different STM-1 signals to form the STM-N signal. Table 1.6 shows commonly used SDH and SONET signal levels, the line rate, and the popular numerical name for that rate. Note that although SONET rates can be defined for data rates beyond 40 Gb/s, emerging requirements for efficient transport of information from high bandwidth data services requires a new technology such as Sect. 1.4.3 describes.

1.4.3 Optical Transport Network (OTN)

In the early days of telecommunications, for decades network traffic consisted principally of voice calls that were transported over networks in which the connections between endpoints followed a predictable connection scheme. When data services emerged, adaptations were developed to map the data traffic onto SONET/SDH networks to provide a single transport network. However, this scheme became increasingly difficult to implement beyond 40 Gb/s because voice and data have inherently different transport requirements. Whereas aggregated voice traffic follows a predictable connection pattern, data services and applications have bursty, unpredictable traffic patterns with widely varying demands on bandwidth and data transmission performance. Consequently, the telecom industry developed a new technology known as the Optical Transport Network for transmission rates of 40 Gb/s and above. The ITU has standardized OTN as G.709. Chapter 13 gives more details on OTN and its applications.

1.5 Multiplexing of Wavelength Channels

The burgeoning development of consumer equipment such as powerful laptop PCs, tablet computers, 4G and 5G smart phones, high-definition TV sets, and control consoles for 3D online games has created a growing demand for more band-width from applications such as video-on-demand, cloud computing, online gaming, music streaming, and social networking. To accommodate this growing bandwidth demand, network providers are continuously seeking new methods for increasing the information-handling capacity of optical links. This is especially important in the links running between a customer and a nearby traffic switching facility (known as the *central office*). These methods include wavelength division, polarization division, mode division, and space division multiplexing techniques that combine many individual information channels onto a single fiber.

1.5.1   Basis of WDM

The basis of *wavelength division multiplexing* (WDM) is to use multiple sources operating at slightly different wavelengths to transmit several independent information streams simultaneously over the same fiber. Figure 1.9 shows the basic WDM concept. Here N independent optically formatted information streams, each trans-mitted at a different wavelength, are combined by means of an optical multiplexer and sent over the same fiber. Note that each of these streams could be at a different data rate. Each information stream maintains its individual data rate after being multiplexed with the other traffic streams, and still operates at its unique wavelength.

Although researchers started looking at WDM techniques in the 1970s, during the ensuing years it generally turned out to be easier to transmit only a single wavelength on a fiber using high-speed electronic and optical devices, than to invoke the greater system complexity called for in WDM. However, a dramatic surge in WDM popu-larity started in the early 1990s owing to several factors. These include new fiber types that provide better performance of multiple-wavelength operation at 1550 nm, advances in producing WDM devices that can combine and separate closely spaced wavelengths, and the development of optical amplifiers that can boost optical signal

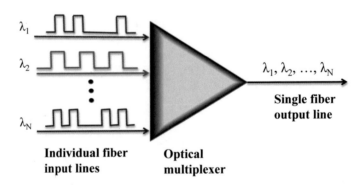

Fig. 1.9 Basic concept of wavelength division multiplexing

levels completely in the optical domain. Chapter 10 presents further details on WDM concepts and components

1.5.2 Polarization Division Multiplexing

As Sect. 3.2.8 describes, signal energy at a given wavelength occupies two orthogonal polarization modes. The basis of polarization division multiplexing (PDM) is to impose independent optical signal streams on the two orthogonal polarization states, thereby doubling the transmission capacity of an optical fiber. The PDM method generally is used with phase modulation or optical quadrature amplitude modulation (QAM) techniques thus allowing data rates of 100 Gb/s or more to be sent over a single optical fiber in a WDM link. A challenge with implementing PDM is to mitigate the problems of polarization-mode dispersion (see Sect. 3.2.8), polarization-dependent loss, and cross-polarization modulation. This challenge is addressed through the use of advanced coding techniques, such as polarization-multiplexed differential quadrature phase-shift keying (PM-DQPSK) modulation formats (see Sect. 13.4.2).

1.5.3 Optical Fibers with Multiple Cores

Another concept for increasing optical fiber capacity is the technique of space division multiplexing (SDM) through the use of fibers with multiple cores. In such fibers, each core provides a spatially isolated transmission path for independent groups of WDM optical signals. SDM simply multiplies the transmission capacity per fiber by the number of fiber cores. This condition holds, provided that each SDM channel (each fiber core) acts independently and has transmission characteristics that are equivalent to the performance of conventional single-core fibers. For example, the capacity of a seven-core fiber would be seven times that of a single-core fiber. Section 3.6 gives some examples of multiple-core optical fibers.

1.6 Basic Elements of Optical Fiber Systems

Similar to electrical communication systems, the basic function of an optical fiber link is to transport a signal from communication equipment (e.g., a computer, telephone, or video device) at one location to corresponding equipment at another location with a high degree of reliability and accuracy. Figure 1.10 shows the main constituents of an optical fiber communications link. The key sections are a transmitter consisting of a light source and its associated drive circuitry, a cable offering mechanical and environmental protection to the optical fibers contained inside, and a receiver

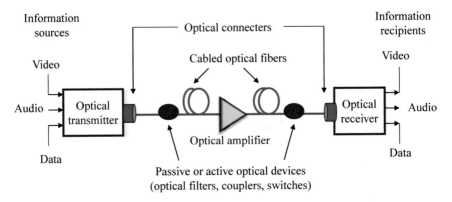

Fig. 1.10 Main constituents of an optical fiber communications link

consisting of a photodetector plus amplification and signal-restoring circuitry. Additional components include optical amplifiers, connectors, splices, couplers, regenerators (for restoring the signal-shape characteristics), and other passive components and active photonic devices.

The cabled fiber is one of the most important elements in an optical fiber link as is described in Chaps. 2 and 3. In addition to protecting the glass fibers during installation and service, the cable may contain copper wires for powering optical amplifiers or signal regenerators, which are needed periodically in long-distance links for amplifying and reshaping the signal. A variety of fiber types with different performance characteristics exist for a wide range of applications. To protect the glass fibers during installation and service, there are many different cable configurations depending on whether the cable is to be installed inside a building, underground in ducts or through direct-burial methods, outside on poles, or under water. Very low-loss optical connectors and splices are needed in all categories of optical fiber networks for joining cables and for attaching one fiber to another.

Analogous to copper cables, the installation of optical fiber cables can be either aerial, in ducts, undersea, or buried directly in the ground, as illustrated in Fig. 1.11. As Chap. 2 describes, the cable structure will vary greatly depending on the specific application and the environment in which it will be installed. Owing to installation and/or manufacturing limitations, individual cable lengths for in-building or terrestrial applications will range from several hundred meters to several kilometers. Practical considerations such as reel size and cable weight determine the actual length of a single cable section. The shorter segments tend to be used when the cables are pulled through ducts. Longer lengths are used in aerial, direct-burial, or underwater applications.

Workers can install optical fiber cables by pulling or blowing them through ducts (both indoor and outdoor), laying them in a trench outside, plowing them directly into the ground, suspending them on poles, or laying or plowing them underwater. Although each method has its own special handling procedures, they all need to adhere to a common set of precautions. These include avoiding sharp bends of the

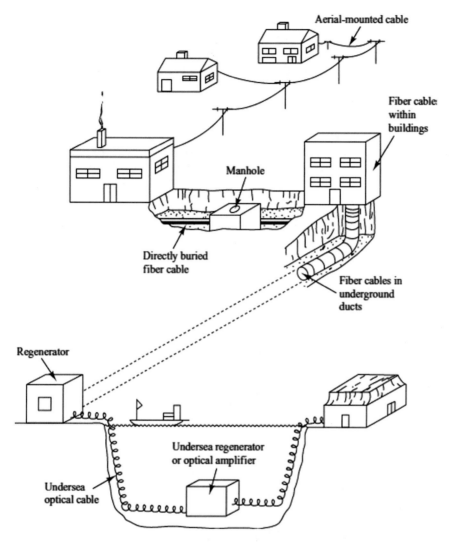

Fig. 1.11 Optical fiber cables can be installed on poles, in ducts, and underwater, or they can be buried directly in the ground

cable, minimizing stresses on the installed cable, periodically allowing extra cable slack along the cable route for unexpected repairs, and avoiding excessive pulling or hard yanks on the cable. For *direct-burial installations* a fiber optic cable can be plowed directly underground or placed in a trench that is filled in later.

Transoceanic cable lengths can be many thousands of kilometers long and include periodically spaced (on the order of 60–100 km) optical repeaters to boost the signal level. The cables are assembled in onshore factories and then are loaded into special cable-laying ships. Splicing together individual cable sections forms continuous transmission lines for these long-distance links.

Once the cable is installed, a transmitter can be used to launch a light signal into the fiber. Chapter 4 describes transmitter configurations and Chap. 5 discusses methods and devices for connecting sources and other photonic devices to fibers. In general, the transmitter consists of a light source that is dimensionally compatible with the fiber core and it contains associated electronic control and modulation circuitry. Semiconductor light-emitting diodes (LEDs) and laser diodes are suitable sources. For these devices the light output amplitude can be modulated rapidly by simply varying the input current at the desired transmission rate, thereby producing a time-varying optical signal. The electric input signals to the transmitter circuitry for driving the optical source can be either of an analog or digital form. The functions of the associated transmitter electronics are to set and stabilize the source operating point and output power level. For high-rate systems (usually greater than about 2.5 Gb/s), direct modulation of the source can lead to unacceptable optical signal distortion. In this case, an external modulator is used to vary the amplitude of a continuous light output from a laser diode source. In the 770-to-910 nm region the light sources are generally alloys of GaAlAs. At longer wavelengths (1260–1675 nm) an InGaAsP alloy is the principal optical source material.

After an optical signal is launched into a fiber, it will become progressively attenuated and distorted with increasing distance because of light scattering, absorption, and dispersion mechanisms in the glass material. As Chap. 6 discusses, at the destination of an optical fiber transmission line, there is a receiving device that interprets the information contained in the optical signal. Inside the receiver is a photodiode that detects the weakened and distorted optical signal emerging from the end of an optical fiber and converts it to an electrical signal (referred to as a *photocurrent*). The receiver also contains electronic amplification devices and circuitry to restore signal fidelity. Silicon photodiodes are used in the 770-to-910 nm region. The primary material in the 1260-to-1675 nm region is an InGaAs alloy.

The design of an optical receiver is inherently complex and can have rather sophisticated functions because it has to interpret the content of the weakened and degraded signal received by the photodetector. Chapters 6–8 discuss basic receivers for digital and analog applications. The principal figure of merit for a receiver is the minimum optical power necessary at the desired data rate to attain either a given error probability for digital systems or a specified signal-to-noise ratio for an analog system. The ability of a receiver to achieve a certain performance level depends on the photodetector type, the effects of noise in the system, and the characteristics of the successive amplification stages in the receiver.

Included in any optical fiber link are various passive and active optical devices that assist in controlling and guiding the light signals. Chapter 10 describes a variety of such components. Passive devices are optical components that require no electronic control for their operation. Among these are optical filters that select only a narrow spectrum of desired light, optical splitters that divide the power in an optical signal into a number of different branches, optical multiplexers that combine signals from two or more distinct wavelengths onto the same fiber (or that separate the wavelengths into individual channels at the receiving end) in multiple-wavelength optical fiber networks, and couplers used to tap off a certain percentage of light, usually for performance monitoring purposes. In addition, modern sophisticated optical fiber networks contain a wide range of active optical components, which require an electronic control for their operation. These include light signal modulators, tunable (wavelength-selectable) optical filters, reconfigurable elements for adding and dropping wavelengths at intermediate nodes, variable optical attenuators, and optical switches.

Chapters 11 through 13 address factors associated with implementing optical telecom networks. After an optical signal has traveled a certain distance along a fiber, it becomes greatly weakened due to power loss along the fiber. Therefore, when setting up an optical link, engineers formulate a power budget and add amplifiers or repeaters when the path loss exceeds the available power margin. The periodically placed amplifiers merely give the optical signal a power boost, whereas a repeater also will attempt to restore the signal to its original shape. Prior to 1990, only repeaters were available for signal amplification. For an incoming optical signal, a repeater performs photon-to-electron conversion, electrical amplification, retiming, pulse shaping, and then electron-to-photon conversion. This process can be fairly complex for high-speed multiple-wavelength systems. Thus researchers expended a great deal of effort to develop all-optical amplifiers, which boost the light power level completely in the optical domain. Optical amplification mechanisms for WDM links include the use of devices based on rare-earth-doped lengths of fiber and distributed amplification by means of a stimulated Raman scattering effect.

The installation and operation of an optical fiber communication system require measurement techniques for verifying that the specified performance characteristics of the constituent components are satisfied. Chapter 14 addresses these techniques. In addition to measuring optical fiber parameters, system engineers are interested in knowing the characteristics of passive splitters, connectors, and couplers, and electro-optic components, such as sources, photodetectors, and optical amplifiers. Furthermore, when a link is being installed and tested, operational parameters that should be measured include bit error rate, timing jitter, and signal-to-noise ratio as indicated by the eye pattern. During actual operation, measurements are needed for maintenance and monitoring functions to determine factors such as fault locations in fibers and the status of remotely located optical amplifiers.

1.7 Evolution of Fiber Optic Networks

Optical networking technology has made tremendous advances since the first basic links were installed to carry live traffic around 1978. The initial installations operated at 6.3 Mb/s over distances of about 10 km using simple on-off keying (OOK) modulation in the transmitter. As shown in Fig. 1.12, there has been a steady 40–50% growth per year in link data rates and transmission distances since then. Until

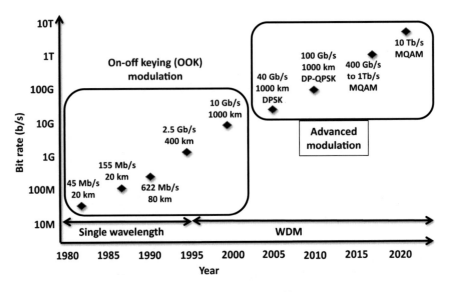

Fig. 1.12 Evolution in optical communication network capacities

about 1995, an optical fiber typically carried data on a single wavelength using OOK modulation. Progressively higher data transmission rates were achieved mainly by developing more efficient data routing and switching equipment and taking advantage of improvements in silicon device technology for laser drivers. Then in the mid-1990s the use of wavelength division multiplexing (WDM) allowed a significant increase in link capacity by sending information on several wavelengths simultaneously over an individual optical fiber. In addition, the development and deployment of optical amplifiers enabled carriers to send information over longer distances without the need for intermediate repeater stations.

Initially, to send data the telecom companies (carriers) used simple OOK modulation at the transmitter and direct detection schemes at the receiver. This changed starting in 2005 with the introduction of advanced modulation techniques to increase spectral efficiency, coherent detection to enable multilevel demodulation, sophisticated digital signal processing (DSP) to compensate for optical signal impairment, and forward error correction (FEC) processing to reduce signal-to-noise ratio requirements. Differential phase-shift keying (DPSK) modulation schemes introduced in 2005 allowed transmissions at 40 Gb/s per wavelength. Dual polarization quadrature phase-shift keying (DP-QPSK) modulation with a coherent receiver started being implemented in 2011 for 100 Gb/s per wavelength transmission. The next step starting in 2017 was to use higher-order modulation techniques, such as 16QAM and 64QAM (quadrature amplitude modulation), to achieve from 400 Gb/s to 10 Tb/s and beyond [36–39] (see Chap. 13 for more details on these techniques).

1.8 Standards for Fiber Optic Communications

To allow components and equipment from different vendors to interface with one another, numerous international standards have been developed [40, 41]. The three basic classes for fiber optics are primary standards, component testing standards, and system standards.

Primary standards refer to measuring and characterizing fundamental physical parameters such as attenuation, bandwidth, operational characteristics of fibers, and optical power levels and spectral widths. In the USA the main organization involved in primary standards is the National Institute of Standards and Technology (NIST). This organization carries out fiber optic and laser standardization work. Other national organizations include the National Physical Laboratory (NPL) in the United Kingdom and the Physikalisch-Technische Bundesanstalt (PTB) in Germany.

Component testing standards define tests for fiber-optic component performance and establish equipment-calibration procedures. Several different organizations are involved in formulating testing standards, some very active ones being the Fiber Optic Association (thefoa.org), the Telecommunication Sector of the International Telecommunication Union (ITU-T), and the International Electrotechnical Commission (IEC).

System standards refer to measurement methods for links and networks. The major organizations are the American National Standards Institute (ANSI), the Institute for Electrical and Electronic Engineers (IEEE), and the ITU-T. Of particular interest for fiber optics system are test standards and recommendations from the ITU-T. Within the G series (in the number range G.650 and higher) the recommendations relate to fiber cables, optical amplifiers, wavelength multiplexing, optical transport networks (OTN), system reliability and availability, and management and control for passive optical networks (PON). The L and O series of the ITU-T address methods and equipment for the construction, installation, maintenance support, monitoring, and testing of cable and other elements in the optical fiber outside plant, that is, the fielded cable system.

1.9 Summary

Following its introduction into telecom networks in the late 1970s, optical fiber communications technology has experienced a dramatic increase in transmission capacities. Starting with a humble 6 Mb/s transmission rate over a 10 km link, forty years later in 2020 optical fiber transmission links carrying information at speeds of 400 Gb/s and 1 Tb/s were being installed on links over hundreds of kilometers long. Many new technology developments were created to achieve such high-speed links and a great deal of effort also was expended in devising installation procedures, network test and monitoring equipment, and a wide variety of international standards.

The following chapters of this book present the fundamental principles for understanding and applying a wide range of optical fiber technologies to modern communication networks. The sequence of topics moves systematically from the underlying principles of components and their interactions with other devices in an optical fiber link, through descriptions of the architectures and performance characteristics of complex optical links and networks, to essential measurement and test procedures required during network installation and maintenance. By mastering these fundamental topics the reader will be prepared not only to contribute to disciplines such as current device, communication link, or equipment designs, but also to understand quickly any further technology developments for future enhanced networks.

Problems

1.1   What are the energies in electron volts (eV) of photons at wavelengths 850, 1310, 1490, and 1550 nm?

1.2   A fundamental analog signal is the *sine wave* shown in Fig. 1.13. Its three main features are its amplitude, period or frequency, and phase. The *amplitude* is the waveform magnitude, which is measured in volts, amperes, or watts depending on the signal type. The *frequency* f is the number of cycles or oscillations per second and is expressed in units of hertz (Hz). The *period* T is the inverse of the frequency, that is, $T = 1/f$. The parameter *phase* describes the position of the waveform relative to time zero and is measured in degrees or radians (rad) with $180° = \pi$ rad. Consider three sine waves that have the following periods: 25 μs, 250 ns, 125 ps. What are their frequencies?

1.3   If the crests and troughs of two sine waves that have the same period occur at the same time, they are *in phase*. Otherwise they are said to be *out of phase*. Consider two waves that have amplitudes $A_1(t)$ and $A_2(t)$, respectively, at a time t. The amplitude of the resulting wave then is $A(t) = A_1(t) + A_2(t)$. (a) If two in-phase waves have the same maximum amplitude A and the same period, what is the maximum amplitude of the combined wave? This case is called *constructive interference*. (b) What is the maximum amplitude of the combined wave if the two waves are $180°$ (π rad) out of phase? This case is called *destructive interference*.

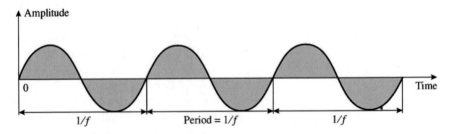

Fig. 1.13 A fundamental sine wave analog signal showing the amplitude, period or frequency, and phase

1.4   Consider two sine wave signals that have the same frequency. Suppose that the second signal is offset by one quarter of a cycle with respect to the first wave. What is the phase shift in degrees between the two signals?

1.5   A common digital binary waveform is represented by a sequence of two types of pulses called *bits*. As Fig. 1.14 shows, the time slot in which a bit occurs is called the *bit interval* or *bit period*. The presence of a pulse in a time slot is a *one bit* or 1 bit, whereas the absence of a pulse represents a *zero bit*. The bit intervals are regularly spaced and occur every 1/R seconds or at a rate of R bits per second. Depending on the signal coding method, a bit can fill the entire bit period (Fig. 1.14a) or only part of it (Fig. 1.14b). The number of photons N in a digital pulse can be found from the relation

$$N = \frac{\text{(Pulse width)(Pulse power)}}{\text{(Energy/photon in eV)}(1.6 \times 10^{-19} \text{ J/eV})}$$

Consider a 1 ns pulse with a 100 nW amplitude at various wavelengths. How many photons are in such a pulse at each of the following wavelength: 850, 1310, 1490, and 1550 nm?

1.6   What is the duration of a bit for each of the following three signals which have bit rates of 64 kb/s, 5 Mb/s, and 10 Gb/s?

1.7   Convert the following absolute power gains to decibel power gains: 10^{-3}, 0.3, 1, 4, 10, 100, 500, 2^n.

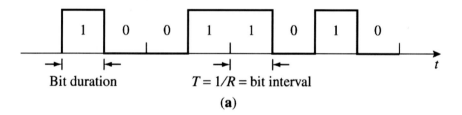

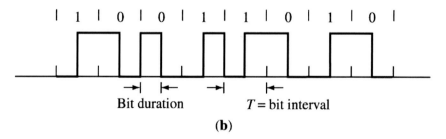

Fig. 1.14 A common digital binary waveform is represented by a sequence of two types of bit pulses

1.8   Convert the following decibel power gains to absolute power gains: -30 dB, 0 dB, 13 dB, 30 dB, $10n$ dB.

1.9   Convert the following absolute power levels to dBm values: 1 pW, 1 nW, 1 mW, 10 mW, 50 mW.

1.10 Convert the following dBm values to power levels in units of mW: -13, -6, 6, 17 dBm.

1.11 A signal travels along a fiber from point A to point B. (a) If the signal power is 1.0 mW at point A and 0.125 mW at point B, what is the fiber attenuation in dB? (b) What is the signal power at point B if the attenuation is 15 dB?

1.12 A signal passes through three cascaded amplifiers, each of which has a 5 dB gain. What is the total gain in dB? By what numerical factor is the signal amplified?

1.13 A 50 km long optical fiber has a total attenuation of 24 dB. If 500 μW of optical power get launched into the fiber, what is the output optical power level in dBm and in μW?

1.14 Based on the Shannon theorem, the maximum data rate R of a channel with a bandwidth B is R = B log2(1 + S/N), where S/N is the signal-to-noise ratio. Suppose a transmission line has a bandwidth of 2 MHz. If the signal-to-noise ratio at the receiving end is 20 dB, what is the maximum data rate that this line can support?

1.15 (a) At the lowest TDM level of the digital service scheme, 24 channels of 64 kb/s each are multiplexed into a 1.544 Mb/s $DS1$ channel. How much is the overhead that is added? (b) The next higher multiplexed level, the $DS2$ rate, is 6.312 Mb/s. How many $DS1$ channels can be accommodated in the $DS2$ rate, and what is the overhead? (c) If the $DS3$ rate that is sent over a $T3$ line is 44.376 Mb/s, how many $DS2$ channels can be accommodated on a $T3$ line, and what is the overhead? (d) Using the above results, find how many $DS0$ channels can be sent over a $T3$ line. What is the total added overhead?

Answers to Selected Problems

1.1   1.46, 0.95, 0.83, and 0.80 eV, respectively
1.2   40 kHz, 4.0 MHz, and 8 GHz, respectively
1.3   (a) 2A; (b) 0
1.4   90°
1.5   428, 657, 753, and 781, respectively
1.6   15.6 μs, 200 ns, and 0.1 ns, respectively
1.7 -30, -5.2, 0, 6, 10, 20, 27, and 3n dB, respectively
1.8 10^{-3}, 1, 20, 1000, and 10^n, respectively
1.9 -90 dBm, -60 dBm, -30 dBm, 10 dBm, and 17 dBm, respectively
1.10 50 μW, 250 μW, 4 mW, and 50 mW, respectively
1.11 (a) 9 dB; (b) 32 μW
1.12 15 dB. The signal is amplified by a factor $10^{1.5} = 31.6$.
1.13 -27 dBm, which is equivalent to 2.0 μW

1.14   13.3 Mb/s

1.15   (a) 8000 bits of overhead; (b) Four DS-1 channels fit into a DS-2 channel. 136 kb/s of overhead are added; (c) A $T3$ line can accommodate seven $DS2$ channels. The overhead is 192 kb/s. (d) 672 $DS0$ channels can be sent over a $T3$ line. The total added overhead is 1.368 Mb/s or 3%.

References

1. R.W. Burns, *Communications: An International History of the Formative Years* (Institution of Electrical Engineers, London, 2004)
2. J. Hecht, *City of Light*, Oxford University Press, revised expanded ed., (2004). This book gives a comprehensive account of the history behind the development of optical fiber communication systems
3. E. Snitzer, Cylindrical dielectric waveguide modes. J. Opt. Soc. Amer. **51**, 491–498 (1961)
4. K.C. Kao, G. A. Hockman, Dielectric-fibre surface waveguides for optical frequencies. Proc. IEE, **113**, 1151–1158 (1966)
5. F.P. Kapron, D.B. Keck, R.D. Maurer, Radiation losses in glass optical waveguides. Appl. Phys. Lett. **17**, 423–425 (1970)
6. A series of books published by Academic Press contains dozens of topics in all areas of optical fiber technology presented by researchers from AT&T Bell Laboratories over a period of forty years. (*a*) S.E. Miller, A.G. Chynoweth (eds.) *Optical Fiber Telecommunications* (1979); (*b*) S.E. Miller, I.P. Kaminow (eds.) *Optical Fiber Telecommunications–II* (1988); (*c*) I.P. Kaminow, T.L. Koch (eds.) *Optical Fiber Telecommunications–III*, vols. A and B (1997); (*d*) I.P. Kaminow, T. Li (eds.) *Optical Fiber Telecommunications–IV*, vols. A and B (2002); (*e*) I.P. Kaminow, T. Li, A.E. Willner (eds.) *Optical Fiber Telecommunications–V*, vols. A and B (2008); (*f*) I.P. Kaminow, T. Li, A. E. Willner (eds.) *Optical Fiber Communications–VI*. vols, A and B (2013); (*g*) A.E. Willner (ed.) *Optical Fiber Telecommunications-VII* (2019)
7. E. Wong, Next-generation broadband access networks and technologies. J. Lightw. Technol. **30**(4), 597–608 (2012)
8. R. Essiambre, R.W. Tkach, Capacity trends and limits of optical communication networks. Proc. IEEE **100**(5), 1035–1055 (2012)
9. C. Kachris, I. Tomkos, A survey on optical interconnects for data centers. *IEEE Commun. Surveys Tutorials* **14**(4), 1021–1036 4th Quarter (2012)
10. C. Lim, K.L. Lee, A.T. Nirmalathas, Review of physical layer networking for optical-wireless integration. Opt. Eng. **55**(3), 031113 (2016)
11. E. Agrell et al., Roadmap of optical communications. J. Opt. **18**, 1–40 (2016)
12. P.J. Winzer, D.T. Neilson, A.R. Chraplyvy, Fiber-optic transmission and networking: the previous 20 and the next 20 years [invited]. Opt. Express **26**(18), 24190–24239 (2018)
13. G. Keiser, *FTTX Concepts and Applications* (Wiley, 2006)
14. A. Brillant, *Digital and Analog Fiber Optic Communication for CATV and FTTx Applications* (Wiley, 2008)
15. J. Chesnoy, *Undersea Fiber Communication Systems* (Academic Press, 2016)
16. A. Paradisi, R. Carvalho Figueirdo, A. Chiuchiarelli, E. de Sousa Rosa (eds.) *Optical Communications* (Springer, Berlin, 2019)
17. R. Hui, *Introduction to Fiber-Optic Communications* (Academic Press, 2019)
18. G.D. Peng, *Handbook of Optical Fibers* (Springer, 2019)
19. M. Ma, *Current Research Progress of Optical Networks* (Springer, Berlin, 2019)
20. B.E.A. Saleh, M.C. Teich, *Fundamentals of Photonics*, 3rd edn. (Wiley, 2019)

21. The Institute of Electrical and Electronic Engineers (IEEE) and the Optical Society (OSA) hold several annual photonics conferences. The main one is the joint *Optical Fiber Communications* (OFC) conference and exposition held in the US
22. The annual *Asia Communications and Photonics Conference* (ACP) held in the Asia-Pacific region is sponsored by IEEE, OSA, SPIE, and the Chinese Optical Society
23. The *European Conference on Optical Fibre Communications (ECOC)* is held annually in Europe. It is sponsored by various European engineering organizations
24. ITU-T Recommendation G.Sup39, *Optical System Design and Engineering Considerations*, ed. 5, Feb. 2016
25. R. Kubo et al., Demonstration of 10-Gbit/s transmission over G.652 fiber for T-band optical access systems using quantum-dot semiconductor devices. IEICE Electron. Express **15**(18), 1–6 (2018)
26. O. Bertran-Pardo, J. Renaudier, G. Charlet, H. Mardoyan, P. Tran, M. Salsi, S. Bigo, Overlaying 10 Gb/s legacy optical networks with 40 and 100 Gb/s coherent terminals. J. Lightw. Technol. **30**(14), 2367–2375 (2012)
27. P.J. Winzer, Optical networking beyond WDM. IEEE Photonics J. **4**(2), 647–651 (2012)
28. IEEE P802.3bs 400 Gb/s Ethernet Task Force. Available online: www.ieee802.org/3/bs/
29. L.W. Couch II, *Digital and Analog Communication Systems*, 8th edn. (Prentice Hall, 2012)
30. B.A. Forouzan, *Data Communications and Networking*, 5th edn., (McGraw-Hill, 2013)
31. B. Sklar, *Digital Communications* (Pearson, 2017)
32. M.S. Alencar, V.C. da Rocha, *Communication Systems* (Springer, 2020)
33. D. Chadha, *Optical WDM Networks* (Wiley, 2019)
34. H. van Helvoort, *The ComSoc Guide to Next Generation Optical Transport: SDH/SONET/ODN* (Wiley/IEEE Press, 2009)
35. R.K. Jain, *Principles of Synchronous Digital Hierarchy* (CRC Press, 2012)
36. I.B. Djordjevic, *Advanced Optical and Wireless Communications Systems* (Springer, Berlin, 2018)
37. M. Mazur, A. Lorences-Riesgo, J. Schroder, P.A. Andrekson, M. Karlsson, 10 Tb/s PM-64QAM self-homodyne comb-based superchannel transmission with 4% shared pilot tone overhead. J. Lightwave Technol. **36**(16), 3176–3184 (2018)
38. Y. Yue, Q. Wang, J. Anderson, Experimental investigation of 400 Gb/s data center interconnect using unamplified high-baud-rate and high-order QAM single-carrier signal. Appl. Sci. **9**, 2455–2464 (2019)
39. J. Yu, Spectrally efficient single carrier 400G optical signal transmission. Front. Optoelectron. **12**, 15–23 (2019)
40. Telecommunications Sector-International Telecommunications Union (ITU-T), https://www.itu.int/en/ITU-YT: This organization publishes various series G.700, G.800, and G.900 Recommendations for various aspects of optical fibers, photonic components, and networking
41. F.D. Wright, The IEEE standards association and its ecosystem. IEEE Commun. Mag. **46**, 32–39 (2008)

Chapter 2
Optical Fiber Structures and Light Guiding Principles

Abstract Photonics technology is the basic indispensible tool and foundation for optical fiber communications. To understand how light signals travel along an optical fiber, this chapter first describes the fundamental nature of light and discusses how light propagates in a dielectric medium such as glass. The discussion then examines the structure of optical fibers and presents two mechanisms that show how light travels along an optical fiber.

The operational characteristics of an optical fiber largely determine the overall performance of a lightwave transmission system. Some of the questions that arise concerning optical fibers are

1. What is the structure of an optical fiber?
2. How does light propagate along a fiber?
3. Of what materials are fibers made?
4. How is the fiber fabricated?
5. How are fibers incorporated into cable structures?
6. What is the signal loss or attenuation mechanism in a fiber?
7. Why and to what degree does a signal get distorted as it travels along a fiber?

The purpose of this chapter is to present some of the fundamental answers to the first five questions in order to attain a good understanding of the physical structure and waveguiding properties of optical fibers. Questions 6 and 7 are answered in Chap. 3. The discussions address both conventional silica and photonic crystal fibers. Chapter 12 addresses additional nonlinear distortion effects that can arise in multiple-wavelength optical networks.

Fiber optics technology involves the emission, transmission, and detection of light, so the discussion first considers the nature of light and then reviews a few basic laws and definitions of optics. Following a description of the structure of optical fibers, two methods are used to describe how an optical fiber guides light. The first approach uses the geometrical or ray optics concept of light reflection and refraction to provide an intuitive picture of the propagation mechanisms. In the second approach, light is treated as an electromagnetic wave that propagates along the optical fiber

© The Author(s), under exclusive license to Springer Nature Singapore Pte Ltd. 2021   31
G. Keiser, *Fiber Optic Communications*,
https://doi.org/10.1007/978-981-33-4665-9_2

waveguide. This involves solving Maxwell's equations subject to the cylindrical boundary conditions of the fiber.

2.1 The Nature of Light

The concepts concerning the nature of light have undergone several variations during the history of physics [1]. Until the early seventeenth century, it was generally believed that light consisted of a stream of minute particles that were emitted by luminous sources. These particles were pictured as traveling in straight lines, and it was assumed that they could penetrate transparent materials but were reflected from opaque ones. This theory adequately described certain large-scale optical effects, such as reflection and refraction, but failed to explain finer-scale phenomena, such as interference and diffraction.

Fresnel gave the correct explanation of diffraction in 1815. He showed that the approximately rectilinear propagation character of light could be interpreted on the assumption that light is a wave motion, and that the diffraction fringes could thus be accounted for in detail. Later, the work of Maxwell in 1864 theorized that light waves must be electromagnetic in nature. Furthermore, observation of polarization effects indicated that light waves are transverse (i.e., the wave motion is perpendicular to the direction in which the wave travels). In this *wave optics* or *physical optics viewpoint*, a series of successive spherical wave fronts (referred to as a *train of waves*) spaced at regular intervals called a *wavelength* can represent the electromagnetic waves radiated by a small optical source with the source at the center as shown in Fig. 2.1. A *wave front* is defined as the locus of all points in the wave train that have the same

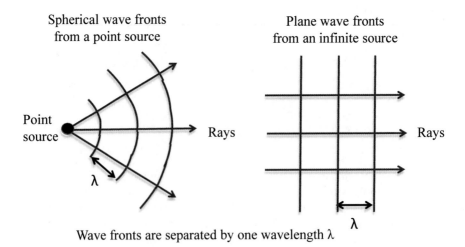

Fig. 2.1 Representations of spherical and plane wave fronts and their associated rays

phase. Generally, one draws wave fronts passing through either the maxima or the minima of the wave, such as the peak or trough of a sine wave, for example. Thus the wave fronts (also called *phase fronts*) are separated by one wavelength.

When the wavelength of the light is much smaller than the object (or opening) that it encounters, the wave fronts appear as straight lines to this object or opening. In this case, the light wave can be represented as a *plane wave,* and its direction of travel can be indicated by a *light ray,* which is drawn perpendicular to the phase front, as shown in Fig. 2.1. The light-ray concept allows large-scale optical effects such as reflection and refraction to be analyzed by the simple geometrical process of *ray tracing.* This view of optics is referred to as *ray* or *geometrical optics.* The concept of light rays is very useful because the rays show the direction of energy flow in the light beam.

2.1.1 Polarization

Light emitted by the sun or by an incandescent lamp is created by electromagnetic waves that vibrate in a variety of directions. This type of light is called *unpolarized light.* Lightwaves in which the vibrations occur in a single plane are known as *polarized light.* The process of transforming unpolarized light into polarized light is known as *polarization.* The polarization characteristics of lightwaves are important when examining the behavior of components such as optical isolators and filters. Polarization-sensitive devices include light signal modulators, polarization filters, Faraday rotators, beam splitters, and beam displacers. Birefringent crystals such as calcite, lithium niobate, rutile, and yttrium vanadate are polarization-sensitive materials used in such components.

Light is composed of one or more *transverse electromagnetic waves* that have both an electric field (called E field) and a magnetic field (called H field) component [2]. In a transverse wave the directions of the vibrating electric and magnetic fields are perpendicular to each other and are at right angles to the direction of propagation of the wave, as Fig. 2.2 shows. The waves are moving in the direction indicated by the *wave vector* **k**. The magnitude of the wave vector **k** is $k = 2\pi/\lambda$, which is known as the *wave propagation constant* with λ being the wavelength of the light. Based on Maxwell's equations, it can be shown that E and H are both perpendicular to the direction of propagation. This condition defines a *plane wave*; that is, the vibrations in the electric field are parallel to each other at all points in the wave. Thus, the electric field forms a plane called the *plane of vibration.* Likewise all points in the magnetic field component of the wave lie in another plane of vibration. Furthermore, **E** and **H** are mutually perpendicular, so that **E**, **H**, and **k** form a set of orthogonal vectors.

An ordinary lightwave is made up of many transverse waves that vibrate in a variety of directions (i.e., in more than one plane) and is referred to as *unpolarized light.* However any arbitrary direction of vibration of a specific transverse wave can be represented as a combination of two orthogonal plane polarization components.

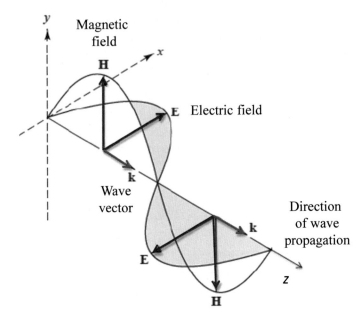

Fig. 2.2 Electric and magnetic field distributions in a train of plane electromagnetic waves at a given instant in time

This concept is important when examining the reflection and refraction of lightwaves at the interface of two different media, and when examining the propagation of light along an optical fiber. In the case when all the electric field planes of the different transverse waves are aligned parallel to each other, then the lightwave is *linearly polarized*. This is the simplest type of polarization.

2.1.2   Linear Polarization

The electric or magnetic field of a train of *plane linearly polarized waves* traveling in a direction $\mathbf{k}$ can be represented in the general form

$$\mathbf{A}(\mathbf{r},t) = \mathbf{e}_i A_0 \exp[j(\omega t - \mathbf{k} \cdot \mathbf{r})] \tag{2.1}$$

with $\mathbf{r} = x\mathbf{e}_x + y\mathbf{e}_y + z\mathbf{e}_z$ representing a general position vector and $\mathbf{k} = k_x\mathbf{e}_x + k_y\mathbf{e}_y + k_z\mathbf{e}_z$ representing the wave propagation vector. Here, A_0 is the maximum amplitude of the wave, $\omega = 2\pi\nu$, where ν is the frequency of the light and $\mathbf{e}_i$ is a unit vector lying parallel to an axis designated by i.

The components of the actual (measurable) electromagnetic field represented by Eq. (2.1) are obtained by taking the real part of this equation. For example, if $\mathbf{k} = k\mathbf{e}_z$, and if $\mathbf{A}$ denotes the electric field $\mathbf{E}$ with the coordinate axes chosen such that

$e_i = e_x$, then the real measurable electric field is given by

$$\mathbf{E}_x(z, t) = \text{Re}(\mathbf{E}) = \mathbf{e}_x E_{0x} \cos(\omega t - kz) = \mathbf{e}_x E_x \qquad (2.2)$$

which represents a plane wave that varies harmonically as it travels in the z direction. Here E_{0x} is the maximum amplitude of the wave along the x axis and E_x is the amplitude at a given value of z. The reason for using the exponential form shown in Eq. (2.1) is that it is more easily handled mathematically than equivalent expressions given in terms of sine and cosine. In addition, the rationale for using harmonic functions is that any waveform can be expressed in terms of sinusoidal waves using Fourier techniques.

The plane wave example given by Eq. (2.2) has its electric field vector always pointing in the $\mathbf{e}_x$ direction. Such a wave is *linearly polarized* with polarization vector $\mathbf{e}_x$. A general state of polarization is described by considering another linearly polarized wave that is independent of the first wave and orthogonal to it. Let this wave be

$$\mathbf{E}_y(z, t) = \mathbf{e}_y E_{0y} \cos(\omega t - kz + \delta) = \mathbf{e}_y E_y \qquad (2.3)$$

where δ is the relative phase difference between the waves. Similar to Eq. (2.2), E_{0y} is the maximum amplitude of the wave along the y axis and E_y is the amplitude at a given value of z. The resultant wave is

$$\mathbf{E}(z, t) = \mathbf{E}_x(z, t) + \mathbf{E}_y(z, t) \qquad (2.4)$$

If δ is zero or an integer multiple of 2π, then the waves are in phase. Equation (2.4) is then also a linearly polarized wave with a polarization vector making an angle

$$\theta = \arctan \frac{E_{0y}}{E_{0x}} \qquad (2.5)$$

with respect to $\mathbf{e}_x$ and having a magnitude

$$E = \left(E_{0x}^2 + E_{0y}^2 \right)^{1/2} \qquad (2.6)$$

This case is shown schematically in Fig. 2.3. Conversely, just as any two orthogonal plane waves can be combined into a linearly polarized wave, an arbitrary linearly polarized wave can be resolved into two independent orthogonal plane waves that are in phase.

Example 2.1 The general form of an electromagnetic wave is

$$y = (\text{amplitude in } \mu m) \times \cos(\omega t - kz) = A \cos[2\pi(\nu t - z/\lambda)]$$

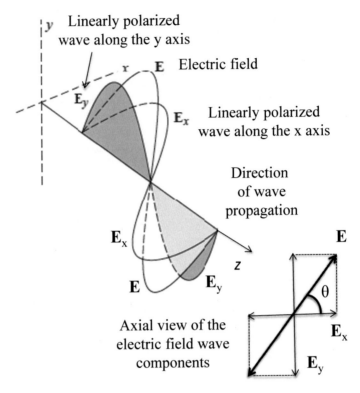

Fig. 2.3 Addition of two linearly polarized waves having a zero relative phase between them

Find (a) the amplitude, (b) the wavelength, (c) the angular frequency, and (d) the displacement at time $t = 0$ and $z = 4$ μm of a given plane electromagnetic wave specified by the equation $y = 12 \cos [2\pi (3t - 1.2z)]$.

Solution From the above general wave equation for y it follows that

(a) Amplitude $= 12$ μm
(b) Wavelength: $1/\lambda = 1.2$ μm^{-1} so that $\lambda = 833$ nm
(c) The angular frequency is $\omega = 2\pi\nu = 2\pi (3) = 6\pi$
(d) At time $t = 0$ and $z = 4$ μm we have that the displacement is

$$y = 12 \cos [2\pi(-1.2 \text{ μm}^{-1})(4 \text{ μm})] = 12 \cos[2\pi(-4.8)] = 10.38 \text{ μm}$$

2.1.3 Elliptical Polarization and Circular Polarization

For general values of δ the wave given by Eq. (2.4) is *elliptically polarized*. The resultant field vector **E** will both rotate and change its magnitude as a function of the angular frequency ω. Eliminating the $(\omega t - kz)$ dependence between Eqs. (2.2) and (2.3) for a general value of δ yields

$$\left(\frac{E_x}{E_{0x}}\right)^2 + \left(\frac{E_y}{E_{0y}}\right)^2 - 2\left(\frac{E_x}{E_{0x}}\right)\left(\frac{E_y}{E_{0y}}\right)\cos\delta = \sin^2\delta \qquad (2.7)$$

which is the general equation of an ellipse. Thus as Fig. 2.4 shows, the endpoint of **E** will trace out an ellipse at a given point in space. The axis of the ellipse makes an angle α relative to the x axis given by

$$\tan 2\alpha = \frac{2E_{0x}E_{0y}\cos\delta}{E_{0x}^2 - E_{0y}^2} \qquad (2.8)$$

Aligning the principal axis of the ellipse with the x axis gives a better picture of Eq. (2.7). In that case $\alpha = 0$, or, equivalently, $\delta = \pm\pi/2, \pm 3\pi/2, ...$, so that Eq. (2.7) becomes

$$\left(\frac{E_x}{E_{0x}}\right)^2 + \left(\frac{E_y}{E_{0y}}\right)^2 = 1 \qquad (2.9)$$

This is the equation of an ellipse with the origin at the center and semi-axes equal to E_{0x} and E_{0y}.

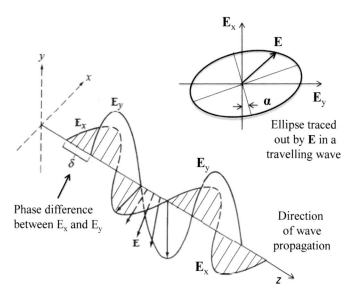

Fig. 2.4 Elliptically polarized light results from the addition of two linearly polarized waves of unequal amplitude having a nonzero phase difference δ between them

When $E_{0x} = E_{0y} = E_0$ and the relative phase difference $\delta = \pm \pi/2 + 2m\pi$, where $m = 0, \pm 1, \pm 2, ...$, then the light is *circularly polarized*. In this case, Eq. (2.9) reduces to

$$E_x^2 + E_y^2 = E_0^2 \qquad (2.10)$$

which defines a circle. Choosing the positive sign for δ, Eqs. (2.2) and (2.3) become

$$\mathbf{E}_x(z, t) = \mathbf{e}_x E_0 \cos(\omega t - kz) \qquad (2.11)$$

$$\mathbf{E}_y(z, t) = -\mathbf{e}_y E_0 \sin(\omega t - kz) \qquad (2.12)$$

In this case, the endpoint of $\mathbf{E}$ will trace out a circle at a given point in space, as Fig. 2.5 illustrates. To see this, consider an observer located at some arbitrary point z_{ref} toward which the wave is moving. For convenience, pick the reference point to be at $z = \pi/k$ at $t = 0$. Then, using Eqs. (2.11) and (2.12) it follows that

$$\mathbf{E}_x(z, t) = -\mathbf{e}_x E_0 \text{ and } \mathbf{E}_y(z, t) = 0$$

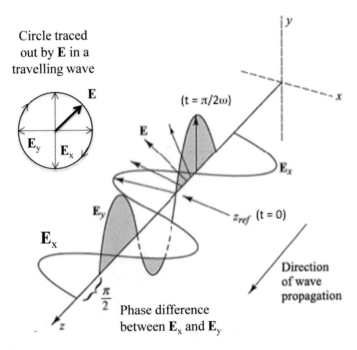

Fig. 2.5 Addition of two equal-amplitude linearly polarized waves with a relative phase difference $\delta = \pi/2 + 2m\pi$ results in a right circularly polarized wave

so that $\mathbf{E}$ lies along the negative x axis as Fig. 2.5 shows. At a later time, say $t = \pi/2\omega$, the electric field vector has rotated through $90°$ and now lies along the positive y axis at z_{ref}. Thus as the wave moves toward the observer with increasing time, the resultant electric field vector $\mathbf{E}$ rotates *clockwise* at an angular frequency ω. It makes one complete rotation as the wave advances through one wavelength. Such a light wave is *right circularly polarized.*

If one chooses the negative sign for δ, then the electric field vector is given by

$$\mathbf{E} = E_0\big[\mathbf{e}_x \cos(\omega t - kz) + \mathbf{e}_y \sin(\omega t - kz)\big] \tag{2.13}$$

Now $\mathbf{E}$ rotates *counterclockwise* and the wave is *left circularly polarized.*

2.1.4 Quantum Aspects of Light

The wave theory of light adequately accounts for all phenomena involving the transmission of light. However, in dealing with the interaction of light and matter, such as occurs in dispersion and in the emission and absorption of light, neither the particle theory nor the wave theory of light is appropriate. Instead, one must turn to quantum theory, which indicates that optical radiation has particle as well as wave properties. The particle nature arises from the observation that light energy is always emitted or absorbed in discrete units called *quanta* or *photons.* In all experiments used to show the existence of photons, the photon energy is found to depend only on the frequency v. This frequency, in turn, must be measured by observing a wave property of light.

As described in Sect. 1.2.1 and illustrated in Fig. 1.2, the physical properties of a photon can be measured in terms of its wavelength, energy, or frequency. The relationship between the energy E and the frequency v of a photon is given by

$$E = hv \tag{2.14}$$

where $h = 6.6256 \times 10^{-34}$ J·s is Planck's constant. When light is incident on an atom, a photon can transfer its energy to an electron within this atom, thereby exciting it to a higher energy level. In this process either all or none of the photon energy is imparted to the electron. The energy absorbed by the electron must be exactly equal to that required to excite the electron to a higher energy level. Conversely, an electron in an excited state can drop to a lower state separated from it by an energy hv by emitting a photon of exactly this energy.

Drill Problem 2.1 Using Eq. (2.14), show that the corresponding wavelengths of photons with energies of 0.95 eV and 0.80 eV are 1310 nm and 1550 nm, respectively.

Example 2.2 (a) Consider an incoming photon that boosts an electron from a ground state level E_1 to an excited level E_2. If the incoming photon energy $E = E_2 - E_1 = 1.512$ eV, what is the wavelength of the incoming photon? (b) Now suppose this excited electron loses some of its energy and moves to a slightly lower energy level E_3. If the electron then drops back to level E_1 thereby emitting a photon of energy $E_3 - E_1 = 1.450$ eV, what is the wavelength of the emitted photon?

Solution (a) Referring back to Sect. 1.2, from Eqs. (1.2) and (1.3), $\lambda_{incident} = 1.2405/1.512$ eV $= 0.820$ μm $= 820$ nm. (b) From Eqs. (1.2) and (1.3), $\lambda_{emitted} = 1.2405/1.450$ eV $= 0.855$ μm $= 855$ nm.

2.2 Basic Laws and Definitions of Optics

This section reviews some of the basic optics laws and definitions relevant to optical fiber transmission technology [1, 3]. These include Snell's law, the definition of the refractive index of a material, and the concepts of reflection, refraction, and polarization.

2.2.1 Concept of Refractive Index

A fundamental optical parameter of a material is the *refractive index* (or *index of refraction*). In free space light travels at a speed $c = 2.99793 \times 10^8$ m/s $\approx 3 \times 10^8$ m/s. The speed of light is related to the frequency v and the wavelength λ by $c = v\lambda$. Upon entering a dielectric or nonconducting medium the wave now travels at a speed s, which is characteristic of the material and is less than c. The ratio of the speed of light in a vacuum to that in matter is the index of refraction n of the material and is given by Eq. (2.15). Representatives values are listed in Table 2.1

$$n = \frac{c}{s} \tag{2.15}$$

2.2.2 Basis of Reflection and Refraction

The concepts of reflection and refraction can be interpreted by considering the behavior of light rays associated with plane waves traveling in a dielectric material. When a light ray encounters a boundary separating two different dielectric media, part of the ray is reflected back into the first medium and the remainder is bent (or refracted) as it enters the second material. This is shown in Fig. 2.6 for the interface

Table. 2.1 Examples of the indices of refraction for various substances

Material	Refractive index
Acetone	1.356
Air	1.000
Diamond	2.419
Ethyl alcohol	1.361
Fused quartz (SiO_2): varies with wavelength	1.453 @ 850 nm
Gallium arsenide (GaAs)	3.299 (infrared region)
Glass, crown	1.52–1.62
Glycerin	1.473
Polymethylmethacrylate (PMMA)	1.489
Silicon (varies with wavelength)	3.650 @ 850 nm
Water	1.333

Fig. 2.6 Refraction and reflection of a light ray at a material boundary

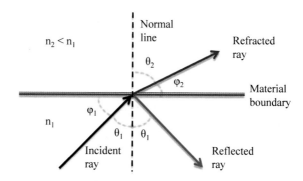

between two materials that have refractive indices n_1 and n_2, where $n_2 < n_1$. The bending or refraction of the light ray at the interface is a result of the difference in the speed of light in two materials that have different refractive indices. The relationship at the interface is known as *Snell's law* and is given by

$$n_1 \sin \theta_1 = n_2 \sin \theta_2 \tag{2.16a}$$

or, equivalently, as

$$n_1 \cos \varphi_1 = n_2 \cos \varphi_2 \tag{2.16b}$$

where the angles are defined in Fig. 2.6. The angle θ_1 between the incident ray and the normal to the surface is known as the *angle of incidence*.

According to the law of reflection, the angle θ_1 at which the incident ray strikes the interface is exactly equal to the angle that the reflected ray makes with the normal line to the same interface. In addition, the incident ray, the normal to the interface, and the

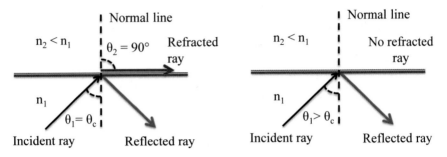

Fig. 2.7 Representation of the critical angle and total internal reflection at a glass-air interface, where n_1 is the refractive index of glass

reflected ray all lie in the same plane, which is perpendicular to the interface plane between the two materials. This plane is called the *plane of incidence*. When light traveling in a certain medium is reflected off an optically denser material (one with a higher refractive index), the process is referred to as *external reflection*. Conversely, the reflection of light off of less optically dense material (such as light traveling in glass being reflected at a glass–air interface) is called *internal reflection*.

As the angle of incidence θ_1 in an optically denser material becomes larger, the refracted angle θ_2 approaches $\pi/2$. Beyond this point no refraction is possible as the incident angle increases and the light rays undergo *total internal reflection*. The conditions required for a light ray to be totally internally reflected can be determined by using Snell's law. Consider Fig. 2.7, which shows a glass surface in air. A light ray gets bent toward the glass surface as it leaves the glass in accordance with Snell's law. If the angle of incidence θ_1 is increased, a point will eventually be reached where the light ray in air is parallel to the glass surface. This point is known as the *critical angle of incidence* θ_c. When the incidence angle θ_1 is greater than the critical angle, the condition for total internal reflection is satisfied; that is, the light is totally reflected back into the glass with no light escaping from the glass surface.

To find the critical angle, consider Snell's law as given by Eq. (2.16). The critical angle is reached when $\theta_2 = 90°$ so that $\sin \theta_2 = 1$. Substituting this value of θ_2 into Eq. (2.16) thus show that the critical angle is thus determined from the condition

$$\sin \theta_c = \frac{n_2}{n_1} \tag{2.17}$$

Example 2.3 Consider the interface between a smooth dielectric material with n_1 = 1.48 and air for which $n_2 = 1.00$. What is the critical angle for light traveling in the dielectric material?

Solution From Eq. (2.17), for light traveling in the dielectric material the critical angle is

$$\theta_c = sin^{-1}\frac{n_2}{n_1} = sin^{-1}0.676 = 42.5°$$

Thus any light ray traveling in the dielectric material that is incident on the material–air interface at an angle θ_1 with respect to the normal (as shown in Fig. 2.7) greater than 42.5° is totally reflected back into the dielectric material.

Example 2.4 A light ray traveling in air ($n_1 = 1.00$) is incident on a smooth, flat slab of crown glass, which has a refractive index $n_2 = 1.52$. If the incoming ray makes an angle of $\theta_1 = 30.0°$ with respect to the normal, what is the angle of refraction θ_2 in the glass?

Solution From Snell's law given by Eq. (2.16),

$$\sin \theta_2 = \frac{n_1}{n_2} \sin \theta_1 = \frac{1.00}{1.52} \sin 30° = 0.658 \times 0.50 = 0.329$$

Solving for θ_2 then yields $\theta_2 = \sin^{-1}(0.329) = 19.2°$.

Drill Problem 2.2 Consider the interface between a GaAs surface with a refractive index $n_1 = 3.299$ and air for which $n_2 = 1.000$. Show that the critical angle is $\theta_c = 17.6°$.

An important consideration for optical communication links is the power reflection for light incident normally at the interfaces between two fibers or between a fiber and a different type of material, such as air, a light source, or a photodetector. This situation is shown in Fig. 2.8 for light that is incident perpendicularly on the interface between materials having refractive indices n_1 and n_2. A typical case in fiber links is the interface between the end of an optical fiber and air. The fraction of the incident power that is reflected at the interface is given by the *reflectance* R

$$R = \left(\frac{n_1 - n_2}{n_1 + n_2}\right)^2 \tag{2.18a}$$

Fig. 2.8 Power reflection for light incident normally at the interface between two different types of material

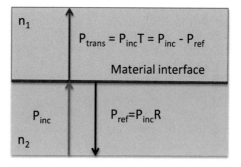

The corresponding fraction of optical power that traverses the material interface is given by the *transmittance* T

$$T = \frac{4n_1 n_2}{(n_1 + n_2)^2} \tag{2.18b}$$

These expressions are derived from the Fresnel reflection coefficient analyses [3] given in the appendix of this chapter. Note that $R + T = 1$. Chapter 5 gives detailed applications and examples of these optical power reflection conditions.

In addition, when light is totally internally reflected, a phase change δ occurs in the reflected wave. This phase change depends on the angle $\theta < \pi/2 - \varphi$ according to the relationships

$$tan \frac{\delta_N}{2} = \frac{\sqrt{n^2 cos^2 \theta_1 - 1}}{n sin \theta_1} \tag{2.19a}$$

$$tan \frac{\delta_p}{2} = \frac{\sqrt{n^2 cos^2 \theta_1 - 1}}{sin \theta_1} \tag{2.19b}$$

Here, δ_N and δ_p are the phase shifts of the electric field wave components normal and parallel to the plane of incidence, respectively, and $n = n_1/n_2$.

2.2.3 Polarization Characteristics of Light

A generic lightwave consists of many transverse electromagnetic waves that vibrate in a variety of directions (i.e., in more than one plane) and is called *unpolarized light*. However, one can represent any arbitrary direction of vibration as a combination of a parallel vibration and a perpendicular vibration, as shown in Fig. 2.9. Therefore, one can consider unpolarized light as consisting of two orthogonal plane polarization components, one that lies in the plane of incidence (the plane containing the incident and reflected rays) and the other of which lies in a plane perpendicular to the plane of incidence. These are the *parallel polarization* and the *perpendicular polarization* components, respectively. In the case when all the electric field planes of the different transverse waves are aligned parallel to each other, then the lightwave is linearly polarized. This is the simplest type of polarization, as Sect. 2.1.1 describes.

Unpolarized light can be split into separate polarization components either by reflection off of a nonmetallic surface or by refraction when the light passes from one material to another. As noted in Fig. 2.10, when an unpolarized light beam traveling in air impinges on a nonmetallic surface such as glass, part of the beam is reflected and part is refracted into the glass. A circled dot and an arrow designate the parallel and perpendicular polarization components, respectively, in Fig. 2.10. The reflected beam is partially polarized and at a specific angle (known as *Brewster's angle*) the reflected light is completely perpendicularly polarized. The parallel component of

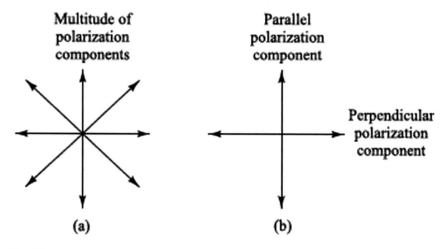

Fig. 2.9 Polarization represented as a combination of a parallel vibration and a perpendicular vibration

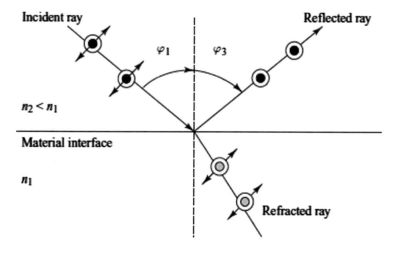

Fig. 2.10 Behavior of an unpolarized light beam at the interface between air and a nonmetallic surface

the refracted beam is transmitted entirely into the glass, whereas the perpendicular component is only partially refracted. How much of the refracted light is polarized depends on the angle at which the light approaches the surface and on the material composition.

2.2.4 Polarization-Sensitive Devices

The polarization characteristics of light are important when examining the behavior of materials for devices such as optical isolators and light filters. Three polarization-sensitive materials or devices that are used in such components are polarizers, Faraday rotators, and birefringent crystals.

A *polarizer* is a material or device that transmits only one polarization component and blocks the other. For example, in the case when unpolarized light enters a polarizer that has a vertical transmission axis as shown in Fig. 2.11, only the vertical polarization component passes through the device. A familiar example of this concept is the use of polarizing sunglasses to reduce the glare of partially polarized sunlight reflections from road or water surfaces. To see the polarization property of the sunglasses, a number of glare spots will appear when users tilt their head sideways. The polarization filters in the sunglasses block out the polarized light coming from these glare spots when the head is held normally.

A *Faraday rotator* is a device that rotates the *state of polarization* (SOP) of light passing through it by a specific amount. For example, a popular device rotates the SOP clockwise by 45° or a quarter of a wavelength, as shown in Fig. 2.12. This rotation is independent of the SOP of input light, but the rotation angle is different depending on the direction in which the light passes through the device. That is, the rotation process is not reciprocal. In this process, the SOP of the input light is

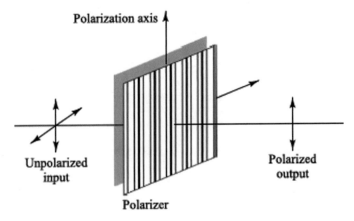

Fig. 2.11 Only the vertical polarization component passes through a vertically oriented polarizer

Fig. 2.12 A Faraday rotator is a device that rotates the state of polarization, for example, clockwise by 45° or a quarter of a wavelength

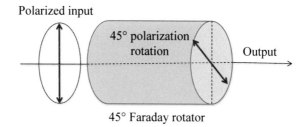

45° Faraday rotator

maintained after the rotation. For example, if the input light to a 45° Faraday rotator is linearly polarized in a vertical direction, then the rotated light exiting the crystal also is linearly polarized at a 45° angle. The Faraday rotator material usually is an asymmetric crystal such as yttrium iron garnet (YIG), and the degree of angular rotation is proportional to the thickness of the device.

Birefringent or *double-refractive crystals* have a property called *double refraction*. This means that the indices of refraction are slightly different along two perpendicular axes of the crystal as shown in Fig. 2.13. A device made from such materials is known as a *spatial walk-off polarizer* (SWP). The SWP splits the light signal entering it into two orthogonally (perpendicularly) polarized beams. One of the beams is called an *ordinary ray* or o-ray because it obeys Snell's law of refraction at the crystal surface. The second beam is called the *extraordinary ray* or e-ray because it refracts at an angle that deviates from the prediction of the standard form of Snell's law. Each of the two orthogonal polarization components thus is refracted at a different angle as shown in Fig. 2.13. For example, if the incident unpolarized light arrives at an angle perpendicular to the surface of the device, the o-ray can pass straight through the device whereas the e-ray component is deflected at a slight angle so it follows a different path through the material.

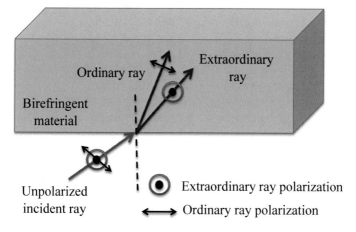

Fig. 2.13 A birefringent crystal splits the light signal entering it into two perpendicularly polarized beams

Table 2.2 Common birefringent crystals and some applications

Crystal name	Symbol	n_o	n_e	Applications
Calcite	CaCO$_3$	1.658	1.486	Polarization controllers and beamsplitters
Lithium niobate	LiNbOj	2.286	2.200	Light signal modulators
Rutile	TiO$_2$	2.616	2.903	Optical isolators and circulators
Yttrium vanadate	YVO$_4$	1.945	2.149	Optical isolators, circulators, and beam displacers

Table 2.2 lists the ordinary index n_o and the extraordinary index n_e of some common birefringent crystals that are used in optical communication components and gives some of their applications.

2.3 Optical Fiber Configurations and Modes

This section first presents an overview of the underlying concepts of optical fiber modes and optical fiber configurations. The discussions in Sects. 2.3 through 2.7 address conventional optical fibers, which consist of solid dielectric structures. Section 2.8 describes the structure of photonic crystal fibers, which can be created to have a variety of internal microstructures. Chapter 3 describes the operational characteristics of both categories of fibers.

2.3.1 Conventional Fiber Types

An optical fiber is a dielectric waveguide that operates at optical frequencies. This fiber waveguide is normally cylindrical in form. It confines electromagnetic energy in the form of light to within its surfaces and guides the light in a direction parallel to its axis. The transmission properties of an optical waveguide are dictated by its structural characteristics, which have a major effect in determining how an optical signal is affected as it propagates along the fiber. The structure basically establishes the information-carrying capacity of the fiber and also influences the response of the waveguide to different kinds of environmental perturbations.

The propagation of light along a waveguide can be described in terms of a set of guided electromagnetic waves called the *modes* of the waveguide. These guided modes are referred to as the *bound* or *trapped* modes of the waveguide. Each guided mode is a pattern of electric and magnetic field distributions that is repeated along the fiber at equal intervals. Only a certain discrete number of modes are capable of propagating along the guide. These modes are those electromagnetic waves that satisfy the homogeneous wave equation in the fiber and the electromagnetic field boundary conditions at the core-cladding interface of the waveguide.

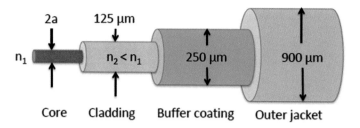

Fig. 2.14 A conventional silica fiber has a circular solid core of refractive index n_1 surrounded by a cladding with a refractive index $n_2 < n_1$; an elastic plastic buffer encapsulates the fiber

Although many different configurations of the optical waveguide have been discussed in the literature, [4] the most widely accepted structure is the single solid dielectric cylinder of radius a and index of refraction n_1 shown in Fig. 2.14. This cylinder is known as the *core* of the fiber. The core is surrounded by a solid dielectric *cladding*, which has a refractive index n_2 that is less than n_1. As described in Sect. 2.3.5, the cladding material is carefully selected to enable light to propagate efficiently along the core of the fiber. In addition, the cladding reduces scattering loss that results from dielectric discontinuities at the core surface, it adds mechanical strength to the fiber, and it protects the core from absorbing surface contaminants with which it could come in contact.

In standard optical fibers the core material is a highly pure silica glass (SiO_2) compound and is surrounded by a glass cladding. Common core sizes are about 9 and 50 μm (micrometers). A standard cladding diameter is 125 μm. Higher-loss plastic-core fibers with plastic claddings are also widely in use. In addition, most fibers are encapsulated in an elastic, abrasion-resistant plastic buffer coating and an outer strengthening jacket that have diameters of 250 μm and 900 μm, respectively. These materials add further strength to the fiber and mechanically isolate or buffer the fibers from small geometrical irregularities, distortions, or roughness of adjacent surfaces. These perturbations could otherwise cause scattering losses induced by random microscopic bends that can arise when the fibers are incorporated into cables or supported by other structures.

Variations in the material composition of the core give rise to the two commonly used fiber types shown in Fig. 2.15. In the first case, which is called a *step-index fiber*, the refractive index of the core is uniform throughout and undergoes an abrupt change (or step) at the cladding boundary. In the other case shown at the bottom of Fig. 2.15, the core refractive index is made to vary as a function of the radial distance from the center of the fiber. This type is a *graded-index fiber.*

Both the step-index and the graded-index fibers can be further divided into single-mode and multimode classes. As the name implies, a single-mode fiber sustains only one mode of propagation, whereas multimode fibers contain many hundreds of modes. A few typical sizes of single-and multimode fibers are given in Fig. 2.15 to provide an idea of the dimensional scale. Multimode fibers offer several advantages compared with single-mode fibers. As shown in Chap. 5, the larger core radii of

Fig. 2.15 Comparison of conventional single-mode and multimode step-index and graded-index optical fibers

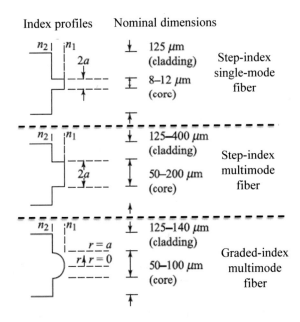

multimode fibers make it easier to launch optical power into the fiber and facilitate the connecting together of similar fibers. Another advantage is that light can be launched into a multimode fiber using a light-emitting-diode (LED) source, whereas single-mode fibers must generally be excited with laser diodes. Although LEDs have less optical output power than laser diodes (as discussed in Chap. 4), they are easier to make, are less expensive, require less complex circuitry, and have longer lifetimes than laser diodes, thus making them more desirable in certain applications.

A limitation of multimode fibers for high-speed long-distance transmission is that the bandwidth is restricted by intermodal dispersion. This effect is described in detail in Chap. 3. Briefly, intermodal dispersion can be defined as follows. When an optical pulse is launched into a fiber, the optical power in the pulse is distributed over all (or most) of the modes of the fiber. Each of the modes that can propagate in a multimode fiber travels at a slightly different velocity. This means that the modes in a given optical pulse arrive at the fiber end at slightly different times, thus causing the pulse to spread out in time as it travels along the fiber. This effect is known as *intermodal dispersion* or *modal delay* and can be reduced by using a graded-index profile in a fiber core. This allows graded-index fibers to have much larger bandwidths (data rate transmission capabilities) than step-index fibers. Even higher bandwidths are possible in single-mode fibers, where intermodal dispersion effects are not present because only one mode travels in the fiber.

2.3.2 Concepts of Rays and Modes

The electromagnetic light field that is guided along an optical fiber can be represented by a superposition of bound or trapped modes. Each of these guided modes consists of a set of simple electromagnetic field configurations. For monochromatic light fields of radian frequency ω, a mode traveling in the positive z direction (i.e., along the fiber axis) has a time and z dependence given by

$$e^{j(\omega t - \beta z)}$$

The factor β is the z component of the wave propagation constant $k = 2\pi/\lambda$ and is the main parameter of interest in describing fiber modes. For guided modes, β can assume only certain discrete values, which are determined from the requirement that the mode field must satisfy Maxwell's equations and the electric and magnetic field boundary conditions at the core-cladding interface.

Another method for theoretically studying the propagation characteristics of light in an optical fiber is the geometrical optics or ray-tracing approach. This method provides a good approximation to the light acceptance and guiding properties of optical fibers when the ratio of the fiber radius to the wavelength is large. This is known as the *small-wavelength limit*. Although the ray approach is strictly valid only in the zero-wavelength limit, it is still relatively accurate and extremely valuable for nonzero wavelengths when the number of guided modes is large; that is, for multimode fibers. The advantage of the ray approach is that, compared with the exact electromagnetic wave (modal) analysis, it gives a more direct physical interpretation of the light propagation characteristics in an optical fiber.

Because the concept of a light ray is very different from that of a mode, it is important to see qualitatively what the relationship is between them. (The mathematical details of this relationship are beyond the scope of this book but can be found in the literature [5–7]). A guided mode traveling in the z direction (along the fiber axis) can be decomposed into a family of superimposed plane waves that collectively form a standing-wave pattern in the direction transverse to the fiber axis. That is, the phases of the plane waves are such that the envelope of the collective set of waves remains stationary. Because with any plane wave one can associate a light ray that is perpendicular to the phase front of the wave, the family of plane waves corresponding to a particular mode forms a set of rays called a *ray congruence*. Each ray of this particular congruent set travels in the fiber at the same angle relative to the fiber axis. It is important to note that, since only a certain number M of discrete guided modes exist in a fiber, the possible angles of the ray congruences corresponding to these modes are also limited to the same number M. Although a simple ray picture appears to allow rays at any angle greater than the critical angle to propagate in a fiber, the allowable quantized propagation angles result when the phase condition for standing waves is introduced into the ray picture. This is discussed further in Sect. 2.3.5.

Despite the usefulness of the approximate geometrical optics method, a number of limitations and discrepancies exist between it and the exact modal analysis. An important case is the analysis of single-mode or few-mode fibers, which must be dealt with by using electromagnetic theory. Problems involving coherence or interference phenomena must also be solved with an electromagnetic approach. In addition, a modal analysis is necessary when knowledge of the field distribution of individual modes is required. This arises, for example, when analyzing the excitation of an individual mode or when analyzing the coupling of power between modes at waveguide imperfections (which is discussed in Sect. 3.1).

Another discrepancy between the ray optics approach and the modal analysis occurs when an optical fiber is uniformly bent with a constant radius of curvature. As shown in Chap. 3, wave optics correctly predicts that every mode of the curved fiber experiences some radiation loss. Ray optics, on the other hand, erroneously predicts that some ray congruences can undergo total internal reflection at the curve and, consequently, can remain guided without loss.

2.3.3 Structure of Step-Index Fibers

To begin the discussion of light propagation in an optical waveguide first consider the step-index fiber illustrated in Fig. 2.15. In practical step-index glass fibers the core of radius a has a refractive index n_1, which is typically equal to 1.48. This is surrounded by a cladding of slightly lower index n_2, where

$$n_2 = n_1(1 - \Delta) \qquad (2.20)$$

The parameter Δ is called the *core-cladding index difference* or simply the *index difference*. Values of n_2 are chosen such that Δ is nominally 1–3% for multimode fibers and from 0.2 to 1.0% for single-mode fibers. Because the core refractive index is larger than the cladding index, electromagnetic energy at optical frequencies is made to propagate along the fiber waveguide through internal reflection at the core-cladding interface.

2.3.4 Ray Optics Representation

Because the core size of multimode fibers is much larger than the wavelength of the light being transmitted (which is approximately 1 μm), an intuitive picture of the propagation mechanism in an ideal multimode step-index optical waveguide is most easily seen by a simple ray (geometrical) optics representation [6–11]. For simplicity, this analysis shall consider only a particular ray belonging to a ray congruence that represents a fiber mode. The two types of rays that can propagate in a fiber are meridional rays and skew rays. *Meridional rays* are confined to the meridian planes of the fiber, which are the planes that contain the axis of symmetry of the fiber (the core axis). Because a given meridional ray lies in a single plane, its path is easy to

Fig. 2.16 Ray optics representation of skew rays traveling in a step-index optical fiber core

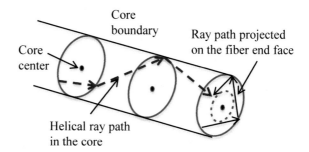

track as it travels along the fiber. Meridional rays can be divided into two general classes: bound rays that are trapped in the core and propagate along the fiber axis according to the laws of geometrical optics, and unbound rays that are refracted out of the fiber core.

Skew rays are not confined to a single plane, but instead tend to follow a helical-type path along the fiber as shown in Fig. 2.16. These rays are more difficult to track as they travel along the fiber because they do not lie in a single plane. Although skew rays constitute a major portion of the total number of guided rays, their analysis is not necessary to obtain a general picture of rays propagating in a fiber. The examination of meridional rays will suffice for this purpose. However, a detailed inclusion of skew rays will change such expressions as the light-acceptance ability of the fiber and power losses of light traveling along a waveguide.

A greater power loss arises when skew rays are included in the analyses because many of the skew rays that geometric optics predicts to be trapped in the fiber are actually leaky rays [6, 12, 13]. These *leaky rays* are only partially confined to the core of the circular optical fiber and attenuate as the light travels along the optical waveguide. This partial reflection of leaky rays cannot be described by pure ray theory alone. Instead, the analysis of radiation loss arising from these types of rays must be described by mode theory.

The meridional ray is shown in Fig. 2.17 for a step-index fiber. The light ray enters the fiber core from a medium of refractive index n at an angle θ_0 with respect to the fiber axis and strikes the core-cladding interface at a normal angle φ. If it strikes this interface at such an angle that it is totally internally reflected, then the meridional ray follows a zigzag path along the fiber core, passing through the axis of the guide after each reflection.

From Snell's law, the minimum or critical angle φ_c that supports total internal reflection for the meridional ray is given by

$$sin\ \phi_c = \frac{n_2}{n_1} \tag{2.21}$$

Rays striking the core-to-cladding interface at angles less than φ_c will refract out of the core and be lost in the cladding, as the dashed line shows. By applying Snell's law to the air–fiber face boundary, the condition of Eq. (2.21) can be related to the

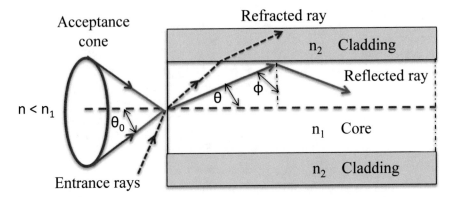

Fig. 2.17 Meridional ray optics representation of the propagation mechanism in an ideal step-index optical waveguide

maximum entrance angle $\theta_{0,\,max}$, which is called the *acceptance angle* θ_A, through the relationship

$$n \sin \theta_{0,max} = n \sin \theta_A = n_1 \sin \theta_c = \left(n_1^2 - n_2^2\right)^{1/2} \qquad (2.22)$$

where $\theta_c = \pi/2 - \varphi_c$. Thus those rays having entrance angles θ_0 less than θ_A will be totally internally reflected at the core–cladding interface. Thus θ_A defines an acceptance cone for an optical fiber.

Equation (2.22) also defines the *numerical aperture* (NA) of a step-index fiber for meridional rays:

$$NA = n \sin \theta_A = \left(n_1^2 - n_2^2\right)^{1/2} \approx n_1 \sqrt{2\Delta} \qquad (2.23)$$

The approximation on the right-hand side is valid for the typical case where Δ, as defined by Eq. (2.20), is much less than 1. Because the numerical aperture is related to the acceptance angle, it is commonly used to describe the light acceptance or gathering capability of a fiber and to calculate source-to-fiber optical power coupling efficiencies. This is detailed in Chap. 5. The numerical aperture is a dimensionless quantity less than unity, with values normally ranging from 0.14 to 0.50.

Example 2.5 Consider a multimode silica fiber that has a core refractive index $n_1 = 1.480$ and a cladding index $n_2 = 1.460$. Find (a) the critical angle, (b) the numerical aperture, and (c) the acceptance angle.

Solution

(a) From Eq. (2.21), the critical angle is given by

$$\varphi_c = \sin^{-1} \frac{n_2}{n_1} = \sin^{-1} \frac{1.460}{1.480} = 80.5°$$

(b) From Eq. (2.23) the numerical aperture is

$$NA = \left(n_1^2 - n_2^2\right)^{1/2} = 0.242$$

(c) From Eq. (2.22) the acceptance angle in air ($n = 1.00$) is

$$\theta_A = \sin^{-1} NA = \sin^{-1} 0.242 = 14°$$

Example 2.6 Consider a multimode fiber that has a core refractive index of 1.480 and a core-cladding index difference of 2.0% ($\Delta = 0.020$). Find (a) the numerical aperture, (b) the acceptance angle, and (c) the critical angle.

Solution From Eq. (2.20), the cladding index is $n_2 = n_1(1 - \Delta) = 1.480(0.980) = 1.450$.

(a) Using Eq. (2.23) then the numerical aperture is

$$NA = n_1\sqrt{2\Delta} = 1.480\sqrt{0.04} = 0.296$$

(b) Using Eq. (2.22) the acceptance angle in air ($n = 1.00$) is

$$\theta_A = \sin^{-1}NA = \sin^{-1}0.296 = 17.2°$$

(c) From Eq. (2.21) the critical angle at the core–cladding interface is

$$\varphi_c = \sin^{-1}\frac{n_2}{n_1} = \sin^{-1}0.980 = 78.5°$$

Drill Problem 2.3 Consider the interface between fiber core and cladding materials that have refractive indices of n_1 and n_2, respectively. If n_2 is smaller than n_1 by 1% and $n_1 = 1.450$, show that $n_2 = 1.435$. Show that the critical angle is $\varphi_c = 81.9°$.

2.3.5 Lightwaves in a Dielectric Slab Waveguide

Referring to Fig. 2.17, the ray theory appears to allow rays at any angle φ greater than the critical angle φ_c to propagate along the fiber. However, when the interference effect due to the phase of the plane wave associated with the ray is taken into account, it is seen that only waves at certain discrete angles greater than or equal to φ_c are capable of propagating along the fiber.

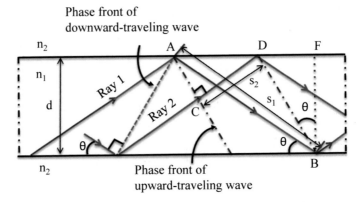

Fig. 2.18 When light propagates along a fiber waveguide, phase changes occur both as the wave travels through the fiber and at the reflection points

To see this, consider wave propagation in an infinite dielectric slab waveguide of thickness d. Its refractive index n_1 is greater than the index n_2 of the material above and below the slab. A wave will thus propagate in this guide through multiple reflections, provided that the angle of incidence with respect to the upper and lower surfaces satisfies the condition given in Eq. (2.22).

Figure 2.18 shows the geometry of the waves reflecting at the material interfaces. Here, first consider two rays, designated ray 1 and ray 2, associated with the same wave. The rays are incident on the material interface at an angle $\theta < \theta_c = \pi/2 - \varphi_c$. In Fig. 2.18 the solid lines denote the ray paths and the dashed lines show their associated constant-phase fronts.

The condition required for wave propagation in the dielectric slab is that all points on the same phase front of a plane wave must be in phase. This means that the phase change occurring in ray 1 when traveling from point A to point B minus the phase change in ray 2 between points C and D must differ by an integer multiple of 2π. As the wave travels through the material, it undergoes a phase shift Δ given by

$$\Delta = k_1 s = n_1 k s = n_1 2\pi s/\lambda$$

where

   $k_1 =$ the propagation constant in the medium of refractive index n_1.
   $k = k_1/n_1$ is the free-space propagation constant.
   $s =$ the distance the wave has traveled in the material.

The phase of the wave changes not only as the wave travels but also upon reflection from a dielectric interface, as described in Sect. 2.2.

In going from point A to point B, ray 1 travels a distance $s_1 = d/\sin \theta$ in the material, and undergoes two phase changes δ at the reflection points. Ray 2 does not incur any reflections in going from point C to point D. To determine its phase change, first note that the distance from point A to point D is $AD = AF - DF = (d/\tan \theta) - d \tan \theta$. Thus, the distance between points C and D is

$$s_2 = AD \cos \theta = (\cos^2 \theta - \sin^2 \theta)d/\sin \theta$$

The requirement for wave propagation can then be written as

$$\frac{2\pi n_1}{\lambda}(s_1 - s_2) + 2\delta = 2\pi m \tag{2.24a}$$

where $m = 0, 1, 2, 3, \ldots.$ Substituting the expressions for s_1 and s_2 into Eq. (2.24a) then yields

$$\frac{2\pi n_1}{\lambda}\left\{\frac{d}{\sin\theta} - \left[\frac{(\cos^2\theta - \sin^2\theta)d}{\sin\theta}\right]\right\} + 2\delta = 2\pi m \tag{2.24b}$$

which can be reduced to

$$\frac{2\pi n_1 d \sin \theta}{\lambda} + \delta = \pi m \tag{2.24c}$$

Considering only electric waves with components normal to the plane of incidence, it follows from Eq. (2.19a) that the phase shift upon reflection is

$$\delta = -2 arctan\left[\frac{\sqrt{\cos^2\theta - (n_2^2/n_1^2)}}{\sin\theta}\right] \tag{2.25}$$

The negative sign is needed here because the wave in the medium must be a decaying and not a growing wave. Substituting this expression into Eq. (2.24c) yields

$$\frac{2\pi n_1 d \sin \theta}{\lambda} - \pi m = 2 arctan\left[\frac{\sqrt{\cos^2\theta - (n_2^2/n_1^2)}}{\sin \theta}\right] \tag{2.26a}$$

or

$$tan\left(\frac{\pi n_1 d \sin \theta}{\lambda} - \frac{\pi m}{2}\right) = \left[\frac{\sqrt{n_1^2\cos^2\theta - n_2^2}}{n_1 \sin \theta}\right] \tag{2.26b}$$

Thus only waves that have those angles θ that satisfy the condition in Eq. (2.26) will propagate in the dielectric slab waveguide.

2.4   Modes in Circular Waveguides

To attain a detailed understanding of the optical power propagation mechanism in a fiber, it is necessary to solve Maxwell's equations subject to their cylindrical boundary conditions of the electric and magnetic fields at the interface between the core and the cladding of the fiber. This has been done in extensive detail in a number of works [6, 10, 14–18]. Because a complete mathematical treatment is beyond the scope of this book, only the results of the analysis of Maxwell's equations is given here.

When solving Maxwell's equations for hollow metallic waveguides, only transverse electric (TE) modes and transverse magnetic (TM) modes are involved. However, in optical fibers the core-cladding boundary conditions lead to a coupling between the electric and magnetic field components. This gives rise to hybrid electromagnetic modes, which makes optical waveguide analysis more complex than metallic waveguide analysis. The hybrid modes are designated as EH or HE modes, depending on whether the transverse electric field (the E field) or the transverse magnetic field (the H field), respectively, is larger for that mode. The two lowest-order modes are called HE_{11} and TE_{01}, where the subscripts refer to possible modes of propagation of the optical field.

2.4.1   Basic Modal Concepts

Before progressing with a discussion of mode theory in circular optical fibers, this section will qualitatively examine the appearance of modal fields in the planar dielectric slab waveguide shown in Fig. 2.19. The core of this waveguide is a dielectric slab of index n_1 that is sandwiched between two dielectric layers that have refractive indices $n_2 < n_1$. These surrounding layers are called the *cladding*. This represents the simplest form of an optical waveguide and can serve as a model to gain an understanding of wave propagation in optical fibers. In fact, a cross-sectional view of the slab waveguide looks the same as the cross-sectional view of an optical fiber cut along its axis. Figure 2.19 shows the field patterns of several of the lower-order transverse electric (TE) modes (which are solutions of Maxwell's equations for the slab waveguide). The *order* of a mode is equal to the number of field zeros across the guide (the locations where the field is zero at the center of the guide or at the axis of the fiber). The order of the mode is also related to the angle that the ray congruence corresponding to this mode makes with the plane of the waveguide (or the axis of a fiber); that is, the steeper the angle, the higher the order of the mode. The plots show that the electric fields of the guided modes are not completely confined to the central dielectric slab (i.e., they do not go to zero at the guide-cladding interface),

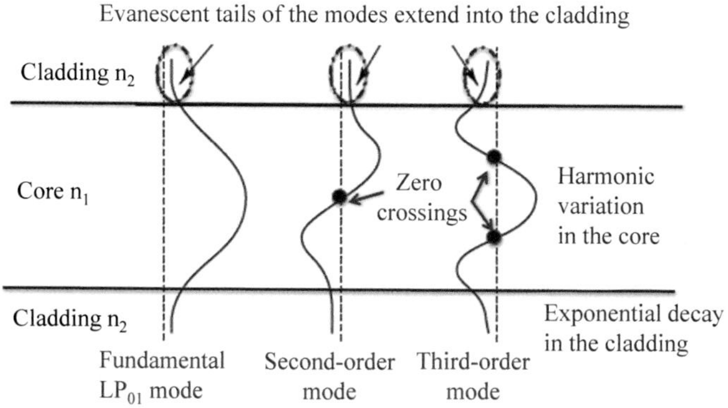

Fig. 2.19 Electric field distributions for several of the lower-order guided modes in a symmetrical-slab waveguide

but, instead, they extend partially into the cladding. The fields vary harmonically in the guiding region of refractive index n_1 and decay exponentially outside of this region. For low-order modes the fields are tightly concentrated near the center of the slab (or the axis of an optical fiber), with little penetration into the cladding region. On the other hand, for higher-order modes the fields are distributed more toward the edges of the guide and penetrate farther into the cladding region.

Solving Maxwell's equations shows that, in addition to supporting a finite number of guided modes, the optical fiber waveguide has an infinite continuum of *radiation modes* that are not trapped in the core and guided by the fiber but are still solutions of the same boundary-value problem. The radiation field basically results from the optical power that is outside the fiber acceptance angle being refracted out of the core. Because of the finite radius of the cladding, some of this radiation gets trapped in the cladding, thereby causing cladding modes to appear. As the core and cladding modes propagate along the fiber, mode coupling occurs between the cladding modes and the higher-order core modes. This coupling occurs because the electric fields of the guided core modes are not completely confined to the core but extend partially into the cladding (see Fig. 2.19) and likewise for the cladding modes. A diffusion of power back and forth between the core and cladding modes thus occurs; this generally results in a loss of power from the core modes.

Guided modes in the fiber occur when the values for β satisfy the condition $n_2 k < \beta < n_1 k$. At the limit of propagation when $\beta = n_2 k$, a mode is no longer properly guided and is called being *cut off*. Thus unguided or radiation modes appear for frequencies below the cutoff point where $\beta < n_2 k$. However, wave propagation can still occur below cutoff for those modes where some of the energy loss due to radiation is blocked by an angular momentum barrier that exists near the core-cladding interface [17]. These propagation states behave as partially confined guided modes rather than radiation modes and are called *leaky modes* [5, 6, 12, 13]. These leaky modes can travel considerable distances along a fiber but lose power through leakage or tunneling into the cladding as they propagate.

2.4.2 Cutoff Wavelength and V Number

An important parameter connected with the cutoff condition is the *normalized frequency V* (also called the V *number* or V *parameter*) defined by

$$V = \frac{2\pi a}{\lambda_c}\left(n_1^2 - n_2^2\right)^{1/2} \approx \frac{2\pi a}{\lambda}n_1\sqrt{2\Delta} \qquad (2.27)$$

where the approximation on the right-hand side comes from Eq. (2.23).

This parameter is a dimensionless number that is related to the wavelength and the numerical aperture and determines how many modes a fiber can support. The number of modes that can exist in a waveguide as a function of V may be conveniently represented in terms of a *normalized propagation constant b* defined by Gloge [19]

$$b = \frac{(\beta/k)^2 - n_2^2}{n_1^2 - n_2^2} \qquad (2.28)$$

Figure 2.20 gives a plot of b (in terms of β/k) as a function of V for a few of the low-order modes. This figure shows that except for the lowest-order HE_{11} mode, each mode can exist only for values of V that exceed a certain limiting value (with each mode having a different V limit). The modes are cut off when $\beta = n_2k$. The wavelength at which all higher-order modes are cut off is called the *cutoff wavelength* λ_c. The HE_{11} mode has no cutoff and ceases to exist only when the core diameter is zero. This is the principle on which single-mode fibers are based. By appropriately choosing a, n_1, and n_2 so that

$$V = \frac{2\pi a}{\lambda}\left(n_1^2 - n_2^2\right)^{1/2} \le 2.405 \qquad (2.29)$$

then all modes except the HE_{11} mode are cut off.

The ITU-T Recommendation G.652 states that an effective cutoff wavelength should be less than or equal to 1260 nm for single-mode fiber operation in the 1310 nm wavelength region [20]. This specification also ensures that the fiber is single-mode for operation in the 1550 nm region.

> **Drill Problem 2.4** Consider a step-index fiber that has a 5 μm core radius, a core index $n_1 = 1.480$, and a cladding index $n_2 = 1.477$. (a) Show that at 820 nm the fiber has $V = 3.514$. (b) From Fig. 2.20 find the four modes that exist at 820 nm. (c) Verify that at 1310 nm only the HE_{11} mode exists.

Example 2.7 A step-index fiber has a normalized frequency $V = 26.6$ at a 1300 nm wavelength. If the core radius is 25 μm, what is the numerical aperture?

Solution From Eq. (2.27) the NA is

$$NA = V\frac{\lambda}{2\pi a} = 26.2\frac{1.30\,\mu m}{2\pi \times 25\,\mu m} = 0.22$$

Drill Problem 2.5 Consider a fiber that has a core refractive index of 1.480, a cladding index of 1.476, and a core radius of 4.4 μm. Using Eq. (2.27), show that the wavelength at which this fiber becomes single-mode with $V = 2.405$ is $\lambda_c = 1250$ nm.

The V number also can be used to express the number of modes M in a multimode step-index fiber when V is large (see Sect. 2.6 for modes in a graded-index multimode fiber). For the multimode step-index case, an estimate of the total number of modes supported in such a fiber is

$$M = \frac{1}{2}\left(\frac{2\pi a}{\lambda}\right)^2 (n_1^2 - n_2^2) = \frac{V^2}{2} \tag{2.30}$$

Example 2.8 Consider a multimode step-index fiber with a 62.5 μm core diameter and a core-cladding index difference of 1.5%. If the core refractive index is 1.480, estimate the normalized frequency of the fiber and the total number of modes supported in the fiber at a wavelength of 850 nm.

Solution From Eq. (2.27) the normalized frequency is

$$V = \frac{2\pi a}{\lambda}n_1\sqrt{2\Delta} = \frac{2\pi \times 31.25\,\mu m \times 1.48}{0.85\,\mu m}\sqrt{2 \times 0.015} = 59.2$$

Using Eq. (2.30), the total number of modes is

$$M = \frac{V^2}{2} = 1752$$

Example 2.9 Consider a multimode step-index optical fiber that has a core radius of 25 μm, a core index of 1.48, and an index difference $\Delta = 0.01$. How many modes are in the fiber at wavelengths 860, 1310, and 1550 nm?

Solution

(a) First, from Eq. (2.27) at an operating wavelength of 860 nm the value of V is

$$V = \frac{2\pi a}{\lambda}n_1\sqrt{2\Delta} = \frac{2\pi \times 25\,\mu m \times 1.48}{0.86\,\mu m}\sqrt{2 \times 0.01} = 38.2$$

Using Eq. (2.30), the total number of modes at 860 nm is

$$M = \frac{V^2}{2} = 729$$

(b)   Similarly, at 1310 nm the parameter $V = 25.1$ and $M = 315$.
(c)   Finally, at 1550 nm the parameter $V = 21.2$ and $M = 224$.

Example 2.10 Consider three multimode step-index optical fibers each of which has a core index of 1.48 and an index difference $\Delta = 0.01$. Assume the three fibers have core diameters of 50, 62.5, and 100 μm. How many modes are in these fibers at a wavelength of 1550 nm?

Solution

(a)   First, from Eq. (2.27) at a core diameter of 50 μm the value of V is

$$V = \frac{2\pi a}{\lambda} n_1 \sqrt{2\Delta} = \frac{2\pi \times 25\,\mu m \times 1.48}{1.55\,\mu m} \sqrt{2 \times 0.01} = 21.2$$

Using Eq. (2.30), the total number of modes in the 50 μm core-diameter fiber is

$$M = \frac{V^2}{2} = 224$$

(b)   Similarly, at 62.5 μm the parameter $V = 26.5$ and $M = 351$.
(c)   Finally at 100 μm the parameter $V = 42.4$ and $M = 898$.

2.4.3   Optical Power in Step-Index Fibers

A final quantity of interest for step-index fibers is the fractional power flow in the core and cladding for a given mode. As illustrated in Fig. 2.19, the electromagnetic field for a given mode does not go to zero at the core-cladding interface, but changes from an oscillating form in the core to an exponential decay in the cladding. Thus the electromagnetic energy of a guided mode is carried partly in the core and partly in the cladding. The farther away a mode is from its cutoff frequency, the more of its energy is concentrated in the core. As cutoff is approached, the field penetrates farther into the cladding region and a greater percentage of the energy travels in the cladding. At cutoff the field no longer decays outside the core and the mode now becomes a fully radiating mode with all the optical power of the mode residing in the cladding.

Far from cutoff, that is, for large values of V, the fraction of the average optical power residing in the cladding can be estimated by

$$\frac{P_{clad}}{P} = \frac{4}{3\sqrt{M}} \tag{2.31}$$

where P is the total optical power in the fiber. Note that because M is proportional to V^2, the power flow in the cladding decreases as V increases. However, this increases the number of modes in the fiber, which is not desirable for a high-bandwidth capability.

Example 2.11 Consider a multimode step-index optical fiber that has a core radius of 25 µm, a core index of 1.48, and an index difference $\Delta = 0.01$. Find the percentage of optical power that propagates in the cladding at 840 nm.

Solution From Eq. (2.27), at an operating wavelength of 840 nm the value of V is

$$V = \frac{2\pi a}{\lambda} n_1 \sqrt{2\Delta} = \frac{2\pi \times 25\,\mu m \times 1.48}{0.84\,\mu m} \sqrt{2 \times 0.01} = 39$$

Using Eq. (2.30), the total number of modes is

$$M = \frac{V^2}{2} = 760$$

From Eq. (2.31) it follows that.

$$\frac{P_{clad}}{P} = \frac{4}{3\sqrt{M}} = 0.048$$

Thus approximately 4.8% of the optical power propagates in the cladding. If Δ is decreased to 0.03 in order to lower the signal dispersion (see Chap. 3), then there are 242 modes in the fiber and about 8.6% of the power propagates in the cladding.

> **Drill Problem 2.6** Consider a multimode step-index optical fiber that has a core diameter of 62.5 µm, a core index of 1.48, and an index difference $\Delta = 0.01$. Show that at 840 nm (a) the value of V is 49, (b) the total number of modes is 1200, and (c) the optical power that propagates in the cladding is 3.8%.

2.4.4 Linearly Polarized Modes

Although the theory of light propagation in optical fibers is well understood, a complete description of the guided and radiation modes requires the use of six-component hybrid electromagnetic fields that have very involved mathematical

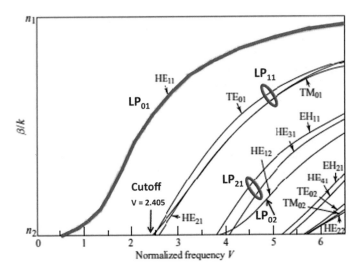

Fig. 2.20 Plots of the propagation constant (in terms of β/k) as a function of V for a few of the lowest-order modes

expressions. A simplification [19, 21–24] of these expressions can be carried out in practice, because fibers usually are constructed so that the difference in the core and cladding indices of refraction is very small; that is, $n_1 - n_2 \ll 1$. With this assumption, only four field components need to be considered and their expressions become significantly simpler. The field components are called *linearly polarized* (LP) modes and are labeled LP_{jm} where j and m are integers designating mode solutions. In this scheme for the lowest-order modes, each LP_{0m} mode is derived from an HE_{1m} mode and each LP_{1m} mode comes from TE_{0m}, TM_{0m}, and HE_{0m} modes. Thus the fundamental LP_{01} mode corresponds to an HE_{11} mode, which is the only mode that propagates in a single-mode fiber. Figure 2.20 shows the LP_{01}, LP_{11}, LP_{21}, and LP_{02} modes and their relations to the lower order HE, EH, TE, and TM modes.

Within the weakly guiding approximation all modes characterized by a common set of j and m satisfy the same characteristic equation. This means that these modes are degenerate. Thus if an $HE_{v+1, m}$ mode is degenerate with an $EH_{v-1, m}$ mode (i.e., if HE and EH modes of corresponding radial order m and equal circumferential order v form degenerate pairs), then any combination of an $HE_{v+1, m}$ mode with an $EH_{v-1, m}$ mode will likewise constitute a guided mode of the fiber.

The corresponding LP modes are designated LP_{jm} modes regardless of their TM, TE, EH, or HE field configuration. In general, the following conditions hold:

1. Each LP_{0m} mode is derived from an HE_{1m} mode.
2. Each LP_{1m} mode comes from TE_{0m}, TM_{0m}, and HE_{2m} modes.
3. Each LP_{vm} mode ($v \geq 2$) is from an $HE_{v+1, m}$ and an $EH_{v-1, m}$ mode.

Table 2.3 Composition of the lower-order linearly polarized modes

LP-mode designation	Traditional-mode designation and number of modes	Number of degenerate modes
LP_{01}	$HE_{11} \times 2$	2
LP_{11}	TE_{01}, TM_{01}, $HE_{21} \times 2$	4
LP_{21}	$EH_{11} \times 2$, $HE_{31} \times 2$	4
LP_{02}	$HE_{12} \times 2$	2
LP_{31}	$EH_{21} \times 2$, $HE_{41} \times 2$	4
LP_{12}	TE_{02}, TM_{02}, $HE_{22} \times 2$	4
LP_{41}	$EH_{31} \times 2$, $HE_{51} \times 2$	4
LP_{22}	$EH_{12} \times 2$, $HE_{32} \times 2$	4
LP_{03}	$HE_{13} \times 2$	2
LP_{51}	$EH_{41} \times 2$, $HE_{61} \times 2$	4

The correspondence between the ten lowest order LP modes (i.e., those having the lowest cutoff frequencies) and the traditional TM, TE, EH, and HE modes is given in Table 2.3. This table also shows the number of degenerate modes.

2.5   Single-Mode Fibers

In multimode fibers the differences in the propagation delays of various modes lead to signal dispersion in an optical fiber link (described in Sect. 3.2). This intermodal delay or modal dispersion effect limits the speed at which information can be transmitted over a fiber. Intermodal signal dispersion can be avoided by designing a fiber such that only the fundamental mode is allowed to propagate. Such a construction forms the basis of a single-mode fiber (SMF).

2.5.1   SMF Construction

Single-mode fibers are constructed by letting the dimensions of the core diameter be a few wavelengths (usually from 8 to 12) and by having small index differences between the core and the cladding. From Eq. (2.27) with $V = 2.4$, it can be seen that single-mode propagation is possible for fairly large variations in values of the physical core size a and the core-cladding index differences Δ. However, in practical designs of single-mode fibers, [25] the core-cladding index difference varies between 0.2 and 1.0%, and the core diameter should be chosen to be just below the cutoff of the first higher-order mode; that is, for V slightly less than 2.4.

Example 2.12 A manufacturing engineer wants to make an optical fiber that has a core index of 1.480 and a cladding index of 1.478. What should the core size be for single-mode operation at 1550 nm?

Solution Using the condition that $V \leq 2.405$ must be satisfied for single-mode operation, then from Eq. (2.27)

$$a = \frac{V\lambda}{2\pi} \frac{1}{\sqrt{n_1^2 - n_2^2}} \leq \frac{2.405 \times 1.55\ \mu m}{2\pi} \frac{1}{\sqrt{(1.480)^2 - (1.478)^2}} = 7.7\mu m$$

If this fiber also should be single-mode at 1310 nm, then the core radius must be less than 6.50 μm.

Example 2.13 An applications engineer has an optical fiber that has a 3.0 μm core radius and a numerical aperture of 0.1. Will this fiber exhibit single-mode operation at 800 nm?

Solution From Eq. (2.27)

$$V = \frac{2\pi a}{\lambda} NA = \frac{2\pi \times 3\ \mu m}{0.80\ \mu m} 0.10 = 2.356$$

Because $V < 2.405$, this fiber will exhibit single-mode operation at 800 nm.

2.5.2 Definition of Mode–Field Diameter

For multimode fibers the core diameter and numerical aperture are key parameters for describing the signal transmission properties. In single-mode fibers the geometric distribution of light in the propagating mode is what is needed when predicting the performance characteristics of these fibers. Thus in a single-mode fiber a fundamental parameter is the *mode-field diameter* (MFD). This parameter can be determined from the mode-field distribution of the fundamental fiber mode and is a function of the optical source wavelength, the core radius, and the refractive index profile of the fiber. The mode-field diameter is analogous to the core diameter in multimode fibers, except that in single-mode fibers not all the light that propagates through the fiber is carried in the core. For example, at $V = 2$ only 75% of the optical power is confined to the core. This percentage increases for larger values of V and is less for smaller V values.

The MFD is an important parameter for single-mode fiber because it is used to predict fiber properties such as splice loss, bending loss, cutoff wavelength, and waveguide dispersion. Chapters 3 and 5 describe these parameters and their effects on fiber performance. A variety of models have been proposed for characterizing and measuring the MFD [26–31]. These include far-field scanning, near-field scanning, transverse offset, variable aperture in the far field, knife-edge, and mask methods. The main consideration of all these methods is how to approximate the optical power distribution.

A standard technique to find the MFD is to measure the far-field intensity distribution $E^2(r)$ and then calculate the MFD using the Petermann II equation [28]

$$\text{MFD} = 2w_0 = 2\left[\frac{2\int_0^\infty E^2(r)r^3 dr}{\int_0^\infty E^2(r)r\,dr}\right]^{1/2} \tag{2.32}$$

where the parameter $2w_0$ (with w_0 being called the *spot size* or the *mode field radius*) is the full width of the far-field distribution. For calculation simplicity the exact field distribution can be fitted to a Gaussian function [22]

$$E(r) = E_0 exp\left(-r^2/w_0^2\right) \tag{2.33}$$

where r is the radial position in the field distribution and E_0 is the field at zero radius, as shown in Fig. 2.21. Then the MFD is given by the $1/e^2$ width of the optical power. The Gaussian pattern given in Eq. (2.33) is a good approximation for values of V lying between 1.8 and 2.4, which designates the operational region of practical single-mode fibers.

Fig. 2.21 Distribution of light in a single-mode fiber above its cutoff wavelength

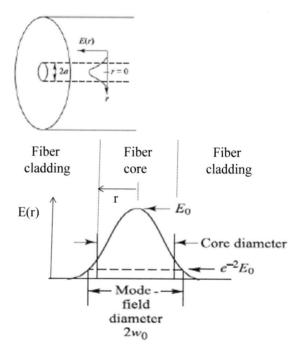

An approximation to the relative spot size w_0/a, which for a step-index fiber has an accuracy better than 1% in the range $1.2 < V < 2.4$, is given by

$$\frac{w_0}{a} = 0.65 + 1.619V^{-3/2} + 2.879V^{-6} \tag{2.34}$$

The condition $V = 2.405$ for single-mode operation yields $w_0/a = 1.1005$. As V decreases from 2.4, the spot size increases. The spot size thus becomes progressively larger than the core radius a and extends farther into the cladding. As a result, when V becomes smaller the optical beam becomes less tightly bound to the core and becomes more susceptible to optical power losses from the cladding. Manufacturers therefore typically design their fibers with V values greater than 2.0 to prevent high cladding losses but somewhat less than 2.4 to avoid the possibility of having more than one mode in the fiber.

Example 2.14 A certain single-mode step-index fiber has an MFD $= 11.2\ \mu$m and $V = 2.25$. What is the core diameter of this fiber?

Solution From Eq. (2.32) $w_0 = $ MFD$/2 = 5.6\ \mu$m. Using Eq. (2.34) then yields

$$a = w_0/\left(0.65 + 1.619V^{-3/2} + 2.879V^{-6}\right) = \frac{5.6\mu m}{0.65 + 1.619V^{-3/2} + 2.879V^{-6}}$$

$$= \frac{5.6\mu m}{1.152} = 4.86\mu m$$

Thus the core diameter is $2a = 9.72\ \mu$m.

Drill Problem 2.7 Compare the relative spot size w_0/a for V values of 1.2, 1.8, and 2.4.

2.5.3   Origin of Birefringence

An important point to keep in mind is that in any ordinary single-mode fiber there are actually two independent, degenerate propagation modes [32–34]. These modes are very similar, but their polarization planes are orthogonal. These may be chosen arbitrarily as the horizontal (H) and the vertical (V) polarizations as shown in Fig. 2.22. Either one of these two polarization modes constitutes the fundamental HE_{11} mode. In general, the electric field of the light propagating along the fiber is a linear superposition of these two polarization modes and depends on the polarization of the light at the launching point into the fiber.

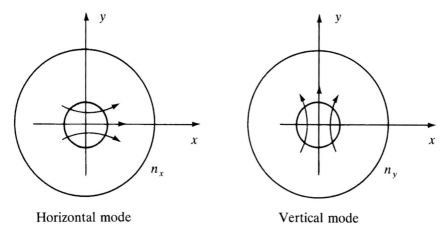

Horizontal mode   Vertical mode

Fig. 2.22 Two polarizations of the fundamental HE_{11} mode in a single-mode fiber

Suppose one of the modes is chosen arbitrarily to have its transverse electric field polarized along the x direction and the other independent, orthogonal mode to be polarized in the y direction as shown in Fig. 2.22. In ideal fibers with perfect rotational symmetry, the two modes are degenerate with equal propagation constants ($\beta_x = \beta_y$), and any polarization state injected into the fiber will propagate unchanged. In actual fibers there are imperfections, such as asymmetrical lateral stresses, noncircular cores, and variations in refractive-index profiles. These imperfections break the circular symmetry of the ideal fiber and lift the degeneracy of the two modes. Then the modes propagate with different phase velocities, and the difference between their effective refractive indices is called the fiber *birefringence*,

$$B_f = \frac{\lambda}{2\pi}(\beta_x - \beta_y) \tag{2.35}$$

If light is injected into the fiber so that both modes are excited, then one mode will be delayed in phase relative to the other as they propagate. When this phase difference is an integral multiple of 2π, the two modes will beat at this point and the input polarization state will be reproduced. The length over which this beating occurs is the *fiber beat length*,

$$L_B = \frac{\lambda}{B_f} = \frac{2\pi}{(\beta_x - \beta_y)} \tag{2.36}$$

Example 2.15 A single-mode optical fiber has a beat length of 8 cm at 1310 nm. What is the birefringence?

Solution From Eq. (2.36) the modal birefringence is

$$B_f = \frac{\lambda}{L_B} = \frac{1.31 \times 10^{-6} \text{ m}}{8 \times 10^{-2} \text{ m}} = 1.64 \times 10^{-5}$$

This is characteristic of an intermediate type fiber, because birefringence can vary from $B_f = 1 \times 10^{-3}$ (for a typical high birefringence fiber) to $B_f = 1 \times 10^{-8}$ (for a typical low birefringence fiber).

2.5.4   Effective Refractive Index

From Maxwell's equations it can be shown that the phase of the fundamental HE_{11} mode (or equivalently, of the LP_{01} mode) is determined by means of the wave propagation constant β for that mode. For various applications, for example, when discussing signal propagation in few-mode fibers or when analyzing fiber Bragg gratings, it is useful to define an *effective refractive index* n_{eff}. The basic definition is that for some fiber mode the propagation constant β is a factor of n_{eff} times the vacuum wave number $k_0 = 2\pi/\lambda$, that is,

$$n_{eff} = \frac{\beta}{k_0} \tag{2.37}$$

In a standard single-mode fiber, because the fundamental guided LP_{01} mode extends significantly beyond the core region, the effective refractive index has a value falling between the refractive indices of the core and the cladding. In a multimode fiber, higher-order modes extend farther into the cladding, and have smaller effective indices than lower-order modes. The exact value of the effective refractive index depends on factors such as the specific mode being considered, the size of the fiber, and the wavelength. It is important to note that the definition of n_{eff} is related to the phase change per unit length along the fiber and not on the intensity distribution of the modes.

2.6   Graded-Index (GI) Fibers

2.6.1   Core Structure of GI Fibers

In the graded-index fiber design the core refractive index decreases continuously with increasing radial distance r from the center of the fiber but is generally constant in the cladding. The most commonly used construction for the refractive-index variation in the core is the power law relationship

$$n(r) = \begin{cases} n_1\left[1 - 2\Delta\left(\frac{r}{a}\right)^\alpha\right]^{1/2} & \text{for } 0 \le r \le a \\ n_1(1 - 2\Delta)^{1/2} \approx n_1(1 - \Delta) = n_2 & \text{for } r \ge a \end{cases} \tag{2.38}$$

Here, r is the radial distance from the fiber axis, a is the core radius, n_1 is the refractive index at the core axis, n_2 is the refractive index of the cladding, and the dimensionless parameter α defines the shape of the index profile. The index difference Δ for the graded-index fiber is given by

$$\Delta = \frac{n_1^2 - n_2^2}{2n_1^2} \approx \frac{n_1 - n_2}{n_1} \tag{2.39}$$

The approximation on the right-hand side of this equation reduces the expression for Δ to that of the step-index fiber given by Eq. (2.20). Thus, the same symbol is used in both cases. For $\alpha = \infty$, inside the core Eq. (2.38) reduces to the step-index profile $n(r) = n_1$.

2.6.2 GI Fiber Numerical Aperture

Determining the NA for graded-index fibers is more complex than for step-index fibers because it is a function of position across the core end face. This is in contrast to the step-index fiber, where the NA is constant across the core. Geometrical optics considerations show that light incident on the fiber core at position r will propagate as a guided mode only if it is within the local numerical aperture NA(r) at that point. The local numerical aperture is defined as [35]

$$NA(r) = \begin{cases} \left[n^2(r) - n_2^2\right]^{1/2} \approx NA(0)\sqrt{1 - (r/a)^\alpha} & \text{for } r \le a \\ 0 & \text{for } r > a \end{cases} \tag{2.40}$$

where the axial numerical aperture is defined as

$$NA(0) = \left[n^2(0) - n_2^2\right]^{1/2} = \left(n_1^2 - n_2^2\right)^{1/2} \approx n_1\sqrt{2\Delta} \tag{2.41}$$

Thus, the NA of a graded-index fiber decreases from NA(0) to zero as r moves from the fiber axis to the core-cladding boundary. A comparison of the numerical apertures for fibers having various α profiles is shown in Fig. 2.23. The number of bound modes in a graded-index fiber is

$$M_g = \frac{\alpha}{\alpha + 2}a^2k^2n_1^2\Delta \approx \frac{\alpha}{\alpha + 2}\frac{V^2}{2} \tag{2.42}$$

where $k = 2\pi/\lambda$ and the right-hand approximation is derived using Eqs. (2.23) and (2.27). Fiber manufacturers typically choose a *parabolic refractive index profile* given

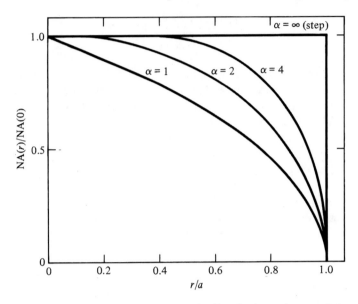

Fig. 2.23 A comparison of the numerical apertures for fibers having various core index profiles

by $\alpha = 2.0$. In this case, $M_g = V^2/4$, which is half the number of modes supported by a step-index fiber (for which $\alpha = \infty$) that has the same V value.

Example 2.16 Consider a 50 μm diameter graded-index fiber that has a parabolic refractive index profile ($\alpha = 2$). If the fiber has a numerical aperture NA $= 0.22$, what is the total number of guided modes at a wavelength of 1310 nm?

Solution First, from Eq. (2.27)

$$V = \frac{2\pi a}{\lambda}NA = \frac{2\pi \times 25\,\mu m}{1.31\,\mu m}0.22 = 26.4$$

Then from Eq. (2.42) the total number of modes for $\alpha = 2$ is

$$M_g = \frac{\alpha}{\alpha + 2}\frac{V^2}{2} = \frac{V^2}{4} = 174$$

2.6.3 Cutoff Condition in GI Fibers

Similar to step-index fibers, in order to eliminate intermodal dispersion graded-index fibers can be designed as single-mode fibers in which only the fundamental mode is allowed to propagate at the desired operational wavelength. An empirical expression of the V parameter at which the second lowest order mode, the L_{11} mode, is cut off for graded-index fibers has been shown to be [3]

$$V = 2.405\sqrt{1 + \frac{2}{\alpha}} \tag{2.43}$$

Equation (2.43) shows that in general for a graded-index fiber the value of V decreases as α increases. It also shows that the critical value of V for the cutoff condition in parabolic graded-index fibers is a factor of $\sqrt{2}$ larger than for a similar-sized step-index fiber. Furthermore, from the definition of V given by Eq. (2.27), the numerical aperture of a graded-index fiber is larger than that of a step-index fiber of comparable size.

2.7   Optical Fiber Materials

In selecting materials for optical fibers, a number of requirements must be satisfied. For example:

1. It must be possible to make long, thin, flexible fibers from the material;
2. The material must have a low loss at a particular optical wavelength in order for the fiber to guide light efficiently;
3. Physically compatible materials that have slightly different refractive indices for the core and cladding must be available.

Materials that satisfy these requirements are glasses and plastics.

The majority of fibers are made of glass consisting of either silica (SiO_2) or a silicate. The variety of available glass fibers ranges from moderate-loss fibers with large cores used for short-transmission distances to very transparent (low-loss) fibers employed in long-haul applications. Plastic fibers typically have a substantially higher attenuation than glass fibers. A main use of plastic fibers is in short-distance applications (several hundred meters) and in abusive environments, where the greater mechanical strength of plastic fibers offers an advantage over the use of glass fibers.

2.7.1   Glass Optical Fibers

Glass is made by fusing mixtures of metal oxides, sulfides, or selenides [36–38]. The resulting material is a randomly connected molecular network rather than an ordered structure as found in crystalline materials. A consequence of this random order is that glasses do not have well defined melting points. When glass is heated up from room temperature, it remains a hard solid up to several hundred degrees centigrade. As the temperature increases further, the glass gradually begins to soften

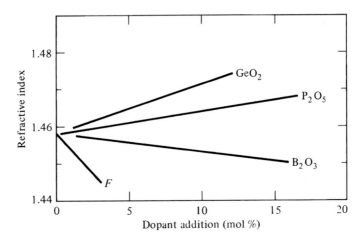

Fig. 2.24 Variation in refractive index as a function of doping concentration in silica glass

until at very high temperatures it becomes a viscous liquid. The expression "melting temperature" is commonly used in glass manufacture. This term refers only to an extended temperature range in which the glass becomes fluid enough to free itself fairly quickly of gas bubbles.

The largest category of optically transparent glasses from which optical fibers are made consists of the oxide glasses. Of these, the most common is silica (SiO_2), which has a refractive index ranging from 1.458 at 850 to 1.444 at 1550 nm. To produce two similar materials that have slightly different indices of refraction for the core and cladding, either fluorine or various oxides (referred to as *dopants*), such as B_2O_3, GeO_2, or P_2O_5, are added to the silica. As shown in Fig. 2.24, the addition of GeO_2 or P_2O_5 increases the refractive index, whereas doping the silica with fluorine or B_2O_3 decreases it. Because the cladding must have a lower index than the core, examples of fiber compositions are

- GeO_2–SiO_2 core; SiO_2 cladding
- P_2O_5–SiO_2 core; SiO_2 cladding
- SiO_2 core; B_2O_3–SiO_2 cladding
- GeO_2–B_2O_3–SiO_2 core; B_2O_3–SiO_2 cladding.

Here, the notation GeO_2–SiO_2, for example, denotes a GeO_2-doped silica glass.

The principal raw material for silica is high-purity sand. Glass composed of pure silica is referred to as *silica glass, fused silica,* or *vitreous silica.* Some of its desirable properties are a resistance to deformation at temperatures as high as 1000 °C, a high resistance to breakage from thermal shock because of its low thermal expansion, good chemical durability, and high transparency in both the visible and infrared regions of interest to fiber optic communication systems.

2.7.2 Standard Fiber Fabrication

Two basic techniques are used in the fabrication of all-glass optical waveguides. These are the vapor-phase oxidation processes and the direct-melt methods. The *direct-melt method* follows traditional glass-making procedures in that optical fibers are made directly from the molten state of purified components of silicate glasses. In the *vapor-phase oxidation process*, highly pure vapors of metal halides (e.g., $SiCl_4$ and $GeCl_4$) react with oxygen to form a white powder of SiO_2 particles. The particles are then collected on the surface of a bulk glass by one of four different commonly used processes and are sintered (transformed to a homogeneous glass mass by heating without melting) by one of a variety of techniques to form a clear glass rod or tube (depending on the process). This rod or tube is called a *preform*. It is typically around 10–25 mm in diameter and 60–120 cm long. Fibers are made from the preform [39] by using the equipment shown in Fig. 2.25. The preform is precision-fed into a circular heater called the *drawing furnace*. Here the preform end is softened to the point where it can be drawn into a very thin filament, which becomes the optical fiber. The turning speed of the takeup drum at the bottom of the draw tower determines how fast the fiber is drawn. This, in turn, will determine the thickness of the fiber, so that a precise rotation rate must be maintained. An optical fiber thickness monitor is used in a feedback loop for this speed regulation. To protect the bare glass fiber from external contaminants, such as dust and water vapor, an elastic coating is applied to the fiber immediately after it is drawn.

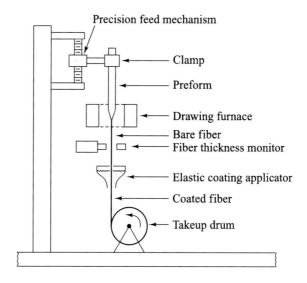

Fig. 2.25 Schematic of a fiber-drawing apparatus

2.7.3 Active Glass Optical Fibers

Incorporating rare-earth elements (atomic numbers 57–71) into a normally passive glass gives the resulting material new optical and magnetic properties. These new properties allow the material to perform amplification, attenuation, and phase retardation on the light passing through it [40, 41]. Doping (i.e., adding impurities) can be carried out for silica, telluride, and halide glasses.

Two commonly used doping materials for fiber lasers are erbium and neodymium. The ionic concentrations of the rare-earth elements are low (on the order of 0.005–0.05 mol %) to avoid clustering effects. To make use of the absorption and fluorescence spectra of these materials, one can use an optical source that emits at an absorption wavelength of the doping material to excite electrons to higher energy levels in the rare-earth dopants. When these excited electrons are stimulated by a signal photon to drop to lower energy levels, the transition process results in the emission of light in a narrow optical spectrum at the fluorescence wavelength. Chapter 11 discusses the applications of fibers doped with rare-earth elements to create optical amplifiers.

2.7.4 Plastic Optical Fibers

The growing demand for delivering high-speed services directly to the workstation has led fiber developers to create high-bandwidth graded-index polymer (plastic) optical fibers (POF) for use on customer premises [42–44]. The core of these fibers is either polymethylmethacrylate or a perfluorinated polymer. These fibers are hence referred to as PMMA POF and PF POF, respectively. Although they exhibit considerably greater optical signal attenuations than glass fibers, plastic fibers are tough and durable. For example, since the modulus of these polymers is nearly two orders of magnitude lower than that of silica, even a 1-mm-diameter graded-index POF is sufficiently flexible to be installed in conventional fiber cable routes. Standard optical connectors can be used on plastic fibers having core sizes that are compatible with the core diameters of standard multimode glass telecom fibers. Thus coupling between similar sized plastic and glass fibers is straightforward. In addition, for the plastic fibers inexpensive plastic injection-molding technologies can be used to fabricate connectors, splices, and transceivers.

Table 2.4 gives sample characteristics of PMMA and PF polymer optical fibers.

2.8 Photonic Crystal Fiber Concepts

In the early 1990s researchers envisioned and demonstrated a new optical fiber structure. Initially this was called a *holey fiber* and later became known as a *photonic*

Table 2.4 Sample characteristics of PMMA and PF polymer optical fibers

Characteristic	PMMAPOF	PFPOF
Core diameter	0.4 mm	0.050–0.30 mm
Cladding diameter	1.0 mm	0.25–0.60 mm
Numerical aperture	0.25	0.20
Attenuation	150 dB/km at 650 nm	<40 dB/km at 650–1300 nm
Bandwidth	2.5 Gb/s over 200 m	2.5 Gb/s over 550 m

crystal fiber (PCF) or a *microstructured* fiber [45–48]. The difference between this new structure and that of a conventional fiber is that the cladding and, in some cases, the core regions of a PCF contain air holes, which run along the entire length of the fiber. Whereas the material properties of the core and cladding define the light transmission characteristics of conventional fibers, the structural arrangement of the hole channels in a PCF creates an internal microstructure, which offers extra dimensions in controlling the optical properties of light, such as the dispersion, nonlinearity, and birefringence effects in optical fibers.

The sizes of the holes and the hole-to-hole spacing (known as the *pitch*) in the microstructure and the refractive index of its constituent material determine the light-guiding characteristics of photonic crystal fibers. The two basic PCF categories are index-guiding fibers and photonic bandgap fibers. The light transmission mechanism in an *index-guiding fiber* is similar to that in a conventional fiber as it has a high-index core surrounded by a lower-index cladding. However, for a PCF the effective refractive index of the cladding depends on the wavelength and the size and pitch of the holes. In contrast, in a *photonic bandgap fiber* light is guided by means of a photonic bandgap effect in either a hollow or microstructured core, which is surrounded by a microstructured cladding.

2.8.1 Index-Guiding PCF

Figure 2.26 shows the two-dimensional cross-sectional end view of a basic structure of an index-guiding PCF. The fibers have a solid core that is surrounded by a cladding region, which contains air holes that run along the length of the fiber and can have a variety of different shapes, sizes, and distribution patterns. As an illustration, in Fig. 2.26 the air holes are arranged in a uniform hexagonal array. The holes all have a diameter d and a hole-to-hole spacing or pitch Λ.

The values of the hole diameter and the pitch are important for determining the operational characteristics of an index-guiding PCF. For a diameter-to-pitch ratio $d/\Lambda < 0.4$ the fiber exhibits single-mode properties over a wide range of wavelengths (from about 300 to 2000 nm). This characteristic is not possible to achieve in standard

Fig. 2.26 Cross-sectional end view of the structure of an index-guiding photonic crystal fiber with air holes of uniform size

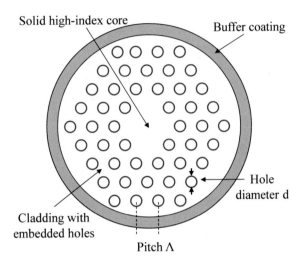

optical fibers and is useful for the simultaneous transmission of multiple wavelengths over the same fiber.

Although the core and the cladding in a PCF are made of the same material (for example, pure silica), the air holes lower the effective index of refraction in the cladding region, because $n = 1.00$ for air and 1.45 for silica. The large difference in refractive indices together with the small dimensions of the microstructures causes the effective index of the cladding to depend strongly on wavelength. The fact that the core can be made of pure silica gives the PCF a number of operational advantages over conventional fibers, which typically have a germanium-doped silica core. These include very low losses, the ability to transmit high optical power levels, and a high resistance to darkening effects from ultraviolet light. The fibers can support single-mode operation over wavelengths ranging from 300 nm to more than 2000 nm. The mode-field area of a PCF can be greater than 300 μm^2 compared to the 80 μm^2 area of conventional single-mode fibers. This allows the PCF to transmit high optical power levels without encountering the nonlinear effects exhibited by standard fibers (see Chap. 12).

2.8.2 Photonic Bandgap Fiber

Photonic bandgap (PBG) fibers have a different light-guiding mechanism, which is based on a two-dimensional photonic bandgap in the transverse plane of the cladding region. This photonic bandgap results from an ordered arrangement of the air holes in the cladding. Wavelengths within this bandgap are prevented from traveling in the cladding and thus are confined to travel in a region where the index is lower than the surrounding material. The functional principle of a photonic bandgap fiber

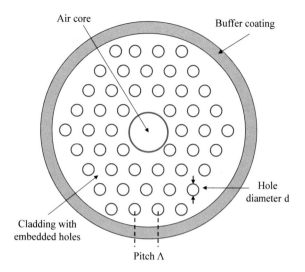

Fig. 2.27 Cross-sectional end view of the structure of one type of photonic bandgap fiber

is analogous to the role of a periodic crystalline lattice in a semiconductor, which blocks electrons from occupying a bandgap region. In a traditional PBG fiber the hollow core acts as a defect in the photonic bandgap structure, which creates a region in which the light can propagate. Whereas modes at all wavelengths can propagate along index-guiding fibers, the guided light in a PBG fiber is allowed in only a relatively narrow wavelength region with a width of approximately 100–200 nm.

Figure 2.27 shows a two-dimensional cross-sectional end view of the structure of an example PBG fiber. Here a large central hollow core is formed by removing material in the center of the fiber from an area that can be occupied by seven air holes. Such a structure is called an *air-guiding* or a *hollow-core* PBG fiber, and allows about 98% of the power in the guided modes to propagate in the air hole regions. Analogous to an index-guiding fiber, the holes in the cladding region have a diameter d and a pitch Λ. Such hollow-core fibers can have a very low nonlinearity and a high damage threshold. Thus PBG fibers can be used for dispersive pulse compression with high optical intensities. In addition, fiber optic sensors or variable power attenuators can be constructed by filling PBG fibers that have larger core holes with gases or liquids.

2.9 Optical Fiber Cables

In any practical application of optical waveguide technology, the fibers need to be incorporated in some type of cable structure [49–53]. The cable structure will vary greatly, depending on whether the cable is to be pulled into underground or intra-building ducts, buried directly in the ground, installed on outdoor poles, or submerged underwater. Different cable designs are required for each type of application, but certain fundamental cable design principles will apply in every case. The objectives of cable manufacturers have been that the optical fiber cables should be installable with the same equipment, installation techniques, and precautions as those used for conventional wire cables. This requires special cable designs because of the mechanical properties of glass fibers.

2.9.1 Fiber Optic Cable Structures

One important mechanical property is the maximum allowable axial load on the cable because this factor determines the length of cable that can be reliably installed. In copper cables the wires themselves are generally the principal load-bearing members of the cable, and elongations of more than 20% are possible without fracture. On the other hand, extremely strong optical fibers tend to break at 4% elongation, whereas typical good-quality fibers exhibit long-length breaking elongations of about 0.5–1.0%. Because fiber ruptures occur very quickly at stress levels above 40% of the permissible elongation and very slowly below 20%, fiber elongations during cable manufacture and installation should be limited to 0.1–0.2%.

Steel wire has been used extensively for reinforcing electric cables and also can be used as a *strength member* for optical fiber cables. For some applications it is desirable to have a completely nonmetallic construction, either to avoid the effects of electromagnetic induction or to reduce cable weight. In such cases, plastic strength members and high- tensile-strength synthetic yarns are used. A popular yarn is Kevlar®, which is a soft but tough yellow synthetic nylon material belonging to a generic yarn family known as *aramids*. Good fabrication practices will isolate the fibers from other cable components, keep them close to the neutral axis of the cable, and allow the fibers to move freely when the cable is flexed or stretched.

The generic cable configuration shown in Fig. 2.28 illustrates some common materials that are used in the optical fiber cabling process. Individual fibers or modules of bundled fiber groupings are wound loosely around the central buffered strength member. A cable wrapping tape and other strength members such as Kevlar then encapsulate and bind these fiber groupings together. Surrounding all these components is a tough polyethylene (PE) *jacket* that provides crush resistance and handles any tensile stresses applied to the cable so that the fibers inside are not damaged. The jacket also protects the fibers inside against abrasion, moisture, oil, solvents, and other contaminants. The jacket type largely defines the application characteristics; for example, heavy-duty outside-plant cables for direct-burial and aerial applications have much thicker and tougher jackets than indoor cables that have lower stress environments. Some cable designs might contain optional copper wires for powering in-line equipment. Other cable components can include steel armoring tapes, water-blocking or water-absorbing materials, optional copper wires for powering in-line equipment, and rip cords that allow the jacket to be cut back easily without damaging the components inside the cable.

To distinguish individual fiber strands within a grouping of fibers, each fiber is designated by a separate and distinct jacket color. The TIA-598-D Optical Fiber

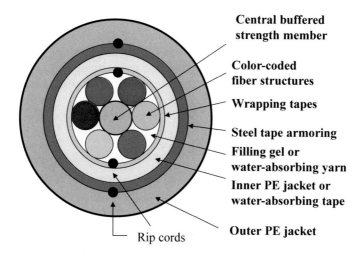

Fig. 2.28 A typical generic six-fiber cable illustrating some common materials used in the optical fiber cabling process

Cable Color Coding standards document prescribes a common set of twelve basic colors. If there are several encapsulated fiber groupings within a cable, then the jacket colors of each fiber grouping follow the same standard color-coding scheme.

The two basic fiber optic cable structures are the *tight-buffered fiber cable* design and the *loose-tube cable* configuration. Cables with tight-buffered fibers nominally are used indoors whereas the loose-tube structure is intended for long-haul outdoor applications. A *ribbon cable* is an extension of the tight-buffered cable. In all cases the fibers themselves consist of the normally manufactured glass core and cladding, which is surrounded by a protective 250 μm diameter coating.

As shown in Fig. 2.29, in the *tight-buffered* design each fiber is individually

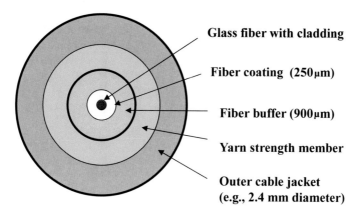

Fig. 2.29 Construction of a simplex tight-buffered fiber cable module

encapsulated within its own 900 μm diameter plastic buffer structure, hence the designation *tight-buffered design*. The 900 μm buffer is nearly four times the diameter and five times the thickness of the 250 μm protective coating material. This construction feature contributes to the excellent moisture and temperature performance of tight-buffered cables and also permits their direct termination with connectors. In a single-fiber module, a layer of amarid strength material surrounds the 900 μm fiber structure. This configuration then is encapsulated within a polyvinyl chloride (PVC) outer jacket.

In the *loose-tube* cable configuration one or more standard-coated fibers are enclosed in a thermoplastic tube that has an inner diameter much larger than the fiber diameter. The fibers in the tube are slightly longer than the cable itself. The purpose of this construction is to isolate the fiber from any stretching of the surrounding cable structure caused by factors such as temperature changes, wind forces, or ice loading. The tube is filled with either a *gel* or a dry *water-blocking material* that acts as a buffer, enables the fibers to move freely within the tube, and prevents moisture from entering the tube. A loose-tube cable typically has a steel armoring layer just inside the jacket to offer crush resistance and protection against gnawing rodents. Such a cable can be used for direct burial or aerial-based outside plant applications.

To facilitate the field operation of splicing cables that contain a large number of fibers, cable designers devised the fiber-ribbon structure. As shown in Fig. 2.30, the *ribbon cable* is an arrangement of fibers that are aligned precisely next to each other and then are encapsulated in a plastic buffer or jacket to form a long continuous ribbon. The number of fibers in a ribbon typically ranges from four to twelve. These ribbons can be stacked on top of each other to form a densely packed arrangement of many fibers (e.g., 144 fibers) within a cable structure.

There are many different ways in which to arrange fibers inside a cable. The particular arrangement of fibers and the cable design itself need to take into account issues such as the physical environment, the services that the optical link will provide, and any anticipated maintenance and repair that may be needed.

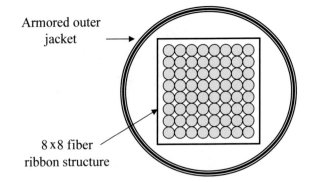

Fig. 2.30 Example of the layered ribbon structure of a 64-fiber cable module

Armored outer jacket

8 x 8 fiber ribbon structure

2.9.2 Designs of Indoor Optical Cables

Indoor cables can be used for interconnecting instruments, for distributing signals among office users, for connections to printers or servers, and for short patch cords in telecommunication equipment racks. The three main types are described here.

Interconnect cable serves light-duty low-fiber count indoor applications such as fiber-to-the-desk links, patch cords, and point-to-point runs in conduits and trays. The cable is flexible, compact, and lightweight with a tight-buffered construction. A popular indoor cable type is the *duplex cable,* which consists of two fibers that are encapsulated in an outer PVC jacket. Fiber optic *patch cords,* also known as *jumper cables,* are short lengths (usually less than 2 m) of simplex or duplex cable with connectors on both ends. They are used to connect lightwave test equipment to a fiber patch panel or to interconnect optical transmission modules within an equipment rack.

Breakout or fanout cable consists of up to 12 tight-buffered fibers stranded around a central strength member. Such cables serve low-to-medium fiber-count applications where it is necessary to protect individual jacketed fibers. The breakout cable allows easy installation of connectors on individual fibers in the cable. With such a cable configuration, routing the individually terminated fibers to separate pieces of equipment can be achieved easily.

Distribution cable consists of individual or small groupings of tight-buffered fibers stranded around a central strength member. This cable serves a wide range of network applications for sending data, voice, and video signals. Distribution cables are designed for use in intra-building cable trays, conduits, and loose placement in dropped-ceiling structures. A main feature is that they enable groupings of fibers within the cable to be branched (distributed) to various locations.

2.9.3 Designs of Outdoor Optical Cables

Outdoor cable installations include aerial, duct, direct burial, and underwater applications. Invariably these cables consist of a loose-tube structure. Many different designs and sizes of outdoor cables are available depending on the physical environment in which the cable will be used and the particular application.

Aerial cable is intended for mounting outside between buildings or on poles or towers. The two popular designs are the self-supporting and the facility-supporting cable structures. The *self-supporting cable* contains an internal strength member that permits the cable to be strung between poles without using any additional support mechanisms for the cable. For the *facility-supporting cable,* first a separate wire or strength member is strung between the poles and then the cable is lashed or clipped to this member.

Armored cable for direct-burial or underground-duct applications has one or more layers of steel-wire or steel-sheath protective armoring below a layer of polyethylene jacketing as shown in Fig. 2.31. This not only provides additional strength to the cable but also protects it from gnawing animals such as squirrels or burrowing rodents, which often cause damage to underground cables. For example, in the United States the plains pocket gopher (*Geomys busarius*) can destroy unprotected cable that is buried less than 2 m (6 ft) deep.

Underwater cable, also known as *submarine cable,* is used in rivers, lakes, and ocean environments. Because such cables normally are exposed to high water pressures, they have much more stringent requirements than underground cables. For example, as shown in Fig. 2.32, cables that can be used in rivers and lakes have various water-blocking layers, one or more protective inner polyethylene sheaths,

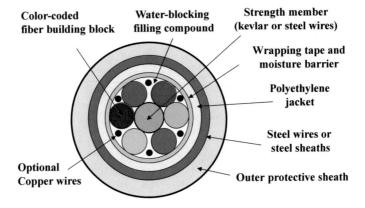

Fig. 2.31 Example configuration of an armored outdoor fiber optic cable

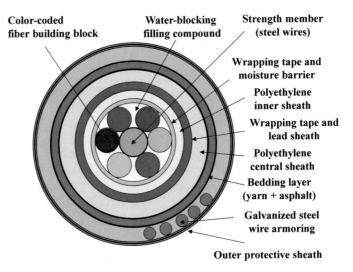

Fig. 2.32 Example configuration of an underwater fiber optic cable

and a heavy outer armor jacket. Cables that run under the ocean have further layers of armoring and contain copper wires to provide electrical power for submersed optical amplifiers or regenerators.

2.10 Summary

This chapter examines the structure of optical fibers and presents two mechanisms that show how light propagates along these fibers. In its simplest form an optical fiber is a coaxial cylindrical arrangement of two homogeneous dielectric (glass or plastic) materials. This fiber type consists of a central core of uniform refractive index n_1 surrounded by a cladding region of refractive index n_2 that is less than n_1. This configuration is referred to as a step-index fiber because the cross-sectional refractive-index profile has a step function at the interface between the core and the cladding.

In a graded-index fiber the refractive-index profile varies as a function of the radial coordinate r in the core but is constant in the cladding. This index profile $n(r)$ often is represented as a r^α power law where α defines the shape of the core index profile. A commonly used value of the power law exponent is $\alpha = 2$. This special case is referred to as a parabolic graded-index profile. A graded-index profile reduces signal dispersion in multimode fibers and thus provides a wider bandwidth than offered by a step-index fiber.

A photonic crystal fiber (PCF) or a microstructured fiber differs from a conventional fiber in that the cladding and, in some cases, the core regions of a PCF contain air holes, which run along the entire length of the fiber. Whereas the material properties of the core and cladding define the light transmission characteristics of conventional fibers, the structural arrangement of holes in a PCF creates an internal microstructure, which offers extra dimensions in controlling the optical properties of light, such as the dispersion, nonlinear effects, and birefringence effects in optical fibers.

A general picture of light propagation in a conventional fiber can be obtained by considering a ray-tracing (or geometrical optics) model in a slab waveguide. The slab consists of a central region of refractive index n_1, which is sandwiched between two material layers having a lower refractive index n_2. Light rays propagate along the slab waveguide by undergoing total internal reflection at the material boundaries.

Although the ray model is adequate for an intuitive picture of how light travels along a fiber, a more comprehensive description of light propagation, signal dispersion, and power loss in a cylindrical optical fiber waveguide requires a wave theory approach. In this approach, electromagnetic fields (at optical frequencies) traveling in the fiber can be expressed as superpositions of elementary field configurations called the modes of the fiber. A mode of monochromatic light of radian frequency ω traveling in the axial (positive z) direction in a fiber can be described by the factor $\exp[j(\omega t - \beta z)]$, where β is the propagation constant of the mode. For guided (bound) modes β can assume only a finite number of possible solutions. These solutions are found by solving Maxwell's equations for a dielectric medium subject to the electromagnetic field boundary conditions at the core-cladding interface of an optical fiber. The analysis is rather complex because the boundary conditions create a coupling between the longitudinal components of the **E** and **H** fields, which leads to hybrid mode solutions.

However, in place of a lengthy exact analysis for the modes of a fiber, a simpler but highly accurate approximation can be used, based on the principle that in a typical step-index fiber the difference between the indices of refraction of the core and cladding is very small. This is the weakly guiding fiber approximation that has been used successfully for evaluating optical fiber waveguide characteristics.

Appendix: The Fresnel Equations

One can consider unpolarized light as consisting of two orthogonal plane polarization components. For analyzing reflected and refracted light, one component can be chosen to lie in the plane of incidence (the plane containing the incident and reflected rays, which here is taken to be the yz-plane) and the other of which lies in a plane perpendicular to the plane of incidence (the xz-plane). For example, these can be the E_x and E_y components of the electric field vector. These then are designated as the *perpendicular polarization* (E_x) and the *parallel polarization* (E_y) components with maximum amplitudes E_{0x} and E_{0y}, respectively.

When an unpolarized light beam traveling in air impinges on a nonmetallic surface such as a glass material, part of the beam (designated by E_{0r}) is reflected and part of the beam (designated by E_{0t}) is refracted and transmitted into the target material. The reflected beam is partially polarized and at a specific angle (known as *Brewster's angle*) the reflected light is completely perpendicularly polarized, so that $(E_{0r})_y = 0$. This condition holds when the angle of incidence is such that $\theta_1 + \theta_2 = 90°$ (see Fig. 2.6 for the angle definitions). The parallel component of the refracted beam is transmitted entirely into the target material, whereas the perpendicular component is only partially refracted. How much of the refracted light is polarized depends on the angle at which the light approaches the surface and on the material composition.

The amount of light of each polarization type that is reflected and refracted at a material interface can be calculated using a set of equations known as the *Fresnel equations*. These field-amplitude ratio equations are given in terms of the perpendicular and parallel *reflection coefficients* r_x and r_y, respectively, and the perpendicular and parallel *transmission coefficients* t_x and t_y, respectively. Given that E_{0i}, E_{0r}, and E_{0t} are the amplitudes of the incident, reflected, and transmitted waves, respectively, then

$$r_\perp = r_x = \left(\frac{E_{0r}}{E_{0i}} \right)_x = \frac{n_1 \cos \theta_1 - n_2 \cos \theta_2}{n_1 \cos \theta_1 + n_2 \cos \theta_2} \tag{2.44}$$

$$r_{\|} = r_y = \left(\frac{E_{0r}}{E_{0i}}\right)_y = \frac{n_2 \cos\theta_1 - n_1 \cos\theta_2}{n_1 \cos\theta_2 + n_2 \cos\theta_1} \qquad (2.45)$$

$$t_{\perp} = t_x = \left(\frac{E_{0t}}{E_{0i}}\right)_x = \frac{2n_1 \cos\theta_1}{n_1 \cos\theta_1 + n_2 \cos\theta_2} \qquad (2.46)$$

$$t_{\|} = t_y = \left(\frac{E_{0t}}{E_{0i}}\right)_y = \frac{2n_1 \cos\theta_1}{n_1 \cos\theta_2 + n_2 \cos\theta_1} \qquad (2.47)$$

If light is incident perpendicularly on the material interface, then the angles are $\theta_1 = \theta_2 = 0$. From Eqs. (2.44) and (2.45) it follows that the reflection coefficients are

$$r_x(\theta_1 = 0) = -r_y(\theta_2 = 0) = \frac{n_1 - n_2}{n_1 + n_2} \qquad (2.48)$$

Similarly, for $\theta_1 = \theta_2 = 0$, the transmission coefficients are

$$t_x(\theta_1 = 0) = t_y(\theta_2 = 0) = \frac{2n_1}{n_1 + n_2} \qquad (2.49)$$

Example 2A.1 Consider the case when light traveling in air ($n_{air} = 1.00$) is incident perpendicularly on a smooth glass surface that has a refractive index $n_{tissue} = 1.48$. What are the reflection and transmission coefficients?

Solution From Eq. (2.48) with $n_1 = n_{air}$ and $n_2 = n_{glass}$ it follows that the reflection coefficient is

$$r_x = -r_x = (1.48 - 1.00)/(1.48 + 1.00) = 0.194$$

and from Eq. (2.49) the transmission coefficient is

$$t_x = t_y = 2(1.00)/(1.48 + 1.00) = 0.806$$

The change in sign of the reflection coefficient r_x means that the field of the perpendicular component shifts by $180°$ upon reflection.

The field amplitude ratios can be used to calculate the *reflectance* R (the ratio of the reflected to the incident flux or power) and the *transmittance* T (the ratio of the transmitted to the incident flux or power). For linearly polarized light in which the vibrational plane of the incident light is perpendicular to the interface plane, the total reflectance and transmittance are

$$R_{\perp} = \left(\frac{E_{0r}}{E_{0i}}\right)_x^2 = R_x = r_x^2 \qquad (2.50)$$

$$R_{\|} = \left(\frac{E_{0r}}{E_{0i}}\right)_y^2 = R_y = r_y^2 \qquad (2.51)$$

$$T_\perp = \frac{n_2 \cos\theta_2}{n_1 \cos\theta_1} \left(\frac{E_{0t}}{E_{0i}}\right)^2_x = T_x = \frac{n_2 \cos\theta_2}{n_1 \cos\theta_1} t_x^2 \tag{2.52}$$

$$T_\parallel = \frac{n_2 \cos\theta_2}{n_1 \cos\theta_1} \left(\frac{E_{0t}}{E_{0i}}\right)^2_y = T_y = \frac{n_2 \cos\theta_2}{n_1 \cos\theta_1} t_y^2 \tag{2.53}$$

The expression for T is a bit more complex compared to R because the shape of the incident light beam changes upon entering the second material and the speeds at which energy is transported into and out of the interface are different.

If light is incident perpendicularly on the material interface, then substituting Eq. (2.48) into Eqs. (2.50) and (2.51) yields the following expression for the reflectance R

$$R = R_\perp(\theta_1 = 0) = R_\parallel(\theta_1 = 0) = \left(\frac{n_1 - n_2}{n_1 + n_2}\right)^2 \tag{2.54}$$

and substituting Eq. (2.49) into Eqs. (2.52) and (2.53) yields the following expression for the transmittance T

$$T = T_\perp(\theta_2 = 0) = T_\parallel(\theta_2 = 0) = \frac{4n_1 n_2}{(n_1 + n_2)^2} \tag{2.55}$$

Example 2A.2 Consider the case described in Example 2A.1 in which light traveling in air ($n_{air} = 1.00$) is incident perpendicularly on a smooth glass sample that has a refractive index $n_{glass} = 1.48$. What are the reflectance and transmittance values?

Solution From Eq. (2.54) and Example 2A.1 the reflectance is

$$R = [(1.48 - 1.00)/(1.48 + 1.00)]^2 = (0.194)^2 = 0.038 \text{ or } 3.8\%$$

From Eq. (2.55) the transmittance is

$$T = 4(1.00)(1.48)/(1.00 + 1.48)^2 = 0.962 \text{ or } 96.2\%$$

Note that $R + T = 1.00$.

Example 2A.3 Consider a plane wave that lies in the plane of incidence of an air-glass interface. What are the values of the reflection coefficients if this lightwave is incident at 30° on the interface? Let $n_{air} = 1.00$ and $n_{glass} = 1.50$.

Solution First from Snell's law, it follows that $\theta_2 = 19.2°$. Substituting the values of the refractive indices and the angles into Eqs. (2.44) and (2.45) then yield $r_x = -0.241$ and $r_y = 0.158$. As noted in Example 2A.1, the change in sign of the reflection coefficient r_x means that the field of the perpendicular component shifts by 180° upon reflection.

Problems

2.1 Suppose a certain wave is specified by $y = 8 \cos 2\pi (2t - 0.8z)$, where y is expressed in micrometers and the propagation constant is given in μm^{-1}. Find (a) the amplitude, (b) the wavelength, (c) the angular frequency, and (d) the displacement at time $t = 0$ and $z = 4 \mu m$. [Answers using Example 2.1: (a) 8 μm; (b) 1.25 μm; (c) 4 π; (d) 7.512 μm.]

2.2 Let $E_{0x} = E_{0y} = 1$ in Eq. (2.7). Write a software program to plot this equation for values of $\delta = (n\pi)/8$, where $n = 0, 1, 2, \ldots. 16$. What does this show about the state of polarization as the angle δ changes?

2.3 Show that any linearly polarized wave may be considered as the superposition of left and right circularly polarized waves that are in phase and have equal amplitudes and frequencies.

2.4 Light traveling in air strikes a glass plate at an angle $\theta_1 = 33°$, where θ_1 is measured between the incoming ray and the normal to the glass surface. Upon striking the glass, part of the beam is reflected and part is refracted. (a) If the refracted and reflected beams make an angle of $90°$ with each other, show that the refractive index of the glass is 1.540. (b) Show that the critical angle for this glass is $40.5°$.

2.5 A point source of light is 12 cm below the surface of a large body of water ($n_{water} = 1.33$). Show that the radius of the largest circle on the water surface through which the light can emerge from the water into air ($n_{air} = 1.000$) is 13.7 cm.

2.6 A right-angle prism (internal angles are $45°$, $45°$, and $90°$) is immersed in alcohol ($n = 1.45$). Show that the refractive index of the prism must be 2.05 if a ray incident normally on one of the short faces is to be totally reflected at the long face of the prism.

2.7 Show that the critical angle at an interface between doped silica with $n_1 = 1.460$ and pure silica with $n_2 = 1.450$ is $83.3°$.

2.8 Consider the power reflection for light incident normally at the interface between an optical fiber end and air. If the refractive indices of the fiber and air are $n_{fiber} = 1.48$ and $n_{air} = 1.000$, show that the reflectance $R = 0.037$ and the transmittance is $T = 0.963$.

2.9 Consider a step-index fiber having $n_1 = 1.48$ and $n_2 = 1.46$. (a) Show that the numerical aperture is 0.243. (b) If the outer medium is air with $n = 1.00$, show that the acceptance angle θ_A for this fiber is $14°$.

2.10 Suppose a certain multimode step-index optical fiber has a core diameter of 62.5 μm, a core index of 1.48, and an index difference $\Delta = 0.015$. (a) Show that at 1310 nm the value of V is 38.4. (b) Show that the total number of modes is 737.

2.11 A given step-index multimode fiber with a numerical aperture of 0.20 supports approximately 1000 modes at an 850 nm wavelength. (a) Show that the diameter of its core is 60.5 μm. (b) Show that the fiber supports 414 modes at 1320 nm. (c) Show that the fiber supports 300 modes at 1550 nm.

2.12 A certain step-index fiber has a 25 μm core radius, $n_1 = 1.48$, and $n_2 = 1.46$. (a) Show that the normalized frequency at 820 nm is $V = 46.5$. (b) Show that 1081 modes propagate in this fiber at 820 nm. (c) Show that 417 modes propagate in this fiber at 1320 nm. (d) Show that 303 modes propagate in this fiber at 1550 nm. (e) Verify that the percent of the optical power flows in the cladding for the different wavelengths is 4.1% at 820 nm, 6.6% at 1320 nm, and 7.8% at 1550 nm.

2.13 Consider a fiber with a 25 μm core radius, a core index $n_1 = 1.48$, and $\Delta = 0.01$. (a) Show that at 1320 nm the parameter $V = 25$ and the number of modes $M = 312$. (b) Verify that 7.5% of the optical power flows in the cladding. (c) If the core-cladding difference is reduced to $\Delta = 0.003$, show that the number of modes $M = 94$ and 13.7% of the optical power flows in the cladding.

2.14 Suppose a certain step-index fiber has a 5 μm core radius, an index difference $\Delta = 0.002$, and a core index $n_1 = 1.480$. (a) By calculating the V number, verify that at 1310 nm this is a single-mode fiber. (b) Verify that at 820 nm the fiber is not single-mode because $V = 3.514$. (c) With the result from (b), verify by observation from Fig. 2.20 that the fiber supports the LP_{01} and LP_{11} modes at 820 nm.

2.15 Consider a 62.5 μm core-diameter graded-index fiber that has a parabolic index profile ($\alpha = 2$). Suppose the fiber has a numerical aperture $NA = 0.275$. (a) Show that the V number for this fiber at 850 nm is 63.5. (b) Verify that 1008 modes propagate in the fiber at 850 nm.

2.16 Consider a 50 μm core diameter graded-index fiber that has a core index $n_1 = 1.480$ and a cladding index $n_2 = 1.465$. (a) Using the exact expression for the index difference given in Eq. (2.39), show that $\Delta = 1.008\%$. (b) Using the approximation for given in the right-hand side of Eq. (2.39), show that $\Delta = 1.014\%$. This shows that the approximation is quite accurate.

2.17 Calculate the number of modes at 820 and 1300 nm in a graded-index fiber having a parabolic-index profile ($\alpha = 2$), a 25 μm core radius, $n_1 = 1.48$, and $n_2 = 1.46$. How does this compare to a step-index fiber? [Answer: At 820 nm for the graded-index fiber, $M = 543$ and at 1300 nm, $M = 216$. For a step-index fiber, at 820 nm, $M = 1078$ and at 1300 nm, $M = 429$.]

2.18 Calculate the numerical apertures of (a) a plastic step-index fiber having a core refractive index of $n_1 = 1.60$ and a cladding index of $n_2 = 1.49$, (b) a step-index fiber having a silica core ($n_1 = 1.458$) and a silicone resin cladding ($n_2 = 1.405$). [Answer: (a) 0.58; (b) 0.39.]

References

1. See any general physics book or introductory optics book; for example: (a) D. Halliday, R. Resnick, and J. Walker, *Fundamentals of Physics*, 11th edn. (Wiley, 2018); (b) E. Hecht, *Optics*, Pearson, 5th edn. (2016); (c) K. Iizuka, *Engineering Optics* (Springer, Berlin, 2019)

2. See any introductory electromagnetics book; for example: (*a*) B.M. Notaros, *Electromagnetics* (Prentice Hall, 2011); (*b*) W.H. Hayt Jr, J.A. Buck, *Engineering Electromagnetics*, 9th edn. (McGraw-Hill, 2019); (*c*) N. Ida, *Engineering Electromagnetics*, 4th edn. (Springer, 2020); (*d*) F.T. Ulaby, *Fundamentals of Applied Electromagnetics*, 7th edn. (Pearson, 2015)

3. B.E.A. Saleh, M.C. Teich, *Fundamentals of Photonics*, 3rd edn. (Wiley, 2019)

4. E.A.J. Marcatili, in *Objectives of Early Fibers: Evolution of Fiber Types*, ed. by S.E. Miller, A.G. Chynoweth, Optical Fiber Telecommunications (Academic Press, 1979)

5. R.J. Black, L. Gagnon, *Optical Waveguide Modes: Polarization, Coupling and Symmetry* (McGraw-Hill, 2010).

6. A.W. Snyder, J.D. Love, *Optical Waveguide Theory* (Chapman and Hall, 1983)

7. C. Yeh, F. Shimabukuro, *The Essence of Dielectric Waveguides* (Springer, Berlin, 2008)

8. K. Okamoto, *Fundamentals of Optical Waveguides*, 2nd edn. (Academic Press, 2006)

9. C.L. Chen, *Foundations of Guided-Wave Optics* (Wiley, 2007)

10. K. Kawano, T. Kitoh, *Introduction to Optical Waveguide Analysis: Solving Maxwell's Equation and the Schrödinger Equation* (Wiley, 2002)

11. J.A. Buck, *Fundamentals of Optical Fibers*, 2nd edn. (Wiley, 2004)

12. J. Hu, C.R. Menyuk, Understanding leaky modes: slab waveguide revisited. Adv. Optics Photon. **1**(1), 58–106 (2009)

13. A.K. Ghatak, Leaky modes in optical waveguides. Opt. Quant. Electron. **17**, 311–321 (1985)

14. E. Snitzer, Cylindrical dielectric waveguide modes. J. Opt. Soc. Amer. **51**, 491–498 (1961)

15. M. Koshiba, *Optical Waveguide Analysis* (McGraw-Hill, 1992)

16. D. Marcuse, *Light Transmission Optics*, 2nd edn. (Van Nostrand-Reinhold, 1982)

17. R. Olshansky, Propagation in glass optical waveguides. Rev. Mod. Phys. **51**, 341–367 (1979)

18. D. Gloge, The optical fiber as a transmission medium. Rep. Progr. Phys. **42**, 1777–1824 (1979)

19. D. Gloge, Weakly guiding fibers. Appl. Opt. **10**, 2252–2258 (1971)

20. ITU-T Recommendation G.652, *Characteristics of a single-mode optical fibre and cable* (2016)

21. A.W. Snyder, Asymptotic expressions for eigenfunctions and eigenvalues of a dielectric or optical waveguide. IEEE Trans. Microwave Theory Tech., **MTT-17**, 1130–1138 (1969)

22. D. Marcuse, Gaussian approximation of the fundamental modes of graded index fibers. J. Opt. Soc. Amer. **68**, 103–109 (1978)

23. H.M. DeRuiter, Integral equation approach to the computation of modes in an optical waveguide. J. Opt. Soc. Amer. **70**, 1519–1524 (1980)

24. A.W. Snyder, Understanding monomode optical fibers. Proc. IEEE **69**, 6–13 (1981)

25. D. Marcuse, D. Gloge, E.A.J. Marcatili, in *Guiding Properties of Fibers*, ed. by S.E. Miller, A.G. Chynoweth, Optical Fiber Telecommunications (Academic Press, 1979)

26. M. Artiglia, G. Coppa, P. DiVita, M. Potenza, A. Sharma, Mode field diameter measurements in single-mode optical fibers. J. Lightw. Technol. **7**, 1139–1152 (1989)

27. T.J. Drapela, D.L. Franzen, A.H. Cherin, R.J. Smith, A comparison of far-field methods for determining mode field diameter of single-mode fibers using both gaussian and Petermann definitions. J. Lightw. Technol. **7**, 1153–1157 (1989)

28. K. Petermann, Constraints for fundamental mode spot size for broadband dispersion-compensated single-mode fibers. Electron. Lett. **19**, 712–714 (1983)

29. (*a*) ITU-T Recommendation G.650.1, *Definitions and Test Methods for Linear, Deterministic Attributes of Single-Mode Fibre and Cable* (2018); (*b*) ITU-T Recommendation G.650.2, *Definitions and Test Methods for Statistical and Nonlinear Related Attributes of Single-Mode Fibre and Cable* (2015); (*c*) ITU-T Recommendation G.650.3, *Test Methods for Installed Single-Mode Optical Fibre Cable Links* (2017)

30. IEC-60793-1-45, *Measurement Methods and Test Procedures—Mode Field Diameter* (2017)

31. R. Hui, M. O'Sullivan, *Fiber Optic Measurement Techniques* (Academic Press, 2009)

32. I.P. Kaminow, Polarization in optical fibers. IEEE J. Quantum Electron. **17**, 15–22 (1981)

33. J.N. Damask, *Polarization Optics in Telecommunications* (Springer, Berlin, 2004)

34. M. Brodsky, N.J. Frigo, M. Tur, in *Polarization Mode Dispersion*, ed. by I.P. Kaminov, T. Li, A.E. Willner, Optical Fiber Telecommunications-V, vol A, chap 17 (Academic Press, 2008), pp. 593–669

35. D. Gloge, E. Marcatili, Multimode theory of graded core fibers. Bell Sys. Tech. J. **52**, 1563–1578 (1973)
36. J.D. Musgraves, J. Hu, L. Calvez, *Springer Handbook of Glass* (Springer, Berlin, 2019)
37. S.R. Nagel, in *Fiber Materials and Fabrication Methods*, ed. by S.E. Miller, I.P. Kaminow, Optical Fiber Telecommunications–II (Academic Press, 1988)
38. B. Mysen, P. Richet, *Silicate Glasses and Melts*, 2nd edn. (Elsevier, 2018)
39. P.C. Schultz, Progress in optical waveguide processes and materials. Appl. Opt. **18**, 3684–3693 (1979)
40. W. Miniscalco, Erbium-doped glasses for fiber amplifiers at 1500 nm. J. Lightw. Technol. **9**, 234–250 (1991)
41. F. Sidiroglou, A. Roberts, G. Baxter, Contributed review: a review of the investigation of rare-earth dopant profiles in optical fibers. Rev. Sci. Instr. **87**(041501), 22 (2016)
42. O. Ziemann, J. Krauser, P.E. Zamzow, W. Daum, *POF Handbook,* 2nd edn. (Springer, Berlin, 2008)
43. R. Nakao, A. Kondo, Y. Koike, Fabrication of high glass transition temperature graded-index plastic optical fiber. J. Lightw. Technol. **30**, 969–973 (2012)
44. C.-A. Bunge, M. Beckers, T. Gries (eds.), *Polymer Optical Fibres* (Woodhead Publishing, 2017)
45. P. St. John Russell, Photonic crystal fibers. J. Lightw. Technol. **24**(12), 4729–4749 (2006)
46. M. Large, L. Poladian, G. Barton, M.A. van Eijkelenborg, *Microstructured Polymer Optical Fibres* (Springer, Berlin, 2008)
47. X. Yu, M. Yan, G. Ren, W. Tong, X. Cheng, J. Zhou, P. Shum, N.Q. Ngo, Nanostructure core fiber with enhanced performances: design, fabrication and devices. J. Lightw. Technol. **27**(11), 1548–1555 (2009)
48. A. Bjarklev, J. Broeng, A.S. Bjarklev, *Photonic Crystal Fibres* (Springer, Berlin, 2003)
49. B. Wiltshire, M.H. Reeve, A review of the environmental factors affecting optical cable design. J. Lightw. Technol. **6**, 179–185 (1988)
50. A. Inoue, Y. Koike, Low-noise graded-index plastic optical fibers for significantly stable and robust data transmission. J. Lightw. Technol. **36**(24), 5887–5892 (2018)
51. X. Jiang, W. Jiang, G. Pan, Y.C. Ye (eds.), *Submarine Optical Cable Engineering* (Elsevier, 2018)
52. M. Charbonneau-Lefort, M. J. Yadlowsky, Optical cables for consumer applications. J. Lightw. Technol. **33**(4), 872–877 (2015)
53. J. Chesnoy, *Undersea Fiber Communication Systems* (Academic Press, 2016)

Chapter 3
Optical Signal Attenuation and Dispersion

Abstract When information signals travel in any type of transmission medium, various signal power losses and signal fidelity distortions are always present. Attenuation of a light signal as it propagates along a fiber is an important consideration in the design of an optical communication system because it plays a major role in determining the maximum transmission distance between a transmitter and a receiver. In addition to being attenuated, an optical signal undergoes continuous broadening and distortion as it travels along a fiber. The signal broadening is a consequence of intramodal and intermodal dispersion effects.

Chapter 2 showed the structure of optical fibers and examined the concepts of how light propagates along a cylindrical dielectric optical waveguide. This chapter continues the discussion of optical fibers by answering two very important questions:

1. What are the optical power loss or signal attenuation mechanisms in a fiber?
2. Why and to what degree do optical signals get distorted as they propagate along a fiber?

Signal attenuation (also known as *fiber attenuation, fiber loss,* or *power level reduction*) is one of the most important properties of an optical fiber because it largely determines the maximum unamplified or repeaterless separation between a transmitter and a receiver. Because amplifiers and repeaters are expensive to fabricate, install, and maintain, the degree of attenuation in a fiber has a large influence on system cost. Of equal importance is signal dispersion. The dispersion mechanisms in a fiber cause optical signal pulses to broaden as they travel along a fiber. If these pulses travel sufficiently far, they will eventually overlap with neighboring pulses, thereby creating errors in the receiver output. The signal dispersion mechanisms thus limit the information-carrying capacity of a fiber.

The original version of this chapter was revised: All the incorrect sentences have been corrected. The correction to this chapter can be found at https://doi.org/10.1007/978-981-33-4665-9_15

© The Author(s), under exclusive license to Springer Nature Singapore Pte Ltd. 2021, 93
corrected publication 2021
G. Keiser, *Fiber Optic Communications*,
https://doi.org/10.1007/978-981-33-4665-9_3

3.1   Fiber Attenuation

Optical power attenuation of a light signal as it propagates along a fiber is an important consideration in the design of an optical communication system; the degree of attenuation plays a major role in determining the maximum transmission distance between a transmitter and a receiver or an in-line amplifier. The basic attenuation mechanisms that cause power level reductions in a fiber are absorption, scattering, and radiative losses of the optical energy [1–3]. Absorption is related to the fiber material, whereas scattering is associated both with the fiber material and with structural imperfections in the optical waveguide. Attenuation owing to radiative loss effects originates from perturbations (both microscopic and macroscopic) of the fiber geometry.

This section first discusses the units in which fiber losses are measured and then presents the physical phenomena that give rise to attenuation.

3.1.1   Units for Fiber Attenuation

As light travels along a fiber, its power decreases exponentially with distance. If $P(0)$ is the optical power in a fiber at the origin (at $z = 0$), then the power $P(z)$ at a distance z farther down the fiber is

$$P(z) = P(0)e^{-\alpha_p z} \tag{3.1}$$

where

$$\alpha_p = \frac{1}{z} \ln \left[\frac{P(0)}{P(z)} \right] \tag{3.2}$$

is the fiber *attenuation coefficient* given in units of, for example, km^{-1}. Note that the units for $2z\alpha_p$ can also be designated by *nepers* (see Appendix B).

For simplicity in calculating optical signal attenuation in a fiber, the common procedure is to express the attenuation coefficient in units of *decibels per kilometer,* denoted by dB/km. Designating this parameter by α yields

$$\alpha(\text{dB/km}) = \frac{10}{z} \log \left[\frac{P(0)}{P(z)} \right] = 4.343 \, \alpha_p \left(\text{km}^{-1} \right) \tag{3.3}$$

This parameter is generally referred to as the *fiber loss* or the *fiber attenuation.* It depends on several variables, as is shown in the following sections, and it is a function of the wavelength.

Example 3.1 An ideal fiber would have no loss so that $P_{out} = P_{in}$. This corresponds to an attenuation of 0 dB/km, which, in practice, is impossible. An actual low-loss fiber might have a 0.35 dB/km loss at 1310 nm and a loss of 0.20 dB/km at 1550 nm, for example. Then at 1310 nm the optical power at 10 km P(10 km) as a fraction of the input power P(0) is

$$\frac{P(10 \text{ km})}{P(0)} = 10^{-\alpha z/10} = 10^{-(0.35)(10)/10} = 0.447 = 44.7\%$$

For operation at 1550 nm, the optical power at 10 km P(10 km) as a fraction of the input power P(0) is

$$\frac{P(10 \text{ km})}{P(0)} = 10^{-\alpha z/10} = 10^{-(0.20)(10)/10} = 0.63 = 63\%$$

This means that after a 10 km transmission distance, at 1310 nm the optical signal power would decrease by 3.5 dB (that is, $10 \log 0.447 = -3.5$ dB) or be 44.7% of the input power. Likewise, at 1550 nm the output optical power is 63% of the input power (a decrease of $10 \log 0.630 = -2.0$ dB). Viewed alternatively, at 1310 nm the loss over 10 km is $(10 \text{ km})(0.35 \text{ dB/km}) = 3.5$ dB loss and at 1550 nm the loss over 10 km is $(10 \text{ km})(0.20 \text{ dB/km}) = 2.0$ dB loss.

Figure 3.1 shows the relationship between decibels and power ratios ranging from 0.1 to 1.0. Thus, as shown in Example 3.1, the dashed lines illustrate that transmission over a 10 km distance using a fiber with an attenuation of 0.5 dB/km results in a 5 dB power attenuation yielding an output-to-input power ratio of 31.6%. Similarly, over a 10 km distance a fiber with an attenuation of 0.3 dB/km results in a 3 dB power attenuation yielding a power ratio of 50%.

Example 3.2 As Sect. 1.3 describes, optical powers are commonly expressed in units of *dBm*, which is the decibel power level referred to 1 mW. Consider a 30 km long

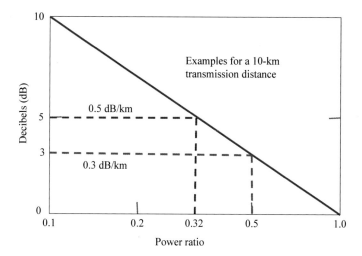

Fig. 3.1 The relationship between decibels and power ratios ranging from 0.1 to 1.0

optical fiber that has an attenuation of 0.4 dB/km at 1310 nm. Suppose an engineer wants to find the optical output power P_{out} if 200 nW of optical power is launched into the fiber. First express the input power in dBm units:

$$P_{in}(dBm) = 10\log\left[\frac{P_{in}(W)}{1\,mW}\right] = 10\log\left[\frac{200 \times 10^{-6}\,W}{1 \times 10^{-3}\,W}\right] = -7.0\,dBm$$

From Eq. (3.3) with $P(0) = P_{in}$ and $P(z) = P_{out}$ the output power level (in dBm) at $z = 30$ km is

$$P_{out}(dBm) = 10\log\left[\frac{P_{out}(W)}{1\,mW}\right] = 10\log\left[\frac{P_{in}(W)}{1\,mW}\right] - \alpha z$$
$$= -7.0\,dBm - (0.4\,dB/km)(30\,km) = -19.0\,dBm$$

In unit of watts, the output power is

$$P(30\,km) = 10^{-19.0/10}(1\,mW) = 12.6 \times 10^{-3}\,mW = 12.6\,\mu W.$$

Drill Problem 3.1 A 50 km long optical fiber has an attenuation of 0.25 dB/km at 1550 nm. If 100 μW of optical power is launched into the fiber, show that the power emerging at the fiber output is -32.5 dBm or 0.56 μW.

Drill Problem 3.2 An optical fiber loses 75% of the optical power traversing the fiber after 25 km. Using the left-hand side of Eq. (3.3) with $z = 25$ km and $P(z) = 0.25\,P(0)$, show that the attenuation is $\alpha = 0.25$ dB/km.

3.1.2 Absorption of Optical Power

Absorption of optical power is caused by three different mechanisms:

1. Absorption by atomic defects in the glass composition.
2. Extrinsic absorption by impurity atoms in the glass material.
3. Intrinsic absorption by the basic constituent atoms of the fiber material.

Atomic defects are imperfections in the atomic structure of the fiber material. Examples of these defects include missing molecules, high-density clusters of atom groups, or oxygen defects in the glass structure. Usually, absorption losses

Table 3.1 Examples of absorption loss in silica glass at different wavelengths due to 1 ppm of water-ions and various transition-metal impurities

Impurity	Loss due to 1 ppm of impurity (dB/km)	Absorption peak (nm)
Iron: Fe^{2+}	0.68	1100
Iron: Fe^{3+}	0.15	400
Copper: Cu^{2+}	1.1	850
Chromium: Cr^{2+}	1.6	625
Vanadium: V^{4+}	2.7	725
Water: OH^-	1.0	950
Water: OH^-	2.0	1240
Water: OH^-	4.0	1380

arising from these defects are negligible compared with intrinsic and impurity absorption effects.

The dominant absorption factor in silica fibers is the presence of minute quantities of impurities in the fiber material. These impurities include OH (water) ions that are dissolved in the glass and transition metal ions such as iron, copper, chromium, and vanadium. Transition metal impurity levels were around 1 part per million (ppm) in glass fibers made in the 1970s, which resulted in losses ranging from 1 to 4 dB/km, as Table 3.1 shows. Impurity absorption losses occur either because of electron transitions between the energy levels within these ions or because of charge transitions between ions. The absorption peaks of the various transition metal impurities tend to be broad, and several peaks may overlap, which further broadens the absorption in a specific region. Modern techniques for producing a fiber preform have reduced the transition-metal impurity levels by several orders of magnitude. Such low impurity levels allow the fabrication of low-loss fibers.

The presence of OH ion impurities in a glass fiber preform results mainly from the elements in the oxyhydrogen flame used in the hydrolysis reaction of the basic $SiCl_4$, $GeCl_4$, and $POCl_3$ starting materials. Water impurity concentrations of less than a few parts per billion (ppb) are required if the attenuation is to be less than 20 dB/km. The high levels of OH ions in early fibers resulted in large absorption peaks at 725, 950, 1240, and 1380 nm. Regions of low attention lie between these absorption peaks.

The peaks and valleys in the attenuation curves resulted in the designation of the various *transmission windows* shown in Fig. 3.2 (see Sect. 1.2.2). By reducing the residual OH content of fibers to below 1 ppb, standard commercially available single-mode fibers have nominal attenuations of below 0.4 dB/km at 1310 nm (in the O-band) and less than 0.25 dB/km at 1550 nm (in the C-band). Further elimination of water ions diminishes the absorption peak around 1440 nm and thus opens up the E-band for data transmission, as indicated by the dashed line in Fig. 3.2. Optical fibers that can be used in the E-band are known by names such as *low-water-peak* or *full-spectrum fibers*.

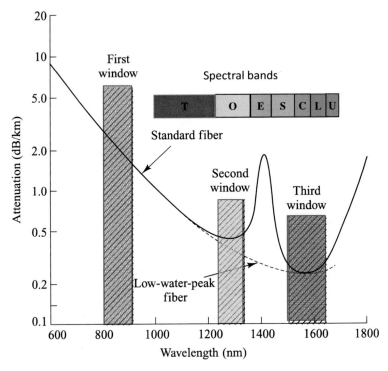

Fig. 3.2 Optical fiber attenuation as a function of wavelength yields nominal values of 0.40 dB/km at 1310 nm and 0.25 dB/km at 1550 nm for standard single-mode fiber; the dashed curve is the attenuation for low-water-peak fiber

Intrinsic absorption is associated with the basic fiber material (e.g., pure SiO_2) and is the principal physical factor that defines the transparency window of a material over a specified spectral region. Intrinsic absorption sets the fundamental lower limit on absorption for any particular material; it is defined as the absorption that occurs when the material is in a perfect state with no density variations, impurities, or material inhomogeneity.

Intrinsic absorption results from electronic absorption bands in the ultraviolet region and from atomic vibration bands in the near-infrared region. The electronic absorption bands are associated with the energy band gaps of the amorphous glass materials. Absorption occurs when a photon interacts with an electron in the valence band and excites it to a higher energy level. The ultraviolet edge of the electron absorption bands of both amorphous and crystalline materials follows the empirical relationship [4]

$$\alpha_{uv} = Ce^{E/E_0} \tag{3.4}$$

which is known as Urbach's rule. Here, C and E_0 are empirical constants and E is the photon energy. The magnitude and characteristic exponential decay of the ultraviolet absorption are shown in Fig. 3.3. Because E is inversely proportional to the wavelength λ, ultraviolet absorption decays exponentially with increasing wavelength. In particular, the ultraviolet loss contribution in dB/km at any wavelength (given in μm) can be expressed empirically (derived from observation or experiment) as a function of the mole fraction x of GeO_2 as [5]

$$\alpha_{uv} = \frac{154.2x}{46.6x + 60} \times 10^{-2} \exp\left(\frac{4.63}{\lambda}\right). \tag{3.5}$$

Example 3.3 Consider two silica fibers that are doped with x = 6% and 18% mole fractions of GeO_2, respectively. Compare the ultraviolet absorptions at wavelengths of 0.7 μm and 1.3 μm.

Solution Using Eq. (3.5) for the ultraviolet absorption, then

(a)   For the fiber with $x = 0.06$ and $\lambda = 0.7$ μm

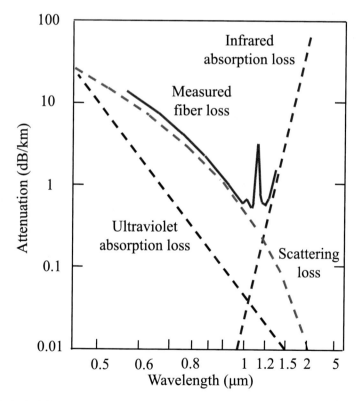

Fig. 3.3 Optical fiber attenuation characteristics and their limiting mechanisms for a GeO_2-doped low-loss low-water-content silica fiber

$$\alpha_{uv} = \frac{154.2(0.06)}{46.6(0.06) + 60} \times 10^{-2} \exp\left(\frac{4.63}{0.7}\right) = 1.10 \text{ dB/km}$$

(b)   For the fiber with $x = 0.06$ and $\lambda = 1.3$ μm

$$\alpha_{uv} = \frac{154.2(0.06)}{46.6(0.06) + 60} \times 10^{-2} \exp\left(\frac{4.63}{1.3}\right) = 0.07 \text{ dB/km}$$

(c)   For the fiber with $x = 0.18$ and $\lambda = 0.7$ μm

$$\alpha_{uv} = \frac{154.2(0.18)}{46.6(0.18) + 60} \times 10^{-2} \exp\left(\frac{4.63}{0.7}\right) = 3.03 \text{ dB/km}$$

(d)   For the fiber with $x = 0.18$ and $\lambda = 1.3$ μm

$$\alpha_{uv} = \frac{154.2(0.18)}{46.6(0.18) + 60} \times 10^{-2} \exp\left(\frac{4.63}{1.3}\right) = 0.19 \text{ dB/km.}$$

Drill Problem 3.3 A silica fiber is doped with a 15% mole fraction of GeO_2. Compare the ultraviolet absorption at 860 nm and 1550 nm.
  [**Answer**: 0.75 dB/km at 860 nm; 0.068 dB/km at 1550 nm.]

As Fig. 3.3 shows, absorption loss is small compared with scattering loss in the ultraviolet, visible, and near-infrared regions ranging from 0.5 to 1.2 μm. In the near-infrared region above 1.2 μm, the optical waveguide loss is predominantly determined by the presence of OH ions and the intrinsic infrared absorption of the constituent material. The intrinsic infrared absorption is associated with the characteristic vibration frequency of the particular chemical bond between the atoms of which the fiber is composed. An interaction between the vibrating bond and the electromagnetic field of the optical signal results in a transfer of energy from the field to the bond, thereby giving rise to absorption. This absorption is quite strong because of the many bonds present in the fiber. An empirical expression for the infrared absorption in dB/km for GeO_2–SiO_2 glass with λ given in μm is [5]

$$\alpha_{IR} = 7.81 \times 10^{11} \times \exp\left(\frac{-48.48}{\lambda}\right) \tag{3.6}$$

These mechanisms result in a wedge-shaped spectral-loss characteristic. Within this wedge, losses as low as 0.148 dB/km at 1.57 μm in a single-mode fiber have been measured [6].

3.1.3 Scattering Losses in Optical Fibers

Scattering losses in glass arise from microscopic variations in the material density, from compositional fluctuations, and from structural inhomogeneity or defects occurring during fiber manufacture. As Sect. 2.7 describes, glass is composed of a randomly connected network of molecules. Such a structure naturally contains regions in which the molecular density is either higher or lower than the average density in the glass. In addition, because glass is made up of several oxides, such as SiO_2, GeO_2, and P_2O_5, compositional fluctuations can occur. These two effects give rise to refractive-index variations that occur within the glass over distances that are small compared with the wavelength. These index variations cause a Rayleigh-type scattering of the light. Rayleigh scattering in glass is the same phenomenon that scatters light from the sun in the atmosphere, thereby giving rise to a blue sky.

The exact expressions for scattering-induced attenuation are fairly complex owing to the random molecular nature and the various oxide constituents of glass. For single-component glass the scattering loss at a wavelength λ (given in μm) resulting from density fluctuations can be approximated by [4, 7] (in base e units)

$$\alpha_{scat} = \frac{8\pi^3}{3\lambda^4}\left(n^2 - 1\right)^2 k_B T_f \beta_T \tag{3.7}$$

Here, n is the refractive index, k_B is Boltzmann's constant, β_T is the isothermal compressibility of the material, and the fictive temperature T_f is the temperature at which the density fluctuations are frozen into the glass as it solidifies (after having been drawn into a fiber). Alternatively, the relation [4, 8] (in base e units)

$$\alpha_{scat} = \frac{8\pi^3}{3\lambda^4} n^8 p^2 k_B T_f \beta_T \tag{3.8}$$

has been derived, where p is the photoelastic coefficient. A comparison of Eqs. (3.7) and (3.8) is given in Problem 3.5. Note that these equations are given in units of *nepers* (that is, base e units). As shown in Eq. (3.1), to change this to decibels for optical power attenuation calculations, multiply these equations by $10 \log e = 4.343$.

Example 3.4 For silica the fictive temperature T_f is 1400 K, the isothermal compressibility β_T is 6.8×10^{-12} cm^2/dyn $= 6.8 \times 10^{-11}$ m^2/N, and the photoelastic coefficient is 0.286. Estimate the scattering loss at a 1.30 μm wavelength where $n = 1.450$.

Solution Using Eq. (3.8)

$$
\begin{aligned}
\alpha_{scat} &= \frac{8\pi^3}{3\lambda^4} n^8 p^2 k_B T_f \beta_T \\
&= \frac{8\pi^3}{3(1.3)^4}(1.45)^8(0.286)^2\left(1.38 \times 10^{-23}\right)(1400)\left(6.8 \times 10^{-12}\right)^{\cdot} \\
&= 6.08 \times 10^{-2} \text{ nepers/km} = 0.26 \text{ dB/km}
\end{aligned}
$$

Example 3.5 For pure silica glass an approximate equation for the Rayleigh scattering loss is given by

$$
\alpha(\lambda) = \alpha_0 \left(\frac{\lambda_0}{\lambda}\right)^4
$$

where $\alpha_0 = 1.64$ dB/km at $\lambda_0 = 850$ nm. This formula predicts scattering losses of 0.291 dB/km at 1310 nm and 0.148 dB/km at 1550 nm.

> **Drill Problem 3.4** Using Eq. (3.8) and the parameter values from Example 3.4, show that the estimated scattering loss in a silica fiber at 850 nm where $n = 1.455$ is 1.49 dB/km.

For multicomponent glasses the scattering at a wavelength λ (measured in μm) is given by [4]

$$
\alpha = \frac{8\pi^3}{3\lambda^4}\left(\delta n^2\right)^2 \delta V \tag{3.9}
$$

where the square of the mean-square refractive-index fluctuation $(\delta n^2)^2$ over a volume of δV is

$$
\left(\delta n^2\right)^2 = \left(\frac{\partial n^2}{\partial \rho}\right)^2 (\delta \rho)^2 + \sum_{i=1}^{m}\left(\frac{\partial n^2}{\partial C_i}\right)^2 (\delta C_i)^2 \tag{3.10}
$$

Here, $\delta\rho$ is the density fluctuation and δC_i is the concentration fluctuation of the ith glass component. Their magnitudes must be determined from experimental scattering data. The factors $\partial n^2/\partial \rho$ and $\partial n^2/\partial C_i$ are the variations of the square of the index with respect to the density and the ith glass component, respectively.

Structural inhomogeneities and defects created during fiber fabrication can also cause scattering of light out of the fiber. These defects may be in the form of trapped gas bubbles, unreacted starting materials, and crystallized regions in the glass. In general, the preform manufacturing methods that have evolved have minimized these extrinsic effects to the point where scattering that results from them is negligible compared with the intrinsic Rayleigh scattering.

Rayleigh scattering follows a characteristic λ^{-4} dependence, so it decreases dramatically with increasing wavelength, as is shown in Fig. 3.3. For wavelengths below about 1 μm it is the dominant loss mechanism in a fiber and gives the attenuation-versus-wavelength plots their characteristic downward trend with increasing wavelength. At wavelengths longer than 1 μm, infrared absorption effects tend to dominate optical signal attenuation.

3.1.4 Fiber Bending Losses

Radiative losses occur whenever an optical fiber undergoes a bend of finite radius of curvature [9, 10]. Fibers can be subject to two types of curvatures: (*a*) macroscopic bends having radii that are large compared with the fiber diameter, such as those that occur when a fiber cable turns a corner, and (*b*) random microscopic bends of the fiber axis that can arise when the fibers are incorporated into cables.

Large-curvature radiation loss is known as *macrobending loss* or simply *bending loss*. For slight bends the excess loss is extremely small and is essentially unobservable. As the radius of curvature decreases, the loss increases exponentially until at a certain critical bend radius the curvature loss becomes observable. If the bend radius is made a bit smaller once this threshold point has been reached, the losses suddenly become extremely large.

The amount of optical radiation from a bent fiber depends on the field strength outside of the fiber core and on the bending radius of curvature R. Because higher-order modes in a multimode fiber are bound less tightly to the fiber core than lower-order modes, the higher-order modes will couple more strongly into the cladding region when the fiber is bent and thus will radiate out of the fiber first. Thus the total number of modes that can be supported by a curved fiber is less than in a straight fiber. The following expression [11] has been derived for the effective number of modes M_{eff} that are guided by a curved multimode fiber of radius a:

$$M_{eff} = M_\infty \left\{ 1 - \frac{\alpha + 2}{2\alpha\Delta} \left[\frac{2a}{R} + \left(\frac{3}{2n_2 kR} \right)^{2/3} \right] \right\} \tag{3.11}$$

where α defines the graded-index profile, Δ is the core-cladding index difference, n_2 is the cladding refractive index, $k = 2\pi/\lambda$ is the wave propagation constant, and

$$M_\infty = \frac{\alpha}{\alpha + 2} (n_1 ka)^2 \Delta \tag{3.12}$$

gives the total number of modes in a straight fiber [see Eq. (2.81)].

Example 3.6 Consider a graded-index multimode fiber for which the index profile $\alpha = 2.0$, the core index $n_1 = 1.480$, the core-cladding index difference $\Delta = 0.01$, and the core radius $a = 25$ μm. If the radius of curvature of the fiber is $R = 1.0$ cm, what percentage of the modes remain in the fiber at a 1300 nm wavelength?

Solution First, from Eq. (2.20) $n_2 = n1(1 - \Delta) = 1.480 \, (1 - 0.01) = 1.465$. Then given that $k = 2\pi/\lambda$, from Eq. (3.7) the percentage of modes at a given curvature R is

$$\frac{M_{eff}}{M_\infty} = 1 - \frac{\alpha + 2}{2\alpha\Delta}\left[\frac{2a}{R} + \left(\frac{3}{2n_2kR}\right)^{2/3}\right]$$

$$= 1 - \frac{1}{0.01}\left[\frac{2(25)}{10000} + \left(\frac{3(1.3)}{2(1.465)2\pi(10000)}\right)^{2/3}\right] = 0.42$$

Thus 42% of the modes remain in this fiber at a 1.0 cm bend radius.

Drill Problem 3.5

(a) Show that for a step-index fiber where the index parameter $\alpha = \infty$, Eq. (3.7) becomes

$$\frac{M_{eff}}{M_\infty} = 1 - \frac{1}{2\Delta}\left[\frac{2a}{R} + \left(\frac{3}{2n_2kR}\right)^{2/3}\right]$$

(b) Consider a step-index multimode fiber for which the core index $n_1 = 1.480$, the index difference $\Delta = 0.01$, and the core radius $a = 25$ μm. If the radius of curvature of the fiber is $R = 1$ cm, show from the above equation that the percentage of the modes remaining in the fiber at 1300 nm is 71%. Note that $k = 2\pi/\lambda$.

Another form of radiation loss in optical waveguide results from mode coupling caused by random microbends of the optical fiber [12]. *Microbends* are repetitive small-scale fluctuations in the radius of curvature of the fiber axis, as is illustrated in Fig. 3.4. They are caused either by nonuniformities in the manufacturing of the fiber or by nonuniform lateral pressures created during the cabling of the fiber. The latter effect is often referred to as *cabling* or *packaging losses*. An increase in attenuation results from microbending because the fiber curvature causes repetitive coupling of energy between the guided modes and the nonguided modes in the fiber.

One method of minimizing microbending losses is by extruding a compressible jacket over the fiber. When external forces are applied to this configuration, the jacket will be deformed but the fiber will tend to stay relatively straight. For a multimode graded-index fiber having a core radius a, outer radius b (excluding the jacket), and index difference Δ, the microbending loss α_M of a jacketed fiber is reduced from that of an unjacketed fiber by a factor [13]

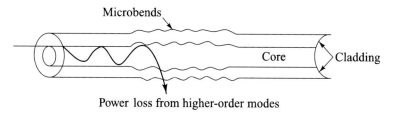

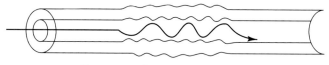

Fig. 3.4 Small-scale fluctuations in the radius of curvature of the fiber axis lead to microbending losses, which can cause power from low-order modes to couple to higher-order modes

$$F(\alpha_M) = \left[1 + \pi \Delta^2 \left(\frac{b}{a}\right)^4 \frac{E_f}{E_j}\right]^{-2} \tag{3.13}$$

Here, E_j and E_f are the Young's moduli of the jacket and fiber, respectively. The Young's modulus of common jacket materials ranges from 20 to 500 MPa. The Young's modulus of fused silica glass is about 65 GPa.

Drill Problem 3.6 Equation (3.13) gives an expression for the factor by which microbending loss is reduced when a compressible jacket is extruded over a fiber. Consider the case when a jacket material that has a Young's modulus $E_j = 58$ MPa is extruded over a glass fiber that has a Young's modulus $E_j = 64$ GPa and a cladding-to-core ratio $b/a = 2.0$. Show that when the refractive index difference $\Delta = 0.01$, the microbending loss reduction factor is $F(\alpha_M) = 0.0233 = 2.33\%$.

3.1.5 Core and Cladding Propagation Losses

Upon measuring the propagation losses in an actual fiber, all the dissipative and scattering losses will be manifested simultaneously. Because the core and cladding have different indices of refraction and therefore differ in composition, generally the core and cladding have different attenuation coefficients denoted by α_1 and α_2, respectively. If the influence of modal coupling is ignored (see Sect. 3.2.4), the loss for a mode of order (v, m) for a step-index waveguide is

$$\alpha_{vm} = \alpha_1 \frac{P_{core}}{P} + \alpha_2 \frac{P_{clad}}{P} \tag{3.14}$$

where P is total power in a given mode and the fractional powers P_{core}/P and P_{clad}/P are shown in Fig. 3.5 as a function of the V number for several low-order modes. From the relation $P = P_{core} + P_{clad}$, it follows that Eq. (3.14) can be written as

$$\alpha_{vm} = \alpha_1 + (\alpha_2 - \alpha_1) \frac{P_{clad}}{P} \tag{3.15}$$

The total loss of the waveguide can be found by summing over all modes weighted by the fractional power in that mode.

For the case of a graded-index fiber the situation is much more complicated. In this case, both the attenuation coefficients and the modal power tend to be functions of the radial coordinate. At a distance r from the core axis the loss is [14]

$$\alpha(r) = \alpha_1 + (\alpha_2 - \alpha_1) \frac{n^2(0) - n^2(r)}{n^2(0) - n_2^2} \tag{3.16}$$

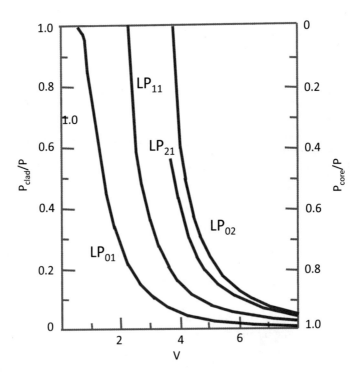

Fig. 3.5 The fractional powers P_{core}/P and P_{clad}/P for several low-order modes

where α_1 and α_2 are the axial and cladding attenuation coefficients, respectively, and the n terms are defined by Eq. (2.38). The loss encountered by a given mode is then

$$\alpha_{gi} = \frac{\int_0^\infty \alpha(r)\, p(r)\, r\, dr}{\int_0^\infty p(r)\, r\, dr} \tag{3.17}$$

where $p(r)$ is the power density of that mode at r. The complexity of the multimode waveguide has prevented an experimental correlation with a model. However, it has generally been observed that the loss increases with increasing mode number.

3.2 Optical Signal Dispersion Effects

As shown in Fig. 3.6, an optical signal weakens from attenuation mechanisms and broadens due to dispersion effects as it travels along a fiber. Eventually these two factors will cause neighboring pulses to overlap. After a certain amount of overlap occurs, the receiver can no longer distinguish the individual adjacent pulses and errors arise when interpreting the received signal.

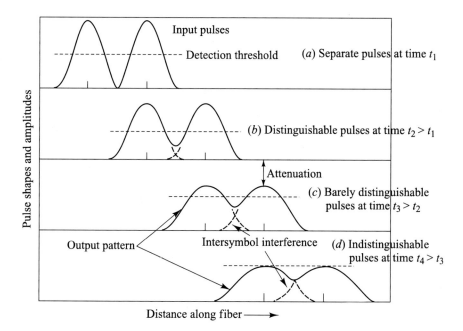

Fig. 3.6 Broadening and attenuation of two adjacent pulses as they travel along a fiber: **a** Originally the pulses are separate; **b** the pulses overlap slightly and are clearly distinguishable; **c** the pulses overlap significantly and are barely distinguishable; **d** eventually the pulses strongly overlap and are indistinguishable

This section first discusses the general factors that cause signal dispersion and then examines the various dispersion mechanisms in more detail. Section 3.2.2 addresses modal delay and shows how this delay is related to the information capacity of a multimode fiber in terms of a transmitted bit rate B. Section 3.2.3 examines the various factors contributing to dispersion in terms of the frequency dependence of the propagation constant β. The next topics include a discussion of group velocity in Sect. 3.2.4 and details of the various dispersion mechanisms in Sects. 3.2.5 through 3.2.8.

3.2.1   Origins of Signal Dispersion

Signal dispersion is a consequence of factors such as intermodal delay (also called intermodal dispersion), intramodal dispersion, polarization-mode dispersion, and higher-order dispersion effects. These effects can be explained by examining the behavior of the group velocities of the guided modes, where the *group velocity* is the speed at which energy in a particular mode travels along the fiber (see Sect. 3.2.4).

Intermodal delay (or simply *modal delay*) appears only in multimode fibers. Modal delay is a result of each mode having a different value of the group velocity at a single frequency. From this effect one can derive an intuitive picture of the information-carrying capacity of a multimode fiber.

Intramodal dispersion or chromatic dispersion is pulse spreading that takes place within a single mode. This spreading arises from the finite spectral emission width of an optical source. The phenomenon also is known as *group velocity dispersion,* because the dispersion is a result of the group velocity being a function of the wavelength. Because intramodal dispersion depends on the wavelength, its effect on signal distortion increases with the enlarging of spectral width of the light source. The spectral width is the band of wavelengths over which the source emits light. This wavelength band normally is characterized by the root-mean-square (rms) spectral width σ_λ. Depending on the device structure of a light-emitting diode (LED), the spectral width is approximately 4–9% of a central wavelength. For example, as Fig. 3.7 illustrates, if the peak wavelength of an LED is 850 nm, a typical source spectral width would be 36 nm; that is, such an LED emits most of its light in the 832–868 nm wavelength band. Laser diode optical sources exhibit much narrower spectral widths, with typical values being 1–2 nm for multimode lasers and 10^{-4} nm for single-mode lasers (see Chap. 4).

The two main causes of intramodal dispersion are as follows:

1. *Material dispersion* arises due to the variations of the refractive index of the core material as a function of wavelength. Material dispersion also is referred to as *chromatic dispersion,* because this is the same effect by which a prism spreads out a spectrum. This refractive index property causes a wavelength dependence of the group velocity of a given mode, because as indicated by

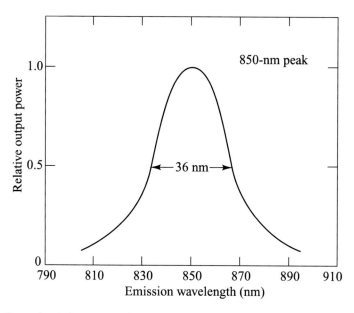

Fig. 3.7 Spectral emission pattern of a representative $Ga_{1-x}Al_xAs$ LED with a peak emission at 850 nm and a half-power width of 36 nm

Eq. (2.15) the velocity of light depends on the value of the refractive index. Thus, pulse spreading occurs even when different wavelengths simultaneously follow the same path, because each wavelength within a pulse travels at a slightly different velocity.

2. *Waveguide dispersion* causes pulse spreading because only part of the optical power propagation along a fiber is confined to the core. Within a single propagating mode, the cross-sectional distribution of light in the optical fiber varies for different wave-lengths. Shorter wavelengths are more completely confined to the fiber core, whereas a larger portion of the optical power at longer wavelengths propagates in the cladding, as shown in Fig. 3.8. The refractive index is lower in the cladding than in the core, so the fraction of light power propagating

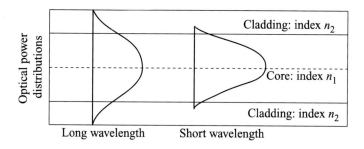

Fig. 3.8 Shorter wavelengths are confined closer to the center of a fiber core than longer wavelengths

in the cladding travels faster than the light confined to the core. In addition, note that the index of refraction depends on the wavelength (see Sect. 3.2.5) so that different spectral components within a single mode have different propagation speeds. Dispersion thus arises because the difference in core-cladding spatial power distributions, together with the speed variations of the various wavelengths, causes a change in propagation velocity for each spectral component. The degree of waveguide dispersion depends on the fiber design (see Sect. 3.3.1). Waveguide dispersion usually can be ignored in multimode fibers, but its effect is significant in single-mode fibers.

Polarization-mode dispersion results from the fact that light-signal energy at a given wavelength in a single-mode fiber actually occupies two orthogonal polarization states or modes (see Sect. 2.5). At the start of the fiber the two polarization states are aligned. However, because fiber material is not perfectly uniform throughout its length, each polarization mode will encounter a slightly different refractive index. Consequently each mode will travel at a slightly different velocity. The resulting difference in propagation times between the two orthogonal polarization modes will cause pulse spreading. Section 3.2.8 gives more details on this effect.

3.2.2   Modal Delay Effects

Intermodal dispersion or *modal delay* appears only in multimode fibers. This signal-distorting mechanism is a result of each mode having a different value of the group velocity at a single frequency. To see why the delay arises, consider the meridional ray picture given in Fig. 2.17 for a multimode step-index fiber. The steeper the angle of propagation of the ray congruence, the higher is the mode number and, consequently, the slower the axial group velocity. This variation in the group velocities of the different modes results in a group delay spread, which is the intermodal dispersion. This dispersion mechanism is eliminated by single-mode operation but is important in multimode fibers. The maximum pulse broadening arising from the modal delay is the difference between the travel time T_{max} of the longest ray congruence paths (the highest-order mode) and the travel time T_{min} of the shortest ray congruence paths (the fundamental mode). This broadening is simply obtained from ray tracing and for a fiber of length L is given by

$$\Delta T = T_{max} - T_{min} = \frac{n_1}{c}\left(\frac{L}{\sin\theta_c} - L\right) = \frac{Ln_1^2}{cn_2}\Delta \approx \frac{Ln_1\Delta}{c} \qquad (3.18)$$

where from Eq. (2.21) $\sin\theta_c = n_2/n_1$ and Δ is the index difference.

Example 3.7 Consider a 1 km long multimode step-index fiber in which $n_1 = 1.480$ and $\Delta = 0.01$, so that $n_2 = 1.465$. What is the modal delay per length in this fiber?

Solution Equation (3.18) yields

$$\frac{\Delta T}{L} = \frac{n_1^2}{c n_2} \Delta = 50 \text{ ns/km}$$

This means that a pulse broadens by 50 ns after traveling a distance of 1 km in this type of fiber.

The question now arises as to what maximum bit rate B can be sent over a multi-mode step-index fiber. Typically the fiber capacity is specified in terms of the *bit rate-distance product BL*, that is, the bit rate B times the possible transmission distance L. In order for neighboring signal pulses to remain distinguishable at the receiver, the pulse spread should be less than $1/B$, which is the width of a bit period. For example, a stringent requirement for a high-performance link might be $\Delta T \leq 0.1/B$. In general, it is necessary to have $\Delta T < 1/B$. Using Eq. (3.18) this inequality gives the bit rate-distance product

$$BL < \frac{n_2}{n_1^2} \frac{c}{\Delta} \tag{3.19}$$

Taking values of $n_1 = 1.480$, $n_2 = 1.465$, and $\Delta = 0.01$, the capacity of this multimode step-index fiber is $BL = 20$ Mb/s-km.

Example 3.8 Viewed alternatively, as illustrated in Example 3.7, for a multimode step-index fiber with a bandwidth-distance value of $BL = 20$ Mb/s km the pulse spreading is 50 ns/km. As an example, suppose the pulse width in a transmission system is allowed to widen by at most 25%. Then for a 10 Mb/s data rate, in which one pulse is transmitted every 100 ns, this limitation allows a spread of at most 25 ns, which occurs in a transmission distance of 500 m. Now, suppose the data rate is increased to 100 Mb/s, which means that one pulse is transmitted every 10 ns. In this case the allowable spreading factor of 50 ns/km will limit the transmission distance to only 50 m in such a multimode step-index fiber.

The root-mean-square (rms) value of the time delay is a useful parameter for assessing the effect of modal delay in a multimode fiber. If it is assumed that the light rays are uniformly distributed over the acceptance angles of the fiber, then the rms impulse response σ_s due to intermodal dispersion in a step-index multimode fiber can be estimated from the expression

$$\sigma_s \approx \frac{L n_1 \Delta}{2\sqrt{3}\, c} \approx \frac{L(NA)^2}{4\sqrt{3}\, n_1 c} \tag{3.20}$$

Here L is the fiber length and NA is the numerical aperture. Equation (3.20) shows that the pulse broadening is directly proportional to the core-cladding index difference and the length of the fiber.

Drill Problem 3.7 A 10 km transmission link consists of a step-index multi-mode fiber that has a core index $n_1 = 1.480$ and a core-cladding refractive index difference $\Delta = 0.01$.

(a) Using the approximation on the right-hand side of Eq. (3.18), show that the delay difference between the fastest and slowest modes is 493 ns.
(b) Using Eq. (3.20), show that the rms pulse broadening resulting from intermodal delay is 142 ns.
(c) Using Eq. (3.18) and the condition that the maximum bit rate B should satisfy the condition $B < 0.1/\Delta T$, show that the maximum bit rate-distance product is $BL = (2.03 \text{ Mb/s}) \text{ km}$.

A successful technique for reducing modal delay in multimode fibers is through the use of a graded refractive index in the fiber core, as shown in Fig. 2.15. In any multimode fiber the ray paths associated with higher-order modes are concentrated near the edge of the core and thus follow a longer path through the fiber than lower-order modes (which are concentrated near the fiber axis). However, if the core has a graded index profile, then the higher-order modes encounter a lower refractive index near the core edge. Because the speed of light in a material depends on the refractive index value, the higher-order modes travel faster in the outer core region than those modes that propagate through a higher refractive index along the fiber center. Consequently this reduces the delay difference between the fastest and slowest modes. A detailed analysis using electromagnetic mode theory gives the following absolute modal delay at the output of a graded-index fiber that has a parabolic ($\alpha = 2$) core index profile:

$$\sigma_s \approx \frac{Ln_1\Delta^2}{20\sqrt{3}\,c} \tag{3.21}$$

Thus for an index difference of $\Delta = 0.01$, the theoretical improvement factor for intermodal rms pulse broadening in a graded-index fiber is 1000.

Example 3.9 Consider the following two multimode fibers: (a) a step-index fiber with a core index $n_1 = 1.458$ and a core-cladding index difference $\Delta = 0.01$; (b) a parabolic-profile ($\alpha = 2$) graded-index fiber with the same values of n_1 and Δ. Compare the rms pulse broadening per kilometer for these two fibers.

Solution

(a) From Eq. (3.20)

$$\frac{\sigma_s}{L} \approx \frac{n_1\Delta}{2\sqrt{3}\,c} = \frac{1.458(0.01)}{2\sqrt{3} \times 3 \times 10^8 \text{ m/s}} = 14.0 \text{ ns/km}$$

(b) From Eq. (3.21)

$$\frac{\sigma_s}{L} \approx \frac{n_1 \Delta^2}{20\sqrt{3}\,c} = \frac{1.458(0.01)^2}{20\sqrt{3} \times 3 \times 10^8 \text{ m/s}} = 14.0 \text{ ps/km}$$

In graded-index fibers, careful selection of the radial refractive-index profile can lead to bit rate-distance products of up to 1 Gb/s km.

3.2.3 Factors Contributing to Dispersion

This section briefly examines the various factors contributing to dispersion. Sections 3.2.4–3.2.8 and Sect. 3.3 describe these factors in more detail.

As Sect. 2.3.2 notes, the z component of the wave propagation constant β is a function of the wavelength or, equivalently, of the angular frequency ω. Because β is a slowly varying function of this angular frequency, one can see where various dispersion effects arise by expanding β in a Taylor series about a central frequency ω_0. Inserting such an expansion into the waveform equation, for example Eq. (2.1), then shows the effects of variations in β due to modal dispersion and delay effects on the frequency components of a pulse during its propagation along a fiber.

Expanding β to third order in a Taylor series yields

$$\beta(\omega) \approx \beta_0(\omega_0) + \beta_1(\omega_0)(\omega - \omega_0) + \frac{1}{2}\beta_2(\omega_0)(\omega - \omega_0)^2 + \frac{1}{6}\beta_3(\omega_0)(\omega - \omega_0)^3$$

(3.22)

where $\beta_m(\omega_0)$ denotes the mth derivative of β with respect to ω evaluated at $\omega = \omega_0$; that is,

$$\beta_m = \left(\frac{\partial^m \beta}{\partial \omega^m}\right)_{\omega=\omega_0}$$

(3.23)

Now consider the different components of the product βz, where z is the distance traveled along the fiber. The resulting first term $\beta_0 z$ describes a phase shift of the propagating optical wave. From the second term of Eq. (3.22), the factor $\beta_1(\omega_0)z$ produces a group delay $\tau_g = z/V_g$, where z is the distance traveled by the pulse and $V_g = 1/\beta_1$ is the group velocity [see Eqs. (3.27) and (3.28)]. Assume β_{1x} and β_{1y} are the propagation constants of the polarization components along the x-axis and y-axis, respectively, of a particular mode. If the corresponding group delays of these two polarization components are $\tau_{gx} = z\,\beta_{1x}$ and $\tau_{gy} = z\,\beta_{1y}$ in a distance z, then the difference in the propagation times of these two modes

$$\Delta\tau_{PMD} = z\big|\beta_{1x} - \beta_{1y}\big|$$

(3.24)

is called the *polarization-mode dispersion* (PMD) of the ideal uniform fiber.

In the third term of Eq. (3.22), the factor β_2 shows that the group velocity of a monochromatic wave depends on the wave frequency. This means that the different group velocities of the frequency components of a pulse cause it to broaden as it travels along a fiber. This spreading of the group velocities is known as *chromatic dispersion* or *group velocity dispersion* (GVD). The factor β_2 is called the *GVD* and the *dispersion D* is related to β_2 through the expression

$$D = -\frac{2\pi c}{\lambda^2}\beta_2 \tag{3.25}$$

In the fourth term of Eq. (3.22), the factor β_3 is known as the *third-order dispersion*. This term is important around the wavelength at which β_2 equals zero. The third-order dispersion can be related to the dispersion D and the *dispersion slope* $S_0 = \partial D/\partial\lambda$ (the variation in the dispersion D with wavelength) by transforming the derivative with respect to ω into a derivative with respect to λ. Thus

$$\beta_3 = \frac{\partial\beta_2}{\partial\omega} = -\frac{\lambda^2}{2\pi c}\frac{\partial\beta_2}{\partial\lambda} = -\frac{\lambda^2}{2\pi c}\frac{\partial}{\partial\lambda}\left[-\frac{\lambda^2}{2\pi c}D\right] = \frac{\lambda^2}{(2\pi c)^2}\left(\lambda^2 S_0 + 2\lambda D\right) \tag{3.26}$$

The procedure for selecting the values of the parameters in Eq. (3.26) are described in ITU-T Recommendation 650.1 (see Sect. 3.3.3 for details).

3.2.4 Group Delay Results

As Example 3.8 mentions, the information-carrying capacity of a fiber link can be determined by examining the deformation of short light pulses propagating along the fiber. The following discussion on signal dispersion thus is carried out primarily from the viewpoint of pulse broadening, which is representative of digital transmission.

First consider an electrical signal that modulates an optical source. For this case, assume that the modulated optical signal excites all modes equally at the input of the fiber. Each waveguide mode thus carries an equal amount of energy through the fiber. Furthermore, each mode contains all the spectral components in the wavelength band over which the source emits. In addition, assume that each of these spectral components is modulated in the same way. As the signal propagates along the fiber, each spectral component can be assumed to travel independently and to undergo a time delay or *group delay* per unit length τ_g/L in the direction of the propagation given by [15]

$$\frac{\tau_g}{L} = \frac{1}{V_g} = \frac{1}{c}\frac{d\beta}{dk} = -\frac{\lambda^2}{2\pi c}\frac{\partial\beta}{\partial\lambda} \tag{3.27}$$

Here, L is the distance traveled by the pulse, β is the propagation constant along the fiber axis, $k = 2\pi/\lambda$, and the *group velocity*

$$V_g = c\left(\frac{d\beta}{dk}\right)^{-1} = \left(\frac{\partial\beta}{\partial\omega}\right)^{-1} \tag{3.28}$$

is the velocity at which the energy in a pulse travels along a fiber.

Because the group delay depends on the wavelength, each spectral component of any particular mode takes a different amount of time to travel a certain distance. As a result of this difference in time delays, the optical signal pulse spreads out with time as it is transmitted over the fiber. Thus the quantity of interest is the amount of pulse spreading that arises from the group delay variation.

If the spectral width of the optical source is not too wide, the delay difference per unit wavelength along the propagation path is approximately $d\tau_g/d\lambda$. For spectral components that are $\delta\lambda$ apart and which lie $\delta\lambda/2$ above and below a central wavelength λ_0, the total delay difference $\delta\tau$ over a distance L is

$$\delta\tau = \frac{d\tau_g}{d\lambda}\delta\lambda = -\frac{L}{2\pi c}\left(2\lambda\frac{d\beta}{d\lambda} + \lambda^2\frac{d^2\beta}{d\lambda^2}\right)\delta\lambda \tag{3.29}$$

In terms of the angular frequency ω, this is written as

$$\delta\tau = \frac{d\tau_g}{d\omega}\delta\omega = \frac{d}{d\omega}\left(\frac{L}{V_g}\right)\delta\omega = L\left(\frac{d^2\beta}{d\omega^2}\right)\delta\omega \tag{3.30}$$

The factor $\beta_2 \equiv d^2\beta/d\omega^2$ is the *GVD parameter*, which determines how much a light pulse broadens as it travels along an optical fiber.

If the spectral width $\delta\lambda$ of an optical source is characterized by its rms value σ_λ (see Fig. 3.7 for a typical LED), then the pulse spreading can be approximated by the rms pulse width,

$$\sigma_g = \left|\frac{d\tau_g}{d\lambda}\right|\sigma_\lambda = \frac{L\sigma_\lambda}{2\pi c}\left|2\lambda\frac{d\beta}{d\lambda} + \lambda^2\frac{d^2\beta}{d\lambda^2}\right| \tag{3.31}$$

The factor

$$D = \frac{1}{L}\frac{d\tau_g}{d\lambda} = \frac{d}{d\lambda}\left(\frac{1}{V_g}\right) = -\frac{2\pi c}{\lambda^2}\beta_2 \tag{3.32}$$

is designated as the *dispersion*. It defines the pulse spread as a function of wavelength and is measured in picoseconds per kilometer per nanometer [ps/(nm km)]. It is a result of material and waveguide dispersion. In many theoretical treatments of intramodal dispersion it is assumed, for simplicity, that material dispersion and waveguide dispersion can be calculated separately and then added to give the total dispersion of the mode. In reality, these two mechanisms are intricately related owing to the fact that the dispersive properties of the refractive index (which give rise to material dispersion) also affect the waveguide dispersion. However, an examination [16] of the interdependence of material and waveguide dispersion has shown that, unless a very precise value to a fraction of a percent is desired, a good estimate of the total intramodal dispersion can be obtained by calculating the effect of signal distortion arising from one type of dispersion in the absence of the other. Thus, to a very good approximation, D can be written as the sum of the material dispersion D_{mat} and the waveguide dispersion D_{wg}. Material dispersion and waveguide dispersion are therefore considered separately in the next two sections.

3.2.5   Material-Induced Dispersion

Material dispersion occurs because the index of refraction varies as a function of the optical wavelength. This is exemplified in Fig. 3.9 for silica. As a consequence, because the group velocity V_g of a mode is a function of the index of refraction, the various spectral components of a given mode will travel at different speeds, depending on the wavelength. Therefore, material dispersion is an intramodal dispersion effect and is of particular importance for single-mode waveguides and for LED systems (because an LED has a broader output spectrum than a laser diode).

To calculate material-induced dispersion, consider a plane wave propagating in an infinitely extended dielectric medium that has a refractive index $n(\lambda)$ equal to that of the fiber core. The propagation constant β is thus given as

Fig. 3.9 Variations in the index of refraction as a function of the optical wavelength for silica

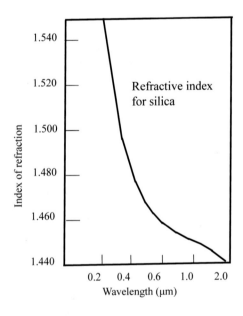

$$\beta = \frac{2\pi n(\lambda)}{\lambda} \tag{3.33}$$

Substituting this expression for β into Eq. (3.27) with $k = 2\pi/\lambda$ yields the group delay τ_{mat} resulting from material dispersion.

$$\tau_{mat} = \frac{L}{c}\left(n - \lambda\frac{dn}{d\lambda}\right) \tag{3.34}$$

Using Eq. (3.31), the pulse spread σ_{mat} for a source of spectral width σ_λ is found by differentiating this group delay with respect to wavelength and multiplying by σ_λ to yield

$$\sigma_{mat} = \left|\frac{d\tau_{mat}}{d\lambda}\right|\sigma_\lambda = \frac{\sigma_\lambda L}{c}\left|\lambda\frac{d^2n}{d\lambda^2}\right| = \sigma_\lambda L|D_{mat}(\lambda)| \tag{3.35}$$

where $D_{mat}(\lambda)$ is the *material dispersion*.

Example 3.10 A manufacturer's data sheet lists the material dispersion D_{mat} of a GeO_2-doped fiber to be 110 ps/(nm km) at a wavelength of 860 nm. Find the rms pulse broadening per kilometer due to material dispersion if the optical source is a GaAlAs LED that has a spectral width σ_λ of 40 nm at a peak output wavelength of 860 nm.

Solution From Eq. (3.35) the rms material dispersion is given by

$$\sigma_{mat}/L = \sigma_\lambda D_{mat} = (40\,nm) \times [110\ ps/(nm \cdot km)] = 4.4\ ns/km$$

Example 3.11 The manufacturer's data shows that the same fiber as in Example 3.10 has a material dispersion D_{mat} of 15 ps/(nm km) at a wavelength of 1550 nm. However, now consider a laser source with a spectral width σ_λ of 0.2 nm at an operating wavelength of 1550 nm. What is the rms pulse broadening per kilometer due to material dispersion in this case?

Solution From Eq. (3.35) the rms material dispersion is given by

$$\sigma_{mat}/L = \sigma_\lambda D_{mat} = (0.2\,nm) \times [15\,ps/(nm \cdot km)] = 3.0\,ps/km$$

This example shows that a dramatic reduction in dispersion can be achieved when operating at longer wavelengths with laser sources that have a narrower spectral width.

3.2.6 Effects of Waveguide Dispersion

The effect of waveguide dispersion on pulse spreading can be approximated by assuming that the refractive index of the material is independent of wavelength. First consider the group delay—that is, the time required for a mode to travel along a fiber of length L. To make the results independent of fiber configuration, [17] the group delay can be expressed in terms of the normalized propagation constant b defined as

$$b = \frac{\beta^2/k^2 - n_2^2}{n_1^2 - n_2^2} \tag{3.36}$$

For small values of the index difference $\Delta = (n_1 - n_2)/n_1$, so Eq. (3.29) can be approximated by

$$b = \frac{\beta/k - n_2}{n_1 - n_2} \tag{3.37}$$

Solving Eq. (3.37) for β then yields

$$\beta \approx n_2 k(b\Delta + 1) \tag{3.38}$$

With this expression for β and using the assumption that n_2 is not a function of wavelength, the group delay τ_{wg} arising from waveguide dispersion is given by

$$\tau_{wg} = \frac{L}{c}\frac{d\beta}{dk} = \frac{L}{c}\left[n_2 + n_2\Delta\frac{d(kb)}{dk}\right] \tag{3.39}$$

The modal propagation constant β is generally given in terms of the normalized frequency V defined by Eq. (2.27). Therefore one can use the approximation

$$V = ka(n_1^2 - n_2^2)^{1/2} \approx ka\,n_1\sqrt{2\Delta} \tag{3.40}$$

which is valid for small values of Δ, to write the group delay in Eq. (3.39) in terms of V instead of k, yielding

$$\tau_{wg} = \frac{L}{c}\left[n_2 + n_2\Delta\frac{d(Vb)}{dV}\right] \tag{3.41}$$

The first term in Eq. (3.41) is a constant and the second term represents the group delay arising from waveguide dispersion. For a fixed value of V, the group delay is different for every guided mode. When a light pulse is launched into a fiber, it is distributed among many guided modes. These various modes arrive at the fiber end at different times depending on their group delay, so that pulse spreading results. For multimode fibers the waveguide dispersion is generally very small compared with material dispersion and can therefore be neglected.

3.2.7 Dispersion Behavior in Single-Mode Fibers

Waveguide dispersion is of importance for single-mode fibers and can be of the same order of magnitude as material dispersion. This can be seen by comparing the two dispersion factors. The pulse spread σ_{wg} occurring over a distribution of wavelengths σ_λ is obtained from the derivative of the group delay with respect to wavelength:

$$\sigma_{wg} = \left|\frac{d\tau_{wg}}{d\lambda}\right|\sigma_\lambda = L\left|D_{wg}(\lambda)\right|\sigma_\lambda = \frac{V}{\lambda}\left|\frac{d\tau_{wg}}{dV}\right|\sigma_\lambda = \frac{n_2 L \Delta \sigma_\lambda}{c\lambda} V \frac{d^2(Vb)}{dV^2} \qquad (3.42)$$

where $D_{wg}(\lambda)$ is the *waveguide dispersion*. From an analysis of Maxwell's equations, for the HE_{11} mode b can be expressed as [17]

$$b(V) = 1 - \frac{\left(1+\sqrt{2}\right)^2}{\left[1 + \left(4 + V^4\right)^{1/4}\right]^2} \qquad (3.43)$$

Figure 3.10 shows plots of this expression for b and its derivatives $d(Vb)/dV$ and $Vd^2(Vb)/dV^2$ as functions of V.

Example 3.12 From Eq. (3.42) the waveguide dispersion is

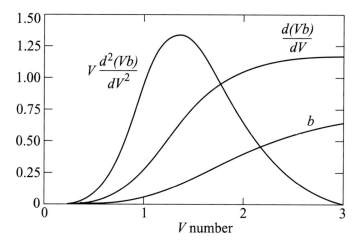

Fig. 3.10 The waveguide parameter b and its derivatives $d(Vb)/dV$ and $Vd^2(Vb)/dV^2$ plotted as a function of the V number for the HE_{11} mode (or the LP_{01} mode)

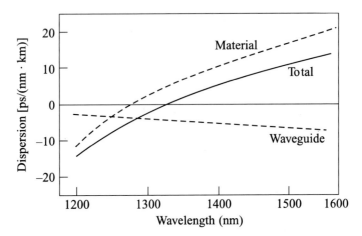

Fig. 3.11 Generic representations of the magnitudes of material and waveguide dispersion as a function of optical wavelength for a single-mode fused-silica-core fiber

$$D_{wg}(\lambda) = -\frac{n_2 \Delta}{c} \frac{1}{\lambda} \left[V \frac{d^2(Vb)}{dV^2} \right]$$

Let $n_2 = 1.48$ and $\Delta = 0.2\%$. Assume that at $V = 2.4$ the expression in square brackets is 0.26. Choosing $\lambda = 1320$ nm, then the waveguide dispersion is $D_{wg}(\lambda) = -1.9$ ps/(nm km).

Figure 3.11 gives generic examples of the magnitudes of material and waveguide dispersion for a fused-silica-core single-mode fiber having $V = 2.4$, such as a G.652 fiber. Comparing the waveguide dispersion with the material dispersion, one can see that for a standard non-dispersion-shifted fiber, waveguide dispersion is important around 1320 nm. At this point, the two dispersion factors cancel to give a zero total dispersion. However, material dispersion dominates waveguide dispersion at shorter and longer wavelengths; for example, at 900 and 1550 nm. This figure used the approximation that material and waveguide dispersions are additive.

3.2.8 Origin of Polarization-Mode Dispersion

The effects of fiber birefringence on the polarization states of an optical signal are another source of pulse broadening. This is particularly critical for high-rate long-haul transmission links (e.g., 10 and 40 Gb/s over tens of kilometers). Birefringence can result from intrinsic factors such as geometric irregularities of the fiber core or internal stresses on it. Deviations of less than 1% in the circularity of the core can already have a noticeable effect in a high-speed lightwave system. In addition, external factors, such as bending, twisting, or pinching of the fiber, can also lead to birefringence.

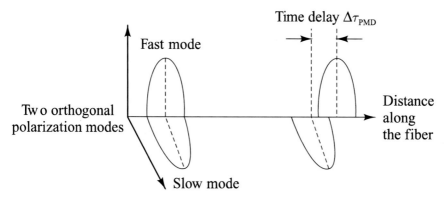

Fig. 3.12 Differences in the polarization-mode propagation times as an optical pulse passes through a fiber with varying birefringence along its length

Because all these mechanisms exist to some extent in any field-installed fiber, there will be a varying birefringence along its length.

A fundamental property of an optical signal is its polarization state. *Polarization* refers to the electric-field orientation of a light signal, which can vary significantly along the length of a fiber. As shown in Fig. 3.12, signal energy at a given wavelength occupies two orthogonal polarization modes. A variation in the birefringence along its length will cause each polarization mode to travel at a slightly different velocity. The resulting difference in propagation time $\Delta\tau_{PMD}$ between the two orthogonal polarization modes will result in pulse spreading. This is the *polarization-mode dispersion* (PMD) [18, 19]. If the group velocities of the two orthogonal polarization modes are V_{gx} and V_{gy}, then the differential time delay $\Delta\tau_{PMD}$ between the two polarization components during propagation of the pulse over a distance L is

$$\Delta\tau_{PMD} = \left| \frac{L}{V_{gx}} - \frac{L}{V_{gy}} \right| \tag{3.44}$$

An important point to note is that, in contrast to chromatic dispersion, which is a relatively stable phenomenon along a fiber, PMD varies randomly along a fiber. A principal reason for this is that the perturbations causing the birefringence effects typically vary with temperature and stress dynamics. In practice, the effect of these perturbations shows up as a random, time-varying fluctuation in the value of the PMD at the fiber output. Thus $\Delta\tau_{PMD}$ given in Eq. (3.39) cannot be used directly to estimate PMD. Instead, statistical estimations are needed to account for its effects.

A useful means of characterizing PMD for long fiber lengths is in terms of the mean value of the differential group delay. This can be calculated according to the relationship

$$\Delta\tau_{PMD} \approx D_{PMD}\sqrt{L} \tag{3.45}$$

where D_{PMD}, which is measured in ps/$\sqrt{km}$, is the average PMD parameter. Typical values of D_{PMD} range from 0.03 to 1.3 ps/$\sqrt{km}$ depending on the type of cable installation. Buried cable typically does not encounter significant variations in the surrounding environments, so the PMD value tends to be low and relatively stable. However, aerial cables experience larger fluctuations in the value of PMD, which result in both gradual and rapid stress variations in the fiber due to temperature fluctuations or from sudden movements of the fiber due to wind.

To keep the probability of errors due to PMD low, a standard limit on the maximum tolerable value of $\Delta \tau_{PMD}$ ranges between 10 and 20% of a bit duration. Thus $\Delta \tau_{PMD}$ should be no more than 10–20 ps for 10 Gb/s data rates and 3 ps at 40 Gb/s. For example, taking the lower tolerance limit, this means that for a 10 Gb/s link that has 20 spans of 80 km each, the PMD of the fiber must be less than 0.2 ps/$\sqrt{km}$.

Example 3.13 Consider a 1600 km fiber link on which data is being sent at a bit rate $B = 10$ Gb/s. Assume that the maximum tolerable delay due to PMD is 10% of a bit period, so that $\Delta \tau_{PMD} < 0.1/B = 0.1/(10 \times 10^9/s) = 10^{-11} s = 10$ ps. Thus for this link, from Eq. (3.45) the PMD must satisfy the condition

$$D_{PMD} < 0.1 \Delta \tau_{PMD} / \sqrt{L} = 10 \text{ ps}/\sqrt{1600 \text{ km}} = 0.25 \text{ ps}/\sqrt{km}$$

Drill Problem 3.8 Consider a 1600 km fiber link on which data is being sent at a bit rate $B = 40$ Gb/s. Assume that the maximum tolerable delay due to PMD is 20% of a bit period. Show that the tolerable PMD of the transmission fiber should <0.13 ps/$\sqrt{km}$.

3.3 Design and Characteristics of SMFs

This section addresses the basic design and operational characteristics of single-mode fibers. These characteristics include index-profile configurations used to produce different fiber types, the concept of cutoff wavelength, signal dispersion designations and calculations, the definition of mode-field diameter, and signal loss due to fiber bending.

3.3.1 Tailoring of Refractive Index Profiles

When creating single-mode fibers, manufacturers pay special attention to how the fiber design affects both chromatic and polarization-mode dispersions. Such considerations are important because these dispersions set the limits on long-distance and high-speed data transmission. As Fig. 3.11 illustrates, the chromatic dispersion of a classic step-index silica fiber is lowest at 1310 nm. However, if the goal is to transmit a signal as far as possible, it is better to operate the link at 1550 nm (in the C-band) where the fiber attenuation is lower. For high-speed links the C-band originally presented a problem for standard single-mode fibers because chromatic dispersion is much larger at 1550 nm than at 1310 nm. Consequently, fiber designers devised methods for adjusting the fiber parameters to shift the zero-dispersion point to longer wavelengths.

The basic material dispersion is hard to alter significantly. However, it is possible to modify the waveguide dispersion by changing from a simple step-index design to more complex index profiles for the cladding, thereby creating different chromatic-dispersion characteristics in single-mode fibers. Figure 3.13 shows representative refractive-index profiles of four fiber-design categories. These are 1310-nm-optimized fibers, dispersion-shifted fibers, dispersion-flattened fibers, and large-effective-core-area fibers.

Popular single-mode fibers that are used widely in telecommunication networks are near-step-index fibers, which are optimized for use in the O-band around 1310 nm. These *1310-nm-optimized single-mode fibers* are of either the *matched-cladding* or the *depressed-cladding* design, as shown in Fig. 3.13a. Matched-cladding fibers have a uniform refractive index throughout the cladding. Typical mode-field diameters are 9.5 μm and the core-to-cladding index differences are around 0.35%. In depressed-cladding fibers the cladding material next to the core has a lower index than the outer cladding region. Mode-field diameters are around 9.0 μm, and typical positive and negative index differences are 0.25 and 0.12%, respectively.

As Eqs. (3.35) and (3.42) show, whereas material dispersion depends only on the composition of the material, waveguide dispersion is a function of the core radius, the refractive index difference, and the shape of the refractive index profile. Thus the waveguide dispersion can vary dramatically with the fiber design parameters. By creating a fiber with a larger negative waveguide dispersion and assuming the same values for material dispersion as in a standard single-mode fiber, the addition of waveguide and material dispersion can then shift the zero dispersion point to longer wavelengths. The resulting optical fiber is known as a *dispersion-shifted fiber* (DSF). Examples of refractive-index profiles for dispersion-shifted fibers are shown in Fig. 3.13b.

Because the zero-dispersion value of a DSF falls at 1550 nm, the chromatic dispersion is negative for wavelengths less than 1550 nm and positive for longer wavelengths. These positive and negative dispersions will seriously affect closely spaced WDM signals within the C-band because of nonlinear effects in the fiber, as Chap. 12 describes. To reduce the effects of fiber nonlinearities, fiber designers developed

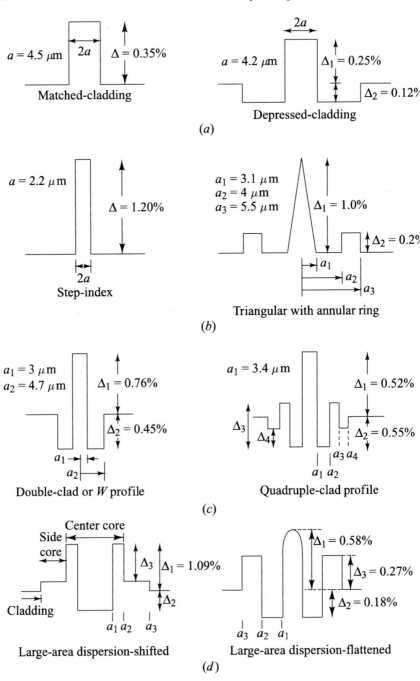

Fig. 3.13 Representative cross sections of index profiles for **a** 1310-nm-optimized, **b** dispersion-shifted, **c** dispersion-flattened, and **d** large-effective-core-area fibers

the *nonzero dispersion-shifted fiber* (NZDSF). These fibers have a small amount of either all positive or all negative dispersion throughout the C-band. A typical positive chromatic dispersion value for a NZDSF is 4.5 ps/(nm km) at 1550 nm.

Among the NZDSF types is a single-mode optical fiber with a larger effective core area. The larger core areas reduce the effects of fiber nonlinearities, which otherwise limit system capacities of transmission systems that have densely spaced WDM channels. Figure 3.13d gives two examples of the index profile for these large-effective-area (LEA) fibers. Whereas standard single-mode fibers have effective core areas of about 55 μm^2, these profiles yield values greater than 100 μm^2.

An alternative fiber design concept is to distribute the dispersion minimum over a wider spectral range. This approach is known as *dispersion flattening*. Dispersion-flattened fibers are more complex to design than dispersion-shifted fibers, because dispersion must be considered over a much broader range of wavelengths. However, they offer desirable characteristics over a wide span of wavelengths. Figure 3.13c shows typical cross-sectional refractive-index profiles. Typical waveguide dispersion curves for three types of fiber are depicted in Fig. 3.14a. Figure 3.14b gives the resultant total material plus waveguide dispersion characteristics.

3.3.2 Concept of Cutoff Wavelength

The cutoff wavelength of the first higher-order mode (LP_{11}) is an important transmission parameter for single-mode fibers because it separates the single- mode from the multimode regions. Recall from Eq. (2.27) that single-mode operation occurs above the theoretical cutoff wavelength given by

$$\lambda_c = \frac{2\pi a}{V}\left(n_1^2 - n_2^2\right)^{1/2} \approx \frac{2\pi a}{V} n_1\sqrt{2\Delta} \tag{3.46}$$

with $V = 2.405$ for step-index fibers. At this wavelength, only the LP_{01} mode (i.e., the HE_{11} mode) should propagate in the fiber.

Example 3.14 A given step-index fiber has a core refractive index of 1.480, a core radius equal to 4.5 μm, and a core-cladding index difference of 0.25%. What is the cutoff wavelength for this fiber?

Solution From Eq. (3.46) for $V = 2.405$

$$\lambda_c = \frac{2\pi a}{V} n_1\sqrt{2\Delta} = \frac{2\pi(4.5)}{2.405}(1.480)\sqrt{2(0.0025)} = 1.23 \ \mu m = 1230 \ nm$$

Because in the cutoff region the field of the LP_{11} mode is widely spread across the fiber cross section (i.e., it is not tightly bound to the core), its attenuation is strongly affected by fiber bends, length, and cabling. Recommendation G.650.1 of the ITU-T specifies methods for determining an effective cutoff wavelength λ_c [20]. The test

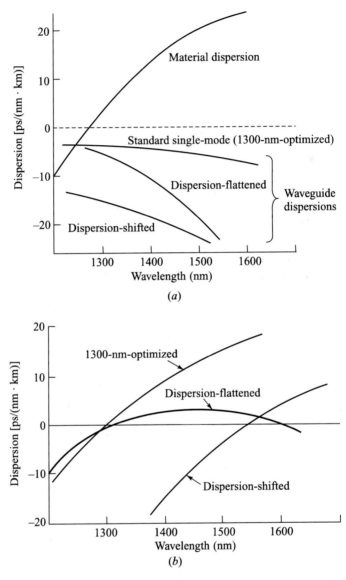

Fig. 3.14 a Typical waveguide dispersions and the common material dispersion for three different single-mode fiber designs; **b** resultant total dispersions

setup consists of a 2 m length of fiber that contains a single 14-cm-radius loop or several 14-cm-radius curvatures that add up to one complete loop. Using a tunable light source that has a full-width half-maximum linewidth not exceeding 10 nm, light is launched into the fiber so that both the LP_{01} and the LP_{11} modes are uniformly excited.

Fig. 3.15 Typical
attenuation-ratio versus
wavelength plot for
determining the cutoff
wavelength using the
bend-reference (or
single-mode-reference)
transmission method

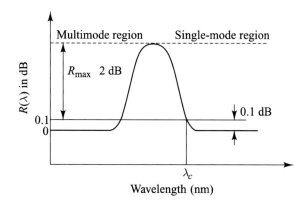

First, the output power $P_1(\lambda)$ is measured as a function of wavelength in a sufficiently wide range around the expected cutoff wavelength. Next, the output power
$P_2(\lambda)$ is measured over the same wavelength range when a loop of sufficiently small
radius is included in the test fiber to filter the LP_{11} mode. A typical radius for this
loop is 30 mm. With this measurement method, the logarithmic ratio $R(\lambda)$ between
the two transmitted powers $P_1(\lambda)$ and $P_2(\lambda)$ is calculated as

$$R(\lambda) = 10\log\left[\frac{P_1(\lambda)}{P_2(\lambda)}\right] \qquad (3.47)$$

Figure 3.15 gives a typical curve of the result. The effective cutoff wavelength
λ_c is defined as the largest wavelength at which the higher-order LP_{11} mode power
relative to the fundamental LP_{01} mode power is reduced to 0.1 dB; that is, when, $R(\lambda)$
= 0.1 dB, as is shown in Fig. 3.15. Recommended values of λ_c range from 1100 to
1280 nm, to avoid modal noise and dispersion problems.

3.3.3 Standards for Dispersion Calculations

As noted in Sect. 3.3.1, the total chromatic dispersion in single-mode fibers consists
mainly of material and waveguide dispersions. The resultant intramodal or chromatic
dispersion is represented by [21]

$$D(\lambda) = \frac{1}{L}\frac{d\tau}{d\lambda} \qquad (3.48)$$

where τ is the group delay. The dispersion is commonly expressed in ps/(nm km).
The broadening σ of an optical pulse over a fiber of length L is given by

$$\sigma = D(\lambda)L\sigma_\lambda \qquad (3.49)$$

where σ_λ is the half-power spectral width of the optical source. To measure the dispersion, one examines the pulse delay over a desired wavelength range.

As illustrated in Fig. 3.14, the dispersion behavior varies with wavelength and also with fiber type. Thus, various standards have recommended different formulas to calculate the chromatic dispersion for specific fiber types operating in a given wavelength region. To calculate the dispersion for a non-dispersion-shifted fiber in the region ranging from 1270 to 1340 nm, the standards recommend fitting the measured group delay per unit wavelength to a three-term Sellmeier equation of the form [20]

$$\tau = A + B\lambda^2 + C\lambda^{-2} \tag{3.50}$$

Here, A, B, and C are the curve-fitting parameters. An equivalent expression is

$$\tau = \tau_0 + \frac{S_0}{8}\left(\lambda - \frac{\lambda_0^2}{\lambda}\right)^2 \tag{3.51}$$

where τ_0 is the relative delay minimum at the zero-dispersion wavelength λ_0, and S_0 is the value of the *dispersion slope* $S(\lambda) = dD/d\lambda$ at λ_0, which is given in ps/(nm^2 km). Using Eq. (3.48), the dispersion for a non-dispersion-shifted fiber is

$$D(\lambda) = \frac{\lambda S_0}{4}\left[1 - \left(\frac{\lambda_0}{\lambda}\right)^4\right] \tag{3.52}$$

To calculate the dispersion for a dispersion- shifted fiber in the 1500–1600 nm region, the standards recommend using the quadratic expression

$$\tau = \tau_0 + \frac{S_0}{2}(\lambda - \lambda_0)^2 \tag{3.53}$$

which results in the dispersion expression

$$D(\lambda) = (\lambda - \lambda_0)S_0 \tag{3.54}$$

Finally, recall from Eq. (3.26) that the *third- order dispersion* β_3 can be given as

$$\beta_3 = \frac{\lambda^2}{(2\pi c)^2}\left(\lambda^2 S_0 + 2\lambda D\right) \tag{3.55}$$

When measuring a set of fibers, one will get values of λ_0 ranging from $\lambda_{0,min}$ to $\lambda_{0,max}$. Figure 3.16 shows the range of the expected dispersion values for a set of non-dispersion-shifted fibers in the 1270–1340 nm region. Typical values of S_0 are 0.092 ps/(nm^2 km) for standard non-dispersion-shifted fibers, and are between 0.06 and 0.08 ps/(nm^2 km) for dispersion-shifted fibers. Alternatively, the ITU-T

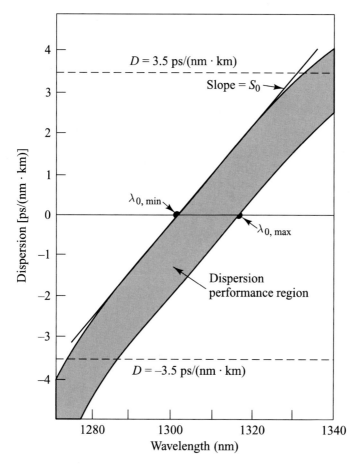

Fig. 3.16 Example of a dispersion performance curve for a set of single-mode fibers, where the two slightly curved lines are found by solving Eq. (3.52)

Rec. G.652 has specified this as a maximum dispersion of 3.5 ps/(nm km) in the 1285–1330 nm region, as denoted by the bounding dashed lines in Fig. 3.16.

Example 3.15 A manufacturer's data sheet states that a non-dispersion-shifted fiber has a zero-dispersion wavelength of 1310 nm and a dispersion slope of 0.092 ps/(nm^2 · km). Compare the dispersions for this fiber at wavelengths of 1280 nm and 1550 nm.

Solution Using Eq. (3.52) the dispersion is

$$D(1280) = \frac{\lambda S_0}{4}\left[1 - \left(\frac{\lambda_0}{\lambda}\right)^4\right] = \frac{(1280)(0.092)}{4}\left[1 - \left(\frac{1310}{1280}\right)^4\right]$$
$$= -2.86 \text{ ps/(nm - km)}$$

$$D(1550) = \frac{\lambda S_0}{4}\left[1 - \left(\frac{\lambda_0}{\lambda}\right)^4\right] = \frac{(1550)(0.092)}{4}\left[1 - \left(\frac{1310}{1550}\right)^4\right]$$

$$= 17.5 \text{ ps/(nm - km)}$$

Drill Problem 3.9 A single-mode optical fiber that is optimized for long-distance high-capacity optically amplified transmission has a dispersion slope at 1550 nm of 0.045 ps/(nm^2 km) and a zero-dispersion wavelength of 1405 nm. Using Eq. (3.52) show that the dispersions at 1310 nm and 1550 nm are − 4.76 ps/(nm km) and +5.67 ps/(nm km), respectively.

In summary for dispersion in single-mode fibers, as optical pulses travel down a fiber, temporal broadening occurs because material and waveguide dispersion cause different wavelengths in the optical pulse to propagate with different velocities. Thus, as Eq. (3.49) implies, the broader the spectral width σ_λ of the source, the greater the pulse dispersion will be.

3.3.4   Definition of Mode-Field Diameter

Section 2.5.2 gives the definition of the mode-field diameter in single-mode fibers. One use of the mode-field diameter is in describing the functional properties of a single-mode fiber, because it takes into account the wavelength-dependent field penetration into the cladding. This is shown in Fig. 3.17 for 1300-nm-optimized, dispersion-shifted, and dispersion-flattened single-mode fibers.

3.3.5   Bending Loss in Single-Mode Fibers

Macrobending and microbending losses are important in the design of single-mode fibers. The bending losses are primarily a function of the mode-field diameter. Generally, the bending losses are less for smaller mode-field diameters, because for smaller mode-field diameters the modes are confined tighter to the core.

By specifying bend-radius limitations when installing standard single-mode fibers, one can largely avoid high microbending losses. Manufacturers usually recommend that a fiber or cable bend diameter should be no smaller than 40–50 mm (1.6–2.0 in.). This is consistent with bend diameter limitations of 50–75 mm specified by installation guides for cable placement in ducts, fiber-splice enclosures, and

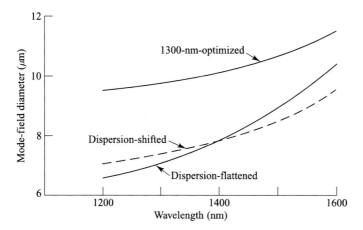

Fig. 3.17 Typical mode-field diameter variations with wavelength for 1300 nm-optimized, dispersion-shifted, and dispersion-flattened single-mode fibers

equipment racks. Furthermore, as Sect. 3.5 describes, the development of bend-insensitive fibers allows much tighter coiling of these fibers in optoelectronic packages. In addition, use of these bend-insensitive fibers in jumper cables between equipment modules greatly reduces bending loss effects when they are installed in highly confined equipment racks.

3.4 ITU-T Standards for Fibers

The ITU-T is a leading organization that develops and publishes a wide range of internationally recognized recommendations and standards. The organization has created a series of recommendations for manufacturing and testing various classes of multimode and single-mode optical fibers used in telecommunications. These documents give guidelines for bounds on fiber parameters, such as core and cladding sizes and circularity, attenuation, cutoff wavelength, and chromatic dispersion. The recommendations allow a reasonable degree of design flexibility, so that fiber manufacturers can improve products and develop new ones within the guidelines given in the performance specifications.

Table 3.2 summarizes the ITU-T recommendations for multimode and single-mode optical fibers used in long-distance, access, and enterprise networks. The following subsections describe the basic characteristics of these fibers. The recommendations are available for downloading at www.itu.int.

Table 3.2 Recommendations for fibers used in telecom, access, and enterprise networks

ITU-T Rec. No.	Title and Description
G.651.1 (Edition 2; Nov. 2018)	Title: Characteristics of a 50/125 μm multimode graded index optical fiber cable for the optical access network Description: Requirements of a silica 50/125 μm multimode graded index fiber cable for the 850 nm or 1300 nm regions
G.652 (Edition 9, Nov. 2016)	Title: Characteristics of a Single-Mode Optical Fiber and Cable Description: Discusses single-mode fiber optimized for O-band (1310 nm) use, but which also can be used in the 1550 nm region
G.653 (Edition 7, July 2010)	Title: Dispersion-Shifted Single-Mode Optical Fiber and Cable Description: Discusses single-mode optical fiber with the zero-dispersion wavelength shifted into the 1550 nm region. Describes chromatic dispersion for the 1460–1625 nm range for CWDM
G.654 (Version E, Mar. 2020)	Title: Cut-Off Shifted Single-Mode Optical Fiber and Cable Description: Undersea single-mode optical fiber applications with a zero-dispersion wavelength around 1300 nm
G.655 (Edition 5, Nov. 2009)	Title: Characteristics of a Non-Zero Dispersion-Shifted Single-Mode Optical Fiber and Cable Description: For applications in long-haul links; describes single-mode optical fiber with chromatic dispersion greater than zero throughout the 1530–1565 nm wavelength range
G.656 (Edition 3, July 2010)	Title: Characteristics of a Fiber and Cable with Non-Zero Dispersion for Wideband Optical Transport Description: Low chromatic dispersion fiber for expanded WDM applications in the wavelength region between 1460 and 1625 nm
G.657 (Edition 4, Nov. 2016)	Title: Characteristics of a bending loss insensitive single mode optical fiber and cable for the access network Description: Addresses use of single-mode fiber for broadband access networks; includes bending conditions for in-building use

3.4.1   Recommendation G.651.1

The economic demand for low-cost installations of high-speed short-distance optical fiber links created an extensive market for multimode fibers. These links use moderately priced light sources that operate in either the short-wavelength region (770–860 nm) or in the O-band (around 1310 nm). Applications include links in locations such as an office or government building, a medical facility, a university campus, or a manufacturing plant, where the desired transmission distance is typically 2 km or less.

Recommendation G.651.1 gives the requirements of a 50/125 μm multimode graded-index silica-optical-fiber cable for use in the 850 nm or 1300 nm regions.

System operation for this fiber is allowed either in each wavelength band individually or simultaneously in both spectral bands. The applications are intended for access and enterprise networks in multiple-tenant building environments in which broadband services have to be delivered to individual apartments or offices of individual businesses. The recommended multimode fiber supports the cost-effective use of 1 Gb/s Ethernet systems over link lengths up to 550 m. The optical fiber attenuation values range from 2.5 dB/km at 850 nm to less than 0.6 dB/km at 1310 nm.

In addition to the recommendations of G.651.1, the IEEE 802.3 series of standards describes the implementation of Ethernet links running at data rates up to 10 Gb/s over distances up to 550 m.

3.4.2   Recommendation G.652

Recommendation G.652 deals with the geometrical, mechanical, and transmission characteristics of a single-mode fiber that has a zero-dispersion value at 1310 nm. Figure 3.18 compares the dispersion of the G.652 fiber with other single-mode fiber types. This fiber consists of a germanium-doped silica core that has a diameter between 5 and 8 μm, and a silica cladding with a 125 μm diameter. The nominal attenuation is 0.4 dB/km at 1310 nm and 0.35 dB/km at 1550 nm. The maximum polarization mode dispersion is 0.2 ps/km. Four subsets ranging from G.652a to G.652d describe different variations of this type of fiber. Because G.652a/b fibers were installed widely in telecommunication networks in the 1990s, they are commonly known as *standard single-mode fibers* or *1310 nm optimized fibers*. The G.652c/d fibers allow operation in the *E*-band and are used widely for *fiber-to-the-premises* (FTTP) installations.

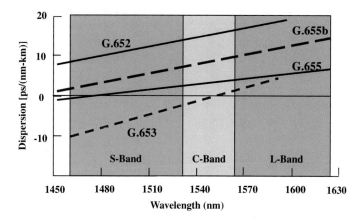

Fig. 3.18 Comparison of the dispersion of the G.652 fiber with other single-mode fiber types

Although many long-distance cable plant installations now are using nonzero-dispersion-shifted fiber, the huge base of G.652 fiber that is installed worldwide will be in service for many years. If the G.652 fiber is used at 1550 nm, the chromatic dispersion value of about 17 ps/(nm km) must be taken into account. This requires implementation of chromatic dispersion-compensation techniques or special data formats at high data rates. As an example, a number of field experiments have demonstrated the ability to transmit 160 Gb/s data rates over long distances of installed G.652a/b fiber.

In G.652c/d fibers the water ion concentration is reduced in order to eliminate the attenuation spike in the 1360–1460 nm E-band. They are called *low-water-peak fiber* and allow operation over the entire wavelength range from 1260 to 1625 nm. One use of this fiber is for low-cost short-reach CWDM (coarse wavelength division multiplexing) applications in the E-band. In CWDM the wavelength channels are spaced by 20 nm, so that minimum wavelength stability control is needed for the optical sources, as described in Chap. 10. Another important application is in a *passive optical network* (PON) for FTTP access networks.

3.4.3 Recommendation G.653

Dispersion-shifted fiber (DSF) was developed for use with 1550 nm lasers. As Fig. 3.18 illustrates, in this fiber type the zero-dispersion point is shifted to 1550 nm where the fiber attenuation is about half that at 1310 nm. Therefore, this fiber allows a high-speed data stream of a single-wavelength channel at or near 1550 nm to maintain its fidelity over long distances. However, it presents problems associated with nonlinear effects in dense wavelength division multiplexing (DWDM) applications in the center of the C-band where many wavelengths are packed tightly into one or more of the operational bands. As noted in Chap. 10, to prevent undesirable nonlinear effects in DWDM systems, the chromatic dispersion values should be positive (or negative) over the entire operational band. Figure 3.18 shows that for G.653 fibers the chromatic dispersion has a different sign above and below 1550 nm. Therefore, the use of G.653 fibers for DWDM should be restricted to either the S-band (wavelengths lower than 1550 nm) or the L-band (wavelengths higher than 1550 nm). These fibers are seldom deployed anymore because G.655 fibers offer a better solution.

3.4.4 Recommendation G.654

This recommendation deals with *cutoff-wavelength-shifted fiber* that is designed for long-distance high-power signal transmission. It describes the geometrical, mechanical, and transmission characteristics of a single-mode optical fiber, which has the zero-dispersion wavelength around 1300 nm. The fiber has a very low loss in the 1550 nm band, which is achieved by using a pure silica core. Because it has a high cutoff wavelength of 1500 nm, this fiber is restricted to operation in the 1500–1600 nm region. It typically is used only in long-distance undersea applications.

3.4.5 Recommendation G.655

Nonzero-dispersion-shifted fiber (NZDSF) was introduced in the mid-1990s for WDM applications. The recommendation has five different versions ranging from G.655.A to G.655.E. Each of these categories has a slightly different value of the dispersion coefficient at 1550 nm. For example, the original G.655.A had a positive dispersion coefficient for wavelengths greater than 1460 nm. This enabled operation of WDM systems in the C-band, which is the spectral operating region for erbium-doped optical fiber amplifiers (see Chap. 11). However, the negative value for lower wavelengths did not allow the use of this fiber in the S-band. Thus, Version G.655.B was introduced to extend WDM applications into the S-band. As shown in Fig. 3.18, the principal characteristic of a G.655.B fiber is that it has a nonzero dispersion value over the entire S-band and the C-band. Version G.655.C specifies a lower PMD value of $0.2\,\text{ps}/\sqrt{\text{km}}$ than the $0.5\,\text{ps}/\sqrt{\text{km}}$ value of G.655.A/B. Recommendations G.655.D and G.655.E have slightly different values of the dispersion coefficient and dispersion slope in order to optimize various tradeoffs among the different G.655 versions in power, channel spacing, amplifier separation, link length and bit rate.

3.4.6 Recommendation G.656

This recommendation describes the characteristics of a single-mode optical fiber that has a positive chromatic dispersion value ranging from 2 to 14 ps/(nm km) in the 1460–1625 nm wavelength band. Whereas the dispersion slope is lower than in G.655 fibers, most G.656 attributes are similar to those of G.655 fibers. For example, the mode-field diameter ranges from 7 to 11 μm (compared to 8–11 μm for G.655 fibers), the maximum PMD value of cabled fiber is $0.2\,\text{ps}/\sqrt{\text{km}}$, and the cutoff wavelength is 1310 nm (the same as for G.655). Thus, because of the close similarity of these two fiber types, a number of optical fiber manufacturers do not bother to distinguish between G.655 and G.656 and simply refer to their offerings as G.655/656 compliant.

3.4.7 Recommendation G.657

The rapidly growing worldwide demand for broadband services in high-capacity access networks and enterprise networks put new demands on the performance characteristics of single-mode fibers that are different from telecom applications. These performance differences are due mainly to the high density of localized distribution and drop cables in the access and enterprise networks compared to metro and long-haul telecom networks. In particular, in building applications the cables can be installed inside moldings that run along the walls and around corners and doors, and in equipment racks where the cables often are tightly coiled or bent. These conditions and the many installation manipulations imposed on the cabling system call for optical fibers with low bending loss sensitivity.

Thus the aim of ITU-T Recommendation G.657 is to describe requirements for a fiber type that exhibits improved bending loss performance compared with existing G.652 single-mode fiber and cables. Two categories of single-mode fibers are specified. Fibers in Category A are fully compliant with the G.652 single-mode fibers and also can be used in other parts of the network. Single-mode fibers in Category B are not necessarily compliant with G.652 but exhibit low values of losses at very small bend radii. Fibers in Category B are predominantly intended for in-building use.

3.5   Designs and Use of Specialty Fibers

Telecommunication fibers, such as those described in Sect. 3.4, are designed to transmit light with minimal change in the signal fidelity. In contrast, *specialty fibers* are designed to interact with light and thereby manipulate or control some characteristics of an optical signal. The light manipulation applications include optical signal amplification, optical power coupling, dispersion compensation, wavelength conversion, and sensing of physical parameters such as temperature, stress, pressure, vibration, and fluid levels. For light-control applications a specialty fiber can be insensitive to bends, maintain polarization states, redirect specific wavelengths, or provide a very high attenuation for fiber terminations.

Specialty fibers can be of either a multimode or a single-mode design. Among the optical devices that may use a specialty fiber are light transmitters, light signal modulators, optical receivers, wavelength multiplexers, light couplers and splitters, optical amplifiers, optical switches, wavelength add/drop modules, and optical power attenuators. Table 3.3 gives a summary of some specialty fibers and their applications.

Rare-Earth Doped Fiber These fibers have small amounts of rare-earth ions (for example, 1000 parts per million weight) added to the silica material to form a basic building block for optical fiber amplifiers. The rare-earth elements could be erbium (Er), ytterbium (Yb), thulium (Tm), or praseodymium (Pr). As described in Chap. 11, a length of such fiber ranging from 10 to 30 m serves as a gain medium for amplifying optical signals in the 1.0 μm region, the C-band (1530–1560 nm), or the L-band (1560–1625 nm). There are many variations on the doping level, cutoff wavelength, mode-field diameter, numerical aperture, and cladding diameter for these fibers. Erbium is the main material used for optical fiber amplifiers operating in the C-band. Specific erbium-doped fiber configurations will yield a variety of optical amplifier designs that can be selected according to pump laser power requirement, noise figure,

Table 3.3 Examples of specialty fibers and their applications

Specialty fiber type	Application
Rare-earth doped fiber	Gain medium for optical fiber amplifiers
Photosensitive fibers	Fabrication of fiber Bragg gratings
Bend-insensitive fibers	Tight loops in device packages
Polarization-preserving fibers	Pump lasers, polarization-sensitive devices
Photonic crystal fibers	Switches; dispersion compensation

signal gain, and flatness of the output spectrum. Higher erbium concentrations allow the use of shorter fiber lengths, smaller claddings are useful for compact packages, and a higher numerical aperture allows the fiber to be coiled tightly in small packages. Table 3.4 lists some generic parameter values of an erbium-doped fiber for use in the C-band.

Photosensitive Sensitive Fiber The refractive index of a photosensitive fiber changes when it is exposed to ultraviolet light. This sensitivity may be provided by doping the fiber material with germanium and boron ions. The main application is to create a fiber Bragg grating, which is a periodic variation of the refractive index along the fiber axis (see Chap. 10). Applications of fiber Bragg gratings include light-coupling mechanisms for pump lasers used in optical amplifiers, wavelength add/drop modules, optical filters, and chromatic dispersion compensation modules.

Bend-Insensitive Fiber In many applications of optical fibers within an indoor telecom facility and along paths inside of homes or businesses, the fibers can experience a sinuous path with sharp bends. Special attention must be paid to this situation, because optical fibers exhibit radiative losses whenever the fiber undergoes a bend with a finite radius of curvature. For slight bends this factor is negligible. However, as the radius of curvature decreases, the losses increase exponentially until at a certain critical radius the losses become extremely large. As shown in Fig. 3.19, for standard

Table 3.4 Generic parameter values of an erbium-doped fiber for the C-band

Parameter	Specification
Peak absorption at 1530 nm	5–10 dB/m
Effective numerical aperture	0.14–0.31
Cutoff wavelength	900 $\pm$ 50 nm; or 1300 nm
Mode-field diameter at 1550 nm	5.0–7.3 μm
Cladding diameter	125 μm standard; 80 μm for tight coils
Coating material	UV-cured acrylic

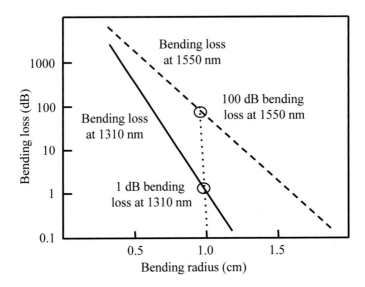

Fig. 3.19 For standard fibers the bending loss becomes more sensitive at longer wavelengths

fibers the bending loss becomes more sensitive at longer wavelengths. A fiber with a small bend radius might be transmitting well at 1310 nm, for example, giving an additional loss of 1 dB at a 1 cm bending radius. However, for this bend radius there could be a significant loss at 1550 nm resulting in an additional loss of about 100 dB for a conventional fiber.

Such situations led the telecom industry to develop bend-loss insensitive fibers that can tolerate numerous sharp bends for indoor installations. Such fibers have a moderately higher numerical aperture (NA) than in a standard single-mode telecom fiber. Increasing the NA reduces the sensitivity of the fiber to bending loss by confining optical power more tightly within the core than in conventional single-mode fibers. Bend-loss insensitive fibers are available commercially from a variety of optical fiber manufacturers. These fibers are offered with either an 80 μm or a 125 μm cladding diameter as standard products. The 80 μm *reduced-cladding fiber* results in a much smaller volume compared with a 125 μm cladding diameter when a fiber length is coiled up within a miniature optoelectronic device package or in a compact test instrument.

For example, various manufacturers offer a bend insensitive fiber that has a lower single-mode cutoff wavelength, a nominally 50% higher index difference value Δ, and a 25% higher NA than conventional telecom fibers. The higher NA of low-bend-loss fibers allows an improved coupling efficiency from laser diode sources to planar waveguides. Generally for bend radii of greater than 20 mm, the bending-induced loss is negligibly small. Fibers are available in which the maximum bending induced loss is less than 0.2 dB due to 100 turns on a 10 mm mandrel. A factor to keep in mind is that at operating wavelengths in the near infrared, the smaller mode field diameter of low-bend-loss fibers can induce a mode-mismatch loss when interconnecting these

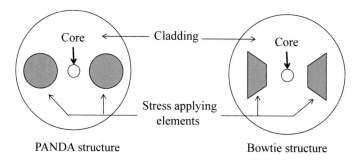

Fig. 3.20 The cross-sectional geometry of two different polarization-maintaining fibers

fibers with standard single-mode fibers. However, carefully made splices between these different fibers typically results in losses less than 0.1 dB.

Polarization-Preserving Fiber In contrast to standard single-mode optical fibers in which the state of polarization fluctuates as a light signal propagates through the fiber, *polarization-preserving fibers* have a special core design that maintains the state of polarization along the fiber. Applications of these fibers include light signal modulators fabricated from lithium niobate, optical amplifiers for polarization multiplexing, light-coupling fibers for pump lasers, and polarization-mode dispersion compensators. Figure 3.20 illustrates the cross-sectional geometry of two different polarization-maintaining fibers, which are known as the PANDA structure and the bowtie structure. The light circles represent the cladding and the dark areas are the core configurations. The goal in each design is to use stress-applying parts to create slow and fast axes in the core. Each of these axes will guide light at a different velocity. Crosstalk between the two axes is suppressed so that polarized light launched into either of the axes will maintain its state of polarization as it travels along the fiber.

3.6 Character of Multicore Optical Fibers

As noted in Sect. 1.5.4, optical fibers with multiple cores have been developed as one solution to increasing the transmission capacity of a fiber [22]. Both multimode and single-mode multicore fibers have been constructed. A common configuration consists of seven pure silica cores, as shown in Fig. 3.21. The material of the silica-based cores also could have a refractive index slightly higher than pure silica. A low refractive-index cladding layer (see the depressed-index structure in Fig. 3.13a) surrounds each core in order to reduce core-to-core crosstalk. A typical core pitch (center-to- center spacing of the cores) is 40 μm and the core diameters are around 8 μm for single-mode operation in the 1310 and 1490 nm regions. A marker can be embedded in the fiber to help identify the different cores.

Fig. 3.21 Example of a multicore optical fiber

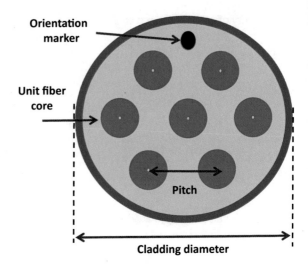

3.7   Summary

Attenuation of a light signal as it propagates along a fiber is an important consideration in the design of an optical communication system because it plays a major role in determining the maximum transmission distance between a transmitter and a receiver. The basic attenuation mechanisms are absorption, scattering, and radiative losses of optical energy. The major causes of absorption are extrinsic absorption by impurity atoms and intrinsic absorption by basic constituent atoms of the fiber material. Intrinsic absorption sets the fundamental lower limit on attenuation for any particular material. Scattering follows a Rayleigh λ^{-4} dependence, which gives attenuation-versus-wavelength plots their characteristic downward trend with increasing wavelength.

Radiative losses occur whenever an optical fiber is bent. These losses can arise from macroscopic bends, such as when an optical fiber turns a corner, or from microscopic bends (microbends) of the fiber axis. Of these various effects, microbends are the most troublesome, so special care must be taken during manufacturing, cabling, and installation to minimize them.

In addition to being attenuated, an optical signal undergoes continuous broadening and distortion as it travels along a fiber. The signal broadening is a consequence of intramodal and intermodal dispersion effects. *Intermodal dispersion* or *modal delay* appears only in multimode fibers. This dispersion mechanism is a result of each mode having a different value of the group velocity at a single frequency. *Intramodal dispersion* is pulse spreading that occurs within an individual mode and thus is of importance in single-mode fibers. Its three main causes are material dispersion, waveguide dispersion, and polarization-mode dispersion.

A variety of multimode and single-mode optical fibers are used in telecommunication, access, and enterprise networks. The ITU-T has created a series of recommendations for manufacturing and testing various classes of multimode and single-mode optical fibers used in telecommunications.

Problems

3.1 Verify the expression given in Eq. (3.3) that relates α, which is in units of dB/km, to α_p, which is in units of km^{-1}.

3.2 A certain optical fiber has an attenuation of 0.6 dB/km at 1310 nm and 0.3 dB/km at 1550 nm. Suppose the following two optical signals are launched simultaneously into the fiber: an optical power of 150 μW at 1310 nm and an optical power of 100 μW at 1550 nm. What are the power levels in μW of these two signals at (a) 8 km and (b) 20 km?

3.3 An optical signal at a specific wavelength has lost 55% of its power after traversing 7.0 km of fiber. What is the attenuation in dB/km of this fiber?

3.4 A continuous 40 km long optical fiber link has a loss of 0.4 dB/km. (a) What is the minimum optical power level that must be launched into the fiber to maintain an optical power level of 2.0 μW at the receiving end? (b) What is the required input power if the fiber has a loss of 0.6 dB/km?

3.5 The optical power loss resulting from Rayleigh scattering in a fiber can be calculated from either Eq. (3.7) or Eq. (3.8). Compare these two equations for silica ($n = 1.460$ at 630 nm), given that the fictive temperature T_f is 1400 K, the isothermal compressibility β_T is 6.8×10^{-12} cm^2/dyn, and the photoelastic coefficient is 0.286. How does this agree with measured values ranging from 3.9 to 4.8 dB/km at 633 nm?

3.6 Consider a graded-index multimode fiber that has a core radius $a = 25$ μm, a refractive index profile $\alpha = 2.0$, a cladding index $n_2 = 1.478$, and an index difference $\Delta = 0.01$. Using Eq. (3.11) compare the ratio M_{eff}/M_∞ for a 1310 nm wavelength when the bending radius $R = 2.5$ cm and when $R = 1.0$ cm.

3.7 Consider a graded-index fiber having an index profile $\alpha = 2.0$, cladding refractive index $n_2 = 1.478$, and an index difference $\Delta = 0.01$. Using Eq. (3.11) compare the ratio M_{eff}/M_∞ for a 1550 nm wavelength for $R = 2.5$ cm when (a) $a = 25$ μm and (b) $a = 50$ μm.

3.8 Equation (3.13) gives an expression for the factor by which microbending loss is reduced when a compressible jacket is extruded over a fiber. Consider the case when a jacket material that has a Young's modulus $E_j = 21$ MPa is extruded over a glass fiber that has a Young's modulus $E_j = 64$ GPa and a cladding-to-core ratio b/a = 2.0.

(a) Show that when the refractive index difference $\Delta = 0.01$, the reduction factor is
$F(\alpha_M) = 0.0038 = 0.38\%$.

(b) Show that when the refractive index difference $\Delta = 0.001$, the reduction factor is $F(\alpha_M) = 0.75 = 75\%$.

3.9   Assume that a step-index fiber has a V number of 6.0. (a) Using Fig. 3.5, estimate the fractional power P_{clad}/P traveling in the cladding for the four lowest-order LP modes. (b) If the fiber in (a) is a glass-core glass-clad fiber having core and cladding attenuations of 3.0 and 4.0 dB/km, respectively, find the attenuations for each of the four lowest-order modes.

3.10   Assume a given mode in a graded-index fiber has a power density $p(r) = P_0 \exp(-Kr^2)$, where the factor K depends on the modal power distribution.

(a)   Letting $n(r)$ in Eq. (3.16) be given by Eq. (2.38) with $\alpha = 2$, show that the loss in this mode is

$$\alpha_{gi} = \alpha_1 + \frac{\alpha_2 - \alpha_1}{Ka^2}$$

Because $p(r)$ is a rapidly decaying function of r and because $\Delta \ll 1$, for ease of calculation assume that the top relation in Eq. (2.38) holds for all values of r.

(b)   Choose K such that $p(a) = 0.1\, P_0$; that is, 10% of the power flows in the cladding. Find α_{gi} in terms of α_1 and α_2.

3.11   A 5 km transmission link consists of a step-index multimode fiber that has a core index $n_1 = 1.480$ and a core-cladding refractive index difference $\Delta = 0.01$.

(a)   Using the approximation on the right-hand side of Eq. (3.18), show that the delay difference between the fastest and slowest modes is 247 ns.

(b)   Using Eq. (3.20), show that the rms pulse broadening resulting from intermodal delay is 71.2 ns.

(c)   Using Eq. (3.18) and the condition that the maximum bit rate B should satisfy the condition $B < 0.1/\Delta T$, show that the maximum bit rate-distance product is $BL = 4.05$ (Mb/s) km.

3.12   For wavelengths less than 1.0 μm the refractive index n satisfies a Sellmeier relation of the form

$$n^2 = 1 + \frac{E_0 E_d}{E_0^2 - E^2}$$

where $E = hc/\lambda$ is the photon energy and E_0 and E_d are, respectively, material oscillator energy and dispersion energy parameters. In SiO_2 glass, $E_0 = 13.4$ eV and $E_d = 14.7$ eV. Show that, for wavelengths between 0.20 and 1.0 μm, the values of n found from the Sellmeier relation are in good agreement with those shown in Fig. 3.9. To make the comparison, select three representative points, for example, at 0.2, 0.6, and 1.0 μm.

3.13   (a) An LED operating at 850 nm has a spectral width of 45 nm. If the material dispersion at that wavelength is 115 ps/(nm km), what is the pulse spreading in ns/km due to material dispersion? (b) Suppose the material dispersion for this fiber is 20 ps/(nm km) at 1550 nm. What is the pulse spreading when a laser diode with a 1 nm spectral width at 1550 nm is used?

3.14 Verify the plots for b, $d(Vb)/dV$, and $Vd^2(Vb)/dV^2$ shown in Fig. 3.10. Use the expression for b given by Eq. (3.43).

3.15 Derive Eq. (3.18) for modal delay by using a ray-tracing method.

3.16 Consider a step-index fiber with core and cladding diameters of 62.5 and 125 μm, respectively. Let the core index $n_1 = 1.48$ and let the index difference $\Delta = 1.5\%$. Compare the modal dispersion in units of ns/km at 1310 nm of this fiber as given by Eq. (3.18) with the more exact expression

$$\frac{\sigma_{\text{mod}}}{L} = \frac{n_1 - n_2}{c}\left(1 - \frac{\pi}{V}\right)$$

where L is the length of the fiber and n_2 is the cladding index.

3.17 Consider a standard G.652 non-dispersion-shifted single-mode optical fiber that has a zero-dispersion wavelength at 1310 nm with a dispersion slope of $S_0 = 0.0970$ ps/(nm^2 km). Plot the dispersion in the wavelength range 1270 nm $\leq 1 \leq$ 1340 nm. Use Eq. (3.52).

3.18 Starting with Eq. (3.50), derive the dispersion expression given in Eq. (3.52).

Answers to Selected Problems

3.2 The input powers in dBm are $P(100\ \mu\text{W}) = -10.0$ dBm; $P(150\ \mu\text{W}) = -8.24$ dBm

(a) $P_{1310}(8\,\text{km}) = -13.0\,\text{dBm} = 50\ \mu\text{W};\ P_{1550}(8\,\text{km}) = -12.4\,\text{dBm} = 57.5\ \mu\text{W}$

(b) $P_{1310}(20\,\text{km}) = -20.2\,\text{dBm} = 9.55\ \mu\text{W};\ P_{1550}(20\,\text{km}) = -16.0\,\text{dBm} = 25.1\ \mu\text{W}$

3.3 $\alpha = 0.5$ dB/km

3.4 (a) $P_{\text{in}} = 79.6\ \mu\text{W} = -11$ dBm;
(b) $P_{\text{in}} = 502\ \mu\text{W} = -3$ dBm

3.5 From Eq. (3.7) $\alpha_{\text{scat}} = 0.0462\ \text{km}^{-1} = 0.40$ dB/km
From Eq. (3.8) $\alpha_{\text{scat}} = 0.0608\ \text{km}^{-1} = 0.26$ dB/km

3.6 0.758 and 0.423 for $R = 2.5$ cm and 1.0 cm, respectively

3.7 0.753 and 0.553 for $a = 25\ \mu$m and 50 μm, respectively

3.9

Mode order	P_{clad}/P	$\alpha_{vm} = \alpha_1 + (\alpha_2 - \alpha_1)P_{\text{clad}}/P$
01	0.02	$3.0 + 0.02$ dB/km
11	0.05	$3.0 + 0.05$
21	0.10	$3.0 + 0.10$
02	0.16	$3.0 + 0.16$

3.10   (b) $p(a) = 0.1 \, P_0 = P_0 \, e^{-Ka^2}$ yields $e^{Ka^2} = 10$.
       This yields $Ka^2 = \ln 10 = 2.3$.
       Thus $\alpha_{gi} = \alpha_1 + \frac{(\alpha_2 - \alpha_1)}{2.3} = 0.57\alpha_1 + 0.43\alpha_2$

3.12

Wavelength λ	Calculated n	n from Fig. 3.9
0.2 μm	1.548	1.550
0.6 μm	1.457	1.458
1.0 μm	1.451	1.450

3.13   (a)   From Eq. (3.35) for the LED $\sigma_{mat}/L = 5.2$ ps/km
       (b)   For the laser diode, $\sigma_{mat}/L = 2.0$ ps/km

References

1. D. Gloge, The optical fibre as a transmission medium. Rpts. Prog. Phys. **42**, 1777–1824 (1979)
2. D.B. Keck, Fundamentals of optical waveguide fibers. IEEE Commun. Mag. **23**, 17–22 (1985)
3. J.D. Musgraves, J. Hu, L. Calvez, *Springer Handbook of Glass* (Springer, Berlin, 2019)
4. R. Olshansky, Propagation in glass optical waveguides. Rev. Mod. Phys. **51**, 341–367 (1979)
5. S.R. Nagel, in *Fiber Materials and Fabrication Methods*, ed. by S.E. Miller, I.P. Kaminow. Optical Fiber Telecommunications–II (Academic, 1988)
6. K. Nagayama, M. Matsui, M. Kakui, T. Saitoh, K. Kawasaki, H. Takamizawa, Y. Ooga, I. Tsuchiya, Y. Chigusa, Ultra low loss (0.1484 dB/km) pure silica core fiber. SEI Tech. Rev. **57**, 3–6 (2003)
7. R. Maurer, Glass fibers for optical communications. Proc. IEEE **61**, 452–462 (1973)
8. D.A. Pinnow, T.C. Rich, F.W. Ostermeyer, M. DiDomenico Jr., Fundamental optical attenuation limits in the liquid and gassy state with application to fiber optical waveguide material. Appl. Phys. Lett. **22**, 527–529 (1973)
9. W.A. Gambling, H. Matsumura, C.M. Ragdale, Curvature and microbending losses in single-mode optical fibers. Opt. Quantum Electron. **11**(1), 43–59 (1979)
10. T. Murao, K. Nagao, K. Saitoh, M. Koshiba, Design principle for realizing low bending losses in all-solid photonic bandgap fibers. J. Lightw. Technol. **29**(16), 2428–2435 (2011)
11. D. Gloge, Bending loss in multimode fibers with graded and ungraded core index. Appl. Opt. **11**, 2506–2512 (1972)
12. J. Sakai, T. Kimura, Practical microbending loss formula for single mode optical fibers. IEEE J. Quantum Electron. **QE-15**, 497–500 (1979)
13. D. Gloge, Optical fiber packaging and its influence on fiber straightness and loss. Bell Sys. Tech. J. **54**, 245–262 (1975)
14. D. Marcuse, *Theory of Dielectric Optical Waveguides*, 2nd edn. (Academic Press, 1991)
15. D. Gloge, E.A.J. Marcatili, D. Marcuse, S.D. Personick, in *Dispersion Properties of Fibers*, ed. by S.E. Miller, A.G. Chynoweth. Optical Fiber Telecommunications (Academic Press, 1979)
16. D. Marcuse, Interdependence of waveguide and material dispersion. Appl. Opt. **18**, 2930–2932 (1979)
17. D. Gloge, Weakly guiding fibers. Appl. Opt. **10**, 2252–2258 (1971); Dispersion in weakly guiding fibers. Appl. Opt. **10**, 2442–2445 (1971)
18. A.E. Willner, S.M.R. Motaghian Nezam, L. Yan, Z. Pan, M.C. Hauer, Monitoring and control of polarization-related impairments in optical fiber systems. J. Lightw. Technol. **22**, 106–125 (2004)

19. P. Barcik, P. Munster, Measurement of slow and fast polarization transients on a fiber-optic testbed. Opt. Exp. **28**(10), 15250–15257 (2020)
20. ITU-T Recommendation G.650.1, *Definitions and Test Methods for Linear, Deterministic Attributes of Single-Mode Fibre and Cable* (2018)
21. R. Hui, M. O'Sullivan, *Fiber Optic Measurement Techniques* (Academic Press, 2009)
22. K. Saitoh, S. Matsua, Multicore fiber technology. J. Lightw. Technol. **34**(1), 55–66 (2016)

Chapter 4
Light Sources for Fiber Links

Abstract Semiconductor-based light-emitting diodes and laser diodes are the two basic types of light sources that are compatible with the dimensions of optical fibers. These components are suitable light sources for optical communications because of their optical power output levels, the ability to directly modulate the light level with an information signal, and their high efficiency. To understand the applications of these optical sources, this chapter first presents some basic concepts of semiconductor physics and then describes the operational characteristics of light-emitting diodes and laser diodes.

Two classes of semiconductor-based light sources that are widely used for fiber optic communications are heterojunction-structured semiconductor *laser diodes* (also referred to as *injection laser diodes* or ILDs) and *light-emitting diodes* (LEDs). A *heterojunction* consists of two adjoining semiconductor materials with different bandgap energies. Note that as described below, the bandgap energy represents the minimum energy that is required to boost an electron in a semiconductor material to a higher energy state in which it can participate in electrical conduction. These devices are suitable for fiber transmission systems because they have adequate output power for a wide range of applications, their optical power output can be directly modulated by varying the input current to the device, they have a high efficiency, and their dimensions are compatible with those of the optical fiber. Comprehensive treatments of the major aspects of LEDs, laser diodes, and their underlying semiconductor principles are presented in various books and review articles [1–6]. Chapter 11 addresses optical fiber lasers and pump lasers used for optical amplifiers.

The intent of this chapter is to give an overview of the pertinent characteristic of fiber-compatible semiconductor luminescent sources. The first section discusses semiconductor material fundamentals that are relevant to light source operation. The next two sections present the output and the operating characteristics of LEDs and laser diodes. Next are sections concerning the temperature responses and linearity characteristics of optical sources.

In this chapter first Sect. 4.1 shows that the light-emitting region of both LEDs and laser diodes consists of a *pn* junction constructed of direct-bandgap III–V semiconductor materials. When this junction is forward biased, electrons and holes are

© The Author(s), under exclusive license to Springer Nature Singapore Pte Ltd. 2021   147
G. Keiser, *Fiber Optic Communications*,
https://doi.org/10.1007/978-981-33-4665-9_4

injected into the p and n regions, respectively. These injected minority carriers can recombine either radiatively, in which case a photon of energy $h\nu$ is emitted, or nonradiatively, whereupon the recombination energy is dissipated in the form of heat. This pn junction is thus known as the *active region* or *recombination region.*

A major difference between LEDs and laser diodes is that the optical output from an LED is incoherent, whereas that from a laser diode is coherent. In a *coherent source*, the optical energy is produced in an optical resonant cavity. The optical energy released from this cavity has spatial and temporal coherence, which means it is highly monochromatic and the output beam is very directional. In an incoherent LED source, no optical cavity exists for wavelength selectivity. The output radiation has a broad spectral width, because the emitted photon energies range over the energy distribution of the recombining electrons and holes, which usually lie between 1 and $2k_BT$ (k_B is Boltzmann's constant and T is the absolute temperature at the pn junction). In addition, the incoherent optical energy is emitted into a broad elliptical region according to a cosine power distribution and thus has a large beam divergence.

In choosing an optical source compatible with the optical waveguide, various characteristics of the fiber, such as its geometry, its attenuation as a function of wavelength, its group delay distortion, and its modal characteristics, must be taken into account. The interplay of these factors with the optical source power, spectral width, radiation pattern, and modulation capability needs to be considered. The spatially directed coherent optical output from a laser diode can be coupled into either single-mode or multimode fibers. In general, LEDs are used with multimode fibers, because normally it is only into a multimode fiber that the incoherent optical power from an LED can be coupled in sufficient quantities to be useful. However, LEDs have been employed in high-speed local-area applications in which one wants to transmit several wavelengths on the same fiber. Here, a technique called *spectral slicing* is used. This entails using a passive device such as a waveguide grating array (see Chap. 10) to split the broad spectral emission of the LED into narrow spectral slices. Because these slices are each centered at a different wavelength, they can be individually modulated externally with independent data streams and simultaneously sent on the same fiber.

4.1   Basic Concepts of Semiconductor Physics

Because the material in this chapter assumes a rudimentary knowledge of semiconductor physics, various relevant definitions are given here for semiconductor material properties, including the concepts of energy bands, intrinsic and extrinsic materials, pn junctions, and direct and indirect bandgaps. Further details can be found in Refs. [4–6].

4.1.1 Semiconductor Energy Bands

Semiconductor materials have conduction properties that lie somewhere between those of metals and insulators. As an example material, consider silicon (Si), which is located in the fourth column (group IV) of the periodic table of elements. A Si atom has four electrons in its outer shell, by which it makes covalent bonds with its neighboring atoms in a crystal. Such outer-shell electrons are called *valence electrons*.

The conduction properties of a semiconductor can be interpreted with the aid of the *energy-band diagrams* shown in Fig. 4.1a. In a semiconductor the valence electrons occupy a band of energy levels called the *valence band* with E_v being the highest energy level. The valence band is the lowest band of allowed states that an electron can occupy. The next higher band of allowed energy levels for the electrons is called the *conduction band* with E_c being the lowest energy level. In a pure crystal at low temperatures, the *conduction band* is completely empty of electrons and the *valence band* is completely full. These two bands are separated by an *energy gap*, or *bandgap*, in which no energy levels exist. As the temperature is raised, some electrons are thermally excited across the bandgap. For Si this excitation energy must be greater than 1.1 eV, which is the bandgap energy. This electron excitation process gives rise to a concentration n of free electrons in the conduction band, which leaves behind an equal concentration p of vacancies (in which there is no electron) or *holes* in the valence band, as is shown schematically in Fig. 4.1b. Both the free electrons and the holes are mobile within the material, so that both can contribute to electrical conductivity; that is, an electron in the valence band can move into a vacant

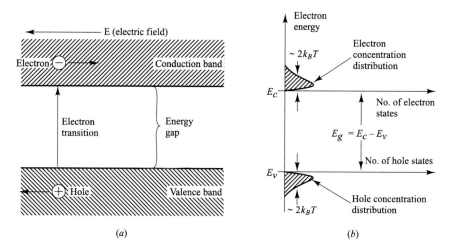

(a)   (b)

Fig. 4.1 a Energy level diagrams showing the excitation of an electron from the valence band to the conduction band; **b** equal electron and hole concentrations in an intrinsic semiconductor created by the thermal excitation of electrons across the bandgap

hole. This action makes the hole move in the opposite direction to the electron flow, as is shown in Fig. 4.1a.

When an electron propagates in a semiconductor, it interacts with the periodically arranged constituent atoms of the material and thus experiences *external forces*. As a result, to describe its acceleration a_{crys} in a semiconductor crystal under an external force F_{ext} its mass needs to be described by a quantum mechanical quantity m_e called the *effective mass*. That is, when using the relationship $F_{ext} = m_e a_{crys}$ (force equals mass times acceleration), the effects of all the forces exerted on the electron within the material are incorporated into m_e.

The concentration of electrons and holes is known as the *intrinsic carrier concentration* n_i, and for a perfect material with no imperfections or impurities it is given by

$$n = p = n_i = K \exp\left(-\frac{E_g}{2k_B T}\right) \tag{4.1}$$

where

$$K = 2\left(2\pi k_B T / h^2\right)^{3/2} (m_e m_h)^{3/4}$$

is a constant that is characteristic of the material. Here, T is the temperature in degrees Kelvin, k_B is Boltzmann's constant, h is Planck's constant, and m_e and m_h are the effective masses of the electrons and holes, respectively, which can be smaller by a factor of 10 or more than the free-space electron rest mass of $m = 9.11 \times 10^{-31}$ kg.

Example 4.1 Consider the following parameter values for GaAs at 300 K:

Electron rest mass $m = 9.11 \times 10^{-31}$ kg.

Effective electron mass $m_e = 0.068\, m = 6.19 \times 10^{-32}$ kg.

Effective hole mass $m_h = 0.56\, m = 5.10 \times 10^{-31}$ kg.

Bandgap energy $E_g = 1.42$ eV.

What is the intrinsic carrier concentration?

Solution First the bandgap energy must be changed to units of joules:

$$E_g = 1.42\ \text{eV} \times 1.60 \times 10 - 19\,\text{J/eV}$$

Then from Eq. (4.1) the intrinsic carrier concentration is

$$n_i = 2\left[\frac{2\pi\left(1.381 \times 10^{-23}\right)300}{\left(6.626 \times 10^{-34}\right)^2}\right]^{3/2} \left[\left(6.19 \times 10^{-32}\right)\left(5.10 \times 10^{-31}\right)\right]^{3/4}$$

$$\exp\left[\frac{1.42 \times 1.6 \times 10^{-19}}{2(1.381 \times 10^{-23})300}\right]$$
$$= 2.62 \times 10^{12} \mathrm{m}^{-3} = 2.62 \times 10^{6}\,\mathrm{cm}^{-3}$$

Drill Problem 4.1 The effective masses for Si are $m_e = 1.09$ m and $m_h = 0.56$ m for electrons and holes, respectively, where m is the electron rest mass given in Example 4.1. Using Eq. (4.1) show that the intrinsic carrier concentration is $n = p = n_i = 1.00 \times 10^{10}\,\mathrm{cm}^{-3}$. To get an accurate value of the ratio $E_g/2k_B T$ for the exponential factor in Eq. (4.1), use the values $E_g = 1.100$ eV and $2k_B T = 0.02586$ eV at $T = 300\,°\mathrm{K}$.

The conduction in Si can be greatly increased by adding traces of impurities from the group V elements (e.g., P, As, Sb). This process is called *doping*, and the doped semiconductor is called an *extrinsic material*. These doping elements have five electrons in the outer shell. When they replace a Si atom, four electrons are used for covalent bonding, and the fifth, loosely bound electron is available for conduction. As shown in Fig. 4.2a, this gives rise to an occupied level E_D, just below the conduction band in the bandgap, called the *donor level*. The impurities are called *donors* because they can give up (donate) an electron to the conduction band. This is reflected by the increase in the free-electron concentration in the conduction band, as shown in Fig. 4.2b. This type of material is called *n-type* material because the current is due to (negative) electrons.

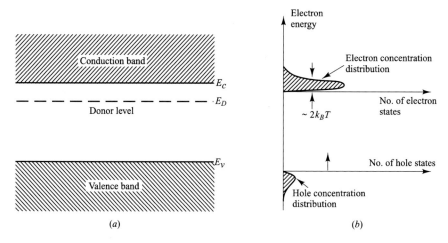

Fig. 4.2 a Donor level in an n-type material; **b** the ionization of donor impurities increases the electron concentration distribution in the conduction band

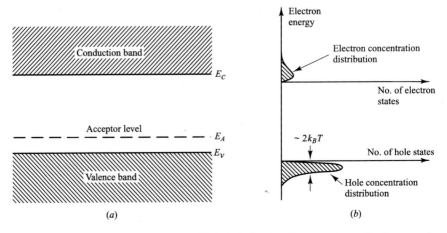

Fig. 4.3   a Acceptor level in a p-type material; **b** the ionization of acceptor impurities increases the hole concentration distribution in the valence band

The conduction also increases by adding group III elements (e.g., Al, Ga, In), which have three electrons in the outer shell. In this case, three electrons make covalent bonds, and a hole with properties identical to that of the donor electron is created. As shown in Fig. 4.3a, this gives rise to an unoccupied level E_A just above the valence band. Conduction occurs when electrons are excited from the valence band to this *acceptor level* (so called because the impurity atoms have accepted electrons from the valence band). Correspondingly, the free-hole concentration increases in the valence band, as shown in Fig. 4.3b. This is called *p*-type material because the conduction is a result of (positive) hole flow.

Drill Problem 4.2   The probability $f(E)$ that an electron occupies a given state at an allowed energy level E is

$$f(E) = \frac{1}{1 + \exp\left[(E - E_f)/k_B T\right]}$$

Here E_f is a reference energy called the *Fermi energy* or *Fermi level*. Consider the case of Si, which has a bandgap energy $E_g = 1.10$ eV at $T = 300°K$. Suppose the Fermi level is at the middle of the energy gap so that $E_c - E_f = E_g/2 = 0.55$ eV for Si. Show that the probability that an electron occupies a state at the bottom of the conduction band (i.e., at $E = E_c$) is $f(E_c) = 5.80 \times 10^{-10}$ at room temperature where $k_B T = 0.02586$ eV. This shows that about one state in two billion is occupied at the bottom of the conduction band.

4.1.2 *Intrinsic and Extrinsic Materials*

A perfect material containing no impurities is called an *intrinsic material*. Because of thermal vibrations of the crystal atoms, some electrons in the valence band gain enough energy to be excited to the conduction band. This *thermal generation process* produces free electron−hole pairs because every electron that moves to the conduction band leaves behind a free hole. Thus for an intrinsic material the number of electrons and holes are both equal to the intrinsic carrier density, as denoted by Eq. (4.1). In the opposite *recombination process*, a free electron in the conduction band releases its energy and drops into a free hole in the valence band. For an extrinsic semiconductor, the increase of one type of carrier reduces the number of the other type. In this case, the product of the two types of carriers remains constant at a given temperature. This gives rise to the *mass-action law*

$$pn = n_i^2 \qquad (4.2)$$

which is valid for both intrinsic and extrinsic materials under thermal equilibrium.

Because the electrical conductivity is proportional to the carrier concentration, two types of charge carriers are defined for this material:

1. *Majority carriers* refer either to electrons in *n*-type material or to holes in *p*-type material;
2. *Minority carriers* refer either to holes in *n*-type material or to electrons in *p*-type material.

The operation of semiconductor devices is essentially based on the *injection* and *extraction* of minority carriers.

Example 4.2 Consider an *n*-type semiconductor that has been doped with a net concentration of N_D donor impurities. Let n_N and p_N be the electron and hole concentrations, respectively, where the subscript N is used to denote n-type semiconductor characteristics. In this case, holes are created exclusively by thermal ionization of intrinsic atoms. This process generates equal concentrations of electrons and holes, so that the concentration of holes in an *n*-type semiconductor is

$$p_N = p_i = n_i$$

Because both impurity and intrinsic atoms generate conduction electrons, the total concentration of conduction electrons n_N is

$$n_N = N_D + n_i = N_D + p_N$$

Substituting Eq. (4.2) for p_N (which states that, in equilibrium, the product of the electron and hole concentrations equals the square of the intrinsic carrier density, so that $p_N = n_i^2/n_N$), it follows that

$$n_N = \frac{N_D}{2}\left(\sqrt{1 + \frac{4n_i^2}{N_D^2}} + 1\right)$$

If $n_i \ll n_D$, which is generally the case, then to a good approximation

$$n_N = N_D \text{ and } p_N = n_i^2/n_D$$

4.1.3   Concept of a pn Junction

Doped n-type or p-type semiconductor material by itself serves only as a conductor. To make devices out of these semiconductors, it is necessary to use both types of materials (in a single, continuous crystal structure). The junction between the two material regions, which is known as the *pn junction,* is responsible for the useful electrical characteristics of a semiconductor device.

When a *pn* junction is created, the majority carriers diffuse across it. This causes electrons to fill holes in the p side of the junction and causes holes to appear on the n side. As a result, an electric field $E(x)$ appears across the junction, as is shown in Fig. 4.4. The variation of this electric field creates a potential $V(x)$ at any point x across the *pn* junction. The potential can be found by integrating the electric field,

Fig. 4.4 Electron diffusion across a pn junction creates a barrier potential (electric field) in the depletion region

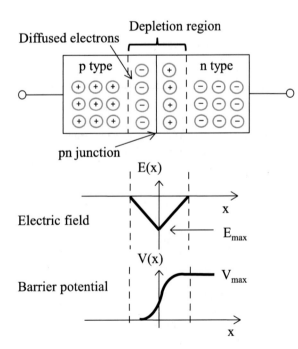

because by definition $E = dV/dx$. The resulting potential across the *pn* junction is the parabolic function shown in Fig. 4.4. The magnitude of the potential varies from 0 to a maximum value V_{max}, which is called the *barrier potential* or the *built-in potential*. This potential prevents further net movements of charges once equilibrium has been established. The junction area now has no mobile carriers because its electrons and holes are locked into a covalent bond structure. This region is called either the *depletion region* or the *space charge region*.

When an external battery is connected to the *pn* junction with its positive terminal to the *n*-type material and its negative terminal to the *p*-type material, the junction is said to be *reverse-biased*. This is shown in Fig. 4.5. As a result of the reverse bias, the width of the depletion region will increase on both the *n* side and the *p* side. This effectively increases the barrier potential and prevents any majority carriers from flowing across the junction. However, minority carriers can move with the field across the junction. The minority carrier flow is small at normal temperatures and operating voltages, but it can be significant when excess carriers are created as, for example, in an illuminated photodiode.

When the *pn* junction is *forward-biased*, as shown in Fig. 4.6, the magnitude of

Fig. 4.5 A reverse bias widens the depletion region but allows minority carriers to move freely with the applied field

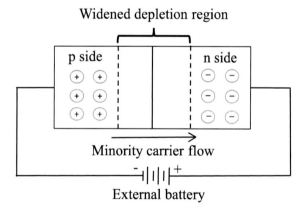

Fig. 4.6 Lowering the barrier potential with a forward bias allows majority carriers to diffuse across the junction

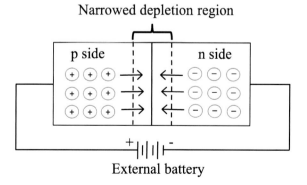

the barrier potential is reduced. Thereby conduction-band electrons on the *n* side and valence-band holes on the *p* side are allowed to diffuse across the junction. Once across, they significantly increase the minority carrier concentrations, and the excess carriers then recombine with the oppositely charged majority carriers. The *recombination of excess minority carriers* is the mechanism by which optical radiation is generated.

4.1.4 Direct Bandgap and Indirect Bandgap

As Sect. 2.1.4 and Eq. (4.3) note, a photon has an energy $E = h\nu = hc/\lambda$, where ν and λ are the frequency and wavelength, respectively, that are associated with the photon. When electron transitions occur between the valence and conduction bands, an electron in the valence band can absorb the energy from an impinging photon and thereby get boosted to the conduction band. In the case when an electron drops from the conduction band and combines with a hole in the valence band, a photon gets emitted in this process as Fig. 4.7 illustrates. In order for electron transitions to take place to or from the conduction band, both energy and momentum must be conserved. Although a photon can have considerable energy, its momentum $h\nu/c$ is very small.

Semiconductors are classified as either *direct-bandgap* or *indirect-bandgap* materials depending on the shape of the bandgap as a function of the momentum k, as shown in Fig. 4.7. Now consider the recombination of an electron and a hole, accompanied by the emission of a photon. The simplest and most probable recombination process will be that where the electron and hole have the same momentum value (see Fig. 4.7a). This is a direct-bandgap material.

For indirect-bandgap materials, the conduction band minimum and the valence band maximum energy levels occur at different values of momentum, as shown in Fig. 4.7b. Here, band-to-band recombination must involve a third particle to conserve momentum because the photon momentum is very small. *Phonons* (i.e., crystal lattice vibrations) serve this purpose.

4.1.5 Fabrication of Semiconductor Devices

In fabricating semiconductor devices, the crystal structure of the various material regions must be carefully taken into account. In any crystal structure, single atoms (e.g., Si or Ge) or groups of atoms (e.g., NaCl or GaAs) are arranged in a repeated pattern in space. This periodic arrangement defines a *lattice,* and the spacing between the atoms or groups of atoms is called the *lattice spacing* or the *lattice constant.* Typical lattice spacings are a few angstroms (1 angstrom $= 0.1$ nm).

Semiconductor devices generally are fabricated by starting with a crystalline substrate that provides mechanical strength for mounting the device and for making

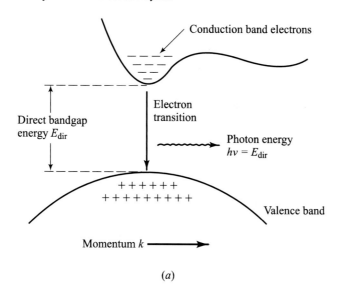

(a)

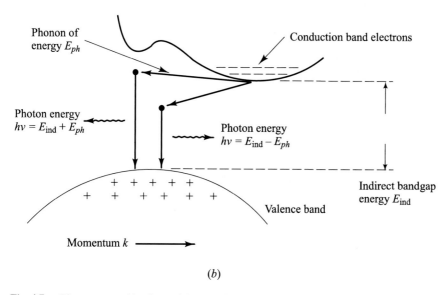

(b)

Fig. 4.7 a Electron recombination and the associated photon emission for a direct-bandgap material; **b** electron recombination for indirect-bandgap materials requires a phonon of energy E_{ph} and momentum k_{ph}

electric contacts. A technique of crystal growth by chemical reaction is then used to grow thin layers of semiconductor materials on the substrate. These materials must have lattice structures that are identical to those of the substrate crystal. In particular, the lattice spacings of adjacent materials should be closely matched to avoid temperature-induced stresses and strains at the material interfaces. This type of growth is called *epitaxial,* which is derived from the Greek words *epi* meaning *on* and *taxis* meaning *arrangement*; that is, it is an arrangement of atoms from one material on another material. An important characteristic of epitaxial growth is that it is relatively simple to change the impurity concentration of successive material layers, so that a layered semiconductor device can be fabricated in a continuous process. Epitaxial layers can be formed by vapor phase, liquid phase, or molecular beam growth techniques [4–6].

4.2 Principles of Light-Emitting Diodes (LEDs)

For optical communication systems requiring bit rates less than approximately 100–200 Mb/s together with multimode fiber-coupled optical power in the tens of microwatts, semiconductor light-emitting diodes (LEDs) are usually the appropriate light source choice. These LEDs require less complex drive circuitry than laser diodes as no thermal or optical stabilization circuits are needed (see Sect. 4.3.6), and they can be fabricated less expensively with higher yields.

4.2.1 LED Structures

To be useful in fiber transmission applications, an LED must have a high radiance output, a fast emission response time, and high quantum efficiency. Its *radiance* is a measure, in watts, of the optical power radiated into a unit solid angle per unit area of the emitting surface. High radiances are necessary to couple sufficiently high optical power levels into a fiber, as shown in detail in Chap. 5. The emission response time is the time delay between the application of a current pulse and the onset of optical emission. As is discussed in Sects. 4.2.4 and 4.3.7, this time delay is the factor limiting the bandwidth (i.e., the data rate) with which the source can be modulated directly by varying the injected current. The quantum efficiency is related to the fraction of injected electron–hole pairs that recombine radiatively. This is defined and described in detail in Sect. 4.2.3.

To achieve a high radiance and high quantum efficiency, the LED structure must provide a means of confining the charge carriers and the stimulated optical emission to the active region of the *pn* junction where radiative recombination takes place. *Carrier confinement* is used to achieve a high level of radiative recombination in the active region of the device, which yields a high quantum efficiency. *Optical confinement* is of importance for preventing absorption of the emitted radiation by the material surrounding the *pn* junction.

To achieve carrier and optical confinement, LED configurations such as homo-junctions and single and double heterojunctions have been widely investigated [7, 8]. The most effective of these structures is the configuration shown in Fig. 4.8. This is referred to as a *double-heterostructure* (or *heterojunction*) device because of the two different semiconductor alloy layers on each side of the active region. This configuration evolved from studies on laser diodes. By means of this sandwich structure of differently composed alloy layers, both the carriers and the optical field are confined

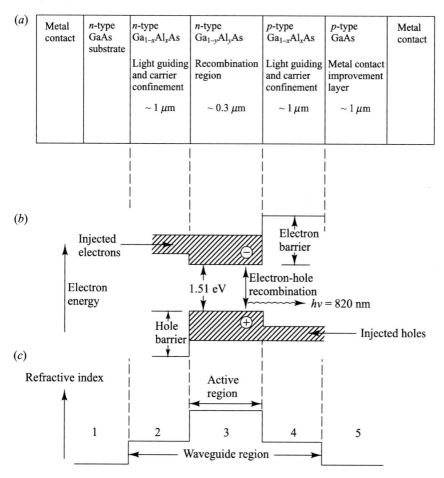

Fig. 4.8 a Cross-sectional drawing (not to scale) of a typical GaAlAs double-heterostructure light emitter in which x > y to provide for both carrier confinement and optical guiding; **b** energy band diagram showing the active region, and the electron and hole barriers that confine the charge carriers to the active layer; **c** variations in the refractive index; the lower index of refraction of the material in regions 1 and 5 creates an optical barrier around the waveguide region

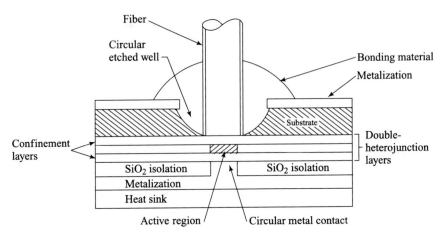

Fig. 4.9 Schematic (not to scale) of a high-radiance surface-emitting LED where the active region is limited to a circular section having an area compatible with the fiber-core end face

in the central active layer. The bandgap differences of adjacent layers confine the charge carriers (Fig. 4.8b), and the differences in the indices of refraction of adjoining layers confine the optical field to the central active layer (Fig. 4.8c). This dual confinement leads to both high efficiency and high radiance. Other parameters influencing the device performance include optical absorption in the active region (self-absorption), carrier recombination at the heterostructure interfaces, doping concentration of the active layer, injection carrier density, and active-layer thickness. The effects of these parameters are discussed in the following sections.

The two basic LED configurations being used for fiber optics are *surface emitters* (also called *Burrus* or *front emitters*) and *edge emitters*. In the surface emitter shown in Fig. 4.9, the plane of the active light-emitting region is oriented perpendicularly to the axis of the fiber [9]. In this configuration, a well is etched through the substrate of the device, into which a fiber is then cemented in order to accept the emitted light. The circular active area in practical surface emitters is nominally 50 μm in diameter and up to 2.5 μm thick. The emission pattern is essentially isotropic with a 120° half-power beam width. This device is useful for coupling light into multimode fibers.

This isotropic pattern from such a surface emitter is called a *Lambertian pattern.* In this pattern (see Fig. 5.2), the source has the same apparent brightness (or luminance) when viewed from any direction, but the power diminishes as cosθ, where θ is the angle between the viewing direction and the normal to the surface (this is because the projected area one sees decreases as cosθ). Thus, the power is down to 50% of its peak when θ = 60°, so that the total half-power beam width is 120°.

The edge emitter depicted in Fig. 4.10 consists of an active junction region, which is the source of the incoherent light, and two guiding layers. The guiding layers both have a refractive index lower than that of the active region but higher than the index of the surrounding material. This structure forms a waveguide channel that directs

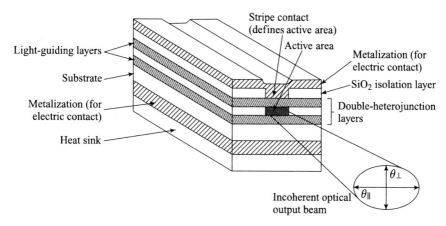

Fig. 4.10 Schematic (not to scale) of an edge-emitting double-heterojunction LED in which the output beam is Lambertian in the plane of the pn junction ($\theta_{||} = 120°$) and highly directional perpendicular to the pn junction ($\theta_\perp \approx 30°$)

the optical radiation toward the fiber core. To match the typical multimode fiber core diameters (50–100 μm), the contact stripes for the edge emitter are 50–70 μm wide. Lengths of the active regions usually range from 100 to 150 μm. The emission pattern of the edge emitter is more directional than that of the surface emitter, as is illustrated in Fig. 4.10. In the plane parallel to the junction, where there is no waveguide effect, the emitted beam is Lambertian (varying as cosθ) with a half-power width of $\theta_{||} = 120°$. In the plane perpendicular to the junction, the half-power beam $\theta_\perp$ has been made as small as 25°–35° by a proper choice of the waveguide thickness.

4.2.2 Semiconductor Materials for Light Sources

The semiconductor material that is used for the active layer of an optical source must have a direct bandgap. In a direct-bandgap semiconductor, electrons and holes can recombine directly across the bandgap without needing a third particle to conserve momentum, as shown in Fig. 4.7a. Only in direct-bandgap material is the radiative recombination sufficiently high to produce an adequate level of optical emission. Although none of the normal single-element semiconductors are direct-bandgap materials, many binary compounds are. The most important of these compounds are made from III-V materials. That is, the compounds consist of selections from a group III element (e.g., Al, Ga, or In) and a group V element (e.g., P, As, or Sb). Various ternary and quaternary combinations of binary compounds of these elements are also direct-gap materials and are suitable candidates for optical sources.

For operation in the spectrum ranging from 800 to 900 nm, the principal material used is the ternary alloy $Ga_{1-x}Al_xAs$. The ratio x of aluminum arsenide to gallium arsenide determines the bandgap of the alloy and, correspondingly, the wavelength

Fig. 4.11 Bandgap energy and output wavelength as a function of aluminum mole fraction x for $Al_xGa_{1-x}As$ at room temperature

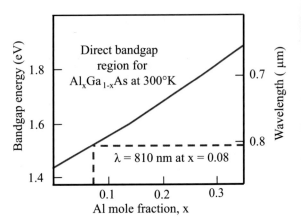

of the peak emitted radiation. This is illustrated in Fig. 4.11. For values of x greater than about 0.37, the bandgap changes from a direct to an indirect bandgap. The value of x for the active-area material is usually chosen to be around 0.1 for an emission wavelength of 800–850 nm. An example of the emission spectrum of a $Ga_{1-x}Al_xAs$ LED with $x = 0.08$ is shown in Fig. 4.12. The peak output power occurs at 810 nm. The width of the spectral pattern at its half-power point is known as the *full-width half-maximum* (FWHM) *spectral width*. For LEDs this parameter varies between 20

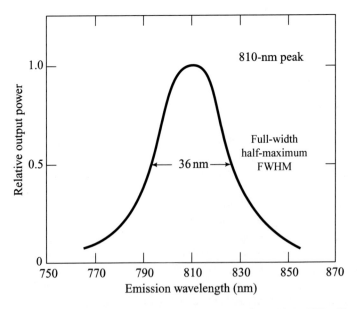

Fig. 4.12 Spectral emission pattern of a representative 810-nm $Ga_{1-x}Al_xAs$ LED with $x = 0.08$, which gives a spectral pattern width at its half-power point of 36 nm

and 50 nm depending on the wavelength. As an example, Fig. 4.12 shows a FWHM spectral width σ_λ of 36 nm for the 810-nm LED.

At longer wavelengths the quaternary alloy $In_{1-x}Ga_xAs_yP_{1-y}$ is one of the primary material candidates. By varying the mole fractions x and y in the active area, LEDs with peak output powers at any wavelength between 1.0 and 1.7 μm can be constructed. For simplicity, the notations GaAlAs and InGaAsP are generally used unless there is an explicit need to know the values of x and y. Other notations such as AlGaAs, (Al, Ga)As, (GaAl)As, GaInPAs, and $In_xGa_{1-x}As_yP_{1-y}$ are also found in the literature. From the last notation, it is obvious that, depending on the preference of the particular author, the values of x and $1 - x$ for the same material could be interchanged in different articles in the literature.

The alloys GaAlAs and InGaAsP are chosen to make semiconductor light sources because it is possible to match the lattice parameters of the heterostructure interfaces by using a proper combination of binary, ternary, and quaternary materials. A very close match between the crystal lattice parameters of the two adjoining heterojunctions is required to reduce interfacial defects and to minimize strains in the device as the temperature varies. These factors directly affect the radiative efficiency and lifetime of a light source. Using the fundamental quantum mechanical relationship between energy E and frequency ν, which is $E = h\nu = hc/\lambda$, the peak emission wavelength λ in micrometers can be expressed as a function of the bandgap energy E_g in electron volts by the equation

$$\lambda(\text{in } \mu m) = \frac{1.240}{E_g(\text{in eV})} \qquad (4.3)$$

One can determine the bandgap of a semiconductor by measuring the energy required to excite electrons from the valence band to the conduction band. Table 4.1 lists the bandgap energies of some common device materials used in various aspects of optical fiber communication applications.

A heterojunction with matching lattice parameters is created by choosing two material compositions that have the same lattice constant but different bandgap energies (the bandgap differences are used to confine the charge carriers). In the ternary alloy $Ga_{1-x}Al_xAs$ the bandgap energy E_g in electron volts for values of x between zero and 0.37 (the direct-bandgap region) can be found from the empirical equation

Table 4.1 Bandgap energies of some common semiconductor materials

Semiconductor material	Bandgap energy (eV)
Silicon (Si)	1.12
GaAs	1.43
Germanium (Ge)	0.67
InP	1.35
$Ga_{0.93}Al_{0.07}As$	1.51
$In_{0.74}Ga_{0.26}As_{0.57}P_{0.43}$	0.97

[1, 9]

$$E_g = 1.424 + 1.266x + 0.266x^2 \tag{4.4}$$

Given the value of E_g in electron volts, the peak emission wavelength in micrometers is found from Eq. (4.3).

Example 4.3 A particular $Ga_{1-x}Al_xAs$ laser is constructed with a material ratio $x = 0.07$. Find (a) the bandgap of this material; (b) the peak emission wavelength.

Solution (a) From Eq. (4.4), $E_g = 1.424 + 1.266(0.07) + 0.266(0.07)^2 = 1.51$ eV.
(b) Using this value of the bandgap energy in Eq. (4.3) yields.

$$\lambda(\mu m) = 1.240/1.51 = 0.82\,\mu m = 820\,nm$$

The bandgap energy and lattice-constant range for the quaternary alloy InGaAsP are much larger. In this case, the compositional parameters x and y follow the relationship $y = 2.20x$ with $0 \leq x \leq 0.47$. For $In_{1-x}Ga_xAs_yP_{1-y}$ compositions that are lattice-matched to InP, the bandgap in eV varies as

$$E_g = 1.35 - 0.72y + 0.12y^2 \tag{4.5}$$

Bandgap wavelengths from 0.92 to 1.65 μm are covered by this material system [8, 10].

Example 4.4 Consider the material alloy $In_{0.74}Ga_{0.26}As_{0.57}P_{0.43}$, that is, $x = 0.26$ and $y = 0.57$ in the general formula $In_{1-x}Ga_xAs_yP_{1-y}$. Find (a) the bandgap of this material; (b) the peak emission wavelength.

Solution (a) From Eq. (4.5), $E_g = 1.35 - 0.72(0.57) + 0.12(0.57)^2 = 0.97$ eV (b) Using this value of the bandgap energy in Eq. (4.3) yields (in micrometers)

$$\lambda(\mu m) = 1.240/0.97 = 1.27\,\mu m = 1270\,nm$$

Whereas the FWHM power spectral widths of LEDs in the 800-nm region are around 35 nm, this increases in longer-wavelength materials. For devices operating in the 1300-to-1600-nm region, the spectral widths vary from around 70 to 180 nm. Figure 4.13 shows examples for devices emitting at 1300 nm. In addition, as Fig. 4.13 shows, the output spectral widths of surface-emitting LEDs tend to be broader than those of edge-emitting LEDs because of different internal absorption effects of the emitted light in the two device structures.

Table 4.2 lists typical characteristics of surface-emitting LEDs (SLED) and edge-emitting LEDS (ELED). The materials used in these examples are GaAlAs for operation at 850 nm and InGaAsP for 1310-nm devices. The fiber-coupled power listing is the amount of light that can be accepted by a 50-μm core diameter multimode fiber.

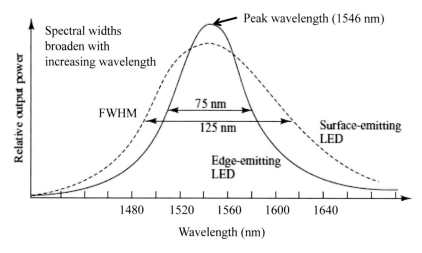

Fig. 4.13 Typical spectral patterns for edge-emitting and surface-emitting LEDs at 1310 nm. The patterns broaden with increasing wavelength and are wider for surface emitters

Table 4.2 Typical characteristics of surface-emitting and edge-emitting LEDs

LED type	Material	Wavelength (nm)	Operating current (mA)	Fiber-coupled power (μW)	Nominal FWHM (nm)
SLED	GaAlAs	850	110	40	35
ELED	InGaAsP	1310	100	15	80
SLED	InGaAsP	1310	110	30	150

Drill Problem 4.3 An important parameter of GaAs is the value of its refractive index as a function of wavelength. For the range $\lambda = 0.89$ to 4.1 μm the refractive index is given by

$$n^2 = 7.10 + \frac{3.78\lambda^2}{\lambda^2 - 0.2767}$$

where λ is given in micrometers. Compare the refractive indices of GaAs at 810 nm and 900 nm [**Answer:** $n = 3.69$ at 810 nm and 3.58 at 900 nm.]

4.2.3   LED Quantum Efficiency and Output Power

An excess of electrons and holes in p-type and n-type material, respectively (referred to as *minority carriers*) is created in a semiconductor light source by carrier injection at the device contacts. The excess densities of electrons n and holes p are equal, because the injected carriers are formed and recombine in pairs in accordance with the requirement for charge neutrality in the crystal. When carrier injection stops, the carrier density returns to the equilibrium value. In general, the excess carrier density decays exponentially with time according to the relation

$$n = n_0 e^{-t/\tau} \tag{4.6}$$

where n_0 is the initial injected excess electron density and the time constant τ is the carrier lifetime (the average time it takes for a minority carrier to recombine). This lifetime is one of the most important operating parameters of an electro-optic device. Its value can range from milliseconds to fractions of a nanosecond depending on material composition and device defects.

The excess carriers can recombine either radiatively or nonradiatively. In radiative recombination a photon of energy $h\nu$, which is approximately equal to the bandgap energy, is emitted. Nonradiative recombination effects include optical absorption in the active region (self-absorption), carrier recombination at the heterostructure interfaces, and the Auger process in which the energy released during an electron-hole recombination is transferred to another carrier in the form of kinetic energy.

When there is a constant current flow into an LED, an equilibrium condition is established. That is, the excess density of electrons n and holes p is equal because the injected carriers are created and recombined in pairs such that charge neutrality is maintained within the device. The total rate at which carriers are generated is the sum of the externally supplied rate and the thermally generated rate. The rate of externally supplied carriers is J/qd, where J is the current density in A/cm^2, q is the electron charge, and d is the thickness of the recombination region. The thermal generation rate is given by n/τ. Then the rate equation for carrier recombination in an LED can be written as

$$\frac{dn}{dt} = \frac{J}{qd} - \frac{n}{\tau} \tag{4.7}$$

The equilibrium condition is found by setting Eq. (4.7) equal to zero, yielding

$$n = \frac{J\tau}{qd} \tag{4.8}$$

This relationship gives the steady-state electron density in the active region when a constant current is flowing through it.

The *internal quantum efficiency* in the active region is the fraction of the electron-hole pairs that recombine radiatively. If the radiative recombination rate is R_r and the nonradiative recombination rate is R_{nr}, then the internal quantum efficiency η_{int} is the ratio of the radiative recombination rate to the total recombination rate:

$$\eta_{int} = \frac{R_r}{R_r + R_{nr}} \tag{4.9}$$

For exponential decay of excess carriers, the radiative recombination lifetime is $\tau_r = n/R_r$ and the nonradiative recombination lifetime is $\tau_{nr} = n/R_{nr}$. Thus the internal quantum efficiency can be expressed as

$$\eta_{int} = \frac{1}{1 + \frac{\tau_r}{\tau_{nr}}} = \frac{\tau}{\tau_r} \tag{4.10}$$

where the *bulk recombination lifetime* τ is

$$\frac{1}{\tau} = \frac{1}{\tau_r} + \frac{1}{\tau_{nr}} \tag{4.11}$$

In general, τ_r and τ_{nr} are comparable for direct-bandgap semiconductors, such as GaAlAs and InGaAsP. This also means that R_r and R_{nr} are similar in magnitude, so that the internal quantum efficiency is about 50 percent for simple homojunction LEDs. However, LEDs having double-heterojunction structures can have quantum efficiencies of 60–80%. This high efficiency is achieved because the thin active regions of these devices mitigate the self-absorption effects, which reduces the nonradiative recombination rate.

If the current injected into the LED is I, then the total number of recombinations per second is

$$R_r + R_{nr} = I/q \tag{4.12}$$

Substituting Eq. (4.12) into Eq. (4.9) then yields $R_r = \eta_{int} I/q$. Noting that R_r is the total number of photons generated per second and that each photon has an energy $h\nu$, then the optical power generated internally to the LED is

$$P_{int} = \eta_{int} \frac{I}{q} h\nu = \eta_{int} \frac{hcI}{q\lambda} \tag{4.13}$$

Example 4.5 Suppose that a double-heterojunction InGaAsP LED emitting at a peak wavelength of 1310 nm has radiative and nonradiative recombination times of 30 and 100 ns, respectively, and let the drive current be 40 mA. Find (*a*) the bulk recombination time; (*b*) the internal quantum efficiency; and (*c*) the internal power level.

Solution (*a*) From Eq. (4.11), the bulk recombination lifetime is

$$\tau = \frac{\tau_r \tau_{nr}}{\tau_r + \tau_{nr}} = \frac{30 \times 300}{30 + 300}\, \text{ns} = 23.1\, \text{ns}$$

(b) Using Eq. (4.10), the internal quantum efficiency is

$$\eta_{int} = \frac{\tau}{\tau_r} = \frac{23.1}{30} = 0.77$$

(c) Substituting this into Eq. (4.13) yields an internal power level of

$$P_{int} = \eta_{int}\frac{hcI}{q\lambda} = 0.77\frac{(6.6256 \times 10^{-34}\, \text{J s})(3 \times 10^8\, \text{m/s})(0.040\, \text{A})}{(1.602 \times 10^{-19}\, \text{C})(1.31 \times 10^{-6}\, \text{m})}$$
$$= 29.2\, \text{mW}$$

Not all internally generated photons will exit the device. To find the emitted power, one needs to consider the *external quantum efficiency* η_{ext}. This is defined as the ratio of the photons emitted from the LED to the number of internally generated photons. To find the external quantum efficiency, one needs to take into account reflection effects at the surface of the LED. As shown in Fig. 4.14 and described in Sect. 2.2, at the interface of a material boundary only that fraction of light falling within a cone defined by the critical angle φ_c will cross the interface. Recall from Eq. (2.18) that $\varphi_c = \sin^{-1}(n_2/n_1)$. Here, n_1 is the refractive index of the semiconductor material and n_2 is the refractive index of the outside material, which nominally is air with $n_2 = 1.0$. The external quantum efficiency can then be calculated from the expression

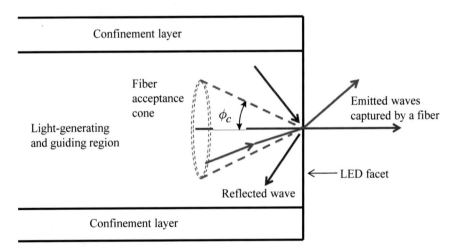

Fig. 4.14 Only light emitted from an optical source that falls within an acceptance cone defined by the critical angle φ_c will be captured by the fiber

$$\eta_{ext} = \frac{1}{4\pi} \int\limits_0^{\phi_c} T(\phi)(2\pi \sin\phi)d\phi \tag{4.14}$$

where $T(\varphi)$ is the *Fresnel transmission coefficient* or *Fresnel transmissivity*. This factor depends on the incidence angle φ, but, for simplicity, one can use the expression for normal incidence, which is (see Sect. 2.2.2)

$$T(0) = \frac{4n_1 n_2}{(n_1 + n_2)^2} \tag{4.15}$$

Assuming the outside medium is air and letting $n_1 = n$, then $T(0) = 4n/(n + 1)^2$. The external quantum efficiency is then approximately given by

$$\eta_{ext} = \frac{1}{n(n + 1)^2} \tag{4.16}$$

From this, it follows that the optical power emitted from the LED is

$$P = \eta_{ext} P_{int} = \frac{P_{int}}{n(n + 1)^2} \tag{4.17}$$

Example 4.6 Assume a typical value of $n = 3.5$ for the refractive index of an LED material. What percent of the internally generated optical power is emitted into an air medium?

Solution Taking the condition for normal incidence, then from Eq. (4.16) the percent of the optical power that is generated internally in the device that is emitted into an air medium is

$$\eta_{ext} = \frac{1}{n(n + 1)^2} = \frac{1}{3.5(3.5 + 1)^2} = 1.41\%$$

This shows that only a small fraction of the internally generated optical power is emitted from the device.

Drill Problem 4.4 (*a*) Verify that 32% of the photons generated inside a GaAs device are reflected when the emitted light is incident normally from the GaAs at an interface with air. The refractive index of air is 1.00 and let that of GaAs be 3.58. (*b*) Show that the reflected fraction of photons changes to 17% when the external material interface is a glass fiber with an index of 1.48.

4.2.4   Response Time of an LED

The *response time* or *frequency response* of an optical source dictates how fast an electrical input drive signal can vary the light output level. The following three factors largely determine the response time: the doping level in the active region, the injected carrier lifetime τ_i in the recombination region, and the parasitic capacitance of the LED. If the drive current is modulated at a frequency ω, the optical output power of the device will vary as

$$P(\omega) = P_0\left[1 + (\omega\tau_i)^2\right]^{-1/2} \tag{4.18}$$

where P_0 is the power emitted at zero modulation frequency. The parasitic capacitance can cause a delay of the carrier injection into the active junction, and, consequently, could delay the optical output [11]. This delay is negligible if a small, constant forward bias is applied to the diode. Under this condition, Eq. (4.18) is valid and the modulation response is limited only by the carrier recombination time.

Example 4.7   A particular LED has a 5-ns injected carrier lifetime. When no modulation current is applied to the device, the optical output power is 0.250 mW for a specified dc bias. Assuming parasitic capacitances are negligible, what are the optical outputs at modulation frequencies f of (a) 10 MHz and (b) 100 MHz? Note: $\omega = 2\pi f$.

Solution   (a) From Eq. (4.18) the optical output at 10 MHz is

$$P(\omega) = \frac{P_0}{\sqrt{1 + (\omega\tau_i)^2}} = \frac{0.250}{\sqrt{1 + \left[2\pi\left(10 \times 10^6\right)\left(5 \times 10^{-9}\right)\right]^2}}$$
$$= 0.239\,\text{mW} = 239\,\mu\text{W}$$

(b) Similarly, the optical output at 100 MHz is

$$P(\omega) = \frac{P_0}{\sqrt{1 + (\omega\tau_i)^2}} = \frac{0.250}{\sqrt{1 + \left[2\pi\left(100 \times 10^6\right)\left(5 \times 10^{-9}\right)\right]^2}}$$
$$= 0.076\,\text{mW} = 76\,\mu\text{W}$$

Thus the output of this particular device decreases at higher modulation rates.

The modulation bandwidth of an LED can be defined in either electrical or optical terms. Normally, electrical terms are used because the bandwidth is actually determined via the associated electrical circuitry. Thus the modulation bandwidth is defined as the point where the electrical signal power, designated by $p(\omega)$, has dropped to half its constant value resulting from the modulated portion of the optical signal. This is the electrical 3-dB point; that is, the frequency at which the output

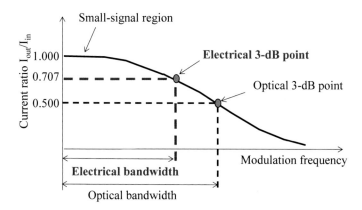

Fig. 4.15 Frequency response of an optical source showing the electrical and optical 3-dB bandwidth points

electrical power is reduced by 3 dB with respect to the input electrical power, as is illustrated in Fig. 4.15.

An optical source exhibits a linear relationship between light power and current, so currents rather than voltages (which are used in electrical systems) are compared in optical systems. Thus, because $p(\omega) = I^2(\omega)/R$, then the ratio of the output electrical power at the frequency ω to the power at zero modulation is

$$\text{Ratio}_{elec} = 10\log\left[\frac{p(\omega)}{p(0)}\right] = 10\log\left[\frac{I^2(\omega)}{I^2(0)}\right] \tag{4.19}$$

where $I(\omega)$ is the electrical current in the detection circuitry and R is the resistance. The electrical 3-dB point occurs at that frequency point where the detected electrical power $p(\omega) = p(0)/2$. This happens when

$$\frac{I^2(\omega)}{I^2(0)} = \frac{1}{2} \tag{4.20}$$

or $I(\omega)/I(0) = 1/\sqrt{2} = 0.707$.

Example 4.8 Consider the particular LED described in Example 4.7, which has a 5-ns injected carrier lifetime. (*a*) What is the 3-dB optical bandwidth of this device? (*b*) What is the 3-dB electrical bandwidth of this device?

Solution (*a*) The 3-dB optical bandwidth occurs at the modulation frequency for which $P(\omega) = 0.5P_0$. Using Eq. (4.18) yields

$$\frac{1}{\sqrt{1 + (\omega\tau_i)^2}} = \frac{1}{2}$$

so that $1 + (\omega \tau_i)^2 = 4$, or $\omega \tau_i = \sqrt{3}$. Solving this expression for the frequency $\omega = 2\pi f$ yields

$$f = \frac{\sqrt{3}}{2\pi \tau_i} = \frac{\sqrt{3}}{2\pi \times 5 \times 10^{-9}} = 55.1 \, \text{MHz}$$

(b) The 3-dB electrical bandwidth is $f/\sqrt{2} = 0.707\,(55.1 \, \text{MHz}) = 39.0 \, \text{MHz}$.

Sometimes, the modulation bandwidth of an LED is given in terms of the 3-dB bandwidth of the modulated optical power $P(\omega)$; that is, it is specified at the frequency where $P(\omega) = P_0/2$. In this case, the 3-dB bandwidth is determined from the ratio of the optical power at frequency ω to the unmodulated value of the optical power P_0. Because the detected current is directly proportional to the optical power, this ratio is

$$\text{Ratio}_{optical} = 10 \log \left[\frac{P(\omega)}{P(0)} \right] = 10 \log \left[\frac{I(\omega)}{I(0)} \right] \tag{4.21}$$

The optical 3-dB point occurs at that frequency where the ratio of the currents is equal to 1/2. As shown in Fig. 4.15, this gives an inflated value of the modulation bandwidth, which corresponds to an electrical power attenuation of 6 dB.

Drill Problem 4.5 A GaAlAs LED with an active region width of 1.0 μm operates at 300°K at a current density level of $J = 100 \, \text{A/cm}^2$. Assuming the steady-state electron density at this current density is $n = 6 \times 10^{16} \, \text{cm}^{-3}$, first use Eq. (4.8) to calculate that the carrier lifetime τ is 9.6 ns. With this value of τ, use the expression given in Example 4.8 to show that the 3-dB cutoff frequency is 28.7 MHz.

4.3   Principles of Laser Diodes

Lasers come in many forms with dimensions ranging from the size of a grain of salt to one that will occupy an entire room. The lasing medium can be a gas, a liquid, an insulating crystal (solid state), or a semiconductor. For optical fiber systems the laser sources used almost exclusively are semiconductor laser diodes. They are similar to other lasers, such as the conventional solid-state and gas lasers, in that the emitted radiation has spatial and temporal coherence; that is, the output radiation is highly monochromatic and the light beam is very directional.

Despite their physical and material differences, the basic principle of operation is the same for each type of laser. Laser action is the result of three key processes:

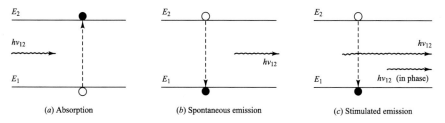

Fig. 4.16 The three key transition processes involved in laser action with the open circle representing the initial state of the electron and the heavy dot representing the final state; incident photons are shown on the left of each diagram and emitted photons are shown on the right

photon absorption, spontaneous emission, and stimulated emission. The simple two-energy-level diagrams in Fig. 4.16 represent these three processes, where E_1 is the ground-state energy and E_2 is the energy at an excited state. According to Planck's law, a transition between these two states involves the absorption or emission of a photon of energy $h\nu_{12} = E_2 - E_1$. Normally, the system is in the ground state. When a photon of energy $h\nu_{12}$ impinges on the system, an electron in state E_1 can absorb the photon energy and be excited to state E_2, as shown in Fig. 4.16a. Because this is an unstable state, the electron will shortly return to the ground state as shown in Fig. 4.16b, thereby emitting a photon of energy $h\nu_{12}$. This occurs without any external stimulation and is called *spontaneous emission*. These emissions are isotropic and of random phase, and thus appear as a narrowband Gaussian output.

The electron can also be induced to make a downward transition from the excited level to the ground-state level by an external stimulation. As shown in Fig. 4.16c, if a photon of energy $h\nu_{12}$ impinges on the system while the electron is still in its excited state, the electron is immediately stimulated to drop to the ground state and give off a photon of energy $h\nu_{12}$. This emitted photon is in phase with the incident photon, and the resultant emission is known as *stimulated emission*.

In thermal equilibrium the density of excited electrons is very small. Most photons incident on the system will therefore be absorbed, so that stimulated emission is essentially negligible. Stimulated emission will exceed absorption only if the population of the excited states is greater than that of the ground state. This condition is known as *population inversion*. Because this is not an equilibrium condition, population inversion is achieved by various externally induced excitation processes, which are known as "pumping" techniques. In a semiconductor laser, population inversion is accomplished by injecting electrons into the material at the device contacts or through an optical absorption method by means of externally injected photons.

4.3.1 Modes and Threshold Conditions in Laser Diodes

For optical fiber communication systems requiring bandwidths greater than approximately 200 MHz, the semiconductor injection laser diode is preferred over the

LED. Laser diodes typically have response times less than 1 ns, can have spectral widths of 1 nm or less, and are capable of coupling tens of milliwatts of useful luminescent power into optical fibers with small cores and small mode-field diameters. The majority of laser diodes in use are multilayered heterojunction devices. As mentioned in Sect. 4.2, the double-heterojunction LED configuration evolved from the successful demonstration of both carrier and optical confinement in heterojunction injection laser diodes. The more rapid evolvement and utilization of LEDs as compared with laser diodes lies in the inherently simpler construction, the smaller temperature dependence of the emitted optical power, and the absence of catastrophic degradation in LEDs. The construction of laser diodes is more complicated, mainly because of the additional requirement of current confinement in a small lasing cavity.

Stimulated emission in semiconductor lasers arises from optical transitions between distributions of energy states in the valence and conduction bands. This differs from gas and solid-state lasers, in which radiative transitions occur between discrete isolated atomic or molecular levels. The radiation in one type of laser diode configuration is generated within a Fabry-Perot resonator cavity [5, 6], shown in Fig. 4.17, as in most other types of lasers. Here the cavity is approximately 250–500 μm long, 5–15 μm wide, and 0.1–0.2 μm thick. These dimensions commonly are referred to as the *longitudinal, lateral,* and *transverse dimensions* of the cavity, respectively. Note from Fig. 4.17 that the light beam emerging from the laser forms a vertical ellipse, even though the lasing spot at the active area facet is a horizontal ellipse. In the lateral dimension the emitted beam has a half-power width of $\theta_{||} \approx 5\text{--}10°$. In the transverse dimension the emitted beam has a half-power width of $\theta_{\perp} \approx 30\text{--}50°$.

As illustrated in Fig. 4.18a, two flat, partially reflecting mirrors are directed toward each other to enclose the Fabry-Perot resonator cavity. The mirror facets are constructed by making two parallel clefts along natural cleavage planes of the

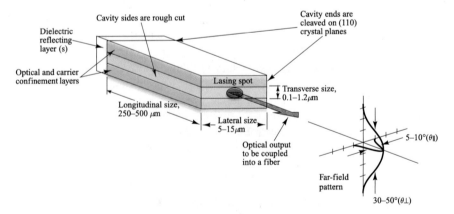

Fig. 4.17 Fabry-Perot resonator cavity for a laser diode where the cleaved crystal ends function as partially reflecting mirrors (note that the light beam emerging from the laser forms a vertical ellipse, even though the lasing spot at the active-area facet is a horizontal ellipse)

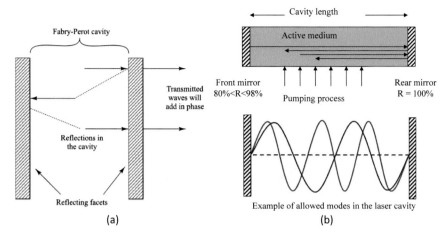

Fig. 4.18 a Two parallel light-reflecting mirrored surfaces define a Fabry-Perot resonator cavity; **b** schematic of a simple laser design and some allowed lasing modes

semiconductor crystal. The purpose of the mirrors is to establish a strong optical feedback in the longitudinal direction. This feedback mechanism converts the device into an oscillator (and hence a light emitter) with a gain mechanism that compensates for optical losses in the cavity at certain resonant optical frequencies. The sides of the cavity are simply formed by roughing the edges of the device to reduce unwanted emissions in the lateral directions.

As the light reflects back and forth within the Fabry-Perot cavity, the electric fields of the light interfere on successive round trips. Figure 4.18b shows that those wavelengths which are integer multiples of the cavity length interfere constructively. Thus their amplitudes add when they exit the device through the right-hand facet. All other wavelengths interfere destructively and therefore cancel themselves out. The optical frequencies at which constructive interference occurs are the *resonant frequencies* of the cavity. Consequently, spontaneously emitted photons that have wavelengths at these resonant frequencies reinforce themselves after multiple trips through the cavity so that their optical field becomes very strong. The resonant wavelengths are called the *longitudinal modes* of the cavity because they resonate along the length of the cavity.

Figure 4.19 illustrates the behavior of the resonant wavelengths for three values of the mirror reflectivity (R = 0.4, 0.7, and 0.9; see Sect. 2.11). The plots give the relative intensity as a function of the wavelength relative to the cavity length. As can be seen from Fig. 4.19, the width of the resonances depends on the value of the reflectivity. The result is that the resonances become sharper as the reflectivity increases. Figure 4.19 also illustrates the *free spectral range* (FSR), which is the spacing in either optical frequency or wavelength between two successive reflected or transmitted optical intensity maxima or minima. Chapter 10 provides further details on the operational theory of Fabry-Perot cavities or etalons.

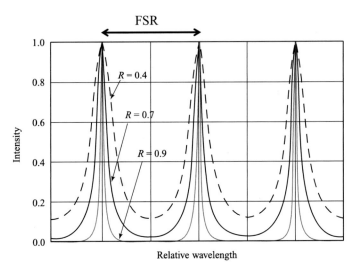

Fig. 4.19 Behavior of the resonant wavelengths in a Fabry-Perot cavity for three values of the mirror reflectivity showing that the distance between adjacent peaks is the free spectral range (FSR)

Example 4.9 As Sect. 10.5 describes, the distance between the adjacent peaks of the resonant wavelengths in a Fabry-Perot cavity, shown in Fig. 4.19, is called the free spectral range (FSR). If D is the distance between the reflecting mirrors in a device of refractive index n, then at a peak wavelength λ the FSR is given by the expression.

$$FSR = \frac{\lambda^2}{2nD}$$

What is the FSR at an 850-nm wavelength for a 0.8-mm long GaAs Fabry-Perot cavity in which the refractive index is 3.5?

Solution From the above expression

$$FSR = \frac{\lambda^2}{2nD} = \frac{\left(0.85 \times 10^{-6}\right)^2}{2(3.5)\left(0.80 \times 10^{-3}\right)} = 0.129\,nm$$

In another laser diode type, commonly referred to as the *distributed-feedback* (DFB) laser [2, 3, 12] the cleaved facets are not required for optical feedback. A typical DFB laser configuration is shown in Fig. 4.20. The fabrication of this device is similar to the Fabry-Perot types, except that the lasing action is obtained from Bragg reflectors (gratings) or periodic variations of the refractive index (called *distributed-feedback corrugations*), which are incorporated into the multilayer structure along the length of the diode. This is discussed in more detail in Sect. 4.3.6.

In general, the full optical output is needed only from the front facet of the laser, that is, the one to be aligned with an optical fiber. In this case, a dielectric reflector

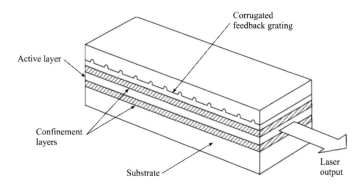

Fig. 4.20 Structure of a distributed-feedback (DFB) laser diode

can be deposited on the rear laser facet to reduce the optical loss in the cavity, to reduce the threshold current density (the point at which lasing starts), and to increase the external quantum efficiency.

The optical radiation within the resonance cavity of a laser diode sets up a pattern of electric and magnetic field lines called the *modes of the cavity* (see Sects. 2.3 and 2.4 for details on modes). These can conveniently be separated into two independent sets of transverse electric (TE) and transverse magnetic (TM) modes. Each set of modes can be described in terms of the longitudinal, lateral, and transverse half-sinusoidal variations of the electromagnetic fields along the major axes of the cavity.

- The *longitudinal modes* are related to the length L of the cavity and determine the principal structure of the frequency spectrum of the emitted optical radiation. Because L is much larger than the lasing wavelength of approximately 1 μm, many longitudinal modes can exist.
- *Lateral modes* lie in the plane of the *pn* junction. These modes depend on the sidewall preparation and the width of the cavity, and determine the shape of the lateral profile of the laser beam.
- *Transverse modes* are associated with the electromagnetic field and beam profile in the direction perpendicular to the plane of the *pn* junction. These modes are of great importance as they largely determine such laser characteristics as the radiation pattern (the transverse angular distribution of the optical output power) and the threshold current density.

To determine the lasing conditions and the resonant frequencies, the electromagnetic wave propagating in the longitudinal direction (along the axis normal to the mirrors) can be expressed in terms of the electric field phasor

$$E(z, t) = I(z)\exp[j(\omega t - \beta z)] \tag{4.22}$$

where $I(z)$ is the optical field intensity, ω is the optical radian frequency, and β is the propagation constant (see Sect. 2.3.2).

Lasing is the condition at which light amplification becomes possible in the laser diode. The requirement for lasing is that a population inversion be achieved. This condition can be understood by considering the fundamental relationship between the optical field intensity I, the absorption coefficient α_λ, and the gain coefficient g in the Fabry-Perot cavity. The stimulated emission rate into a given mode is proportional to the intensity of the radiation in that mode. The radiation intensity at a photon energy $h\nu$ varies exponentially with the distance z that it traverses along the lasing cavity according to the relationship

$$I(z) = I(0)\exp\{[\Gamma g(h\nu) - \alpha_{mat}(h\nu)]z\} \tag{4.23}$$

where α_{mat} is the effective absorption coefficient of the material in the optical path and Γ is the *optical-field confinement factor*, i.e., the fraction of optical power in the active layer (see Problem 4.11 concerning details of transverse and lateral optical-field confinement factors).

The feedback mechanism of the optical cavity provides optical amplification of selected modes. In the repeated passes between the two partially reflecting parallel mirrors, a portion of the radiation associated with those modes that have the highest optical gain coefficient is retained and further amplified during each trip through the cavity.

Lasing occurs when the gain of one or several guided modes is sufficient to exceed the optical loss during one round trip through the cavity; that is, $z = 2L$. During this round trip, only the fractions R_1 and R_2 of the optical radiation are reflected from the two laser ends 1 and 2, respectively, where R_1 and R_2 are the mirror reflectivities or Fresnel reflection coefficients, which are given by

$$R = \left(\frac{n_1 - n_2}{n_1 + n_2}\right)^2 \tag{4.24}$$

for the reflection of light at an interface between two materials having refractive indices n_1 and n_2. From this lasing condition, Eq. (4.23) becomes

$$I(2L) = I(0)R_1R_2\exp\{2L[\Gamma g(h\nu) - \alpha_{mat}(h\nu)]\} \tag{4.25}$$

Example 4.10 Assume that the cleaved mirror end faces of a GaAs laser are uncoated and that the outside medium is air. What is the reflectivity for normal incidence of a plane wave on the GaAs-air interface if the GaAs refractive index is 3.6?

Solution From Eq. (4.24), with $n_1 = 3.6$ for GaAs and $n_2 = 1.0$ for air, for both interfaces the reflectivity is

$$R_1 = R_2 = \left(\frac{3.6 - 1}{3.6 + 1}\right)^2 = 0.32$$

For an uncoated cleaved facet the reflectivity is only about 30%. To reduce the loss in the cavity and to make the optical feedback stronger, the facets typically are coated with a dielectric material. This can produce a reflectivity of about 99 percent for the rear facet and 90% for the front facet through which the lasing light emerges.

At the lasing threshold, a steady-state oscillation takes place, and the magnitude and phase of the returned wave must be equal to those of the original wave. This gives the conditions

$$I(2L) = I(0) \tag{4.26}$$

for the amplitude and

$$e^{-j2\beta L} = 1 \tag{4.27}$$

for the phase. Equation (4.27) gives information concerning the resonant frequencies of the Fabry-Perot cavity. This is discussed further in Sect. 4.3.2. From Eq. (4.26) one can determine which modes have sufficient gain for sustained oscillation, and one can find the amplitudes of these modes. The condition to just reach the lasing threshold is the point at which the optical gain is equal to the total loss α_t in the cavity. From Eq. (4.26), this condition is

$$g_{th} = \alpha_t = \alpha_{mat} + \frac{1}{2L} \ln\left(\frac{1}{R_1 R_2}\right) = \alpha_{mat} + \alpha_{end} \tag{4.28}$$

where α_{end} is the end mirror loss in the lasing cavity. Thus, for lasing to occur it is necessary to have the gain $g \geq g_{th}$. This means that the pumping source that maintains the population inversion must be sufficiently strong to support or exceed all the energy-consuming mechanisms within the lasing cavity.

The mode that satisfies Eq. (4.28) reaches threshold first. Theoretically, at the onset of this condition, all additional energy introduced into the laser should augment the growth of this particular mode. In practice, various phenomena lead to the excitation of more than one mode. Studies on the conditions needed for longitudinal single-mode operation show that important factors are thin active regions and a high degree of temperature stability.

Example 4.11 Assume for GaAs that $R_1 = R_2 = R = 0.32$ for uncoated facets (i.e., 32 percent of the radiation is reflected at a facet) and $\alpha_{mat} \approx 10 \text{ cm}^{-1}$. What is the gain threshold for a 500-μm long laser diode (Note: $L = 500 \times 10^{-4} \text{ cm}^{-1}$)?

Solution From Eq. (4.28)

$$g_{th} = \alpha_{mat} + \frac{1}{2L} \ln\left(\frac{1}{R^2}\right)$$

$$= 10 + \frac{1}{2(500 \times 10^{-4})} \ln\left(\frac{1}{(0.32)^2}\right) = 33 \text{ cm}^{-1}$$

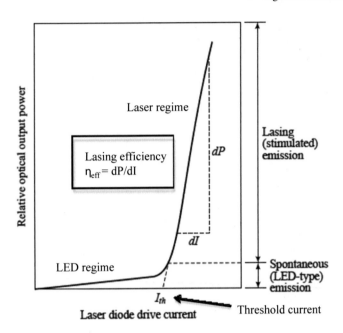

Fig. 4.21 Relationship between optical output power and laser diode drive current

The relationship between optical output power and diode drive current is presented in Fig. 4.21. At low diode currents, only spontaneous radiation is emitted. Both the spectral range and the lateral beam width of this emission are broad like that of an LED. A dramatic and sharply defined increase in the power output occurs at the lasing threshold. As this transition point is approached, the spectral range and the beam width both narrow with increasing drive current. The final spectral width of approximately 1 nm and the fully narrowed lateral beam width of nominally $5 - 10°$ are reached just past the threshold point. The *threshold current* I_{th} is conventionally defined by extrapolation of the lasing region of the power-versus-current curve, as shown in Fig. 4.21. At high power outputs, the slope of the curve decreases because of junction heating.

For laser structures that have strong carrier confinement, the *threshold current density* for stimulated emission J_{th} can to a good approximation be related to the lasing-threshold optical gain by

$$g_{th} = \beta_{th} J_{th} \tag{4.29}$$

where the gain factor β_{th} is a constant that depends on the specific device construction.

Example 4.12 Suppose that a given GaAlAs laser has an optical cavity length of 300 μm and a 100-μm width. Let the gain factor $\beta_{th} = 21 \times 10^{-3}$ A·cm^3 and take the loss coefficient to be $\alpha_{mat} \approx 10$ cm^{-1} at a normal operating temperature. Assume the reflectivity is $R_1 = R_2 = R = 0.32$ for each end face. Find (a) the threshold current density and (b) the threshold current for this device.

Solution (a) From Eqs. (4.28) and (4.29)

$$J_{th} = \frac{1}{\beta_{th}} \left[\alpha_{mat} + \frac{1}{L} ln \left(\frac{1}{R} \right) \right] = \frac{1}{21 \times 10^{-3}} \left[10 + \frac{1}{300 \times 10^{-4}} ln \left(\frac{1}{0.32} \right) \right]$$

$$= 2.28 \times 10^3 \text{ A}km^2$$

(b) The threshold current I_{th} is given by.
$I_{th} = J_{th} \times$ cross-sectional area of the optical cavity $= (2.28 \times 10^3 \text{ A/cm}^2) \times$
$(300 \times 10^{-4} \text{ cm}) \times (100 \times 10^{-4} \text{ cm}) = 684 \text{ mA}$.

Drill Problem 4.6 Consider a Fabry-Perot laser cavity in which the absorption loss coefficient is 20 cm^{-1}. If the cavity mirror reflection coefficient is 0.33 at both ends, use Eq. (4.28) to verify that the cavity length at which the absorption loss and the cavity loss become equal is $L = 554$ μm.

4.3.2 Laser Diode Rate Equations

The relationship between optical output power and the diode drive current can be determined by examining the rate equations that govern the interaction of photons and electrons in the active region. As noted earlier, the total carrier population is determined by carrier injection, spontaneous recombination, and stimulated emission. For a *pn* junction with a carrier-confinement region of depth d, the *rate equations* are given by

$$\frac{d\Phi}{dt} = Cn\Phi + R_{sp} - \frac{\Phi}{\tau_{ph}}$$

$$= \text{stimulated emission} + \text{spontaneous emission} + \text{photon loss} \qquad (4.30)$$

which governs the number of photons Φ, and

$$\frac{dn}{dt} = \frac{J}{qd} - \frac{n}{\tau_{sp}} - Cn\Phi$$

$$= \text{injection} + \text{spontaneous recombination} + \text{stimulated emission} \qquad (4.31)$$

which governs the number of electrons n. Here, C is a coefficient describing the strength of the optical absorption and emission interactions, R_{sp} is the rate of spontaneous emission into the lasing mode (which is much smaller than the total spontaneous-emission rate), τ_{ph} is the photon lifetime, τ_{sp} is the spontaneous-recombination lifetime, and J is the injection-current density.

Equations (4.30) and (4.31) may be balanced by considering all the factors that affect the number of carriers in the laser cavity. The first term in Eq. (4.30) is a source of photons resulting from stimulated emission. The second term, describing the number of photons produced by spontaneous emission, is relatively small compared with the first term. The third term in Eq. (4.30) indicates the decay in the number of photons caused by loss mechanisms in the lasing cavity. In Eq. (4.31), the first term represents the increase in the electron concentration in the conduction band as current flows into the device. The second and third terms give the number of electrons lost from the conduction band owing to spontaneous and stimulated transitions, respectively.

Solving these two equations for a steady-state condition will yield an expression for the output power. The steady state is characterized when the left-hand sides of Eq. (4.30) and Eq. (4.31) are equal to zero. First, from Eq. (4.30), assuming R_{sp} is negligible and noting that $d\Phi/dt$ must be positive when Φ is small, it follows that

$$Cn - \frac{1}{\tau_{ph}} \geq 0 \qquad (4.32)$$

This shows that n must exceed a threshold value n_{th} in order for Φ to increase. Using Eq. (4.31), this threshold value can be expressed in terms of the threshold current J_{th} needed to maintain an inversion level $n = n_{th}$ in the steady state when the number of photons $\Phi = 0$:

$$\frac{n_{th}}{\tau_{sp}} = \frac{J_{th}}{qd} \qquad (4.33)$$

This expression defines the current required to sustain an excess electron density in the laser when spontaneous emission is the only decay mechanism.

Next, consider the photon and electron rate equations in the steady-state condition at the lasing threshold. Respectively, Eqs. (4.30) and (4.31) become

$$0 = Cn_{th}\Phi_s + R_{sp} - \frac{\Phi_s}{\tau_{ph}}$$

$$= \text{stimulated emission} + \text{spontaneous emission} + \text{photon loss} \qquad (4.34)$$

and

$$0 = \frac{J}{qd} - \frac{n_{th}}{\tau_{sp}} - Cn_{th}\Phi_s$$

$$= \text{injection} + \text{spontaneous recombination} + \text{stimulated emission} \qquad (4.35)$$

where Φ_s is the steady-state photon density. Adding Eqs. (4.34) and (4.35), using Eq. (4.33) for the term n_{th}/τ_{sp}, and solving for Φ_s yields the number of photons per unit volume:

$$\Phi_s = \frac{\tau_{ph}}{qd}(J - J_{th}) + \tau_{ph} R_{sp} \tag{4.36}$$

The first term in Eq. (4.36) is the number of photons resulting from stimulated emission. The power from these photons is generally concentrated in one or a few modes. The second term gives the spontaneously generated photons. The power resulting from these photons is not mode-selective, but is spread over all the possible modes of the volume, which are on the order of 10^8 modes.

4.3.3 External Differential Quantum Efficiency

The *external differential quantum efficiency* η_{ext} is defined as the number of photons emitted per radiative electron-hole pair recombination above threshold. Under the assumption that above threshold the gain coefficient remains fixed at g_{th}, then η_{ext} is given by [2, 3]

$$\eta_{ext} = \frac{\eta_i (g_{th} - \alpha_{mat})}{g_{th}} \tag{4.37}$$

Here η_i is the internal quantum efficiency. This is not a well-defined quantity in laser diodes, but most measurements show that η_i is $0.6 - 0.7$ at room temperature. Experimentally, η_{ext} is calculated from the straight-line portion of the curve for the emitted optical power P versus drive current I, which gives

$$\eta_{ext} = \frac{q}{E_g} \frac{dP}{dI} = 0.8065\lambda(\mu m)\frac{dP \text{ (mW)}}{dI \text{ (mA)}} \tag{4.38}$$

where E_g is the bandgap energy in electron volts, dP is the incremental change in the emitted optical power in milliwatts for an incremental change dI in the drive current (in milliamperes), and λ is the emission wavelength in micrometers. For standard semiconductor lasers, external differential quantum efficiencies of 15−20% per facet are typical. High-quality devices have differential quantum efficiencies of 30−40%.

4.3.4 Laser Resonant Frequencies

Now reconsider Eq. (4.27) to examine the resonant frequencies of the laser. The condition in Eq. (4.27) holds when

$$2\beta L = 2\pi m \tag{4.39}$$

where m is an integer. Using $\beta = 2\pi n/\lambda$ for the propagation constant from Eq. (2.46), it follows that

$$m = \frac{L}{\lambda/2n} = \frac{2Ln}{c}v \tag{4.40}$$

where $c = v\lambda$. This states that the cavity resonates (i.e., a standing-wave pattern exists within it) when an integer number m of half-wavelengths spans the region between the mirrors.

Because in all lasers the gain is a function of frequency (or wavelength, because $c = v\lambda$), there will be a range of frequencies (or wavelengths) for which Eq. (4.40) holds. Each of these frequencies corresponds to a mode of oscillation of the laser. Depending on the laser structure, any number of frequencies can satisfy Eqs. (4.26) and (4.27). Thus some lasers are single-mode and some are multimode. The relationship between gain and frequency can be assumed to have the Gaussian form

$$g(\lambda) = g(0)\exp\left[-\frac{(\lambda - \lambda_0)^2}{2\sigma^2}\right] \tag{4.41}$$

where λ_0 is the wavelength at the center of the spectrum, σ is the spectral width of the gain, and the maximum gain $g(0)$ is proportional to the population inversion.

An important point is the frequency, or wavelength, spacing between the modes of a multimode laser. Here, only the longitudinal modes will be considered. Note, however, that for each longitudinal mode there may be several transverse modes that arise from one or more reflections of the propagating wave at the sides of the resonator cavity [2, 3]. To find the frequency spacing, consider two successive modes of frequencies v_{m-1} and v_m represented by the integers $m - 1$ and m. From Eq. (4.40), it follows that

$$m - 1 = \frac{2Ln}{c}v_{m-1} \tag{4.42}$$

and

$$m = \frac{2Ln}{c}v_m \tag{4.43}$$

Subtracting these two equations yields

$$1 = \frac{2Ln}{c}(v_m - v_{m-1}) = \frac{2Ln}{c}\Delta v \tag{4.44}$$

which gives rise to the following expression for the frequency spacing

Fig. 4.22 Typical spectrum from a Fabry-Perot GaAlAs/GaAs laser diode

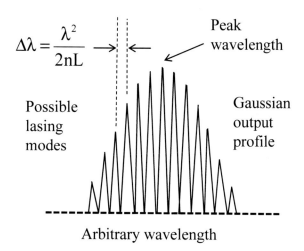

$$\Delta\lambda = \frac{\lambda^2}{2nL}$$

Peak wavelength

Possible lasing modes

Gaussian output profile

Arbitrary wavelength

$$\Delta\nu = \frac{c}{2Ln} \tag{4.45}$$

This can be related to the wavelength spacing $\Delta\lambda$ through the relationship $\Delta\nu/\nu = \Delta\lambda/\lambda$, yielding

$$\Delta\lambda = \frac{\lambda^2}{2Ln} \tag{4.46}$$

Thus, given Eqs. (4.41) and (4.46), the output spectrum of a multimode laser follows the typical gain-versus-frequency plot given in Fig. 4.22, where the exact number of modes, their heights, and their spacing depend on the laser construction.

Example 4.13 A GaAs laser operating at 850 nm has a 500-μm length and a refractive index $n = 3.7$. (a) What are the frequency spacing and the wavelength spacing? (b) If, at the half-power point, $\lambda - \lambda_0 = 2$ nm, what is the spectral width σ of the gain?

Solution (a) From Eq. (4.45) the frequency spacing is

$$\Delta\nu = \frac{3 \times 10^8 \text{ m/s}}{2(500 \times 10^{-6} \text{ m})(3.7)} = 81 \text{ GHz}$$

From Eq. (4.46) the wavelength spacing is

$$\Delta\lambda = \frac{(850 \times 10^{-9} \text{ m})^2}{2(500 \times 10^{-6} \text{ m})(3.7)} = 0.195 \text{ nm}$$

(b) Using Eq. (4.41) with $g(\lambda) = 0.5\, g(0)$ and then solving for σ with $\lambda - \lambda_0 = \Delta\lambda = 0.195$ nm yields

$$\sigma = \frac{\lambda - \lambda_0}{\sqrt{2ln2}} = \frac{0.195 \text{ nm}}{\sqrt{2ln2}} = 0.166 \text{ nm}$$

Example 4.14 Consider a double-heterostructure edge-emitting Fabry-Perot AlGaAs laser, which emits at 900 nm. Suppose that the laser chip is 300 μm long and the refractive index of the laser material is 4.3. (*a*) How many half-wavelengths span the region between the Fabry-Perot mirror surfaces? (*b*) What is the spacing between the lasing modes?

Solution (*a*) From Eq. (4.40) the number of half-wavelengths that span the region between the Fabry-Perot mirror surfaces is

$$m = \frac{2nL}{\lambda} = \frac{2(4.3)(300\,\mu\text{m})}{0.90\,\mu\text{m}} = 2866$$

(*b*) From Eq. (4.46) the spacing between the lasing modes is

$$\Delta\lambda = \frac{(900 \times 10^{-9}\,\text{m})^2}{2(300 \times 10^{-6}\,\text{m})(4.3)} = 0.314\,\text{nm}$$

Drill Problem 4.7 A GaAs laser operating at 900 nm has a 300-μm length and a refractive index $n = 3.58$. Use Eq. (4.45) to show that the frequency separation of the resonant modes is 140 GHz.

4.3.5   Structures and Radiation Patterns of Laser Diodes

A basic requirement for efficient operation of laser diodes is that, in addition to transverse optical confinement and carrier confinement between heterojunction layers, the current flow must be restricted laterally to a narrow stripe along the length of the laser. Numerous novel methods of achieving this, with varying degrees of success, have been proposed, but all strive for the same goals of limiting the number of lateral modes so that lasing is confined to a single filament, stabilizing the lateral gain, and ensuring a relatively low threshold current. The most common devices are called *index-guided lasers*. If a particular index-guided laser supports only the fundamental transverse mode and the fundamental longitudinal mode, it is known as a *single-mode laser*. Such a device emits a single, well-collimated beam of light that has an intensity profile that is a bell-shaped Gaussian curve.

When designing the width and thickness of the optical cavity, a tradeoff must be made between current density and output beam width. As either the width or the thickness of the active region is increased, a narrowing occurs of the lateral or transverse beam widths, respectively, but at the expense of an increase in the threshold current density. Most waveguide lasers have a lasing spot 3 μm wide by 0.6 μm high. This is significantly greater than the active-layer thickness, because about half the light travels in the confining layers. Such lasers can operate reliably only up to continuous-wave (CW) output powers of 3−5 mW.

Although the active layer in a standard double-heterostructure laser is thin enough (1−3 μm) to confine electrons and the optical field, the electronic and optical properties remain the same as in the bulk material. This limits the achievable threshold current density, modulation speed, and linewidth of the device. *Quantum-well lasers* overcome these limitations by having an active-layer thickness around 10 nm [13]. This changes the electronic and optical properties dramatically, because the dimensionality of the free-electron motion is reduced from three to two dimensions. As shown in Fig. 4.23, the restriction of the carrier motion normal to the active layer results in a quantization of the energy levels. The possible energy-level transitions that lead to photon emission are designated by ΔE_{ij} (see Problem 4.16). Both single quantum-well (SQW) and multiple quantum-well (MQW) lasers have been fabricated. These structures contain single and multiple active regions, respectively. The layers separating the active regions are called *barrier layers*. The MQW lasers have a better optical-mode confinement, which results in a lower threshold current density. The wavelength of the output light depends the layer thickness d. For example, in an

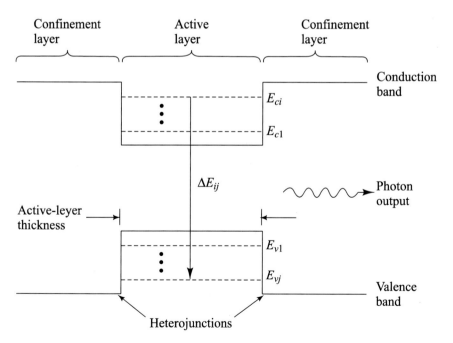

Fig. 4.23 Energy-band diagram for a quantum layer in a multiple quantum-well (MQW) laser where the parameters ΔE_{ij} represents the allowed energy-level transitions

InGaAs quantum-well laser, the peak output wavelength moves from 1550 nm when $d = 10$ nm to a peak of 1500 nm when $d = 8$ nm.

4.3.6 Lasers Operating in a Single Mode

For high-speed long-distance communications one needs single-mode lasers, which must contain only a single longitudinal mode and a single transverse mode. Consequently, the spectral width of the optical emission is very narrow.

One way of restricting a laser to have only one longitudinal mode is to reduce the length L of the lasing cavity to the point where the frequency separation $\Delta \nu$ of the adjacent modes given in Eq. (4.45) is larger than the laser transition line width; that is, only a single longitudinal mode falls within the gain bandwidth of the device. For example, for a Fabry-Perot cavity, all longitudinal modes have nearly equal losses and are spaced by about 1 nm in a 250-μm-long cavity at 1300 nm. By reducing L from 250 to 25 μm, the mode spacing increases from 1 to 10 nm. However, these lengths make the device hard to handle, and they are limited to optical output powers of only a few milliwatts.

Alternative devices were thus developed. Among these are vertical-cavity surface-emitting lasers, structures that have a built-in frequency-selective grating, and tunable lasers. This section discusses the first two structures. The special feature of a *vertical-cavity surface-emitting laser* (VCSEL) [14, 15] is that the light emission is perpendicular to the semiconductor surface, as shown in Fig. 4.24. This feature facilitates the integration of multiple lasers onto a single chip in one-dimensional or two-dimensional arrays, which makes them attractive for wavelength-division multiplexing applications. The active-region volume of these devices is very small, which leads to very low threshold currents (<100 μA). In addition, for an equivalent output power compared to edge-emitting lasers, the modulation bandwidths are

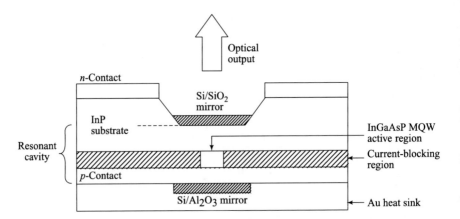

Fig. 4.24 Basic architecture of a vertical-cavity surface-emitting laser (VCSEL)

much greater, because the higher photon densities reduce radiative lifetimes. The mirror system used in VCSELs to form the resonant cavity is of critical importance, because maximum reflectivity is needed for efficient operation. Figure 4.24 shows one mirror system that consists of a semiconductor material, such as Si/SiO_2, as one material and an oxide layer, such as Si/Al_2O_3, as the other material.

Three types of laser configurations using a built-in *frequency-selective reflector* are shown in Fig. 4.25. In each case, the frequency-selective reflector is a corrugated grating that is a passive waveguide layer adjacent to the active region. The optical wave propagates parallel to this grating. The operation of these types of lasers is based on the use of a distributed Bragg phase-grating reflector. A phase grating is essentially a region of periodically varying refractive index that causes two counter propagating traveling waves to couple. The coupling is at a maximum for wavelengths close to the Bragg wavelength λ_B, which is related to the period Λ of the corrugations by

$$\lambda_B = \frac{2n_e\Lambda}{k} \tag{4.47}$$

where n_e is the effective refractive index of the mode (see Sect. 2.5.4) and k is the order of the grating. First-order gratings ($k = 1$) provide the strongest coupling, but sometimes second-order gratings are used because their larger corrugation period makes fabrication easier. Lasers based on this architecture exhibit good single-mode longitudinal operation with low sensitivity to drive-current and temperature variations.

In the *distributed-feedback* (DFB) laser the grating for the wavelength selector is formed over the entire active region (Fig. 4.25a) [2, 3, 12]. As shown in Fig. 4.26, in an ideal DFB laser the longitudinal modes are spaced symmetrically around λ_B at wavelengths given by

$$\lambda = \lambda_B \pm \frac{\lambda_B^2}{2n_e L_e}\left(m + \frac{1}{2}\right) \tag{4.48}$$

where $m = 0, 1, 2, \ldots$ is the mode order and L_e is the effective grating length. The amplitudes of successively higher-order lasing modes are greatly reduced from the zero-order amplitude; for example, the first-order mode ($m = 1$) is usually more than 30 dB down from the zero-order amplitude ($m = 0$).

Theoretically, in a DFB laser that has both ends antireflection-coated, the two zero-order modes on either side of the Bragg wavelength should experience the same lowest threshold gain and would lase simultaneously in an idealized symmetrical structure. However, in practice, the randomness of the cleaving process of the laser crystal material lifts the degeneracy in the modal gain and results in single-mode operation. This facet asymmetry can be further increased by putting a high-reflection coating on one end and a low-reflection coating on the other; for example, around 2 percent on the front facet and 30% on the rear facet.

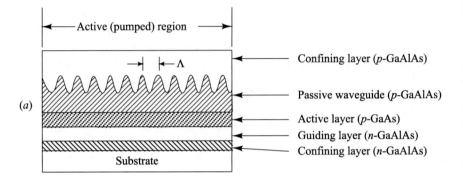

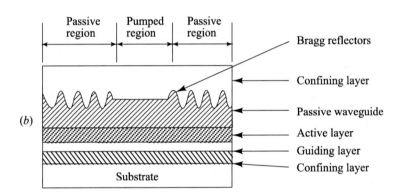

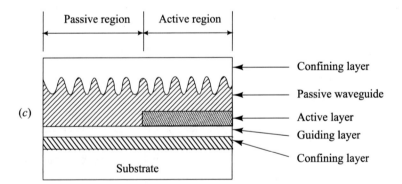

Fig. 4.25 Three types of laser structures using built-in frequency-selective resonator gratings: **a** distributed-feedback (DFB) laser, **b** distributed-Bragg-reflector (DBR) laser, and **c** distributed-reflector (DR) laser

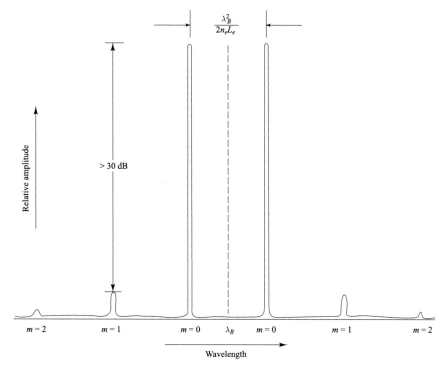

Fig. 4.26 Output spectrum symmetrically distributed around λ_B in an idealized distributed-feedback (DFB) laser diode

For the *distributed-Bragg-reflector* (DBR) laser the gratings are located at the ends of the normal active layer of the laser to replace the cleaved end mirrors used in the Fabry-Perot optical resonator (Fig. 4.25b). The *distributed-reflector* laser consists of active and passive distributed reflectors (Fig. 4.25c). This structure improves the lasing properties of conventional DFB and DBR lasers, and has a high efficiency and high output capability.

4.3.7   Modulation of Laser Diodes

The process of putting information onto a lightwave is called *modulation*. For data rates of less than approximately 10 Gb/s (typically 2.5 Gb/s), the process of imposing information on a laser-emitted light stream can be realized by *direct modulation*. This involves directly varying the laser drive current with the electrically formatted information stream to produce a correspondingly varying optical output power. For higher data rates one needs to use a device called an *external modulator* to temporally modify a steady optical power level emitted by the laser (see Sect. 4.3.9). A variety of external modulators are available commercially either as a separate device or as an integral part of the laser transmitter package.

The basic limitation on the direct modulation rate of laser diodes depends on the spontaneous and stimulated emission carrier lifetimes and on the photon lifetime. The *spontaneous carrier lifetime* τ_{sp} is a function of the semiconductor band structure and the carrier concentration. At room temperature this lifetime is about 1 ns in GaAs-based materials for dopant concentrations on the order of 10^{19} cm^{-3}. The *stimulated carrier lifetime* τ_{st} depends on the optical density in the lasing cavity and is on the order of 10 ps. The *photon lifetime* τ_{ph} is the average time that the photon resides in the lasing cavity before being lost either by absorption or by emission through the facets. In a Fabry-Perot cavity, the photon lifetime is[1-3]

$$\tau_{ph}^{-1} = \frac{c}{n}\left(\alpha_{mat} + \frac{1}{2L}ln\frac{1}{R_1 R_2}\right) = \frac{c}{n}g_{th} \qquad (4.49)$$

For a typical value of $g_{th} = 50$ cm^{-1} and a refractive index in the GaAs lasing material of $n = 3.5$, the photon lifetime is approximately $\tau_{ph} = 2$ ps. This value sets the upper limit to the direct modulation capability of the laser diode.

A laser diode can readily be pulse modulated because the photon lifetime is much smaller than the carrier lifetime. If the laser is completely turned off after each pulse, the spontaneous carrier lifetime will limit the modulation rate. This is because, at the onset of a current pulse of amplitude I_p, a period of time t_d given by (see Problem 4.19)

$$t_d = \tau ln\frac{I_p}{I_p + (I_B - I_{th})} \qquad (4.50)$$

is needed to achieve the population inversion necessary to produce a gain that is sufficient to overcome the optical losses in the lasing cavity. In Eq. (4.50) the parameter I_B is the bias current, which is a fixed dc current applied to the laser. The parameter τ is the average lifetime of the carriers in the combination region when the total current $I = I_p + I_B$ is close to the threshold current I_{th}. Equation (4.50) shows that by dc-biasing the diode at the lasing threshold current the delay time can be eliminated. Pulse modulation is then carried out by modulating the laser only in the operating region above the threshold current (see Fig. 4.21). In this region, the carrier lifetime is now shortened to the stimulated emission lifetime, so that high modulation rates are possible.

When using a directly modulated laser diode for high-speed transmission systems, the modulation frequency can be no larger than the frequency of the relaxation oscillations of the laser field. The relaxation oscillation depends on both the spontaneous lifetime and the photon lifetime. Theoretically, assuming a linear dependence of the optical gain on carrier density, the relaxation oscillation occurs approximately at

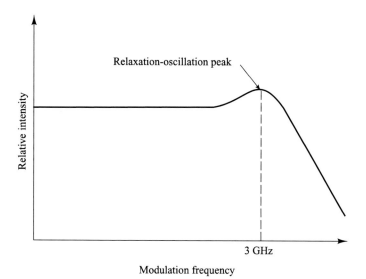

Fig. 4.27 Example of the relaxation–oscillation peak of a laser diode

$$f = \frac{1}{2\pi} \frac{1}{\left(\tau_{sp}\tau_{ph}\right)^{1/2}} \left(\frac{I}{I_{th}} - 1\right)^{1/2} \tag{4.51}$$

Because τ_{sp} is about 1 ns and τ_{ph} is on the order of 2 ps for a 300-μm-long laser, then when the injection current is about twice the threshold current, the maximum modulation frequency is a few gigahertz. An example of a laser that has a relaxation–oscillation peak at 3 GHz is shown in Fig. 4.27.

4.3.8 Laser Output Spectral Width

In non-semiconductor lasers, such as in solid state lasers, it can be shown that noise arising from spontaneous emission effects results in a finite spectral width or *linewidth* $\Delta\nu$ for the lasing output. However, a semiconductor laser has a significantly higher linewidth than what is predicted by this simple theory. In a semiconductor material both the optical gain and the refractive index depend on the actual carrier density in the medium. This relationship leads to an index-gain coupling mechanism; that is, it gives rise in an interaction between phase noise and the light intensity. The theoretically calculated result is [10]

$$\Delta\nu = \frac{R_{sp}}{4\pi I}\left(1 + \alpha^2\right) \tag{4.52}$$

Here I is the average number of photons in the lasing cavity, R_{sp} is the spontaneous emission rate [see Eq. (4.30)], and the parameter α is the *linewidth enhancement factor*. Basically this shows that in semiconductor lasers the linewidth is increased by a factor $(1 + \alpha^2)$.

The linewidth expression in Eq. (4.52) can be rewritten in terms of the optical output power P_{out} as

$$\Delta \nu = \frac{V_g^2 h\nu \, g_{th} \, n_{sp} \, \alpha_t}{8\pi \, P_{out}} (1 + \alpha^2) \qquad (4.53)$$

where V_g is the group velocity of light, $h\nu$ is the photon energy, g_{th} is the threshold gain, α_t is the cavity loss [see Eq. (4.28)], and n_{sp} is the *spontaneous emission factor* (the ratio of spontaneous emission coupled into the lasing mode to the total spontaneous emission).

Equation (4.53) shows that a number of variables influence the magnitude of the laser linewidth. For example, typically $\Delta \nu$ decreases as the laser output power increases. The value of the α-factor also impacts the linewidth. Common values of the dimensionless α-factor range from 2.0 to 6.0 with calculated numbers being in good agreement with experimental measurements. In addition, the laser construction can influence the linewidth because the α-factor values are different depending on the material type and the laser diode structure. For example, the α-factor is smaller in MQW laser structures than in bulk material, and even smaller values are exhibited in devices such as quantum-dot lasers. For DFB lasers the linewidth ranges from 5 to 10 MHz (or, equivalently, around 10^{-4} nm).

The spectral width of a laser also can increase significantly when direct modulation is used to vary the light output level. This line broadening is referred to as a *chirping effect*, which is explained in more detail in Sect. 8.2.6.

4.3.9 External Laser Light Modulation

When direct modulation is used in a laser transmitter, the process of turning the laser on and off with an electrical drive current produces a widening of the laser linewidth. This phenomenon is referred to as *chirp* and makes directly modulated lasers undesirable for operation at data rates greater than about 2.5 Gb/s. For these higher-rate applications it is preferable to use an *external modulator*, as shown in Fig. 4.28. In such a configuration, the optical source emits a constant-amplitude light signal, which enters the external modulator. In this case, instead of varying the amplitude of the light coming out of the laser, the electrical driving signal dynamically changes the optical power level that exits the external modulator. This process thus produces a time-varying optical signal. The external modulator either can be integrated physically in the same package with the light source or it can be a separate device. The two

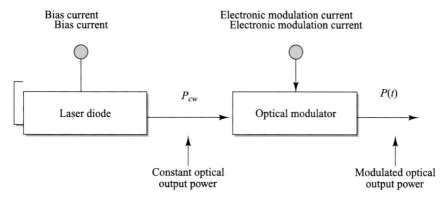

Fig. 4.28 Operational concept of a generic external modulator

main device types are the electro-optical phase modulator and the electro-absorption modulator [16, 17].

The *electro-optical* (EO) phase modulator (also called a *Mach-Zehnder Modulator* or MZM) typically is made of lithium niobate (LiNbO$_3$). In an EO modulator the light beam is split in half and then sent through two separate paths, as shown in Fig. 4.29. A high-speed electric signal then changes the phase of the light signal in one of the paths. This is done in such a manner that when the two halves of the signal meet again at the device output, they will recombine either constructively or destructively. The constructive recombination produces a bright signal and corresponds to a 1 pulse. On the other hand, destructive recombination results in the two signal halves canceling each other so there is no light at the output of the beam combiner. This corresponds to a 0 pulse. LiNbO$_3$ modulators are separately packaged devices and can be up to 12 cm (about 5 inches) long.

The *electro-absorption modulator* (EAM) typically is constructed on an electro-optic substrate material, such as *indium phosphide* (InP). As shown in Fig. 4.30, an EAM operates by having an electric signal change the transmission properties of the material in the light path to make it either transparent during a 1 pulse or opaque during a 0 pulse. Because InP is used as the material for an EAM, the device can

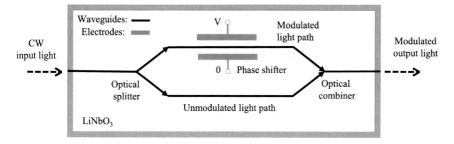

Fig. 4.29 Operational concept of an electro-optical lithium niobate external modulator

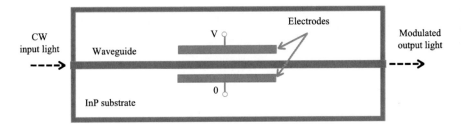

Fig. 4.30 Operational concept of an electro-absorption modulator (EAM)

be integrated onto the same substrate as a DFB laser diode chip. The complete laser plus modulator unit then can be put in a standard butterfly package, thereby reducing drive voltage, power, and space requirements compared to having separate laser and LiNbO$_3$ modulator packages.

4.3.10   Lasing Threshold Temperature Effects

An important factor to consider in the application of laser diodes is the temperature dependence of the threshold current $I_{th}(T)$. This parameter increases with temperature in all types of semiconductor lasers because of various temperature-dependent factors. The complexity of these factors prevents the formulation of a single equation that holds for all devices and temperature ranges. However, the temperature variation of I_{th} can be approximated by the empirical expression

$$I_{th}(T) = I_z \exp(T/T_0) \tag{4.54}$$

where T_0 is a measure of the threshold temperature coefficient and I_z is a constant. For a conventional stripe-geometry GaAlAs laser diode, T_0 is typically $120-165$ °C in the vicinity of room temperature. An example of a laser diode with $T_0 = 135$ °C and $I_z = 52$ mA is shown in Fig. 4.31. The threshold current increases by a factor of about 1.4 between 20 and 60 °C. The variation in I_{th} with temperature is 0.8 percent/°C, as is shown in Fig. 4.32. Smaller dependences of I_{th} on temperature have been demonstrated for GaAlAs quantum-well heterostructure lasers. For these lasers, T_0 can be as high as 437 °C. The temperature dependence of I_{th} for this device is also shown in Fig. 4.32. The threshold variation for this particular laser type is 0.23%/°C.

In addition to being affected by temperature, the lasing threshold can change as the laser ages. Consequently, if a constant optical output power level is to be maintained as the temperature of the laser changes or as the laser ages, it is necessary to adjust the dc-bias current level. One possible method for achieving this automatically is an optical feedback scheme.

A photodetector can be used to achieve optical feedback either by sensing the variation in optical power emitted from the rear facet of the laser or by tapping

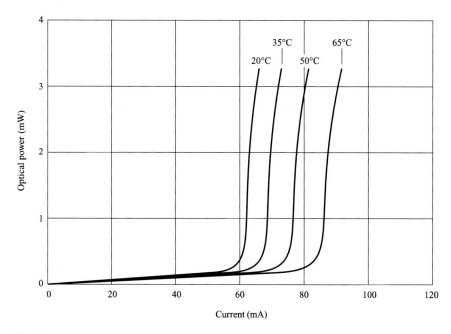

Fig. 4.31 Temperature-dependent behavior of the optical output power as a function of the bias current for a particular laser diode with $T_0 = 135\,°C$ and $I_z = 52$ mA

off and monitoring a small portion of the fiber-coupled power emitted from the front facet. The photodetector compares the optical power output with a reference level and adjusts the dc-bias current level automatically to maintain a constant peak light output relative to the reference. The photodetector used must have a stable long-term responsivity that remains constant over a wide temperature range. For operation in the 800-to-900-nm region, a silicon *pin* photodiode generally exhibits these characteristics (see Chap. 6).

Another standard method of stabilizing the optical output of a laser diode is to use a miniature thermoelectric cooler [18]. This device maintains the laser at a constant temperature and thus stabilizes the output level. Normally, a thermoelectric cooler is used in conjunction with a rear-facet detector feedback loop, as is shown in Fig. 4.33.

Example 4.15 An engineer has a GaAlAs laser that has a threshold temperature coefficient $T_0 = 135\,°C$ and an InGaAsP laser with $T_0 = 55\,°C$. Compare the percent change in the threshold current for each of these lasers when the temperature increases from 20 to 65 °C.

Solution (*a*) Letting $T_1 = 20\,°C$ and $T_2 = 65\,°C$, then from Eq. (4.54) when the temperature changes from T_1 to T_2 for the GaAlAs laser the threshold current increases by

$$\frac{I_{th}(65\,°C)}{I_{th}(20\,°C)} = \exp[(T_2 - T_1)/T_0]$$

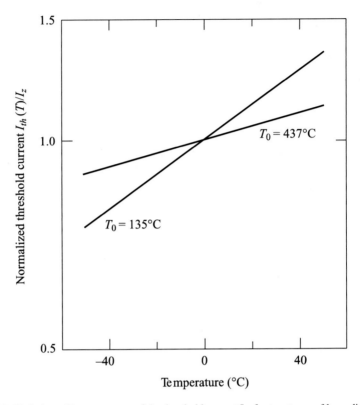

Fig. 4.32 Variation with temperature of the threshold current I_{th} for two types of laser diodes

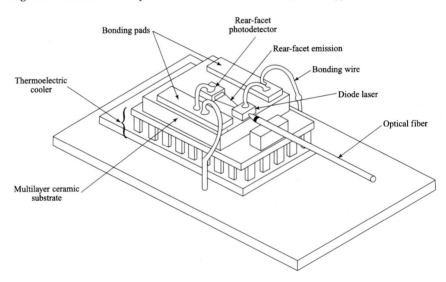

Fig. 4.33 Construction of a laser transmitter that uses a rear-facet photodiode for output monitoring and a thermoelectric cooler for temperature stabilization

$$= \exp[(65 - 20)/135]$$
$$= 1.40 = 140\%$$

(*b*) Similarly, letting $T_1 = 20$ °C and $T_2 = 65$ °C, then from Eq. (4.54) for the InGaAsP laser the threshold current increases by

$$\frac{I_{th}(65\,°C)}{I_{th}(20\,°C)} = \exp[(T_2 - T_1)/T_0]$$
$$= \exp[(65 - 20)/55]$$
$$= 2.27 = 227\%$$

4.4 Output Linearity of Light Sources

High-radiance LEDs and laser diodes are well-suited optical sources for wideband analog applications provided a method is implemented to compensate for any nonlinearity of these devices. In an analog system, a time-varying electric analog signal $s(t)$ is used to modulate an optical source directly about a bias current point I_B, as shown in Fig. 4.34. With no signal input, the optical power output is P_t. When the signal $s(t)$ is applied, the time-varying (analog) optical output power $P(t)$ is

$$P(t) = P_t[1 + m\, s(t)] \tag{4.55}$$

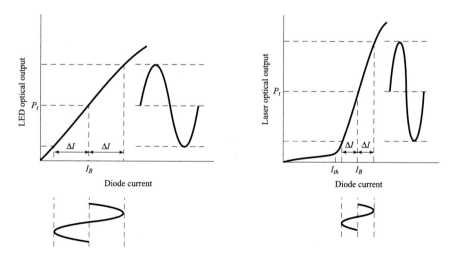

Fig. 4.34 Bias point and amplitude modulation range for analog applications of LEDs (left) and laser diodes (right)

Here, m is the *modulation index* (or *modulation depth*) defined as

$$m = \frac{\Delta I}{I'_B} \tag{4.56}$$

where $I'_B = I_B$ for LEDs and $I'_B = I_B - I_{th}$ for laser diodes. The parameter ΔI is the variation in current about the bias point. To prevent distortions in the output signal, the modulation must be confined to the linear region of the curve for optical output versus drive current. Furthermore, if ΔI is greater than I'_B (i.e., m is greater than 100 percent), the lower portion of the signal gets cut off and severe distortion will result. Typical m values for analog applications range from 0.25 to 0.50.

In analog applications, any device nonlinearities will create frequency components in the output signal that were not present in the input signal. Two important nonlinear effects are harmonic and intermodulation distortions. If the signal input to a nonlinear device is a simple cosine wave $x(t) = A \cos \omega t$ with frequency ω and amplitude A, the output will be

$$y(t) = A_0 + A_1\cos \omega t + A_2\cos 2\omega t + A_3\cos 3\omega t + \cdots \tag{4.57}$$

where the factors A_j are the amplitudes of the jth harmonic. That is, the output signal will consist of a component at the input frequency ω plus spurious components at zero frequency, at the second harmonic frequency 2ω, at the third harmonic frequency 3ω, and so on. This effect is known as *harmonic distortion*. The amount of nth-order distortion in decibels is given by

$$n\text{th-order harmonic distortion} = 20 \log\frac{A_n}{A_1} \tag{4.58}$$

To determine *intermodulation distortion*, the modulating signal of a nonlinear device is taken to be the sum of two cosine waves $x(t) = A_1 \cos \omega_1 t + A_2 \cos \omega_2 t$. The output signal will then be of the form

$$y(t) = \sum_{m,n} B_{mn}\cos(m\omega_1 + n\omega_2) \tag{4.59}$$

where m and $n = 0, \pm 1, \pm 2, \pm 3, \ldots$. This signal includes all the harmonics of ω_1 and ω_2 plus cross-product terms such as $\omega_2 - \omega_1$, $\omega_2 + \omega_1$, $\omega_2 - 2\omega_1$, $\omega_2 + 2\omega_1$, and so on. The sum and difference frequencies give rise to the intermodulation distortion. The sum of the absolute values of the coefficients m and n determines the order of the intermodulation distortion. For example, the second-order intermodulation products are at $\omega_1 \pm \omega_2$ with amplitude B_{11}, the third-order intermodulation products are at $\omega_1 \pm 2\omega_2$ and $2\omega_1 \pm \omega_2$ with amplitudes B_{12} and B_{21}, and so on. (Harmonic distortions are also present wherever either $m \neq 0$ and $n = 0$ or when $m = 0$ and $n \neq 0$. The corresponding amplitudes are B_{m0} and B_{0n}, respectively.) In general, the odd-order intermodulation products with $m = n \pm 1$ (such as $2\omega_1 - \omega_2$, $2\omega_2 - \omega_1$, $3\omega_1 -$

$2\omega_2$, etc.) are the most troublesome because they may fall within the bandwidth of the channel. Of these, usually only the third-order terms are important, because the amplitudes of higher-order terms tend to be significantly smaller. If the operating frequency band is less than an octave, all other intermodulation products will fall outside the passband and can be eliminated with appropriate filters in the receiver.

4.5 Summary

This chapter examines the basic operating characteristics of heterojunction-structured light-emitting diodes (LEDs) and laser diodes. The first topic deals with the basic structure of these sources, which is a sandwich type construction of different semiconductor materials. These layers serve to confine the electrical and optical carriers to yield optical sources with high outputs and high efficiencies. The principal materials of which these layers are composed include the ternary alloy GaAlAs for operation in the 800-nm to 900-nm wavelength region and the quaternary alloy InGaAsP for use between 1100 and 1700 nm.

An important source characteristic associated with the quantum efficiency of a light source is the modulation capability and the response to transient current pulses. By applying a small dc bias to the source, the time delay between the application of a current pulse and the onset of optical power output can be reduced. This bias reduces the parasitic diode space charge capacitance that could cause a delay of the carrier injection into the active region. Such an injection delay otherwise could postpone the start of the optical output.

When deciding whether to choose an LED or a laser diode source, a tradeoff must be made between the advantages and limitations of each type of device. The advantages that a laser diode has over an LED are as follows:

1. A faster response time, so that much higher modulation rates (higher data transmission rates) are possible with a laser diode.
2. A narrower spectral width of the laser output, which implies less dispersion-induced pulse broadening during data transmission.
3. A much higher optical power level that can be coupled from a laser diode into a fiber, thus allowing longer transmission distances.

Compared to LED sources some application issues of laser diodes are as follows:

1. The construction of laser diodes is more complex, mainly because of the requirement of current confinement in a small lasing cavity. This makes the laser diode more expensive than an LED.
2. The optical output power level of a laser is strongly dependent on temperature. If a laser diode is to be used over a wide external temperature range, then either a cooling mechanism (such as a thermoelectric cooler) must be used to maintain the laser at a constant temperature or a circuit that senses the lasing threshold can be implemented to adjust the bias current with changes in temperature. Using a thermoelectric cooler typically is the preferred temperature control method.

For high-speed long-distance communications one needs single-mode lasers, which must contain only a single longitudinal mode and a single transverse mode. Consequently, the spectral width of the optical emission pattern is very narrow. Popular single-mode lasers include the distributed-feedback (DFB) laser and the vertical-cavity surface-emitting laser (VCSEL).

Problems

4.1 Measurements show that the bandgap energy E_g for GaAs varies with temperature according to the empirical formula

$$E_g(T) \approx 1.55 - 4.3 \times 10^{-4} T$$

where E_g is given in electron volts (eV). Using this expression, show that the temperature dependence of the intrinsic electron concentration n_i is

$$n_i = 5 \times 10^{15} T^{3/2} e^{-8991/T}$$

4.2 Repeat the steps given in Example 4.2 for a p-type semiconductor. In particular, show that when the net acceptor concentration is much greater than n_i, then $p_p = N_A$ and $n_p = n_i^2/N_A$.

4.3 An engineer has two $Ga_{1-x}Al_xAs$ LEDs: one has a bandgap energy of 1.540 eV and the other has $x = 0.015$.

(a) Find the aluminum mole fraction x and the emission wavelength for the first LED.

(b) Find the bandgap energy and the emission wavelength of the other LED.

[Answer:
(a) $x = 0.090$, $\lambda = 805$ nm; (b) $E_g = 1.620$ eV, $\lambda = 766$ nm.]

4.4 The lattice spacing of $In_{1-x}Ga_xAs_yP_{1-y}$ has been shown to obey Vegard's law. This states that for quaternary alloys of the form $A_{1-x}B_xC_yD_{1-y}$, where A and B are group III elements (e.g., Al, In, and Ga) and C and D are group V elements (e.g., As, P, and Sb), the lattice spacing $a(x, y)$ of the quaternary alloy can be approximated by

$$a(x, y) = xya(BC) + x(1 - y)a(BD)$$
$$+ (1 - x)ya(AC) + (1 - x)(1 - y)a(AD)$$

where the $a(IJ)$ parameters are the lattice spacings of the binary compounds IJ.

(a) Show that for $In_{1-x}Ga_xAs_yP_{1-y}$ with

$$a(GaAs) = 5.6536 \,\text{Å}$$
$$a(GaP) = 5.4512 \,\text{Å}$$

$$a(\text{InAs}) = 6.0590 \,\text{Å}$$

$$a(\text{InP}) = 5.8696 \,\text{Å}$$

where $1 \,\text{Å} = 10^{-10}$ m, the quaternary lattice spacing becomes

$$a(x, y) = 0.1894y - 0.4184x + 0.0130xy + 5.8696 \,\text{Å}$$

(b) For quaternary alloys that are lattice-matched to InP, the relation between x and y can be determined by letting $a(x, y) = a(\text{InP})$. Show that because $0 \le x \le 0.47$, the resulting expression can be approximated by $y \approx 2.20x$.

(c) A simple empirical relation that gives the bandgap energy in eV in terms of x and y is

$$E_g(x, y) = 1.35 + 0.668x - 1.17y + 0.758x^2 + 0.18y^2$$
$$- 0.069xy - 0.322x^2 y + 0.33xy^2.$$

Find the bandgap energy and the peak emission wavelength of $\text{In}_{0.74}\text{Ga}_{0.26}\text{As}_{0.56}\text{P}_{0.44}$. [Answer: (c) $E_g = 0.956$ eV; $\lambda(\mu\text{m}) = 1.240/ E_g(\text{eV}) = 1.297\ \mu\text{m}$.]

4.5 Using the expression $E = hc/\lambda$, show why the FWHM power spectral width of LEDs becomes wider at longer wavelengths. [Answer: Differentiating the expression for E yields $\Delta\lambda = \frac{\lambda^2}{hc}\Delta E$. For the same energy difference ΔE, the spectral width $\Delta\lambda$ is proportional to the wavelength squared. Thus, for example, $\frac{\Delta\lambda_{1550}}{\Delta\lambda_{1310}} = \left(\frac{1550}{1310}\right)^2 = 1.40$.]

4.6 A double-heterojunction InGaAsP LED emitting at a peak wavelength of 1310 nm has radiative and nonradiative recombination times of 25 and 90 ns, respectively. The drive current is 35 mA.

(a) Find the internal quantum efficiency and the internal power level.

(b) If the refractive index of the light source material is $n = 3.5$, find the power emitted from the device.

[Answer: (a) $\eta_{\text{int}} = 0.783$, $P_{\text{int}} = 26$ mW; (b) 0.37 mW].

4.7 Assume the injected minority carrier lifetime of an LED is 5 ns and that the device has an optical output of 0.30 mW when a constant dc drive current is applied. Using Eq. (4.18), plot the optical output power when the LED is modulated at frequencies ranging from 20 to 100 MHz. Note what happens to the LED output power at higher modulation frequencies.

4.8 Consider an LED having a minority carrier lifetime of 5 ns. Find the 3-dB optical bandwidth and the 3-dB electrical bandwidth. [Answer: The 3-dB optical bandwidth = 9.5 MHz; the 3-dB electrical bandwidth = 6.7 MHz.]

4.9 (a) A GaAlAs laser diode has a 500-μm cavity length, which has an effective absorption coefficient of 10 cm^{-1}. For uncoated facets the reflectivities are

0.32 at each end. What is the optical gain at the lasing threshold?

(b) If one end of the laser is coated with a dielectric reflector so that its reflectivity is now 90%, what is the optical gain at the lasing threshold?

(c) If the internal quantum efficiency is 0.65, what is the external quantum efficiency in cases (a) and (b)? [Answers: (a) 55.6 cm^{-1}; (b) 34.9 cm^{-1}; (c) η_{ext} = 0.53 and 0.46, respectively.]

4.10 Find the external quantum efficiency for a Ga$_{1-x}$Al$_x$As laser diode (with x = 0.03) that has an optical-power-versus-drive-current relationship of 0.5 mW/mA. [Hint: Use Eq. (4.3) to find E_g = 1.462 eV and then find λ = 848 nm. Then from Eq. (4.38) η_{ext} = 0.342.]

4.11 Approximate expressions for the transverse and lateral optical-field confinement factors Γ_T and Γ_L, respectively, in a Fabry-Perot lasing cavity are

$$\Gamma_T = \frac{D^2}{2 + D^2} \text{ with } D = \frac{2\pi d}{\lambda}\left(n_1^2 - n_2^2\right)^{1/2}$$

and

$$\Gamma_L = \frac{W^2}{2 + W^2} \text{ with } W = \frac{2\pi w}{\lambda}\left(n_{eff}^2 - n_2^2\right)^{1/2}$$

where

$$n_{eff}^2 = n_2^2 + \Gamma_T\left(n_1^2 - n_2^2\right)$$

Here w and d are the width and thickness, respectively, of the active layer, and n_1 and n_2 are the refractive indices inside and outside the cavity, respectively.

(a) Consider a 1300-nm InGaAsP laser diode in which the active region is 0.1 μm thick, 1.0 μm wide, and 250 μm long with refractive indices n_1 = 3.55 and n_2 = 3.20. What are the transverse and lateral optical-field confinement factors?

(b) Given that the total confinement factor is.

$\Gamma = \Gamma_T\Gamma_L$, what is the gain threshold if the effective absorption coefficient is α_{mat} = 30 cm^{-1} and the facet reflectivities are $R_1 = R_2 = 0.31$?

[Answer: (a) Γ_T= 0.216 and Γ_L= 0.856; (b) Γ = 0.185].

4.12 A GaAs laser emitting at 800 nm has a 400-μm cavity length with a refractive index n = 3.6. If the gain g exceeds the total loss α_t throughout the range 750 nm < λ < 850 nm, how many modes will exist in the laser? [Answer: $\Delta\lambda$ = 0.22 nm yielding 455 modes].

4.13 A laser emitting at λ_0 = 850 nm has a gain-spectral width of σ = 32 nm and a peak gain of $g(0)$ = 50 cm^{-1}. Plot $g(\lambda)$ from Eq. 4.41. If α_t= 32.2 cm^{-1}, show the region where lasing takes place. If the laser is 400 μm long and n = 3.6, how many modes will be excited in this laser? [Answer: $\Delta\lambda$ = 0.25 nm yielding 240 modes].

4.14 The derivation of Eq. (4.46) assumes that the refractive index n is independent of wavelength.

(a) Show that when n depends on λ, we have

$$\Delta\lambda = \frac{\lambda^2}{2L\left(n - \lambda\frac{dn}{d\lambda}\right)}$$

(b) If the group refractive index $(n - \lambda dn/d\lambda)$ is 4.5 for GaAs at 850 nm, what is the mode spacing for a 400-μm-long laser? [Answer: (b) $\Delta\lambda = 0.20$ nm].

4.15 For laser structures that have strong carrier confinement, the threshold current density for stimulated emission J_{th} can to a good approximation be related to the lasing-threshold optical gain g_{th} by $g_{th} = \beta_{dev}J_{th}$ where β_{dev} is a constant that depends on the specific device construction. Consider a GaAs laser with an optical cavity of length 250 μm and width 100 μm. At the normal operating temperature, the gain factor $\beta_{dev} = 21 \times 10^{-3}$ A/cm^3 and the effective absorption coefficient $\alpha_{mat} = 10$ cm^{-1}.

(a) If the refractive index is 3.6, find the threshold current density and the threshold current I_{th}. Assume the laser end faces are uncoated and the current is restricted to the optical cavity.

(b) What is the threshold current if the laser cavity width is reduced to 10 μm?

[Answers: (a) $J_{th} = 2.65 \times 10^3$ A/cm^2 and $I_{th} = J_{th} \times 1 \times w = 663$ mA.

(b) $I_{th} = 66.3$ mA].

4.16 From quantum mechanics, the energy levels for electrons and holes in the quantum-well laser structure shown in Fig. 4.23 are given by

$$E_{ci} = E_c + \frac{h^2}{8d^2}\frac{i^2}{m_e} \quad \text{with } i = 1, 2, 3, \dots \text{ for electrons}$$

and

$$E_{vj} = E_v - \frac{h^2}{8d^2}\frac{j^2}{m_h} \quad \text{with } j = 1, 2, 3, \dots \text{ for holes}$$

where E_c and E_v are the conduction-band and valence-band energies (see Fig. 4.1), d is the active layer thickness, h is Planck's constant, and m_e and m_h are the electron and hole masses as defined in Example 4.1. The possible energy-level transitions that lead to photon emission are given by

$$\Delta E_{ij} = E_{ci} - E_{vj} = E_g + \frac{h^2}{8d^2}\left(\frac{i^2}{m_e} + \frac{j^2}{m_v}\right)$$

If $E_g = 1.43$ eV for GaAs, what is the emission wavelength between the $i = j = 1$ states if the active layer thickness is $d = 5$ nm?

[Answer: $\lambda = 739$ nm].

4.17   In a multiple quantum-well laser the temperature dependence of the differential or external quantum efficiency can be described by

$$\eta_{ext}(T) = \eta_i(T) \frac{\alpha_{end}}{N_w[\alpha_w + \gamma(T - T_{th})] + \alpha_{end}}$$

where $\eta_i(T)$ is the internal quantum efficiency, α_{end} is the mirror loss of the lasing cavity as given in Eq. (4.28), N_w is the number of quantum wells, T_{th} is the threshold temperature, α_w is the internal loss of the wells at $T = T_{th}$ and γ is a temperature-dependent internal-loss parameter. Consider a six-well, 350-μm-long MQW laser having the following characteristics: $\alpha_w = 1.25$ cm^{-1}, $\gamma = 0.025$ cm^{-1}/K, and $T_{th} = 303$ K. The lasing cavity has a standard uncoated facet on the front ($R_1 = 0.31$) and a high-reflection coating on the near facet ($R_2 = 0.96$).

(a) Assuming that the internal quantum efficiency is constant, plot the external quantum efficiency as a function of temperature over the range 303 K $\leq T \leq$ 375 K. Let $\eta_{ext}(T) = 0.8$ at $T = 303$ K.
   (b) Given that the optical output power at $T = 303$ K is 30 mW at a drive current of $I_d = 50$ mA, plot the power output as a function of temperature over the range 303 K $\leq T \leq$ 375 K at this fixed bias current.

4.18   A distributed-feedback laser has a Bragg wavelength of 1570 nm, a second-order grating with $\Lambda = 460$ nm, and a 300-μm cavity length. Assuming a perfectly symmetrical DFB laser, find the zeroth-order, first-order, and second-order lasing wavelengths to a tenth of a nanometer. Draw a relative amplitude-versus-wavelength plot.

4.19   When a current pulse is applied to a laser diode, the injected carrier pair density n within the recombination region of width d changes with time according to the relationship

$$\frac{\partial n}{\partial t} = \frac{J}{qd} - \frac{n}{\tau}$$

Assume τ is the average carrier lifetime in the recombination region when the injected carrier pair density is n_{th} near the threshold current density J_{th}. That is, in the steady state $\partial n/\partial t = 0$, so that

$$n_{th} = \frac{J_{th}\tau}{qd}$$

If a current pulse of amplitude I_p is applied to an unbiased laser diode, show that the time needed for the onset of stimulated emission is

$$t_d = \tau \ln \frac{I_p}{I_p - I_{th}}$$

Assume the drive current $I = JA$, where J is the current density and A is the area of the active region.

(b) If the laser is now prebiased to a current density $J_B = I_B/A$, so that the initial excess carrier pair density is $n_B = J_B \tau/qd$, then the current density in the active region during a current pulse I_p is $J = J_B + J_p$. Show that in this case Eq. (4.50) results.

4.20 A laser diode has a maximum average output of 1 mW (0 dBm). The laser is to be amplitude-modulated with a signal $x(t)$ that has a dc component of 0.2 and a periodic component of ± 2.56. If the current-input to optical-output relationship is $P(t) = i(t)/10$, find the values of I_0 and m if the modulating current is $i(t) = I_0 [1 + mx(t)]$.

[Answer: $i(t) = I_0 [1 + mx(t)]$ mA $= 9.2 [1 + 0.42 \, x(t)]$ mA].

4.21 Consider the following taylor series expansion of the optical-power-versus-drive-current relationship of an optical source about a given bias point:

$$y(t) = a_1 x(t) + a_2 x^2(t) + a_3 x^3(t) + a_4 x^4(t)$$

Let the modulating signal $x(t)$ be the sum of two sinusoidal tones at frequencies ω_1 and ω_2 given by

$$x(t) = b_1 \cos \omega_1 t + b_2 \cos \omega_2 t$$

(a) Find the second-, third-, and fourth-order intermodulation distortion coefficients B_{mn} (where m and $n = \pm 1, \pm 2, \pm 3,$ and ± 4) in terms of b_1, b_2, and the a_i.

(b) Find the second-, third-, and fourth-order harmonic distortion coefficients A_2, A_3 and A_4 in terms of b_1, b_2, and the a_i.

References

1. L.A. Coldren, S.W. Corzine, M.L. Mashanovitch, *Diode Lasers and Photonic Integrated Circuits*, 2nd edn. (Wiley, 2012)
2. T. Numai, *Fundamentals of Semiconductor Lasers*, 2nd edn. (Springer, 2015)
3. D.J. Klotzkin, *Introduction to Semiconductor Lasers for Optical Communications* (Springer, 2020)
4. D.A. Neaman, 4th edn. *Semiconductor Physics and Devices* (McGraw-Hill, 2012)

5. B.L. Anderson, R.L. Anderson, *Fundamentals of Semiconductor Devices*, 2nd edn. (McGraw-Hill, 2018)
6. S.O. Kasap, *Principles of Electronic Materials and Devices*, 4th edn. (McGraw-Hill, 2018)
7. J.J. Coleman, A.C. Bryce, C. Jagadish (eds.) *Advances in Semiconductor Lasers* (Academic Press, 2012)
8. S. Adachi, III–V ternary and quaternary compounds, Chap. 1, in *Springer Handbook of Electronic and Photonic Materials*, ed. by S. Kasap, P. Capper (Springer 2017)
9. C.A. Burrus, B.I. Miller, Small-area double heterostructure AlGaAs electroluminescent diode sources for optical fiber transmission lines. Opt. Commun. **4**, 307–309 (1971)
10. B.E.A. Saleh, M.C. Teich, *Fundamentals of Photonics*, 3rd edn. (Wiley, 2019)
11. T.P. Lee, Effects of junction capacitance on the rise time of LEDs and the turn-on delay of injection lasers. Bell Sys. Tech. J. **54**, 53–68 (1975)
12. M.N. Zervas, Chap. 1, Advances in fiber distributed-feedback lasers, in *Optical Fiber Telecommunications Vol. VIA: Components and Subsystems*, ed. by I. Kaminow, T. Li, A.E. Willner (Academic Press, 2013)
13. E.O. Odoh, A.S. Njapba, A review of semiconductor quantum well devices. Adv. Physics Theories Appl. **46**, 26–32 (2015)
14. R. Michalzik, *VCSELs: Fundamentals, Technology and Applications of Vertical-Cavity Surface-Emitting Lasers* (Springer 2013)
15. C.C. Shen et al., Design, modeling, and fabrication of high-speed VCSEL with data rate up to 50 Gb/s. Nanoscale Res. Lett. **14**, 276 (2019)
16. M. Mohsin, D. Schall, M. Otto, A. Noculak, D. Neumaier, H. Kurz, Graphene based low insertion loss electroabsorption modulator on SOI waveguide. Opt. Express, **22**(12), 15292–15297 (2014)
17. M. He et al., High-performance hybrid silicon and lithium niobate Mach-Zehnder modulators for 100 Gb/s and beyond. Nat. Photon. **13**(5), 359–364 (2019)
18. W. Zhu, Y. Deng, Y. Wang, A. Wang, Finite element analysis of miniature thermoelectric coolers with high cooling performance and short response time. Microelectron. J. **44**, 860–868 (2013)

Chapter 5
Optical Power Coupling

Abstract Having examined the characteristics of optical fibers and the associated light sources used with them, the next step is to investigate how to launch optical power into a particular fiber efficiently from some type of luminescent source. In conjunction with this topic is the question is how to couple optical power from one fiber into another. Each jointing technique is subject to certain conditions, which can cause varying degrees of optical power loss at the joint. The purpose of this chapter is to highlight these conditions and determine methods for minimizing the joint loss.

In implementing an optical fiber link, two of the major system questions are how to launch optical power into a particular fiber from some type of luminescent source and how to couple optical power from one fiber into another. Launching optical power from a source into a fiber entails considerations such as the numerical aperture, core size, refractive-index profile, and core-cladding index difference of the fiber, plus the size, radiance, and angular power distribution of the optical source.

A measure of the amount of optical power emitted from a source that can be coupled into a fiber is usually given by the *coupling efficiency* η defined as

$$\eta = \frac{P_F}{P_S}$$

Here P_F is the power coupled into the fiber and P_S is the power emitted from the light source. The launching or coupling efficiency depends on the type of fiber that is attached to the source and on the coupling process; for example, whether lenses or other coupling improvement schemes are used.

In practice, many source suppliers offer devices with a short length of optical fiber (1 m or less) already attached in an optimum power-coupling configuration. This section of fiber is generally referred to as a *flylead* or a *pigtail*. The power-launching problem for these pigtailed sources thus reduces to a simpler one of coupling optical power from one fiber into another. The effects to be considered in this case include fiber misalignments; different core sizes, numerical apertures,

The original version of this chapter was revised: All the incorrect sentences have been corrected. The correction to this chapter can be found at https://doi.org/10.1007/978-981-33-4665-9_15

209
G. Keiser, *Fiber Optic Communications*,
https://doi.org/10.1007/978-981-33-4665-9_5

and core refractive-index profiles; plus the need for clean and smooth fiber end faces that are either perpendicular to the axis or that are polished at a slight angle to prevent back reflections of optical power into the device.

An alternate arrangement consists of light sources and optical fiber receptacles that are integrated within a transceiver package. To achieve fiber-to-fiber coupling in this case, the fiber connector from a cable is simply mated to the built-in connector in the transceiver package. Among the various commercially available configurations are the popular *small-form-factor* (SFF) and the SFF pluggable (SFP) devices.

5.1 Source-to-Fiber Power Coupling

A convenient and useful measure of the optical output of a luminescent source is its radiance at a given diode drive current. *Radiance* is traditionally designated by the symbol L and is the optical power radiated into a unit solid angle per unit emitting surface area. It is generally specified in terms of watts per square centimeter per steradian. Because the optical power that can be coupled into a fiber depends on the radiance (i.e., on the spatial distribution of the optical power), the radiance of an optical source rather than the total output power is the important parameter when considering source-to-fiber coupling efficiencies.

5.1.1 Light Source Emission Patterns

To determine the optical power-accepting capability of a fiber, the spatial radiation pattern of the source must first be known. This pattern can be fairly complex. Consider Fig. 5.1, which shows a spherical coordinate system characterized by R, θ, and φ, with the normal to the emitting surface being the polar axis. The radiance may be a function of both θ and φ, and can also vary from point to point on the emitting surface. A reasonable assumption for simplicity of analysis is to take the emission to be uniform across the source area.

Fig. 5.1 Spherical coordinate systems for characterizing the emission pattern from an optical source

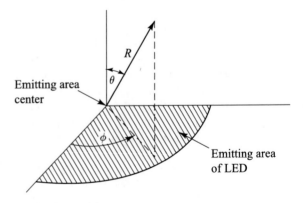

Surface-emitting LEDs are characterized by their Lambertian output pattern, which means the source has the same luminous intensity (or colloquially "is equally bright") when viewed from any direction. The power delivered at an angle θ, measured relative to a normal to the emitting surface, varies as $\cos \theta$ because the projected area of the emitting surface varies as $\cos \theta$ with viewing direction. The emission pattern for a Lambertian source thus follows the relationship.

$$L(\theta, \varphi) = L_0 \cos \theta \tag{5.1}$$

where L_0 is the radiance along the normal to the radiating surface. The radiance pattern for this type of source is shown in Fig. 5.2.

Drill Problem 5.1 (*a*) At what angle from the normal in a Lambertian LED is the power level 50 percent of the normal level? This is the half-power point. (*b*) At what angle from its axis does the light appear only 40 percent as intense as it does when viewed down its centerline? **Answers:** [(*a*) 60°; (*b*) 67°.]

Edge-emitting LEDs and laser diodes have a more complex emission pattern. These devices have different radiances $L(\theta, 0°)$ and $L(\theta, 90°)$ in the planes parallel and normal, respectively, to the emitting-junction plane of the device. These radiances can be approximated by the general form [1, 2]

Fig. 5.2 Radiance patterns for a Lambertian source and the lateral output of a highly directional laser diode with L_0 normalized to unity for both sources

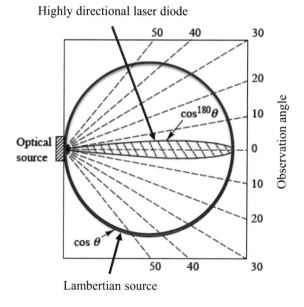

Highly directional laser diode

Lambertian source

$$\frac{1}{L(\theta, \phi)} = \frac{\sin^2\phi}{L_0\cos^{d1}\theta} + \frac{\cos^2\phi}{L_0\cos^{d2}\theta} \tag{5.2}$$

The integers d_1 and d_2 are the transverse and lateral power distribution coefficients, respectively.

Example 5.1 Figure 5.2 compares a Lambertian pattern with a laser diode that has a lateral ($\varphi = 0°$) half-power beam width of $2\theta = 10°$. What is the lateral power distribution coefficient of the laser diode?

Solution From Eq. (5.2), it follows that $L(\theta = 5°, \varphi = 0°) = L_0(\cos 5°)^{d_2} = 0.5L_0$. Solving for d_2,

$$d_2 = \frac{\log 0.5}{\log(\cos 5°)} = \frac{\log 0.5}{\log 0.9962} = 182$$

The much narrower output beam from a laser diode allows significantly more optical power to be coupled into a fiber.

Drill Problem 5.2 Verify that the relative radiance at the angular position $\varphi = 45°$ and $\theta = 5°$ for a laser diode with $d_2 = 1$ and $d_1 = 150$ is $L(\theta, \varphi) = 0.721\, L_0$.

Drill Problem 5.3 For a laser diode, consider a modified Lambertian approximation to the emission pattern of the form.

$$L = L_0\cos^m\theta$$

Suppose that in a certain laser diode the -3-dB power level occurs at an angle of $15°$ from the normal to the emitting surface. Show that the value of m is 20.

In general, for edge emitters, $d_2 = 1$ (which is a Lambertian distribution with a $120°$ half-power beam width) and d_1 is significantly larger. For laser diodes, d_1 can take on values over 100.

5.1.2 Calculation of Power Coupling

To calculate the maximum optical power coupled into a fiber, consider first the case shown in Fig. 5.3 for a symmetric source of radiance $L(A_s, \Omega_s)$, where A_s and Ω_s are

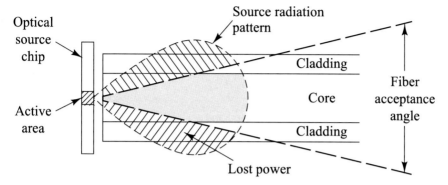

Fig. 5.3 Schematic diagram of a light source coupled to an optical fiber; light falling outside of the acceptance angle is lost

the area and solid emission angle of the source, respectively. Here, the fiber end face is centered over the emitting surface of the source and is positioned as close to it as possible. The coupled power can be found using the relationship

$$P = \int\limits_{A_f} dA_s \int\limits_{\Omega_f} d\Omega_s L(A_s, \Omega_s)$$

$$= \int\limits_{0}^{r_m} \int\limits_{0}^{2\pi} \left[\int\limits_{0}^{2\pi} \int\limits_{0}^{\Theta_A} L(\theta, \phi) \sin\theta \, d\theta \, d\phi \right] d\theta_s \, r \, dr \qquad (5.3)$$

where the area A_f and solid acceptance angle Ω_f of the fiber define the limits of the integrals. In this expression, first the radiance $L(\theta, \varphi)$ from an individual radiating point source on the emitting surface is integrated over the solid acceptance angle of the fiber. This is shown by the expression in square brackets, where θ_A is the acceptance angle of the fiber, which is related to the numerical aperture NA through Eq. (2.23). The total coupled power is then determined by summing up the contributions from each individual emitting point source of incremental area $d\theta_s r \, dr$, that is, integrating over the emitting area. For simplicity, here the emitting surface is taken as being circular. If the source radius r_s is less than the fiber core radius a, then the upper integration limit $r_m = r_s$; for source areas larger than the fiber-core area, $r_m = a$.

As an example, assume a surface-emitting LED has a radius r_s that is less than the fiber core radius a. Because this is a Lambertian emitter, Eq. (5.1) applies and Eq. (5.3) becomes

$$P = \int\limits_{0}^{r_s} \int\limits_{0}^{2\pi} \left[2\pi L_0 \int\limits_{0}^{\Theta_A} \cos\theta \, \sin\theta \, d\theta \right] d\theta_s \, r \, dr$$

$$= \pi L_0 \int_0^{r_s} \int_0^{2\pi} \sin^2\theta_A d\theta_s \, r \, dr$$

$$= \pi L_0 \int_0^{r_s} \int_0^{2\pi} NA^2 d\theta_s \, r \, dr \qquad (5.4)$$

where the numerical aperture NA is defined by Eq. (2.23). For step-index fibers the numerical aperture is independent of the positions θ_s and r on the fiber end face, so that Eq. (5.4) becomes (for $r_s < a$)

$$P_{LED,step} = \pi^2 r_s^2 L_0 (NA)^2 \approx 2\pi^2 r_s^2 L_0 n_1^2 \Delta \qquad (5.5)$$

Consider now the total optical power P_s that is emitted from the source of area A_s into a hemisphere (2π sr). This is given by

$$P_s = A_s \int_0^{2\pi} \int_0^{\pi/2} L(\theta, \phi) \sin\theta \, d\theta \, d\phi$$

$$= \pi r_s^2 2\pi L_0 \int_0^{\pi/2} \cos\theta \sin\theta \, d\theta$$

$$= \pi^2 r_s^2 L_0 \qquad (5.6)$$

Equation (5.5) can, therefore, be expressed in terms of P_s:

$$P_{LED,step} = P_s (NA)^2 \text{ for } r_s \leq a \qquad (5.7)$$

When the radius of the emitting area is larger than the fiber core radius, only the power from the fractional area $\pi a^2 / \pi r_s^2$ is coupled into the fiber, so Eq. (5.7) becomes

$$P_{LED,step} = \left(\frac{a}{r_s}\right)^2 P_s (NA)^2 \text{ for } r_s > a \qquad (5.8)$$

Example 5.2 Consider an LED that has a circular emitting area of radius 35 μm and a Lambertian emission pattern with 150 W/(cm^2-sr) axial radiance at a given drive current. Compare the optical powers coupled into two step-index fibers, one of which has a core radius of 25 μm with NA $= 0.20$ and the other which has a core radius of 50 μm with NA $= 0.20$.

Solution For the larger core fiber, use Eqs. (5.6) and (5.7) to get

$$P_{LED,step} = P_s (NA)^2 = \pi^2 r_s^2 L_0 (NA)^2$$

$$= \pi^2 (0.0035 \text{ cm})^2 [150 \text{ W}/(\text{cm}^2 \text{ sr})](0.20)^2$$
$$= 0.725 \text{ mW}$$

For the case when the fiber end-face area is smaller than the emitting surface area, one needs to use Eq. (5.8). Thus the coupled power is less than the above case by the ratio of the radii squared:

$$P_{LED,step} = \left(\frac{25 \ \mu\text{m}}{35 \ \mu\text{m}}\right)^2 P_s (NA)^2 = \left(\frac{25 \ \mu\text{m}}{35 \ \mu\text{m}}\right)^2 (0.725 \text{ mW})$$
$$= 0.37 \text{ mW} = -4.32 \text{ dBm}$$

In the case of a graded-index fiber, the numerical aperture depends on the distance r from the fiber axis through the relationship defined by Eq. (2.40). Thus using Eqs. (2.40) and (2.41), the power coupled from a surface-emitting LED into a graded-index fiber becomes (for $r_s < a$)

$$P_{LED,graded} = 2\pi^2 L_0 \int_0^{r_s} [n^2(r) - n_2^2] r \, dr$$

$$= 2\pi^2 r_s^2 L_0 n_1^2 \Delta \left[1 - \frac{2}{\alpha + 2}\left(\frac{r_s}{a}\right)^\alpha\right] = 2P_s n_1^2 \Delta \left[1 - \frac{2}{\alpha + 2}\left(\frac{r_s}{a}\right)^\alpha\right]$$
$$\tag{5.9a}$$

where the last expression was obtained from Eq. (5.6). When the source radius is larger than the fiber core radius, the upper limit of integration becomes $r = a$ and only the radiance from the fractional area $(a/r_s)^2$ is coupled into the fiber. Then (for $r_s > a$)

$$P_{LED,graded} = 2\pi^2 a^2 L_0 n_1^2 \Delta \frac{\alpha}{\alpha + 2} = \pi^2 a^2 L_0 NA(0)^2 \frac{\alpha}{\alpha + 2} \tag{5.9b}$$

where Eq. (2.41) is used for the approximation of the axial numerical aperture NA(0) on the right-hand side. The power launched into a fiber from an edge-emitting LED that has a noncylindrical distribution is a bit complex [3].

Drill Problem 5.4 Consider an LED that has a circular emitting area of radius $r_s = 35 \ \mu\text{m}$ and a Lambertian emission pattern with an axial radiance $L_0 = 150 \text{ W}/(\text{cm}^2\cdot\text{sr})$ at a given drive current. (a) Show that the optical power coupled into a graded-index fiber which has a core radius of 50 μm with NA(0) = 0.20 and $\alpha = 2.0$ is 0.55 mW = −2.62 dBm. (b) Show that the optical power coupled into a graded-index fiber which has a core radius of 25 μm with NA(0) = 0.20

and $\alpha = 2.0$ is 0.185 mW $= -7.33$ dBm. (c) How do these numbers compare to the values for similar sized step-index fibers described in Example 5.2?

These analyses assumed perfect coupling conditions between the source and the fiber. This can be achieved only if the refractive index of the medium separating the source and the fiber end matches the index n_1 of the fiber core. If the refractive index n of this medium is different from n_1, then for perpendicular fiber end faces the power coupled into the fiber reduces by the factor

$$R = \left(\frac{n_1 - n}{n_1 + n}\right)^2 \tag{5.10}$$

where R is the *Fresnel reflection* or the *reflectivity* at the fiber-core end face (see Sect. 2.11). The ratio $r = (n_1 - n)/(n_1 + n)$, which is known as the *reflection coefficient*, relates the amplitude of the reflected wave to the amplitude of the incident wave.

Example 5.3 A GaAs optical source with a refractive index of 3.6 is coupled to a silica fiber that has a refractive index of 1.48. What is the power loss between the source and the fiber?

Solution If the fiber end and the source are in close physical contact, then, from Eq. (5.10), the Fresnel reflection at the interface is

$$R = \left(\frac{n_1 - n}{n_1 + n}\right)^2 = \left(\frac{3.60 - 1.48}{3.60 + 1.48}\right)^2 = 0.174$$

This value of R corresponds to a reflection of 17.4% of the emitted optical power back into the source. Given that

$$P_{\text{coupled}} = (1 - R) P_{\text{emitted}}$$

the power loss L_{power} in decibels is found from:

$$L_{power} = -10\log\left(\frac{P_{\text{coupled}}}{P_{\text{emitted}}}\right) = -10\log(1 - R)$$
$$= -10\log(0.826) = 0.83 \,\text{dB}$$

An index-matching material between the source and the fiber end can reduce this loss number.

Example 5.4 An InGaAsP light source that has a refractive index of 3.540 is coupled to a step-index fiber that has a core refractive index of 1.480. Assume that the source size is smaller than the fiber core and that there is a small gap between the source and the fiber. (*a*) If the gap is filled with a gel that has a refractive index of 1.520, what is the power loss in decibels from the source into the fiber? (*b*) What is the power loss if no gel is used to fill the small gap?

Solution (*a*) Here we need to consider the reflectivity at two interfaces. First, using Eq. (5.10) the reflectivity R_{sg} at the source-to-gel interface is

$$R_{sg} = \left(\frac{n_{source} - n_{gel}}{n_{source} + n_{gel}}\right)^2 = \left(\frac{3.540 - 1.520}{3.540 + 1.520}\right)^2 = 0.159$$

Similarly, using Eq. (5.10) the reflectivity R_{gf} at the gel-to-fiber interface is

$$R_{gf} = \left(\frac{n_{fiber} - n_{gel}}{n_{fiber} + n_{gel}}\right)^2 = \left(\frac{1.480 - 1.520}{1.480 + 1.520}\right)^2 = 1.777 \times 10^{-4}$$

The transmission T through the matching gel then is the product of two individual transmissions:

$$\begin{aligned} T &= \left(1 - R_{sg}\right) \times \left(1 - R_{gf}\right) \\ &= (1 - 0.159) \times \left(1 - 1.777 \times 10^{-4}\right) \\ &= 0.841 \end{aligned}$$

The power lost (as a ratio) is

$$L_{power} = 1 - T = 0.159 \text{ (i.e., 15.9\% of the power is lost)}$$

The power loss in decibels is

$$L_{power}(\text{dB}) = -10\log(1 - R) = -10\log 0.841 = 0.752 \text{ dB}$$

(*b*) Similarly, the transmission T through the air gap is the product of two individual transmissions:

$$\begin{aligned} T &= \left[1 - \left(\frac{n_{source} - n_{air}}{n_{source} + n_{air}}\right)^2\right] \times \left[1 - \left(\frac{n_{fiber} - n_{air}}{n_{fiber} + n_{air}}\right)^2\right] \\ &= (1 - 0.313) \times (1 - 0.037) = 0.662 \end{aligned}$$

The power lost (as a ratio) is

$$L_{power} = 1 - T = 0.338 \text{ (i.e., 33.8\% of the power is lost)}$$

In this case the power loss in decibels is

$$L_{power}(\text{dB}) = -10 \log(1 - R) - 10 \log 0.662 = 1.791 \, \text{dB}$$

5.1.3 Optical Coupling Versus Wavelength

It is of interest to note that the optical power launched into a fiber does not depend on the wavelength of the source but only on its radiance. To explore this concept a little further, Eqs. (2.30) and (2.42) show that the number of modes that can propagate in a multimode fiber (either step-index or graded-index) is proportional to the inverse wavelength squared:

$$M \propto \lambda^{-2} \tag{5.11}$$

Thus, for example, twice as many modes propagate in a given fiber at 900 nm than at 1300 nm.

The radiated power per mode, P_s/M, from a source at a particular wavelength is given by the radiance multiplied by the square of the nominal source wavelength [4],

$$\frac{P_s}{M} = L_0 \lambda^2 \tag{5.12}$$

Thus twice as much power is launched into a given mode at 1300 nm than at 900 nm. Hence, two identically sized sources operating at different wavelengths but having identical radiances will launch equal amounts of optical power into the same fiber.

> **Drill Problem 5.5** Consider two identically sized optical sources, one emitting at 895 nm and the other at 1550 nm. (a) Verify that about three times as many modes propagate in the same multimode fiber at 895 nm than at 1550 nm. (b) Show that if the two sources have equal radiances then about three times as much power is launched into a given mode at 1550 nm compared to 895 nm. (c) Using Eqs. (5.11) and (5.12) show that these two sources will launch equal amounts of power into the same fiber.

5.1.4 Equilibrium Numerical Aperture

As noted earlier, a light source may be supplied with a short (typically 1 m long) fiber flylead attached to it in order to facilitate coupling the source to a system fiber. To achieve a low coupling loss, this flylead should be connected to a system fiber that has a nominally identical NA and core diameter. A certain amount of optical power (ranging from 0.1 to 1 dB) is lost at this junction, the exact loss depending on the connecting mechanism and on the fiber type; this is discussed in Sect. 5.3. In addition to the coupling loss, an excess power loss will occur in the first few tens of meters of a multimode system fiber. This excess loss is a result of nonpropagating modes scattering out of the fiber as the launched modes come to an equilibrium condition. This loss is of particular importance for surface-emitting LEDs, which tend to launch power into all modes of the fiber. Fiber-coupled lasers are less prone to this effect because they tend to excite fewer nonpropagating fiber modes.

The excess power loss must be analyzed carefully in any system design as it can be significantly higher for some types of fibers than for others [5]. An example of the excess power loss is shown in Fig. 5.4 in terms of the fiber numerical aperture. At the input end of the fiber, the light acceptance is described in terms of the launch numerical aperture NA_{in}. If the light-emitting area of the LED is less than the cross-sectional area of the fiber core, then at this point the power coupled into the fiber is given by Eq. (5.7), where $NA = NA_{in}$.

However, when the optical power is measured in long multimode fibers after the launched modes have come to equilibrium (which is often taken to occur at 50 m), the effect of the equilibrium numerical aperture NA_{eq} becomes apparent. At this point, the optical power in the fiber scales as

$$P_{eq} = P_{50} \left(\frac{NA_{eq}}{NA_{in}} \right)^2 \tag{5.13}$$

Fig. 5.4 Example of the change in NA as a function of distance along a multimode fiber

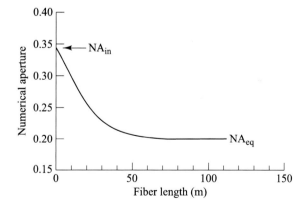

where P_{50} is the power expected in the fiber at the 50-m point based on the launch NA. The degree of mode coupling occurring in a fiber is primarily a function of the core-cladding index difference. It can thus vary significantly among different fiber types. Because most optical fibers attain 80–90% of their equilibrium NA after about 50 m, it is the value of NA_{eq} that is important when calculating launched optical power in multimode fibers.

5.2   Coupling Improvement with Lensing Schemes

The optical power-launching analysis given in Sect. 5.1 is based on centering a flat fiber end face directly over the light source as close to it as possible. If the source-emitting area is larger than the fiber-core area, then the resulting optical power coupled into the fiber is the maximum that can be achieved. This is a result of fundamental energy and radiance conservation principles [6]. However, if the emitting area of the source is smaller than the core area, a miniature lens may be placed between the source and the fiber to improve the power-coupling efficiency.

The function of the microlens is to magnify the emitting area of the source to match the core area of the fiber end face exactly. If the emitting area is increased by a magnification factor M, the solid angle within which optical power is coupled to the fiber from the source is increased by the same factor.

Several possible lensing schemes [1, 2, 7–12] are shown in Fig. 5.5. These include a rounded-end fiber, a small glass sphere (nonimaging microsphere) in contact with both the fiber and the source, a larger spherical lens used to image the source on the core area of the fiber end, a cylindrical lens generally formed from a short section of fiber, a system consisting of a spherical-surfaced LED and a spherical-ended fiber, and a taper-ended fiber.

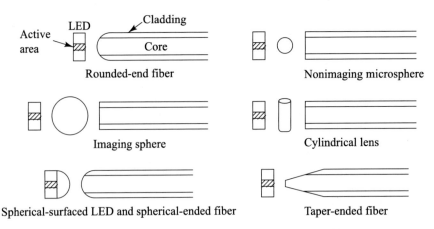

Fig. 5.5 Examples of possible lensing schemes used to improve optical source-to-fiber coupling efficiency

Fig. 5.6 Schematic diagram of an LED emitter with a microsphere lens

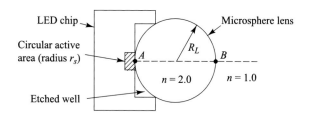

Although these techniques can improve the source-to-fiber coupling efficiency, they also create additional complexities. One problem is that the lens size is similar to the source and fiber-core dimensions, which introduces fabrication and handling difficulties. In the case of the taper-ended fiber, the mechanical alignment must be carried out with greater precision because the coupling efficiency becomes a more sharply peaked function of the spatial alignment. However, alignment tolerances are increased for other types of lensing systems.

One of the most efficient lensing methods is the use of a nonimaging microsphere. Its use for a surface emitter is shown in Fig. 5.6. For carrying out image position calculations, first make the following practical assumptions: the spherical lens has a radius R_L and a refractive index of about 2.0, the outside medium is air ($n = 1.0$), and the emitting area is circular. To collimate the output from the LED, the emitting surface should be located at the focal point of the lens. The focal point can be found from the Gaussian lens formula [13]

$$\frac{n}{s} + \frac{n'}{q} = \frac{n' - n}{r} \tag{5.14}$$

where s and q are the object and image distances, respectively, as measured from the lens surface, n is the refractive index of the lens, n' is the refractive index of the outside medium, and r is the radius of curvature of the lens surface.

The following sign conventions are used with Eq. (5.14):

1. Light travels from left to right.
2. Object distances are measured as positive to the left of a vertex and negative to the right.
3. Image distances are measured as positive to the right of a vertex and negative to the left.
4. All convex surfaces encountered by the light have a positive radius of curvature, and concave surfaces have a negative radius.

Example 5.5 Using the sign conventions for Eq. (5.14), find the focal point for the right-hand surface shown in Fig. 5.6.

Solution To find the focal point, set $q = \infty$ and solve for s in Eq. (5.14), where s is measured from point B. With $n = 2.0$, $n' = 1.0$, $q = \infty$, and $r = -R_L$, Eq. (5.14) yields.

$$s = f = 2R_L$$

Thus the focal point is located on the lens surface at point A.

Placing the LED close to the lens surface thus results in a magnification M of the emitting area. This is given by the ratio of the cross-sectional area of the lens to that of the emitting area:

$$M = \frac{\pi R_L^2}{\pi r_s^2} = \left(\frac{R_L}{r_s}\right)^2 \tag{5.15}$$

Using Eq. (5.4) one can show that, with the lens the optical power P_L that can be coupled into a full aperture angle 2θ is given by

$$P_L = P_S \left(\frac{R_L}{r_s}\right)^2 \sin^2\theta \tag{5.16}$$

where P_s is the total output power from the LED without the lens. Note that the maximum magnification occurs when the magnified source area is equal to the fiber core area. Thus $M_{max} = (a/r_s)^2$.

The theoretical coupling efficiency that can be achieved is based on energy and radiance conservation principles. This efficiency is usually determined by the size of the fiber. For a fiber of radius a and numerical aperture NA, the maximum coupling efficiency η_{max} for a Lambertian source is given by [14]

$$\begin{aligned}
\eta_{max} &= \left(\frac{a}{r_s}\right)^2 (NA)^2 \quad &\text{for } \frac{r_s}{a} > 1 \\
&= (NA)^2 \quad &\text{for } \frac{r_s}{a} \leq 1
\end{aligned} \tag{5.17}$$

Thus when the radius of the emitting area is larger than the fiber radius, no improvement in coupling efficiency is possible with a lens. In this case, the best coupling efficiency is achieved by a direct-butt method.

Example 5.6 An optical source with a circular output pattern is closely coupled to a step-index fiber that has a numerical aperture of 0.22. If the source radius is $r_s = 50 \,\mu$m and the fiber core radius $a = 25 \,\mu$m, what is the maximum coupling efficiency from the source into the fiber?

Solution Because the ratio $r_s/a > 1$, the maximum coupling efficiency η_{max} can be found from the top expression in Eq. (5.17):

$$\eta_{max} = \left(\frac{a}{r_s}\right)^2 (NA)^2 = \left(\frac{25}{50}\right)^2 (0.22)^2$$
$$= 0.25(0.22)^2 = 0.012 = 1.2\%$$

Thus the coupling efficiency is reduced to 25 percent compared to the case in which the source and fiber radii are equal, that is, when $\eta_{max} = (NA)^2$.

5.3 Losses Between Fiber Joints

A significant factor in any fiber optic system installation is the requirement to interconnect fibers in a low-loss manner. These interconnections occur at the optical source, at the photodetector, at intermediate points within a cable where two fibers are joined, and at intermediate points in a link where two cables are connected. The particular technique selected for joining the fibers depends on whether a permanent bond or an easily demountable connection is desired. A permanent bond is generally referred to as a *splice*, whereas a demountable joint is known as a *connector*.

Every joining technique is subject to certain conditions that can cause various amounts of optical power loss at the joint. The loss at a particular junction or through a component is called the *insertion loss*. These losses depend on parameters such as the input power distribution to the joint, the length of the fiber between the optical source and the joint, the geometrical and waveguide characteristics of the two fiber ends at the joint, and the fiber end-face qualities.

The number of modes that can propagate in each fiber limits the coupling of optical power from one fiber to another. For example, if a fiber in which 500 modes can propagate is connected to a fiber in which only 400 modes can propagate, then, at most, 80 percent of the optical power from the first fiber can be coupled into the second fiber (if we assume that all modes are equally excited). For a graded-index fiber with a core radius a and a cladding index n_2, and with $k = 2\pi/\lambda$, the total number of modes can be found from the expression [6]

$$M = k^2 \int_0^a \left[n^2(r) - n_2^2 \right] r \, dr \qquad (5.18)$$

where $n(r)$ defines the variation in the refractive-index profile of the core. This can be related to a general local numerical aperture $NA(r)$ through Eq. (2.40) to yield

$$M = k^2 \int_0^a NA(r)^2 r \, dr = k^2 NA(0)^2 \int_0^a \left[1 - \left(\frac{r}{a} \right)^\alpha \right] r \, dr \qquad (5.19)$$

In general, any two fibers that are to be joined will have varying degrees of differences in their radii a, axial numerical apertures $NA(0)$, and index profiles α. Thus the fraction of energy coupled from one fiber to another is proportional to the number of modes common to both fibers M_{comm} (if a uniform distribution of energy over the modes is assumed). The fiber-to-fiber coupling efficiency η_F is given by

$$\eta_F = \frac{M_{comm}}{M_E} \tag{5.20}$$

where M_E is the number of modes in the *emitting fiber* (the fiber that launches power into the next fiber).

The fiber-to-fiber coupling loss L_F is given in terms of η_F as

$$L_F = -10 \log \eta_F \tag{5.21}$$

Drill Problem 5.6 When the number of modes M_R in a receiving fiber is less than the number of modes M_E in an emitting fiber, Eq. (5.20) can be written as $\eta_F = M_R/M_E$. From Eq. (2.30) the number of modes in a step-index multimode fiber is

$$M = \left(\frac{2\pi a n_1}{\lambda} \right)^2 \Delta$$

(a) If two joined fibers are identical except for their radii, where $a_R = 0.90 a_E$, show that the fiber-to-fiber coupling efficiency is $\eta_F = 0.81$. (b) Show that the coupling loss is $L_F = 0.92$ dB.

An analytical estimate of the optical power loss at a joint between multimode fibers is difficult to make because the loss depends on the power distribution among the modes in the fiber. For example, consider first the case where all modes in a fiber are equally excited, as shown in Fig. 5.7a. The emerging optical beam thus fills the entire exit numerical aperture of this emitting fiber. Suppose now that a second identical fiber, called the *receiving fiber*, is to be joined to the emitting fiber. For the receiving fiber to accept all the optical power emitted by the first fiber, there must be perfect mechanical alignment between the two optical waveguides, and their geometric and waveguide characteristics must match precisely.

On the other hand, if steady-state modal equilibrium has been established in the emitting fiber, most of the energy is concentrated in the lower-order fiber modes. This means that the optical power is concentrated near the center of the fiber core, as shown in Fig. 5.7b. The optical power emerging from the fiber then fills only the equilibrium numerical aperture (see Fig. 5.4). In this case, because the input NA of the receiving fiber is larger than the equilibrium NA of the emitting fiber, slight mechanical misalignments of the two joined fibers and small variations in their geometric characteristics do not contribute significantly to joint loss.

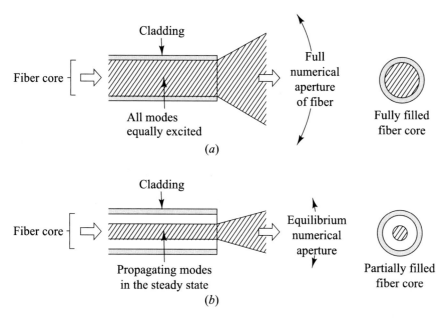

Fig. 5.7 Different modal distributions of the optical beam emerging from a fiber result in different degrees of coupling loss: **a** When all modes are equally excited, the output beam fills the entire output NA; **b** for a steady-state modal distribution, only the equilibrium NA is filled by the output beam

Steady-state modal equilibrium is generally established in long fiber lengths. Thus, when estimating joint losses between long fibers, calculations based on a uniform modal power distribution tend to lead to results that may be too pessimistic. However, if a steady-state equilibrium modal distribution is assumed, the estimate may be too optimistic because mechanical misalignments and fiber-to-fiber variations in operational characteristics cause a redistribution of power among the modes in the second fiber. As the power propagates along the second fiber, an additional loss will thus occur when a steady-state distribution is again established.

An exact calculation of coupling loss between different optical fibers, which takes into account nonuniform distribution of power among the modes and propagation effects in the second fiber, is lengthy and involved [15]. Therefore, here the assumption will be made that all modes in the fiber are equally excited. Although this gives a somewhat pessimistic prediction of joint loss, it will allow an estimate of the relative effects of losses resulting from mechanical misalignments, geometrical mismatches, and variations in the waveguide properties between two joined fibers.

5.3.1 Mechanical Misalignment Effects

Mechanical alignment is a major problem when joining two fibers, owing to their microscopic size [16–19]. A standard multimode graded-index fiber core is 50–100 μm in diameter, which is roughly the thickness of a human hair, whereas single-mode fibers have core diameters on the order of 9 μm. Radiation losses result from mechanical misalignments because the radiation cone of the emitting fiber does not match the acceptance cone of the receiving fiber. The magnitude of the radiation loss depends on the degree of misalignment. The three fundamental types of misalignment between fibers are shown in Fig. 5.8.

Axial displacement (which is also often called *lateral displacement*) results when the axes of the two fibers are separated by a distance d. *Longitudinal separation* occurs when the fibers have the same axis but have a gap s between their end faces. *Angular misalignment* results when the two axes form an angle so that the fiber end faces are no longer parallel.

The most common misalignment occurring in practice, which also causes the greatest power loss, is axial displacement. This axial offset reduces the overlap area of the two fiber-core end faces, as illustrated in Fig. 5.9, and consequently reduces the amount of optical power that can be coupled from one fiber into the other.

To illustrate the effects of axial misalignment, first consider the simple case of two identical step-index fibers of radii a. Suppose that their axes are offset by a separation d as is shown in Fig. 5.9, and assume there is a uniform modal power distribution in the emitting fiber. Because the numerical aperture is constant across

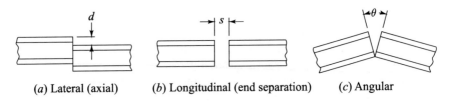

(a) Lateral (axial)   (b) Longitudinal (end separation)   (c) Angular

Fig. 5.8 Three types of mechanical misalignments that can occur between two joined fibers

Fig. 5.9 Axial offset reduces the shaded common core area of the two fiber end faces

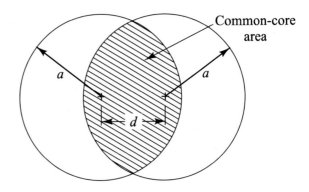

Common-core area

the end faces of the two fibers, the optical power coupled from one fiber to another is simply proportional to the common area A_{comm} of the two fiber cores. Using the standard geometry formula for the area of a circular segment, it is straightforward to show that this is

$$A_{comm} = 2a^2 \arccos\frac{d}{2a} - d\left(a^2 - \frac{d^2}{4}\right)^{1/2} \tag{5.22}$$

For the step-index fiber, the coupling efficiency is simply the ratio of the common-core area to the core end-face area,

$$\eta_{F,step} = \frac{A_{comm}}{\pi a^2} = \frac{2}{\pi}\arccos\frac{d}{2a} - \frac{d}{\pi a}\left[1 - \left(\frac{d}{2a}\right)^2\right]^{1/2} \tag{5.23}$$

The calculation of power coupled from one graded-index fiber into another identical one is more involved, because the numerical aperture varies across the fiber end face. Because of this, the numerical aperture of the transmitting or receiving fiber limits the total power coupled into the receiving fiber at a given point in the common-core area, depending on which NA is smaller at that point.

Example 5.7 An engineer makes a joint between two identical step-index fibers. Each fiber has a core diameter of 50 µm. If the two fibers have an axial (lateral) misalignment of 5 µm, what is the insertion loss at the joint?

Solution Using Eq. (5.23) the coupling efficiency is

$$\eta_{F,step} = \frac{2}{\pi}\arccos\left(\frac{5}{50}\right) - \frac{5}{\pi 25}\left[1 - \left(\frac{5}{50}\right)^2\right]^{1/2} = 0.873$$

From Eq. (5.21) the fiber-to-fiber insertion loss L_F is

$$L_F = -10\log\eta_F = -10\log 0.873 = 0.590 \ \text{dB}$$

If the end face of a graded-index fiber is uniformly illuminated, the optical power accepted by the core will be that power which falls within the numerical aperture of the fiber. The optical power density $p(r)$ at a point r on the fiber end is proportional to the square of the local numerical aperture $NA(r)$ at that point: [20]

$$p(r) = p(0)\frac{NA^2(r)}{NA^2(0)} \tag{5.24}$$

where $NA(r)$ and $NA(0)$ are defined by Eqs. (2.40) and (2.41), respectively. The parameter $p(0)$ is the power density at the core axis, which is related to the total power P in the fiber by

$$P = \int_0^{2\pi} \int_0^a p(r)\,r\,dr\,d\theta \tag{5.25}$$

For an arbitrary index profile, the double integral in Eq. (5.25) must be evaluated numerically. However, an analytic expression can be found by using a fiber with a parabolic index profile ($\alpha = 2.0$). Using Eq. (2.40), the power density expression at a point r given by Eq. (5.24) becomes

$$p(r) = p(0)\left[1 - \left(\frac{r}{a}\right)^2\right] \tag{5.26}$$

Using Eqs. (5.25) and (5.26), the relationship between the axial power density $p(0)$ and the total power P in the emitting fiber is

$$P = \frac{\pi a^2}{2} p(0) \tag{5.27}$$

To calculate the power transmitted across the butt joint of the two parabolic graded-index fibers with an axial offset d, consider the diagram shown in Fig. 5.10. The overlap region must be considered separately for the areas A_1 and A_2. In area A_1 the emitting fiber limits the numerical aperture, whereas in area A_2 the numerical aperture of the receiving fiber is smaller than that of the emitting fiber. The vertical dashed line separating the two areas is the locus of points where the numerical apertures are equal.

To determine the power coupled into the receiving fiber, the power density given by Eq. (5.26) is integrated separately over areas A_1 and A_2. Because the numerical aperture of the emitting fiber is smaller than that of the receiving fiber in area A_1, all of the power emitted in this region will be accepted by the receiving fiber. The received power P_1 in area A_1 is thus

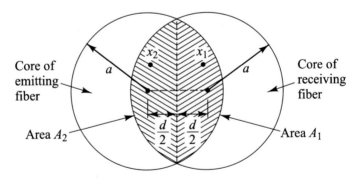

Fig. 5.10 Core overlap region for two identical parabolic graded-index fibers with an axial separation d, where points x_1 and x_2 are arbitrary points of symmetry in areas A_1 and A_2

Fig. 5.11 Area and limits of integration for the common core area of two parabolic graded-index fibers

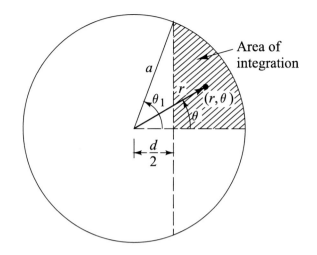

Area of integration

$$P_1 = 2 \int_0^{\theta_1} \int_{r_1}^a p(r)\, r\, dr\, d\theta$$

$$= 2p(0) \int_0^{\theta_1} \int_{r_1}^a \left[1 - \left(\frac{r}{a}\right)^2 \right] r\, dr\, d\theta \tag{5.28}$$

where the limits of integration, shown in Fig. 5.11, are

$$r_1 = \frac{d}{2\cos\theta}$$

and

$$\theta_1 = \arccos\frac{d}{2a}$$

Carrying out the integration yields

$$P_1 = \frac{a^2}{2} p(0) \left\{ \arccos\frac{d}{2a} - \left[1 - \left(\frac{d}{2a}\right)^2 \right]^{1/2} \frac{d}{6a} \left(5 - \frac{d^2}{2a^2} \right) \right\} \tag{5.29}$$

where $p(0)$ is given by Eq. (5.27).

In area A_2 the emitting fiber has a larger numerical aperture than the receiving fiber. This means that the receiving fiber will accept only that fraction of the emitted optical power that falls within its own numerical aperture. This power can be found from symmetry considerations. The numerical aperture of the receiving fiber at a point x_2 in area A_2 is the same as the numerical aperture of the emitting fiber at the symmetrical point x_1 in area A_1. Thus the optical power accepted by the receiving fiber at any point x_2 in area A_2 is equal to that emitted from the symmetrical point x_1 in area A_1. The total power P_2 coupled across area A_2 is thus equal to the power P_1 coupled across area A_1. Combining these results, then yields that the total power P_T accepted by the receiving fiber is

$$P_T = 2P_1 = \frac{2}{\pi} P \left\{ \arccos \frac{d}{2a} - \left[1 - \left(\frac{d}{2a} \right)^2 \right]^{1/2} \frac{d}{6a} \left(5 - \frac{d^2}{2a^2} \right) \right\} \qquad (5.30)$$

Example 5.8 Suppose two identical graded-index fibers are misaligned with an axial offset of $d = 0.3a$. What is the power coupling loss between these two fibers?

Solution From Eq. (5.30), the fraction of optical power coupled from the first fiber into the second fiber is

$$\frac{P_T}{P} = 0.748$$

Or, in decibels,

$$10 \log \left(\frac{P_T}{P} \right) = -1.26 \text{ dB}$$

When the axial misalignment d is small compared with the core radius a, Eq. (5.30) can be approximated by

$$P_T \approx P \left(1 - \frac{8d}{3\pi a} \right) \qquad (5.31)$$

This is accurate to within 1 percent for $d/a < 0.4$. The coupling loss for the offsets given by Eqs. (5.30) and (5.31) is

$$L_F = -10 \log \eta_F = -10 \log \frac{P_T}{P} \qquad (5.32)$$

The effect of separating the two fiber ends longitudinally by a gap s is shown in Fig. 5.12. Not all the higher-mode optical power emitted in the ring of width x will be intercepted by the receiving fiber. The fraction of optical power coupled into the receiving fiber is given by the ratio of the cross-sectional area of the receiving fiber (πr^2) to the area $\pi(a + x)^2$ over which the emitted power is distributed at a distance s. From Fig. 5.12 it follows that $x = s \tan \theta_A$, where θ_A is the acceptance angle of the fibers, as defined in Eq. (2.2). From this ratio the loss for an offset joint between two

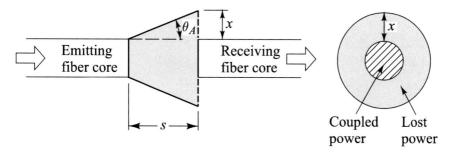

Fig. 5.12 Optical power loss when fiber ends are separated longitudinally by a gap s

identical step-index fibers is found to be

$$L_F = -10 \log\left(\frac{a}{a+x}\right)^2 = -10 \log\left(\frac{a}{a+s\tan\theta_A}\right)^2$$

$$= -10 \log\left[1 + \frac{s}{a}\sin^{-1}\left(\frac{NA}{n}\right)\right]^{-2} \tag{5.33}$$

where a is the fiber radius, NA is the numerical aperture of the fiber, and n is the refractive index of the material between the fiber ends (usually either air or an index matching gel).

Example 5.9 Two identical step-index fibers each have a 25-μm core radius and an acceptance angle of 14°. Assume the two fibers are perfectly aligned axially and angularly. What is the insertion loss for a longitudinal separation of 0.025 mm?

Solution The insertion loss due to a fiber gap between can be found by using Eq. (5.33). For a 0.025-mm = 25-μm gap

$$L_F = -10 \log\left(\frac{25}{25 + 25\tan 14°}\right)^2 = 1.93 \, \text{dB}$$

When the axes of two joined fibers are angularly misaligned at the joint, the optical power that leaves the emitting fiber outside of the solid acceptance angle of the receiving fiber will be lost. For two step-index fibers that have an angular misalignment θ, the optical power loss at the joint has been shown to be [21]

$$L_F = -10 \log\left(\cos\theta\left\{\frac{1}{2} - \frac{1}{\pi}p\left(1-p^2\right)^{1/2} - \frac{1}{\pi}\arcsin p - q\right.\right.$$
$$\left.\left.\left[\frac{1}{\pi}y\left(1-y^2\right)^{1/2} + \frac{1}{\pi}\arcsin y + \frac{1}{2}\right]\right\}\right) \tag{5.34}$$

where

$$p = \frac{\cos \theta_A (1 - \cos \theta)}{\sin \theta_A \sin \theta}$$

$$q = \frac{\cos^3 \theta_A}{\left(\cos^2 \theta_A - \sin^2 \theta\right)^{3/2}}$$

$$y = \frac{\cos^2 \theta_A (1 - \cos \theta) - \sin^2 \theta}{\sin \theta_A \cos \theta_A \sin \theta}$$

The derivation of Eq. (5.34) again assumes that all modes are uniformly excited.

Of the three mechanical misalignments, the dominant loss arises from lateral displacement. In practice, angular misalignments of less than 1° are readily achievable in splices and connectors. Experimental data show that these misalignments result in losses of less than 0.5 dB.

For splices, the separation losses are normally negligible as the fibers should be in relatively close contact. In most connectors, the fiber ends are intentionally separated by a small gap. This prevents them from rubbing against each other and becoming damaged during connector engagement. Typical gaps in these applications range from 0.025 to 0.10 mm, which results in losses of less than 0.8 dB for a 50-μm-diameter fiber.

5.3.2   Fiber Variation Losses

In addition to mechanical misalignments, differences in the geometrical and waveguide characteristics of any two waveguides being joined can have a profound effect on fiber-to-fiber coupling loss. These include variations in core diameter, core-area ellipticity, numerical aperture, refractive-index profile, and core-cladding concentricity of each fiber. Because these are manufacturer-related variations, the user generally has little control over them. Theoretical and experimental studies of the effects of these variations have shown that, for a given percentage mismatch, differences in core radii and numerical apertures have a significantly larger effect on joint loss than mismatches in the refractive-index profile or core ellipticity.

The joint losses resulting from core diameter, numerical aperture, and core refractive-index-profile mismatches can be found from Eqs. (5.19) and (5.20). For simplicity, let the subscripts E and R refer to the emitting and receiving fibers, respectively. If the radii a_E and a_R are not equal but the axial numerical apertures and the index profiles are equal [$NA_E(0) = NA_R(0)$ and $\alpha_E = \alpha_R$], then the coupling loss is

$$L_F(a) = \begin{cases} -10 \log\left(\frac{a_R}{a_E}\right)^2 & \text{for } a_R < a_E \\ 0 \text{ for } a_R \geq a_E \end{cases} \tag{5.35}$$

Drill Problem 5.7 Consider the case when light from a multimode step-index fiber that has a core radius of 62.5 μm is coupled into a similar fiber that has a core radius of 50 μm. Show that the coupling loss in going from the larger to the smaller fiber is 1.94 dB.

If the radii and the index profiles of the two coupled fibers are identical but their axial numerical apertures are different, then

$$L_F(NA) = \begin{cases} -10 \log\left[\frac{NA_R(0)}{NA_E(0)}\right]^2 & \text{for } NA_R(0) < NA_E(0) \\ 0 & \text{for } NA_R(0) \geq NA_E(0) \end{cases} \tag{5.36}$$

Example 5.10 Consider two joined step-index fibers that are perfectly aligned. What is the coupling loss if the numerical apertures are $NA_R = 0.20$ for the receiving fiber and $NA_E = 0.22$ for the emitting fiber?

Solution From Eq. (5.36)

$$L_F(NA) = -10 \log\left(\frac{0.20}{0.22}\right)^2 = -10 \log 0.826 = -0.828 \text{ dB}$$

Finally, if the radii and the axial numerical apertures are the same but the core refractive-index profiles differ in two joined fibers, then the coupling loss is

$$L_F(\alpha) = \begin{cases} -10 \log\left[\frac{\alpha_R(\alpha_E+2)}{\alpha_E(\alpha_R+2)}\right] & \text{for } \alpha_R < \alpha_E \\ 0 & \text{for } \alpha_R \geq \alpha_E \end{cases} \tag{5.37}$$

This results because for $\alpha_R < \alpha_E$ the number of modes that can be supported by the receiving fiber is less than the number of modes in the emitting fiber. If $\alpha_R > \alpha_E$ then all modes in the emitting fiber can be captured by the receiving fiber.

Example 5.11 Consider two joined graded-index fibers that are perfectly aligned. What is the coupling loss if the refractive index profiles are $\alpha_R = 1.98$ for the receiving fiber and $\alpha_E = 2.20$ for the emitting fiber?

Solution From Eq. (5.37)

$$L_F(\alpha) = -10 \log\left[\frac{\alpha_R(\alpha_E + 2)}{\alpha_E(\alpha_R + 2)}\right] = -10 \log 0.950 = -0.22 \text{ dB}$$

5.3.3   Single-Mode Fiber Losses

As is the case in multimode fibers, in single-mode fibers the lateral (axial) offset misalignment presents the most serious loss. This loss depends on the shape of the propagating mode. For Gaussian-shaped beams the loss between identical fibers is [22]

$$L_{SM,lat} = -10 \log \left\{ exp \left[-\left(\frac{d}{w} \right)^2 \right] \right\} \tag{5.38}$$

where the spot size w is the mode-field radius defined in Eq. (2.34), and d is the lateral displacement shown in Fig. 5.9. Because the spot size is only a few micrometers in single-mode fibers, low-loss coupling requires a very high degree of mechanical precision in the axial dimension.

Example 5.12 A single-mode fiber has a normalized frequency $V = 2.20$, a core refractive index $n_1 = 1.47$, a cladding refractive index $n_2 = 1.465$, and a core diameter $2a = 9$ μm. What is the insertion loss of a fiber joint having a lateral offset of $d = 1$ μm? For the mode-field diameter use the expression

$$w = a(0.65 + 1.619V^{-3/2} + 2.879V^{-6}).$$

Solution First, using the above expression for the mode-field diameter,

$$w = 4.5[0.65 + 1.619(2.20)^{-3/2} + 2.879(2.20)^{-6}] = 5.27 \, μm$$

w.
Then, from Eq. (5.38),

$$L_{SM,lat} = -10 \log \left\{ exp \left[-(1/5.27)^2 \right] \right\} = 0.156 \ dB$$

For angular misalignment in single-mode fibers, the loss at a wavelength λ is [22]

$$L_{SM,ang} = -10 \log \left\{ exp \left[-\left(\frac{\pi n_2 w \theta}{\lambda} \right)^2 \right] \right\} \tag{5.39}$$

where n_2 is the refractive index of the cladding, θ is the angular misalignment in radians shown in Fig. 5.9, and w is the mode-field radius.

For a gap s with a material of index n_3, and letting $G = s/kw^2$, the gap loss for identical single-mode fiber splices is

$$L_{SM,gap} = -10 \log \frac{64n_1^2 n_3^2}{(n_1 + n_3)^4 (G^2 + 4)} \tag{5.40}$$

Example 5.13 Consider the single-mode fiber described in Example 5.12. Find the loss at a joint having an angular misalignment of $1° = 0.0175$ radians at a 1300-nm wavelength.

Solution From Eq. (5.39)

$$L_{SM,ang} = -10 \log \left\{ exp \left[-\left(\frac{\pi (1.465)(5.27)(0.0175)}{1.3} \right)^2 \right] \right\} = 0.46 \, \text{dB}$$

5.3.4 Preparation of Fiber End Faces

One of the first steps that must be followed before fibers are connected or spliced to each other is to prepare the fiber end faces properly. In order not to have light deflected or scattered at the joint, the fiber ends must be flat, perpendicular to the fiber axis, and smooth. End-preparation techniques that have been extensively used include sawing, grinding and polishing, controlled fracture, and laser cleaving.

Conventional grinding and polishing techniques can produce a very smooth surface that is perpendicular to the fiber axis. However, this method is quite time-consuming and requires a fair amount of operator skill. Although it is often implemented in a controlled environment such as a laboratory or a factory, it is not readily adaptable for field use. The procedure employed in the grinding and polishing technique is to use successively finer abrasives to polish the fiber end face. The end face is polished with each successive abrasive until the finer scratches of the present abrasive replace the scratches from the previous abrasive material. The number of abrasives used depends on the degree of smoothness that is desired.

Controlled-fracture techniques are based on score-and-break methods for cleaving fibers. In this operation, the fiber to be cleaved is first scratched to create a stress concentration at the surface. The fiber is then bent over a curved form while tension is simultaneously applied, as shown in Fig. 5.13. This action produces a stress distribution across the fiber. The maximum stress occurs at the scratch point so that a crack starts to propagate through the fiber.

One can produce a highly smooth and perpendicular end face in this way. A number of different tools based on the controlled-fracture technique have been developed and

Fig. 5.13
Controlled-fracture
procedure for fiber end
preparation

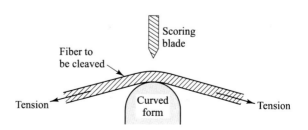

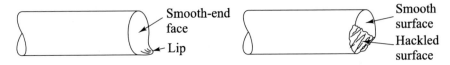

Fig. 5.14 Two examples of improperly cleaved fiber ends

are being used both in the field and in factory environments. However, the controlled-fracture method requires careful control of the curvature of the fiber and of the amount of tension applied. If the stress distribution across the crack is not properly controlled, the fracture propagating across the fiber can fork into several cracks. This forking produces defects such as a lip or a hackled portion on the fiber end, as shown in Fig. 5.14. Common end-face defects include:

Lip. This is a sharp protrusion from the edge of a cleaved fiber that prevents the cores from coming in close contact. Excessive lip height can cause fiber damage.

Rolloff. This rounding-off of the edge of a fiber is the opposite condition to lipping. It is also known as *breakover* and can cause high insertion or splice loss.

Chip. A chip is a localized fracture or break at the end of a cleaved fiber.

Hackle. Figure 5.14 shows this as severe irregularities across a fiber end face.

Mist. This is similar to hackle but much less severe.

Spiral or step. These are abrupt changes in the end-face surface topology.

Shattering. This is the result of an uncontrolled fracture and has no definable cleavage or surface characteristics.

An alternative to a mechanical score-and-break method is the use of a laser to cleave fibers [23, 24].

5.4 Summary

This chapter addresses the problem of launching optical power from a light source into a fiber and the factors involved in coupling light from one fiber into another. The coupling of optical power from a light source into a fiber is influenced by the following considerations:

1. The numerical aperture of the fiber, which defines the light acceptance cone of the fiber.
2. The cross-sectional area of the fiber core compared to the source emitting area. If the emitting area is smaller than the fiber core, then lensing schemes can be used to improve
    the coupling efficiency.

3. The radiance of the light source, which is the optical power radiated into a solid angle per emitting area (measured in watts per square centimeter per steradian).
4. The spatial radiation pattern of the source. The incompatibility between the wide beam divergence of LEDs and the narrow acceptance cone of the fiber is a major contributor to coupling loss. This holds to a lesser extent for laser diodes.

In practice, many suppliers offer optical sources that have a short length of optical fiber (nominally 1–2 m) already attached in an optimum power-coupling configuration. This fiber, which is referred to as a flylead or a pigtail, makes it easier for the user to couple the source to a system fiber. The power-launching problem now becomes a simpler one of coupling optical power from one fiber into another. To achieve a low coupling loss, the fiber flylead should be connected to a system fiber having a nominally identical numerical aperture and core diameter.

Fiber-to-fiber joints can exist between the source flylead and the system fiber, at the photodetector, at intermediate points in a link where two cable sections are interconnected, or at the fiber to component junctions in a communication link. The two principal types of joints are splices, which are permanent bonds between two fibers, and optical connectors, which are used when an easily demountable connection between fibers or between a fiber and an optical component is desired.

Each jointing technique is subject to certain conditions, which can cause varying degrees of optical power loss at the joint. These parameters depend on factors such as the following:

1. *The geometrical characteristics of the fibers.* For example, optical power will be lost because of area mismatches if an emitting fiber has a larger core diameter than the receiving fiber.
2. *The waveguide characteristics of the fibers.* For example, if an emitting fiber has a larger numerical aperture than the receiving fiber, all optical power falling outside of the acceptance cone of the receiving fiber is lost.
3. *The various mechanical misalignments between the two fiber ends at the joint.* These misalignments include longitudinal separation, angular misalignment, and axial (or lateral) displacement. The most common misalignment occurring in practice, which also causes the greatest power loss, is axial displacement.
4. *The input power distribution to the joint.* If all the modes of an emitting fiber are equally excited, there must be perfect mechanical alignment between the two optical waveguides, and their geometric and waveguide characteristics must match precisely in order for no optical power loss to occur at the joint. On the other hand, if steady-state modal equilibrium has been established in the emitting fiber (which happens in long fiber lengths), most of the energy is concentrated in the lower-order fiber modes. In this case, slight mechanical misalignments of two joined fibers and small variations in their geometric and waveguide characteristics do not contribute significantly to joint loss.
5. *The fiber end face quality.* One criterion for low-low joints is that the fiber end faces be clean and smooth. End preparation techniques include sawing, grinding and polishing, and controlled fracture.

Problems

5.1   Analogous to Fig. 5.2, plot and compare the emission patterns from a Lamber-
      tian source and a source with an emission pattern given by $L(\theta) = L_0 \cos^3 \theta$.
      Assume both sources have the same peak radiance L_0, which is normalized
      to unity in each case.

5.2   Consider light sources where the emission pattern is given by $L(\theta) = L_0$
      $\cos^m \theta$. Plot $L(\theta)$ as a function of m in the range $1 \le m \le 20$ at viewing angles
      of $10°$, $20°$, and $45°$. Assume all sources have the same peak radiance L_0.

5.3   A laser diode has lateral ($\varphi = 0°$) and transverse ($\varphi = 90°$) half-power beam
      widths of $2\theta = 60°$ and $30°$, respectively. Show that the transverse and lateral
      power distribution coefficients for this device are T = 20.0 and L = 4.82.

5.4   An LED with a circular emitting area of radius 20 μm has a Lambertian
      emission pattern with a 100-W/(cm² - sr) axial radiance at a 100-mA drive
      current. (a) How much optical power can be coupled into a step-index fiber
      having a 100-μm core diameter and NA = 0.22? (b) How much optical power
      can be coupled from this source into a 50-μm core-diameter graded-index
      fiber having $\alpha = 2.0$, $n_1 = 1.48$, and $\Delta = 0.01$? [Answer: (a) $P_{LED;\ step} =$
      191 μW; (b) $P_{LED;\ graded} = 159$ μW]

5.5   A GaAs optical source that has a refractive index of 3.600 is closely coupled
      to a step-index fiber that has a core refractive index of 1.465. If the source
      size is smaller than the fiber core, and the small gap between the source and
      the fiber is filled with a gel that has a refractive index of 1.305, verify the
      following parameters (see Example 5.4): (a) The reflectivity at the source-
      to-gel interface $R_{s-g} = 0.219$; (b) The reflectivity at the gel-to-fiber interface
      $R_{g-f} = 3.34 \times 10^{-3}$; (c) The total transmission through the gel is T = 0.778
      (d) The power loss from the source into the fiber is L = 1.09 dB.

5.6   Consider a Lambertian LED source with a 50-μm diameter emitting area.
      (*a*) If this source is connected to an optical fiber that has a 62.5-μm core
      diameter with NA = 0.18, show that the coupling efficiency is 3.24% = −
      14.9 dB. (*b*) If a spherical microlens is used to improve the coupling efficiency,
      show that the maximum magnification is $M_{max} = 1.56$. (*c*) Show that with
      this lens the coupling efficiency is 5.06% = −13.0 dB.

5.7   The end faces of two optical fibers with core refractive indices of 1.485 are
      perfectly aligned and have a small gap between them. (a) If this gap is filled
      with a gel that has a refractive index of 1.305, show that the reflectivity at a
      gel-fiber interface is 4.16×10^{-3}; (b) Show that the power loss at a gel-fiber
      interface is L = −10 log (1 − R) = 0.018 dB. (c) If the gap is very small,
      show that the power loss in decibels through the joint when no index-matching
      material is used is 0.17 dB. Note that $n = 1.0$ for air.

5.8   Consider the three fibers having the properties listed in Table 5.1. Use
      Eqs. (5.23) and (5.21) to verify the entries in this table for connector losses
      (in decibels) due to the indicated axial misalignments.

5.9   Consider Eq. (5.35) that gives the coupling loss for two fibers with unequal
      core radii. Show that the coupling losses in decibels as a function of a_R/a_E for

Table 5.1 For Problem 5.8

	Coupling losses (in dB) for given axial misalignments			
Core/cladding diameters (μm)	1	3	5	10
50/125	0.112	0.385	0.590	0.266
62.5/125	0.089	0.274	0.465	0.985
100/140	0.56	0.169	0.286	0.590

Table 5.2 For Problem 5.12

Parameter	Fiber 1	Fiber 2
Core index n_1	1.46	1.48
Index difference Δ	0.010	0.015
Core radius a	50 μm	62.5 μm
Profile factor	2.00	1.80

the values $a_R/a_E = 0.5, 0.7$, and 0.9 are -6.02 dB, -3.10 dB, and -0.92 dB, respectively.

5.10 Consider Eq. (5.36) that gives the coupling loss for two fibers with unequal axial numerical apertures. Show that the coupling losses in decibels for the values $NA_R(0)/NA_E(0) = 0.5, 0.7$, and 0.9 are -6.02 dB, -3.10 dB, and -0.92 dB, respectively.

5.11 Consider Eq. (5.37) that gives the coupling loss for two fibers with different core refractive-index profiles. If $\alpha_E = 2.20$, what are the coupling losses when (a) $\alpha_R = 1.80$, (b) $\alpha_R = 2.00$? [Answers: (a) 0.44 dB; (b) 0.20 dB].

5.12 Consider two multimode graded-index fibers that have the characteristics given in Table 5.2. If these two fibers are perfectly aligned with no gap between them, calculate the splice losses and the coupling efficiencies for the following cases:
Light going from fiber 1 to fiber 2.
Light going from fiber 2 to fiber 1.
[Answers: (a) $L_1 \rightarrow {}_2(\alpha) = 0.24$ dB; (b) $L_2 \rightarrow {}_1(a) = 1.94$ dB, $L_2 \rightarrow {}_1(NA) = 1.89$ dB].

References

1. Y. Uematsu, T. Ozeki, Y. Unno, Efficient power coupling between an MH LED and a taper-ended multimode fiber. IEEE J. Quantum Electron. **15**, 86–92 (1979)
2. H. Kuwahara, M. Sasaki, N. Tokoyo, Efficient coupling from semiconductor lasers into single-mode fibers with tapered hemispherical ends. Appl. Opt. **19**, 2578–2583 (1980)
3. D. Marcuse, LED fundamentals: comparison of front and edge-emitting diodes. IEEE J. Quantum Electron. **13**, 819–827 (1977)
4. B.E.A. Saleh, M. Teich, *Fundamentals of Photonics*, 3rd edn. (Wiley, 2019)

5. TIA-455–54B, *Mode Scrambler Requirements for Overfilled Launching Conditions to Multimode Fibers* (1998)
6. M, Bass (ed.), *Handbook of Optics*, vol II (McGraw-Hill, 2010)
7. M. Forrer et al., High precision automated tab assembly with micro optics for optimized high-power diode laser collimation, in Paper 11261-10, *SPIE Photonics West*, Feb. 2020
8. K. Sakai, M. Kawano, H. Aruga, S.-I. Takagi, S.-I. Kaneko, J. Suzuki, M. Negishi, Y. Kondoh, K.-I. Fukuda, Photodiode packaging technique using ball lens and offset parabolic mirror. J. Lightw. Technol. **27**(17), 3874–3879 (2009)
9. A. Nicia, Lens coupling in fiber-optic devices: efficiency limits. Appl. Opt. **20**, 3136–3145 (1981)
10. K. Keränen, J.T. Mäkinen, K.T. Kautio, J. Ollila, J. Petäjä, V. Heikkinen, J. Heilala, P. Karioja, Fiber pigtailed multimode laser module based on passive device alignment on an LTCC substrate. IEEE Trans. Adv. Packag. **29**, 463–472 (2006)
11. G. Jiang, L. Diao, K. Kuang, Understanding lasers, laser diodes, laser diode packaging and their relationship to tungsten copper, in *Advanced Thermal Manage. Materials* (Springer, 2013)
12. C. Tsou, Y.S. Huang, Silicon-based packaging platform for light-emitting diode. IEEE Trans. Adv. Pack. **29**, 607–614 (2006)
13. See any general physics or introductory optics book; for example: (a) H.D. Young, R.A. Freedman, *University Physics with Modern Physics*, 15th edn. (Pearson, 2020); (b) K. Lizuka, *Engineering Optics* (Springer, 2019); (c) C.A. Diarzio, *Optics for Engineers* (Apple Academic Press, 2021)
14. M.C. Hudson, Calculation of the maximum optical coupling efficiency into multimode optical waveguides. Appl. Opt. **13**, 1029–1033 (1974)
15. P. Di Vita, U. Rossi, Realistic evaluation of coupling loss between different optical fibers. J. Opt. Common. **1**, 26–32 (1980); Evaluation of splice losses induced by mismatch in fiber parameters. Opt. Quantum Electron. **13**, 91–94 (1981)
16. M.J. Adams, D.N. Payne, F.M.E. Staden, Splicing tolerances in graded index fibers. Appl. Phys. Lett. **28**, 524–526 (1976)
17. D. Gloge, Offset and tilt loss in optical fiber splices. Bell Sys. Tech. J. **55**, 905–916 (1976)
18. T.C. Chu, A.R. McCormick, Measurement of loss due to offset, end separation and angular misalignment in graded index fibers excited by an incoherent source. Bell Sys. Tech. J. **57**, 595–602 (1978)
19. C.M. Miller, Transmission vs. transverse offset for parabolic-profile fiber splices with unequal core diameters. Bell Sys. Tech. J. **55**, 917–927 (1976)
20. D. Gloge, E.A.J. Marcatili, Multimode theory of graded-core fibers. Bell Sys. Tech. J. **52**, 1563–1578 (1973)
21. F.L. Thiel, D.H. Davis, Contributions of optical-waveguide manufacturing variations to joint loss. Electron. Lett. **12**, 340–341 (1976)
22. D. Marcuse, D. Gloge, E.A.J. Marcatili, Guiding properties of fibers, in ed. by S.E. Miller, A.G. Chynoweth. *Optical Fiber Telecommunications* (Academic Press, 1979)
23. G. Van Steenberge, P. Geerinck, S. Van Put, J. Watté, H. Ottevaere, H. Thienpont, P. Van Daele, Laser cleaving of glass fibers and glass fiber arrays. J. Lightw. Technol. **23**, 609–614 (2005)
24. W.H. Wu, C.L. Chang, C.H. Hwang, A study on cutting glass fibers by CO_2 laser. Appl. Mech. Mater. **590**, 192–196 (2014)

Chapter 6
Photodetection Devices

Abstract The two main photodetection devices used in optical fiber communications are semiconductor-based *pin* photodiodes and avalanche photodiodes. The key advantages of these devices are their size compatibility with optical fibers, their high sensitivities at the desired operational wavelengths, and their fast response times that allow accurate signal tracking. Because optical signals generally are weakened and distorted when they emerge from the end of a fiber, the photodiodes must meet specific stringent performance requirements. The purpose of this chapter is to describe the performance characteristics and show how the photodiodes are used in optical fiber links.

At the output end of an optical transmission line, there must be a receiving device that interprets the information contained in the optical signal. The first component of this receiver is a photodetector. The photodetector senses the luminescent power falling upon it and converts the variation of this optical power into a correspondingly varying electric current. The optical signal is generally weakened and distorted when it emerges from the end of the fiber, the photodetector must meet very high performance requirements. Among the foremost of these requirements are a high response or sensitivity in the emission wavelength range of the optical source being used, a minimum addition of noise to the system, and a fast response speed or sufficient bandwidth to handle the desired data rate. The photodetector should also be insensitive to variations in temperature, be compatible with the physical dimensions of the optical fiber, have a reasonable cost in relation to the other components of the system, and have a long operating life.

Among the different photodetectors are photomultipliers, pyroelectric detectors, and semiconductor-based photoconductors, phototransistors, and photodiodes. Of these detectors, the semiconductor-based photodiode is used almost exclusively for fiber optic systems because of its small size, suitable material, high sensitivity, and fast response time. The two types of photodiodes used are the *pin* photodetector and the avalanche photodiode (APD). The following sections examine the fundamental characteristics of these two device types. The elementary principles of semiconductor device physics given in Sect. 4.1 will be used in describing these components. Basic

G. Keiser, *Fiber Optic Communications*,
https://doi.org/10.1007/978-981-33-4665-9_6

discussions of semiconductor photodetectors and photodetection processes can be found in the literature [1–12].

6.1 Operation of Photodiodes

6.1.1 The pin Photodetector

The basic semiconductor photodetector is the *pin* photodiode, shown schematically in Fig. 6.1. The device structure consists of p and n regions separated by a very lightly n-doped intrinsic (i) region. In normal operation a sufficiently large reverse-bias voltage is applied across the device through a load resistor R_L so that the intrinsic region is fully depleted of carriers. That is, the intrinsic n and p carrier concentrations are negligibly small in comparison with the impurity concentration in this region.

As a photon flux Φ penetrates into a semiconductor, the photons will be absorbed and converted to charge carriers as they progresses through the material. Suppose P_{in} is the optical power level falling on the photodetector at $x = 0$ and $P(x)$ is the power level at a distance x into the material. Then the incremental change $dP(x)$ in

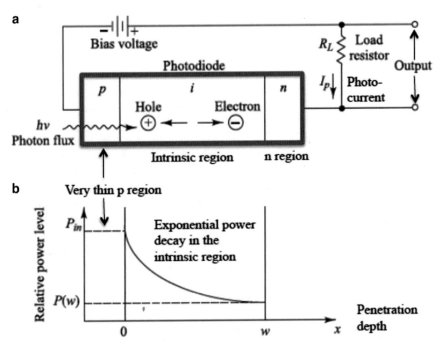

Fig. 6.1 a Representation of a *pin* photodiode circuit with an applied reverse bias. **b** An incident optical power level decays exponentially inside the device

the optical power level as this photon flux passes through an incremental distance dx in the semiconductor is given by $dP(x) = -\alpha_s(\lambda)P(x)dx$, where $\alpha_s(\lambda)$ is the *photon absorption coefficient* at a wavelength λ. Integrating this relationship gives the power level at a distance x into the material as

$$P(x) = P_{in} \exp(-\alpha_s x) \tag{6.1}$$

Figure 6.1 gives an example of the power level as a function of the penetration depth into the intrinsic region, which has a width w. The width of the p region typically is very thin so that little radiation is absorbed there.

Example 6.1 If the absorption coefficient of $In_{0.53}Ga_{0.47}As$ is $0.8\ \mu m^{-1}$ at 1550 nm, what is the penetration depth at which $P(x)/P_{in} = 1/e = 0.368$?

Solution From Eq. (6.1)

$$\frac{P(x)}{P_{in}} = \exp(-\alpha_s x) = \exp(-0.8x) = 0.368$$

Therefore

$$-0.8x = \ln 0.368 = -0.9997$$

which yields $x = 1.25\ \mu m$.

Example 6.2 A high-speed $In_{0.53}Ga_{0.47}As$ *pin* photodetector is made with a depletion layer thickness of $0.15\ \mu m$. What percent of incident photons are absorbed in this photodetector at 1310 nm if the absorption coefficient is $1.5\ \mu m^{-1}$ at this wavelength?

Solution From Eq. (6.1), the optical power level at $x = 0.15\ \mu m$ relative to the incident power level is

$$\frac{P(x)}{P_{in}} = \exp(-\alpha_s x) = \exp[(-1.50)0.15] = 0.80$$

Therefore only 20% of the incident photons are absorbed.

Drill Problem 6.1 An InGaAs *pin* photodetector has an absorption coefficient of $1.0\ \mu m^{-1}$ at 1550 nm. Show that the penetration depth at which 50% of the photons are absorbed is $0.69\ \mu m$.

When the energy of an incident photon is greater than or equal to the bandgap energy E_g of the semiconductor material, the photon can give up its energy and excite an electron from the valence band to the conduction band. This absorption process

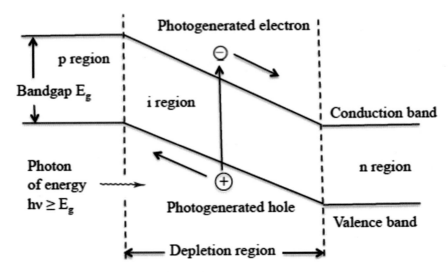

Fig. 6.2 Simple energy-band diagram for a *pin* photodiode showing that photons with energies greater than or equal to the bandgap energy E_g can generate free electron–hole pairs that act as photocurrent carriers

generates mobile electron–hole pairs, as Fig. 6.2 shows. These electrons and holes are known as *photocarriers*, because they are photon-generated charge carriers that are available to produce a current flow when a bias voltage is applied across the device. The concentration level of impurity elements that are intentionally added to the material controls the number of charge carriers (see Sect. 4.1). The photodetector normally is designed so that these carriers are generated mainly in the depletion region (the depleted intrinsic region) where most of the incident light is absorbed. The high electric field present in the depletion region causes the carriers to separate and be collected across the reverse-biased junction. This gives rise to a current flow in an external circuit, with one electron flowing for every carrier pair generated. This current flow is known as the *photocurrent*.

As the charge carriers flow through the material, some electron–hole pairs will recombine and hence disappear. On the average, the charge carriers move a distance L_n or L_p for electrons and holes, respectively. This distance is known as the *diffusion length*. The time it takes for an electron or hole to recombine is known as the *carrier lifetime* and is represented by τ_n and τ_p, respectively. The lifetimes and the diffusion lengths are related by the expressions.

$$L_n = (D_n \tau_n)^{1/2} \quad \text{and} \quad L_p = (D_p \tau_p)^{1/2}$$

where D_n and D_p are the electron and hole diffusion coefficients (or constants), respectively, which are expressed in units of centimeters squared per second.

The dependence of the optical absorption coefficient on wavelength is shown in Fig. 6.3 for several photodiode materials [13]. As the curves clearly show, α_s depends

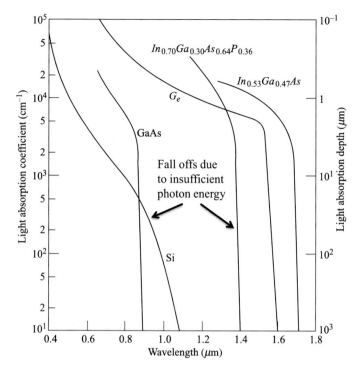

Fig. 6.3 Optical absorption coefficient as a function of wavelength of several different photodetector materials

strongly on the wavelength. Thus a particular semiconductor material can be used only over a limited wavelength range. The upper wavelength cutoff λ_c is determined by the bandgap energy E_g of the material. If E_g is expressed in units of electron volts (eV), then λ_c is given in units of micrometers (μm) by

$$\lambda_c(\mu m) = \frac{hc}{E_g} = \frac{1.2406}{E_g(eV)} \tag{6.2}$$

The cutoff wavelength is about 1.06 μm for Si, 1.6 μm for Ge, and 1.7 μm for InGaAs. For longer wavelengths, the photon energy is not sufficient to excite an electron from the valence to the conduction band.

At the lower-wavelength end, the photoresponse cuts off as a result of the very large values of α_s at the shorter wavelengths. In this case, the photons are absorbed very close to the photodetector surface, where the recombination time of the generated electron–hole pairs is very short. The generated carriers thus recombine before they can be collected by the photodetector circuitry.

Example 6.3 A photodiode is constructed of GaAs, which has a bandgap energy of 1.43 eV at 300 K. What is the cutoff wavelength of this device?

Solution From Eq. (6.2), the long-wavelength cutoff is.

$$\lambda_c = \frac{hc}{E_g} = \frac{(6.625 \times 10^{-34} \text{ J-s})(3 \times 10^8 \text{ m/s})}{(1.43 \text{ eV})(1.625 \times 10^{-19} \text{ J/eV})} = 869 \text{ nm}$$

This GaAs photodiode will not operate for photons of wavelength greater than 869 nm.

Drill Problem 6.2 A particular InGaAs *pin* photodiode has a bandgap energy of 0.74 eV. Show that the cutoff wavelength of this device is 1675 nm.

If the depletion region has a width w, then from Eq. (6.1) the total power absorbed in the distance w is

$$P_{absorbed}(w) = \int_0^w \alpha_s P_{in} \exp(-\alpha_s x)\, dx = P_{in}\left(1 - e^{-\alpha_s w}\right) \qquad (6.3)$$

When taking into account a reflectivity R_f at the entrance face of the photodiode, then the primary photocurrent i_p resulting from the power absorption of Eq. (6.3) is given by

$$i_p = \frac{q}{h\nu} P_{in}\left[1 - \exp(-\alpha_s w)\right]\left(1 - R_f\right) \qquad (6.4)$$

where P_{in} is the optical power incident on the photodetector, q is the electron charge, and $h\nu$ is the photon energy.

Two important characteristics of a photodetector are its quantum efficiency and its response speed. These parameters depend on the material bandgap, the operating wavelength, and the doping and thickness of the p, i, and n regions of the device. The *quantum efficiency* η is the number of the electron–hole carrier pairs generated divided by the number of absorbed incident photons of energy $h\nu$. This parameter is given by

$$\eta = \frac{\text{number of electron-hole pairs generated}}{\text{number of absorbed incident photons}} = \frac{i_p/q}{P_{in}/h\nu} \qquad (6.5)$$

Here, i_p is the photocurrent generated by an optical power P_{in} incident on the photodetector.

In a practical photodiode, 100 photons will create between 30 and 95 electron–hole pairs, thus giving a detector quantum efficiency ranging from 30 to 95%. To achieve a high quantum efficiency, the depletion layer must be thick enough to permit a large fraction of the incident light to be absorbed. However, the thicker the depletion layer, the longer it takes for the photogenerated carriers to drift across the reverse-biased junction. Because the carrier drift time determines the response speed of the photodiode, a compromise has to be made between response speed and quantum efficiency. This relationship is discussed further in Sect. 6.3.

The performance of a photodiode is often characterized by the *responsivity* $\mathscr{R}$. This is related to the quantum efficiency by

$$\mathscr{R} = \frac{i_p}{P_{in}} = \frac{\eta q}{h\nu} \tag{6.6}$$

The responsivity parameter is quite useful because it specifies the photocurrent generated per unit of optical power. Typical *pin* photodiode responsivities as a function of wavelength are shown in Fig. 6.4. Representative values are 0.65 A/W for silicon at 900 nm and 0.45 A/W for germanium at 1.3 μm. For InGaAs, typical responsivity values are 0.9 A/W at 1.3 μm and 1.0 A/W at 1.55 μm.

Example 6.4 In a 100-ns pulse, 6×10^6 photons at a wavelength of 1300 nm fall on an InGaAs photodetector. On the average, 5.4×10^6 electron–hole (e–h) pairs are generated. The quantum efficiency is found from Eq. (6.5) as

$$\eta = \frac{\text{number of electron-hole pairs generated}}{\text{number of absorbed incident photons}} = \frac{5.4 \times 10^6}{6 \times 10^6} = 0.90$$

Thus here the quantum efficiency at 1300 nm is 90%.

Example 6.5 Photons of energy 1.53×10^{-19} J are incident on a photodiode which has a responsivity of 0.65 A/W.

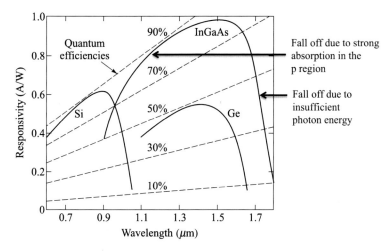

Fig. 6.4 Typical *pin* photodiode responsivities as a function of wavelength of three different materials

If the optical power level is $10\,\mu W$, then from Eq. (6.6) the photocurrent generated is.

$$i_p = \mathcal{R}P_{in} = (0.65\ A/W)(10\,\mu W) = 6.5\,\mu A$$

In most photodiodes the quantum efficiency is independent of the power level falling on the detector at a given photon energy. Thus the responsivity is a linear function of the optical power. That is, the photocurrent i_p is directly proportional to the optical power P_{in} incident upon the photodetector, so that the responsivity $\mathcal{R}$ is constant at a given wavelength (at a given value of $h\nu$). Note, however, that the quantum efficiency is not a constant at all wavelengths because it varies according to the photon energy. Consequently, the responsivity is a function of the wavelength and of the photodiode material (because different materials have different bandgap energies). For a given material, as the wavelength of the incident photon becomes longer, the photon energy becomes less than that required to excite an electron from the valence band to the conduction band. The responsivity thus falls off rapidly beyond the cutoff wavelength, as shown in Fig. 6.4.

Example 6.6 As shown in Fig. 6.4, for the wavelength range $1300\,nm < \lambda < 1600\,nm$, the quantum efficiency for InGaAs is around 90%. Thus in this wavelength range the responsivity is.

$$\mathcal{R} = \frac{\eta q}{h\nu} = \frac{\eta q \lambda}{hc} = \frac{(0.90)\left(1.6 \times 10^{-19}\,C\right)\lambda}{\left(6.625 \times 10^{-34}\,J\text{-}s\right)\left(3 \times 10^8\,m/s\right)} = 7.25 \times 10^5 \lambda$$

For example, at 1300 nm.

$$\mathcal{R} = [7.25 \times 10^5 (A/W)/m]\,(1.30 \times 10^{-6}m) = 0.92\ A/W$$

At wavelengths higher than 1600 nm, the photon energy is not sufficient to excite an electron from the valence band to the conduction band. For example, $In_{0.53}Ga_{0.47}As$ has an energy gap $E_g = 0.73\ eV$, so that from Eq. (6.2) the cutoff wavelength is.

$$\lambda_c = \frac{1.2406}{E_g(eV)} = \frac{1.2406}{0.73} = 1.7\,\mu m$$

At wavelengths less than 1100 nm for InGaAs, the photons are absorbed very close to the photodetector surface, where the recombination rate of the generated electron–hole pairs is very short. The responsivity in this material and in other photodetectors thus decreases rapidly for smaller wavelengths, because many of the generated carriers do not contribute to the photocurrent.

6.1.2 Basics of Avalanche Photodiodes

Avalanche photodiodes (APDs) internally multiply the primary signal photocurrent before it enters the input circuitry of the following amplifier [14–16]. This multiplication action increases receiver sensitivity because the photocurrent is multiplied before encountering the thermal noise associated with the receiver circuit. As shown in Fig. 6.5, *photocurrent multiplication* takes place when the photon-generated carriers traverse a multiplication region where a very high electric field is present. In this high-field region, a photon-generated electron or hole can gain enough energy so that it ionizes bound electrons in the valence band upon colliding with them. This carrier multiplication mechanism is known as *impact ionization*. The newly created carriers are also accelerated by the high electric field, thus gaining enough energy to cause further impact ionization. This phenomenon is the *avalanche effect*.

The multiplication M for all carriers generated in the photodiode is defined by

$$M = \frac{i_M}{i_p} \tag{6.7}$$

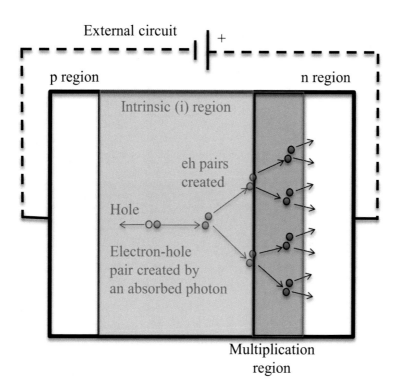

Fig. 6.5 Concept of photocurrent multiplication through an avalanche effect in an APD

where i_M is the average value of the total multiplied output current and i_p is the primary unmultiplied photocurrent defined in Eq. (6.4). In practice, the avalanche mechanism is a statistical process, because not every carrier pair generated in the diode experiences the same multiplication. Thus, the measured value of M is expressed as an average quantity.

Analogous to the *pin* photodiode, the performance of an APD is characterized by its responsivity $\mathscr{R}_{APD}$, which is given by

$$\mathscr{R}_{APD} = \frac{\eta q}{h \nu} M = \mathscr{R} M \tag{6.8}$$

where $\mathscr{R}$ is the unity gain responsivity.

Example 6.7 A given silicon avalanche photodiode has a quantum efficiency of 65% at a wavelength of 900 nm. Suppose 0.5 μW of optical power produces a multiplied photocurrent of 10 μA. What is the multiplication M?

Solution First from Eq. (6.6) the primary photocurrent generated is.

$$i_p = \mathscr{R} P_{in} = \frac{\eta q \lambda}{hc} P_{in} = \frac{(0.65)(1.6 \times 10^{-19}\,\text{C})(9 \times 10^{-7}\,\text{m})}{(6.625 \times 10^{-34}\,\text{J} \cdot \text{s})(3 \times 10^8\,\text{m/s})}(5 \times 10^{-7}\,\text{W})$$

$$= 0.235\,\mu\text{A}$$

Then from Eq. (6.7) the multiplication is.

$$M = \frac{i_M}{i_p} = \frac{10\,\mu\text{A}}{0.235\,\mu\text{A}} = 43$$

Thus the primary photocurrent is multiplied by a factor of 43.

Drill Problem 6.3 A given InGaAs avalanche photodiode has a quantum efficiency of 90% at a wavelength of 1310 nm. Suppose 0.5 μW of optical power produces a multiplied photocurrent of 8 μA. Show that the multiplication $M = 16$.

6.2 Noise Effects in Photodetectors

6.2.1 Signal-to-Noise Ratio

When detecting a weak optical signal, the photodetector and its following amplification circuitry need to be designed so that a desired signal-to-noise ratio is maintained for proper signal interpretation. The power *signal-to-noise ratio* (designated by SNR or S/N) at the output of an optical receiver is defined by

$$SNR = \frac{S}{N} = \frac{\text{signal power from photocurrent}}{\text{photodetector noise power} + \text{amplifier noise power}}$$

$$= \frac{\text{mean square signal current}}{\sum \text{mean square noise currents}}$$

$$= \frac{\langle i_s^2(t)\rangle}{\langle i_{th}^2\rangle + \langle i_{shot}^2\rangle + \langle i_{dark}^2\rangle} = \frac{\langle i_s^2(t)\rangle}{\langle i_{th}^2\rangle + \langle i_N^2\rangle} \qquad (6.9)$$

The noise currents in the receiver arise from the shot noise $\langle i_{shot}^2\rangle$ and dark current noise $\langle i_{dark}^2\rangle$ of the photodetector and the thermal noise $\langle i_{th}^2\rangle$ associated with the combined resistance of the photodiode and the amplifier circuitry. The noise sources are described in Sect. 6.2.2. In most applications, it is the noise currents that determine the minimum optical power level that can be detected, since the photodiode quantum efficiency responsible for the signal current is normally close to its maximum possible value. Note that in the analyses of the noise sources, although the carrier generation in a photodetector follows a Poisson statistics process, to a good approximation Gaussian statistics can be used describe the statistical nature of the shot and dark current noises. Thus their noise powers can be represented by the variances of the noise currents. In addition, the thermal noise also follows Gaussian statistics [17]. These approximations simplify the receiver SNR anaysis.

6.2.2 Sources of Photodetector Noise

To see the interrelationship of the different types of noises affecting the signal-to-noise ratio, consider the circuit of a simple receiver model and its equivalent circuit shown in Fig. 6.6. The photodiode has a small series resistance R_s, a total capacitance C_d consisting of junction and packaging capacitances, and a bias (or load) resistor R_L. The amplifier following the photodiode has an input capacitance C_a and a resistance R_a. For practical purposes, R_s typically is much smaller than the load resistance R_L and can be neglected.

For *pin* photodiodes the mean square value of the signal current i_s is given by

$$\langle i_s^2\rangle_{pin} = \sigma_{s,pin}^2 = \langle i_p^2(t)\rangle \qquad (6.10)$$

where $i_p(t)$ is the primary time varying current resulting from a time varying optical power $P_{in}(t)$ falling on the photodetector and σ is the variance. For avalanche photodiodes

$$\langle i_s^2\rangle_{APD} = \sigma_{s,APD}^2 = \langle i_p^2(t)\rangle M^2 \qquad (6.11)$$

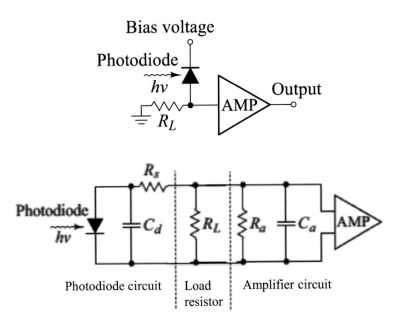

Fig. 6.6 Simple model of a photodetector receiver and its equivalent circuit

where M is the average of the statistically varying avalanche gain as defined in Eq. (6.7).

The principal noise sources associated with photodetectors are shot noise (also called *quantum noise*) and dark-current noise generated in the photodiode material. The *shot noise* arises from the statistical nature of the production and collection of photoelectrons when an optical signal is incident on a photodetector. The fluctuations in the number of photocarriers created from the photoelectric effect are a fundamental property of the photodetection process, so that the shot noise sets the lower limit on the receiver sensitivity when all other conditions are optimized. The *shot noise current* i_{shot} has a mean-square value in a receiver electrical bandwidth B_e that is proportional to the average value of the photocurrent i_p, that is,

$$\langle i_{shot}^2 \rangle = \sigma_{shot}^2 = 2qi_p B_e M^2 F(M) \qquad (6.12)$$

where $F(M)$ is a *noise figure* associated with the random nature of the avalanche process. For an APD the noise figure is typically 3 to 6 dB. From experimental results, it has been found that to a reasonable approximation $F(M) \approx M^x$, where x (with $0 \leq x \leq 1.0$) depends on the material. The parameter x takes on values of 0.3 for Si, 0.7 for InGaAs, and 1.0 for Ge avalanche photodiodes. For *pin* photodiodes M and F(M) are unity.

The photodiode *dark current* is the current i_D that continues to flow through the bias circuit of the device when no light is incident on the photodiode. This is a combination of bulk and surface dark currents, but in general the surface dark current is negligible. The bulk dark current i_{dark} arises from electrons and/or holes that are thermally generated in the pn junction of the photodiode. In an APD, these liberated carriers also get accelerated by the high electric field present at the pn junction, and are therefore multiplied by the avalanche gain mechanism. The mean-square value of this dark current is given by

$$\langle i_{dark}^2 \rangle = \sigma_{dark}^2 = 2qi_D M^2 F(M) B_e \tag{6.13}$$

where i_D is the primary (unmultiplied) detector bulk dark current, which is listed on component data sheets.

Drill Problem 6.4 Suppose a Si APD with $x \approx 0.3$ is biased to operate at $M = 100$. (a) If no signal falls on the photodetector and the unmultiplied dark current is $i_D = 10$ nA, using Eq. (6.13) show that the APD noise current per square root of bandwidth is $\left[\langle i_{dark}^2 \rangle\right]^{1/2} = \left[2qi_D M^2 F(M) B_e\right]^{1/2} = 11.3\ B_e^{1/2}$ pA/ $Hz^{1/2}$. (b) If the receiver bandwidth is 50 MHz, show that the APD dark noise current is 79.9 nA.

Because the dark currents and the signal current are uncorrelated, the total mean-square photodetector noise current $\langle i_N^2 \rangle$ can be written as

$$\langle i_N^2 \rangle = \sigma_N^2 = \langle i_{shot}^2 \rangle + \langle i_{dark}^2 \rangle = \sigma_{shot}^2 + \sigma_{dark}^2$$
$$= 2q(i_p + i_D) M^2 F(M) B_e \tag{6.14}$$

To simplify the analysis of the receiver circuitry, one can assume that the amplifier input impedance is much greater than the load resistance, so that the thermal noise from R_a is much smaller than that of R_L. The photodetector load resistor then dominates and contributes a mean-square *thermal noise current*

$$\langle i_{th}^2 \rangle = \sigma_{th}^2 = \frac{4k_B T}{R_L} B_e \tag{6.15}$$

where k_B is Boltzmann's constant and T is the absolute temperature. Using a load resistor that is large but still consistent with the receiver bandwidth requirements can reduce this noise.

Example 6.8 An InGaAs *pin* photodiode has the following parameters at a wavelength of 1300 nm: $i_D = 4$ nA, $\eta = 0.90$, and $R_L = 1000\ \Omega$. Assume the incident optical power is 300 nW (−35 dBm), the temperature is 293 K, and the receiver bandwidth is 20 MHz. Find (a) The primary photocurrent; (b) The mean-square shot noise current; (c) The mean-square dark current noise; and (d) The mean-square thermal noise current.

Solution

(a) From Eq. (6.6) the primary photocurrent is

$$i_p = \mathscr{R} P_{in} = \frac{\eta q \lambda}{hc} P_{in} = \frac{(0.90)\left(1.6 \times 10^{-19}\,\text{C}\right)\left(1.3 \times 10^{-6}\,\text{m}\right)}{\left(6.625 \times 10^{-34}\,\text{J} \cdot \text{s}\right)\left(3 \times 10^{8}\,\text{m/s}\right)}\left(3 \times 10^{-7}\,\text{W}\right)$$
$$= 0.282\,\mu\text{A}$$

(b) From Eq. (6.12) the mean-square shot noise current for a *pin* photodiode is.

$$\left\langle i_{shot}^2 \right\rangle = 2q i_p B_e = 2\left(1.6 \times 10^{-19}\,\text{C}\right)\left(0.282 \times 10^{-6}\,\text{A}\right)\left(20 \times 10^{6}\,\text{Hz}\right)$$
$$= 1.80 \times 10^{-18}\,\text{A}^2$$

or $\left\langle i_{shot}^2 \right\rangle^{1/2} = 1.34$ nA.

(c) From Eq. (6.13) the mean-square dark current is

$$\left\langle i_{dark}^2 \right\rangle = 2q i_D B_e = 2\left(1.6 \times 10^{-19}\,\text{C}\right)\left(4 \times 10^{-9}\,\text{A}\right)\left(20 \times 10^{6}\,\text{Hz}\right)$$
$$= 2.56 \times 10^{-20}\,\text{A}^2$$

or $\left\langle i_{dark}^2 \right\rangle^{1/2} = 0.16$ nA.

(d) From Eq. (6.15) the mean-square thermal noise current for the receiver is

$$\left\langle i_{th}^2 \right\rangle = \frac{4k_B T}{R_L} B_e = \frac{4\left(1.38 \times 10^{-23}\,\text{J/K}\right)(293\,\text{K})}{1000\,\Omega}\left(20 \times 10^{6}\,\text{Hz}\right)$$
$$= 323 \times 10^{-18}\,\text{A}^2$$

or $\left\langle i_{th}^2 \right\rangle^{1/2} = 18$ nA.

Thus for this receiver the rms thermal noise current is about 14 times greater than the rms shot noise current and about 100 times greater than the rms dark current.

6.2.3 Signal-to-Noise Ratio Limits

By examining the general magnitudes of the various noises, a simplification of the SNR can be made for certain limiting conditions. First consider the full expression of the SNR at the amplifier input. The SNR can be found by substituting Eqs. (6.11), (6.14), and (6.15) into Eq. (6.9). This yields

$$SNR = \frac{\left\langle i_p^2 \right\rangle M^2}{2q\left(i_p + i_D\right)M^2 F(M)B_e + 4k_B T B_e / R_L} \tag{6.16}$$

In general, the term involving i_D can be dropped when the average signal current is much larger than the dark current. The SNR then becomes

$$SNR = \frac{\langle i_p^2 \rangle M^2}{2qi_p M^2 F(M) B_e + 4k_B T B_e / R_L} \tag{6.17}$$

Example 6.9 Consider the InGaAs *pin* photodiode described in Example 6.8. What is the SNR in decibels?

Solution Because the dark current noise is negligible compared to the shot noise and thermal noise, then substituting the numerical results from Example 6.8 into Eq. (6.17) yields.

$$SNR = \frac{\left(0.282 \times 10^{-6}\right)^2}{1.80 \times 10^{-18} + 323 \times 10^{-18}} = 245$$

In decibels the SNR is

$$SNR = 10 \log 245 = 23.9$$

When the optical signal power is relatively high, then the shot noise power is much greater than the thermal noise power. In this case the SNR is called *shot noise limited* or *quantum noise limited*. When the optical signal power is low, then thermal noise usually dominates over the shot noise. In this case the SNR is referred to as being *thermal noise limited*.

When *pin* photodiodes are used, the dominating noise currents are those of the detector load resistor (the thermal current i_T) and the active elements of the amplifier circuitry (i_{amp}). For avalanche photodiodes, the thermal noise is of lesser importance and the photodetector noises usually dominate.

From Eq. (6.16), it can be seen that the signal power is multiplied by M^2 and the shot noise plus bulk dark current is multiplied by $M^2 F(M)$. Because the noise figure $F(M)$ increases with M, there always exists an optimum value of M that maximizes the signal-to-noise ratio. The optimum gain at the maximum signal-to-noise ratio can be found by differentiating Eq. (6.16) with respect to M, setting the result equal to zero, and solving for M. Doing so for a sinusoidally modulated signal, with $m = 1$ and $F(M)$ approximated by M^x, yields

$$M_{opt}^{x+2} = \frac{4k_B T / R_L}{xq(i_p + i_D)} \tag{6.18}$$

Example 6.10 Consider a Si APD operating at 300°K and with a load resistor $R_L = 1000 \, \Omega$. For this APD assume the responsivity $\mathcal{R} = 0.65$ A/W and let $x = 0.3$. (a) If dark current is neglected and 100 nW of optical power falls on the photodetector, what is the optimum avalanche gain? (b) What is the SNR if $B_e = 100$ MHz? (c) How does the SNR of this APD compare with the corresponding SNR of a Si *pin* photodiode?

Solution

(a) Neglecting dark current and with $i_p = \mathscr{R}P = (0.65)(100 \times 10^{-9})$, Eq. (6.18) yields the following optimum gain.

$$M_{opt} = \left(\frac{4k_BT}{xqR_Li_P}\right)^{1/(x+2)} = \left(\frac{4\left(1.38 \times 10^{-23}\right)(300)}{0.3\left(1.60 \times 10^{-19}\right)(1000)(0.65)\left(100 \times 10^{-9}\right)}\right)^{1/2.3} = 42$$

(b) Neglecting dark current and with $F(M) = M^x = (42)^{0.3}$, yields the following SNR value.

$$SNR = \frac{\left(i_pM\right)^2}{\left[2qi_pM^{2.3} + \left(\frac{4k_BT}{R_L}\right)\right]B_e}$$

$$= \frac{\left[(0.65)\left(100 \times 10^{-9}\right)42\right]^2}{\left[2\left(1.6 \times 10^{-19}\right)(0.65)\left(100 \times 10^{-9}\right)42^{2.3} + \left(\frac{4(1.38 \times 10^{-23})300}{1000}\right)\right]\left(100 \times 10^6\right)}$$

$$= 659$$

or in decibels, SNR(APD) $= 10 \log 659 = 28.2$ dB.

(c) For a *pin* photodiode with $M = 1$, the above equation yields SNR(*pin*) $= 2.3$ $= 3.5$ dB. Thus, compared to a *pin* photodiode, the APD improves the SNR by 24.7 dB.

6.2.4 Noise-Equivalent Power and Detectivity

The sensitivity of a photodetector is describable in terms of the *minimum detectable optical power*. This is the optical power necessary to produce a photocurrent of the same magnitude as the root-mean-square (rms) of the total noise current, or equivalently, a signal-to-noise ratio of 1. This optical signal power is referred to as the *noise equivalent power* or NEP, which is designated in units of $W/\sqrt{Hz}$.

As an example, consider the thermal-noise-limited case for a *pin* photodiode. A thermal-limited SNR occurs when the optical signal power is low so that thermal noise dominates over shot noise. Then the SNR becomes

$$SNR = \mathscr{R}^2P^2/(4k_BTB_e/R_L) \tag{6.19}$$

To find the NEP, set the SNR equal to 1 and solve for P to obtain

$$NEP = \frac{P_{min}}{\sqrt{B_e}} = \sqrt{4k_BT/R_L}/\mathscr{R} \tag{6.20}$$

Example 6.11 Let the responsivity $\mathscr{R} = 0.90$ A/W for an InGaAs photodetector operating at 1550 nm. What is the NEP in the thermal-noise-limited case if the load resistor $R_L = 1000\ \Omega$ and T = 300 K?

Solution From Eq. (6.17) the value for NEP is

$$NEP = [4(1.38 \times 10^{-23}\,J/K)(300K)/1000\,\Omega]^{1/2}/(0.90\,A/W)$$
$$= 4.52 \times 10^{-12}\,W\sqrt{Hz}$$

Drill Problem 6.5 A particular silicon *pin* photodiode has an NEP $= 1 \times 10^{-13}$ W/Hz$^{1/2}$. If the receiver operating bandwidth is 1 GHz, show that the required optical signal power is 3.16 nW for a signal-to-noise ratio equal to 1.

The parameter *detectivity*, or D*, is a figure of merit for a photodetector used to characterize its performance. The detectivity is equal to the reciprocal of NEP normalized per unit area A.

$$D* = A^{1/2}/NEP \tag{6.21}$$

Its units commonly are expressed in cm·$\sqrt{Hz}$/W.

6.3 Response Times of Photodiodes

6.3.1 *Photocurrent in the Depletion Layer*

To understand the frequency response of photodiodes, first consider the schematic representation of a reverse-biased *pin* photodiode shown in Fig. 6.7. Light enters the device through the *p* layer and produces electron–hole pairs as it is absorbed in the semiconductor material. Those electron–hole pairs that are generated in the depletion region or within a diffusion length of it will be separated by the reverse-bias-voltage-induced electric field, thereby leading to a current flow in the external circuit as the carriers drift across the depletion layer.

Under steady-state conditions, the total current density J_{tot} flowing through the reverse-biased depletion layer is

$$J_{tot} = J_{dr} + J_{diff} \tag{6.22}$$

Here, J_{dr} is the drift current density resulting from carriers generated inside the depletion region, and J_{diff} is the diffusion current density arising from the carriers

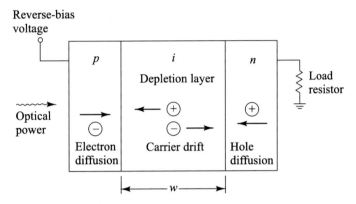

Fig. 6.7 Schematic representation of a reverse-biased *pin* photodiode

that are produced outside of the depletion layer in the bulk of the semiconductor (i.e., in the n and p regions) and diffuse into the reverse-biased junction. The drift current density can be found from Eq. (6.4):

$$J_{dr} = \frac{i_p}{A} = q\Phi_0\left(1 - e^{-\alpha_s w}\right) \tag{6.23}$$

where A is the photodiode area and Φ_0 is the incident photon flux per unit area given by

$$\Phi_0 = \frac{P_{in}\left((1 - R_f)\right)}{Ah\nu} \tag{6.24}$$

The surface p layer of a *pin* photodiode is normally very thin. The diffusion current is thus principally determined by hole diffusion from the bulk n region. The diffusion of holes in this material can be determined by the one-dimensional diffusion equation [10]

$$D_p\frac{\partial^2 p_n}{\partial x^2} - \frac{p_n - p_{n0}}{\tau_p} + G(x) = 0 \tag{6.25}$$

where D_p is the hole diffusion coefficient, p_n is the hole concentration in the n-type material, τ_p is the excess hole lifetime, p_{n0} is the equilibrium hole density, and $G(x)$ is the electron–hole generation rate given by

$$G(x) = \Phi_0\alpha_s e^{-\alpha_s x} \tag{6.26}$$

From Eq. (6.25), the diffusion current density J_{diff} is found to be

$$J_{diff} = q\Phi_0 \frac{\alpha_s L_p}{1 + \alpha_s L_p} e^{-\alpha_s w} + q p_{n0} \frac{D_p}{L_p} \qquad (6.27)$$

Substituting Eqs. (6.23) and (6.27) into Eq. (6.22), then the total current density J_{tot} through the reverse-biased depletion layer is given by

$$J_{tot} = q\Phi_0 \left(1 - \frac{e^{-\alpha_s w}}{1 + \alpha_s L_p}\right) + q p_{n0} \frac{D_p}{L_p} \qquad (6.28)$$

The term involving p_{n0} is normally small, so that the total photogenerated current is proportional to the photon flux Φ_0.

6.3.2 Response Time Characteristics

The response time of a photodiode together with its output circuit (see Fig. 6.6) depends mainly on the following three factors:

1. The transit time of the photocarriers in the depletion region;
2. The diffusion time of the photocarriers generated outside the depletion region;
3. The RC time constant of the photodiode and its associated circuit.

The photodiode parameters responsible for these three factors are the absorption coefficient α_s, the depletion region width w, the photodiode junction and package capacitances, the amplifier capacitance, the detector load resistance, the amplifier input resistance, and the photodiode series resistance. The photodiode series resistance is generally only a few ohms and can be neglected in comparison with the large load resistance and the amplifier input resistance.

The first step is to look at the transit time of the photocarriers in the depletion region. The response speed of a photodiode is fundamentally limited by the time it takes photogenerated carriers to travel across the depletion region. This transit time t_d depends on the carrier drift velocity v_d and the depletion layer width w, and is given by

$$t_d = \frac{w}{v_d} \qquad (6.29)$$

In general, the electric field in the depletion region is large enough so that the carriers have reached their scattering-limited velocity. For silicon, the maximum velocities for electrons and holes are 8.4×10^6 and 4.4×10^6 cm/s, respectively, when the field strength is on the order of 2×10^4 V/cm. A typical high-speed silicon photodiode with a depletion layer width of 10 μm thus has a response time limit of about 0.1 ns.

The diffusion processes are slow compared with the drift of carriers in the high-field region. Therefore, to have a high-speed photodiode, the photocarriers should be generated in the depletion region or so close to it that the diffusion times are less than or equal to the carrier drift times. The effect of long diffusion times can be seen by considering the photodiode response time. This response time is described by the rise time and fall time of the detector output when the detector is illuminated by a step input of optical radiation. The rise time τ_r is typically measured from the 10-to-90% points of the leading edge of the output pulse, as is shown in Fig. 6.8. For fully depleted photodiodes the rise time τ_r and fall time τ_f are generally the same. However, they can be different at low bias levels where the photodiode is not fully depleted, because the photon collection time then starts to become a significant contributor to the rise time. In this case, charge carriers produced in the depletion region are separated and collected quickly. On the other hand, electron–hole pairs generated in the n and p regions must slowly diffuse to the depletion region before they can be separated and collected. A typical response time of a partially depleted photodiode is shown in Fig. 6.9. The fast carriers allow the device output to rise to 50% of its maximum value in approximately 1 ns, but the slow carriers cause a relatively long delay before the output reaches its maximum value.

To achieve a high quantum efficiency, the depletion layer width must be much larger than $1/\alpha_s$ (the inverse of the absorption coefficient), so that most of the light will be absorbed. Figure 6.10b shows the response to the rectangular input pulse

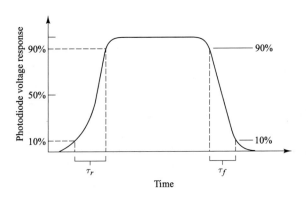

Fig. 6.8 Photodiode response to an optical input pulse showing the 10-to-90% rise time and the 10-to-90% fall time

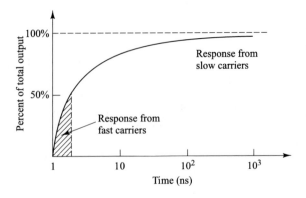

Fig. 6.9 Typical response time of a photodiode that is not fully depleted

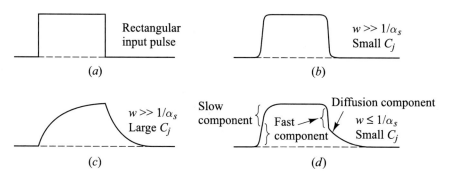

Fig. 6.10 Photodiode pulse responses under various detector parameters

shown in Fig. 6.10a of a low-capacitance photodiode having $w \gg 1/\alpha_s$. The rise and fall times of the photodiode follow the input pulse quite well. If the photodiode capacitance is larger, the response time becomes limited by the RC time constant of the load resistor R_L and the photodiode capacitance. The photodetector response then begins to appear like that shown in Fig. 6.10c.

If the depletion layer is too narrow, any carriers created in the undepleted material would have to diffuse back into the depletion region before they could be collected. Devices with very thin depletion regions thus tend to show distinct slow and fast response components, as shown in Fig. 6.10d. The fast component in the rise time is due to carriers generated in the depletion region, whereas the slow component arises from the diffusion of carriers that are created with a distance L_n from the edge of the depletion region. At the end of the optical pulse, the carriers in the depletion region are collected quickly, which results in the fast detector response component in the fall time. The diffusion of carriers that are within a distance L_n of the depletion region edge appears as the slowly decaying tail at the end of the pulse. Also, if w is too thin, the junction capacitance will become excessive. The junction capacitance C_j is

$$C_j = \frac{\varepsilon_s A}{w} \tag{6.30}$$

where

ε_s = the permittivity of the semiconductor material = $\varepsilon_0 K_s$.
K_s = the semiconductor dielectric constant.
$\varepsilon_0 = 8.8542 \times 10^{-12}$ F/m is the free-space permittivity.
A = the diffusion layer area.

This excessiveness will then give rise to a large RC time constant, which limits the response time of the detector. A reasonable compromise between high-frequency response and high quantum efficiency is found for absorption region thicknesses between $1/\alpha_s$ and $2/\alpha_s$.

If R_T is the combination of the load and amplifier input resistances and C_T is the sum of the photodiode and amplifier capacitances, as shown in Fig. 6.6, the detector behaves approximately like a simple RC low-pass filter with a passband B_c given by

$$B_c = \frac{1}{2\pi R_T C_T} \tag{6.31}$$

Example 6.12 If the photodiode capacitance is 3 pF, the amplifier capacitance is 4 pF, the load resistor is 1 kΩ, and the amplifier input resistance is 1 MΩ, then $C_T = 7$ pF and $R_T = 1$ kΩ, so that the circuit bandwidth is.

$$B_c = \frac{1}{2\pi R_T C_T} = 23 \text{ MHz}$$

If the photodector load resistance is reduced to 50 Ω, then the circuit bandwidth becomes $B_c = 455$ MHz.

6.4 Comparisons of Photodetectors

This section summarizes some generic operating characteristics of Si, Ge, and InGaAs photodiodes. Tables 6.1 and 6.2 list the performance values for *pin* and avalanche photodiodes, respectively. The values were derived from various vendor data sheets and from performance numbers reported in the literature. They are given as guidelines for comparison purposes. Detailed values on specific devices for particular applications can be obtained from suppliers of photodetectors and receiver modules.

For short-distance applications, Si devices operating around 850 nm provide relatively inexpensive solutions for most links. Longer links usually require operation in the 1300-nm and 1550-nm windows; here, one normally uses InGaAs-based devices.

Table 6.1 Generic operating parameters of Si, Ge, and InGaAs pin photodiodes

Parameter	Symbol	Unit	Si	Ge	InGaAs
Wavelength range	λ	nm	400–1100	800–1650	1100–1700
Responsivity	$\mathcal{R}$	A/W	0.4–0.6	0.4–0.5	0.75–0.95
Dark current	i_D	nA	1–10	50–500	0.5–2.0
Rise time	τ_r	ns	0.5–1	0.1–0.5	0.05–0.5
Modulation (bandwidth)	B_m	GHz	0.3–0.7	0.5–3	1–2
Bias voltage	V_B	V	5	5–10	5

Table 6.2 Generic operating parameters of Si, Ge, and InGaAs avalanche photodiodes

Parameter	Symbol	Unit	Si	Ge	InGaAs
Wavelength range	λ	nm	400–1100	800–1650	1100–1700
Avalanche gain	M	–	20–400	50–200	10–40
Dark current	i_D	nA	0.1–1	50–500	10–50 @M = 10
Rise time	τ_r	ns	0.1–2	0.5–0.8	0.1–0.5
Gain bandwidth	$M \cdot B_m$	GHz	100–400	2–10	20–250
Bias voltage	V_B	V	150–400	20–40	20–30

6.5 Summary

Semiconductor *pin* and avalanche photodiodes are the principal devices used as light signal detectors in optical fiber links because of their size compatibility with fibers, their high sensitivities at the desired optical wavelengths, and their fast response times. For short-distance applications at relatively low data rates, Si devices operating around 850 nm provide relatively inexpensive solutions for most links. Longer transmissions or very high-speed short distance links usually require operation in the 1300-nm and 1550-nm windows, where InGaAs-based devices normally are used.

When light falls on a photodetector with photon energies greater than or equal to the bandgap energy of the semiconductor material, the photons can give up their energy and excite electrons from the valence band to the conduction band. This process generates free electron–hole pairs, called photocarriers, in the photodetector. When a reverse-bias voltage is applied across the photodetector, the resultant electric field in the device causes the carriers to separate. This carrier separation gives rise to a current flow in an external circuit, which is known as the photocurrent.

The quantum efficiency η is an important photodetector performance parameter. This parameter is defined as the number of electron–hole carrier pairs generated per incident photon of energy $h\nu$. In practice, quantum efficiencies range from 30 to 95%. Another important parameter is the responsivity $\mathcal{R}$, which is related to the quantum efficiency by

$$\mathcal{R} = \frac{\eta q}{h\nu}$$

This parameter specifies the photocurrent generated per unit optical power. Representative responsivities for *pin* photodiodes are 0.65 A/W for Si at 800 nm, 0.45 A/W for Ge at 1300 nm, and 0.95 A/W for InGaAs at 1550 nm.

An avalanche photodiode (APD) internally multiplies the primary signal photocurrent. This action increases receiver sensitivity because the photocurrent is amplified before encountering the thermal noise associated with the receiver circuitry. The carrier multiplication M is a result of impact ionization in the device. Because the amplification mechanism is a statistical process, not every carrier pair that is generated in the photodiode experiences the same multiplication. Thus the measured value of M is expressed as an average quantity. Analogous to the *pin* photodiode, the performance of an APD is characterized by its responsivity $\mathscr{R}_{APD}$

$$\mathscr{R}_{APD} = \frac{\eta q}{h\nu} M = \mathscr{R} M$$

where $\mathscr{R}$ is the unity gain responsivity.

The sensitivity of a photodetector and its associated receiver essentially is determined by the photodetector noises resulting from the statistical nature of the photon-to-electron conversion process and the thermal noises in the amplifier circuitry. The main noise currents of photodetectors are as follows:

1. Quantum or shot noise current arising from the statistical nature of the production and collection of photoelectrons
2. Dark current arising from electrons and/or holes which are thermally generated in the *pn* junction of the photodiode

In general, for *pin* photodiode receivers the thermal noise currents of the detector load resistor and the active elements of the amplifier circuitry are the dominant noise sources. For avalanche photodiode receivers the thermal noise is of lesser importance and the photodetector noises usually dominate.

The usefulness of a given photodiode in a particular application depends on the required response time. To reproduce the incoming signal faithfully, the photodiode must be able to track the variations in this signal accurately. The response time depends on the absorption coefficient of the material at the desired operating wavelength, the photodiode depletion layer width, and the various capacitances and resistances of the photodiode and its associate receiver circuitry.

Because the multiplication process in an avalanche photodiode is statistical in nature, an additional noise parameter is introduced which is not present in a *pin* photodiode. A measure of this noise increase is given by the excess noise factor $F(M)$.

Problems

6.1 An InGaAs pin photodetector has an absorption coefficient of 1.0 μm^{-1} at 1550 nm. Show that the penetration depth at which 50% of the photons are absorbed is 0.69 μm.

6.2 If an optical power level P_{in} is incident on a photodiode, the electron–hole generation rate $G(x) = \Phi_0 \alpha_s \exp(-\alpha_s x)$. Here Φ_0 is the incident photon flux per unit area given by Eq. (6.24). From this, use the expression

$$i_p = q A \int_0^w G(x) dx$$

to show that the primary photocurrent in the depletion region of width w is given by Eq. (6.4).

6.3 If the absorption coefficient of silicon is 0.05 μm^{-1} at 860 nm, show that the penetration depth x at which $P(x)/P_{in} = 1/e = 0.368$ is equal to 20 μm.

6.4 A particular InGaAs pin photodiode has a bandgap energy of 0.74 eV. Show that the cutoff wavelength of this device is 1678 nm. [Thus, this GaAs photodiode will not respond to photons that have a wavelength greater than 1678 nm.]

6.5 An InGaAs *pin* photodiode has the following parameters at 1550 nm: $i_D = 1.0$ nA, $\eta = 0.95$, and $R_L = 500$ Ω. The incident optical power is 500 nW (−33 dBm) and the receiver bandwidth is 150 MHz. (a) Show from Eq. (6.6) that the primary photocurrent is 0.593 μA. (b) Show that the noise currents given by Eqs. (6.12), (6.13), and (6.15) are as follows:

$$\sigma_{shot}^2 = 2qi_p B_e = 2.84 \times 10^{-17}\ A^2$$
$$\sigma_{dark}^2 = 2qi_D B_e = 4.81 \times 10^{-20}\ A^2$$
$$\sigma_{th}^2 = \frac{4k_B T}{R_L} B_e = 4.85 \times 10^{-15}\ A^2$$

6.6 Consider an avalanche photodiode receiver that has the following parameters: dark current $i_D = 1$ nA, quantum efficiency $\eta = 0.85$, gain $M = 100$, excess noise factor $F = M^{1/2}$, load resistor $R_L = 10^4$ Ω, and bandwidth $B_e = 10$ kHz. Suppose a sinusoidally varying 850-nm signal having a modulation index $m = 0.85$ falls on the photodiode, which is at room temperature ($T = 300$ K). Show that the responsivity is 0.58 A/W.

6.7 Consider an avalanche photodiode receiver that has the parameters listed in Problem 6.6. To compare the contributions from the individual noise terms to the SNR given by Eq. (6.9), examine each of the noises independently in terms of the incident optical power P_{in}. Show that for this particular set of parameters, the relative contributions to the SNR are as follows:

$$SNR_{shot} = \frac{\langle i_s^2(t) \rangle}{\langle i_{shot}^2(t) \rangle} = 6.565 \times 10^{12} P_{in}$$

$$SNR_{dark} = \frac{\langle i_s^2(t) \rangle}{\langle i_{dark}^2(t) \rangle} = 3.798 \times 10^{22} P_{in}^2$$

$$SNR_{th} = \frac{\langle i_s^2(t) \rangle}{\langle i_{th}^2(t) \rangle} = 7.333 \times 10^{22} P_{in}^2$$

6.8 A given InGaAs avalanche photodiode has a quantum efficiency of 90% at a wavelength of 1310 nm. Suppose 0.5 μW of optical power produces a multiplied photocurrent of 8 μA. Show that the multiplication M = 16.

6.9 Suppose an avalanche photodiode has the following parameters: $i_D = 1$ nA, $\eta = 0.85$, $F = M^{1/2}$, $R_L = 10^3$ Ω, and $B_e = 1$ kHz. Consider a sinusoidal 850-nm signal, which has a modulation index $m = 0.85$ and an average power level $P_{in} = -50$ dBm, to fall on the detector at room temperature. (a) Using Eq. (6.16) show that

$$SNR = \frac{1.215 \times 10^{-17} M^2}{2.176 \times 10^{-24} M^{5/2} + 1.656 \times 10^{-20}}$$

(b) Plot the SNR as a function of M for gains ranging from 20 to 100. Show from the plot that the optimum value of M occurs at M = 62.

6.10 The optimum gain at the maximum signal-to-noise ratio can be found by differentiating Eq. (6.16) with respect to M, setting the result equal to zero, and solving for M. Show that doing this procedure yields Eq. (6.18).

6.11 Consider a silicon *pin* photodiode that has a depletion layer width $w = 20 \, \mu m$, an area $A = 0.05 \, mm^2$, and a dielectric constant $K_s = 11.7$. If the photodiode is to operate with a 10-kΩ load resistor at 800 nm, where the absorption coefficient $\alpha_s = 10^3 \, cm^{-1}$, compare the RC time constant and the carrier drift time of this device. Is carrier diffusion time of importance in this photodiode? [Answer: From Eq. (6.30) $t_{RC} = 2.59$ ns and from Eq. (6.29) $t_d = 0.45$ ns. Thus most carriers are absorbed in the depletion region, so the carrier diffusion time is not important here. The detector response time is dominated by the RC time constant.]

References

1. P.C. Eng, S. Song, B. Ping, State-of-the-art photodetectors for optoelectronic integration at telecommunication wavelength. Nanophotonics (2015). https://doi.org/10.1515/nanoph-2015-0012(Reviewarticle)
2. A. Beling, J.C. Campbell, InP-based high-speed photodetectors: tutorial. J. Lightw. Technol. **27**(3), 343–355 (2009)
3. M. Casalino, G. Coppola, R.M. De La Rue, D.F. Logan, State-of-the-art all-silicon sub-bandgap photodetectors at telecom and datacom wavelengths. Laser Photonics Rev. **10**(6), 895–921 (2016)
4. M.J. Deen, P.K. Basu, *Silicon Photonics: Fundamentals and Devices* (Wiley, 2012)
5. S. Donati, *Photodetectors: Devices* (Prentice Hall, Circuits and Applications, 2000)
6. Z. Zhao, J. Liu, Y. Liu, N. Zhu, High-speed photodetectors in optical communication system. J. Semicond. **38**(12), 121001 (2017) (review article)
7. H. Schneider, H.C. Liu, *Quantum Well Infrared Photodetectors* (Springer, 2006)
8. B.E.A. Saleh, M. Teich, *Fundamentals of Photonics*, 3rd edn. (Wiley, 2019)
9. B.L. Anderson, R.L. Anderson, *Fundamentals of Semiconductor Devices*, 2nd edn. (McGraw-Hill, 2018)
10. D.A. Neaman, *Semiconductor Physics and Devices*, 4th edn. (McGraw-Hill, 2012)
11. S.O. Kasap, *Principles of Electronic Materials and Devices*, 4th edn. (McGraw-Hill, 2018)
12. O. Manasreh, *Semiconductor Heterojunctions and Nanostructures* (McGraw-Hill, 2005)
13. S.E. Miller, E.A.J. Marcatili, T. Li, Research toward optical-fiber transmission systems. Proc. IEEE **61**, 1703–1751 (1973)
14. P.P. Webb, R.J. McIntyre, J. Conradi, Properties of avalanche photodiodes. RCA Rev. **35**, 234–278 (1974)
15. D.S.G. Ong, M.M. Hayat, J.P.R. David, J.S. Ng, Sensitivity of high-speed lightwave system receivers using InAlAs avalanche photodiodes. IEEE Photonics Technol. Lett. **23**(4), 233–235 (2011)
16. S. Cao, Y. Zhao, S. ur Rehman, S. Feng, Y. Zuo, C. Li, L. Zhang, B. Cheng, Q. Wang, *Theoretical studies on InGaAs/InAlAs SAGCM avalanche photodiodes*. Nanoscale Res. Lett. **13**, 158 (2018)
17. B.M. Oliver, Thermal and quantum noise. IEEE Proc. **53**, 436–454 (1965)

Chapter 7
Optical Receiver Operation

Abstract The design of an optical receiver can be quite sophisticated because the receiver must be able to detect weak, distorted signals and make decisions on what type of data was sent based on an amplified and reshaped version of this distorted signal. In the photodetection processes, various noises and distortions will unavoidably be introduced, which can cause signal interpretation errors. Noise considerations are thus important in the design of optical receivers, because the noise sources operating in the receiver generally set the lowest limit for the signal levels that can be processed. This chapter describes the origins of these noises and their effect on link performance.

Having discussed the characteristics and operation of photodetectors in the previous chapter, the next step is to consider features of the optical receiver. An optical receiver consists of a photodetector, an amplifier, and signal-processing circuitry. The receiver has the task of first converting the optical energy emerging from the end of a fiber into an electric signal, and then amplifying this signal to a large enough level so that it can be processed by the electronics following the receiver amplifier.

In these processes, various noises and distortions will unavoidably be introduced, which can lead to errors in the interpretation of the received signal. Depending on the magnitude of the received optical signal, the current generated by the photodetector could be very weak and is adversely affected by the random noises associated with the photodetection process. When this electric signal output from the photodiode is amplified, additional noises arising from the amplifier electronics will further corrupt the signal. Noise considerations are thus important in the design of optical receivers, because the noise sources operating in the receiver generally set the lowest limit for the signals that can be processed.

In designing a receiver, it is desirable to predict its performance based on mathematical models of the various receiver stages. These models must take into account the noises and distortions added to the signal by the components in each stage, and they must show the designer which components to choose so that the desired performance criteria of the receiver are met.

The average error probability is an especially meaningful criterion for measuring the performance of a digital communication system. In an analog system the fidelity

© The Author(s), under exclusive license to Springer Nature Singapore Pte Ltd. 2021   267
G. Keiser, *Fiber Optic Communications*,
https://doi.org/10.1007/978-981-33-4665-9_7

criterion normally is specified in terms of a peak signal-to-rms-noise ratio. The calculation of the error probability for a digital optical communication receiver differs from that of conventional electric systems. This is due to the discrete quantum nature of the photon arrival and detection processes and the random gain fluctuations when an avalanche photodiode is used. Researchers have used a variety of analytical methods to derive approximate predictions for receiver performance. In carrying out these analyses, one needs to make a tradeoff between calculation simplicity and accuracy of the approximation. General reviews and detailed concepts of optical receiver designs are given in the literature [1–12].

First Sect. 7.1 presents an overview of the fundamental operational characteristics of the various stages of an optical receiver. This consists of tracing the path of a digital signal through the receiver and showing what happens at each step along the way. Section 7.2 then outlines the fundamental methods for determining the *bit-error rate* or probability of error (the chance that a bit is corrupted and received in error) of a digital receiver based on signal-to-noise considerations. This section also discusses the concept of *receiver sensitivity*, which is an important parameter for estimating the minimum received optical power that is needed to achieve a specific probability of error.

The *eye diagram* is a common measurement tool to determine the fidelity of the received signal. Eye diagrams have been used extensively for all types of communication links, including wire lines, wireless systems, and optical links. Section 7.3 describes the method for generating an eye diagram and how to interpret various signal fidelity parameters with it.

For *passive optical network* (PON) applications, the operational characteristics of an optical receiver located at the central telecommunications switching office differ significantly from receivers used in conventional point-to-point links. Section 7.4 gives a brief description of these devices, which are known as *burst-mode receivers*. Finally, Sect. 7.5 describes another optical receiver type that is used for analog links.

7.1 Basic Receiver Operation

The design of an optical receiver can be quite sophisticated because the receiver must be able to detect and interpret what type of data was sent based on an amplified and reshaped version of the weakened and distorted received signal. To get an appreciation of the function of an optical receiver, first consider what happens to a signal as it is sent through an optical fiber link. Because traditionally fiber optic communication links are intensity-modulated direct-detection (IM-DD) systems that use a binary *on-off keyed* (OOK) digital signal, Sect. 7.1.1 analyzes direct-detection receiver performance by using this OOK signal format.

7.1.1 Transmitting Digital Signals

Figure 7.1 illustrates the shape of a digital signal at different points along an optical link. The transmitted signal is a two-level binary data stream consisting of either a 0 or a 1 in a time slot of duration T_b. This time slot is referred to as a *bit period*. Electrically there are many ways of sending a given digital message [13–16]. One of the simplest techniques for sending binary data is *amplitude-shift keying* (ASK) or *on–off keying* (OOK), wherein a voltage level is switched between two values, which are usually *on* or *off*. The resultant signal wave thus consists of a voltage pulse of amplitude V relative to the zero voltage level when a binary 1 occurs and a zero-voltage-level space when a binary 0 occurs. Depending on the coding scheme to be used, a binary 1 may or may not fill the time slot T_b. For simplicity, in this chapter it is assumed that when 1 is sent, a voltage pulse of duration T_b occurs, whereas for 0 the voltage remains at its zero level.

The function of the optical transmitter is to convert the electric signal to an optical signal. As shown in Sect. 4.3.7, one way of doing this is by directly modulating the light source drive current with the information stream to produce a varying optical output power $P(t)$. Thus in the optical signal emerging from the LED or laser transmitter, 1 is represented by a pulse of optical power (light) of duration T_b, whereas 0 is the absence of any light.

The optical signal that is coupled from the light source to the fiber becomes attenuated and distorted as it propagates along the fiber waveguide. Upon arriving at the end of a fiber, a receiver converts the optical signal back to an electrical format.

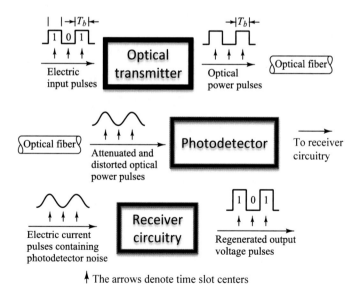

Fig. 7.1 Signal path through an optical data link

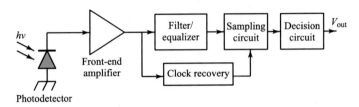

Fig. 7.2 The basic constituents of an optical receiver

Figure 7.2 shows the basic components of an optical receiver. The first element is either a *pin* or an avalanche photodiode, which produces an electric current that is proportional to the received optical power level. Because this electric current typically is very weak, a *front-end amplifier* boosts the signal to a level that can be used by the following electronics.

After the electric signal produced by the photodiode is amplified, it passes through a *low-pass filter* to reduce the noise that is outside of the signal bandwidth. This filter thus defines the receiver bandwidth. In addition, to minimize the effects of inter-symbol interference (ISI) the filter can reshape the pulses that have become distorted as they traveled through the fiber. This function is called *equalization* because it equalizes or cancels pulse-spreading effects.

In the final optical receiver module shown (on the right) in Fig. 7.2, a sampling and decision circuit samples the signal level at the midpoint of each time slot and compares it with a certain reference voltage known as the *threshold level*. If the received signal level is greater than the threshold level, a 1 pulse was received. If the voltage is below the threshold level, a 0 pulse was received. To accomplish this bit interpretation, the receiver must know where the bit boundaries are. This is done with the assistance of a periodic waveform called a *clock*, which has a periodicity equal to the bit interval. Thus this function is called *clock recovery* or *timing recovery* [17, 18].

In some cases, an optical preamplifier is placed ahead of the photodiode to boost the optical signal level before photodetection takes place. This is done so that the signal-to-noise ratio degradation caused by thermal noise in the receiver electronics can be suppressed. Compared with other front-end devices, such as avalanche photo-diodes or optical heterodyne detectors, an optical preamplifier provides a larger gain factor and a broader bandwidth. However, this process also introduces additional noise to the optical signal. Chapter 11 addresses optical amplifiers and their effects on system performance.

7.1.2 Sources of Detection Errors

Errors in the detection mechanism can arise from various noises and disturbances associated with the signal detection system that are shown in Fig. 7.3. The term

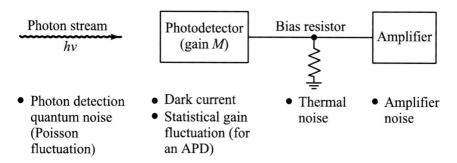

Fig. 7.3 Noise sources and disturbances in an optical pulse detection mechanism

noise is used customarily to describe unwanted components of an electric signal that tend to disturb the transmission and processing of the signal in a physical system, and over which there is incomplete control. The noise sources can be either external to the system (e.g., from electric power lines, motors, radio transmitters, lightning) or internal to the system (e.g., from switch and power supply transients). Here, the concern is mainly with internal noise, which is present in every communication system and represents a basic limitation on the transmission or detection of signals. This noise is caused by the spontaneous fluctuations of current or voltage in electric circuits. The two most common examples of these spontaneous fluctuations are shot or quantum noise and thermal noise. Shot noise arises in electronic devices because of the discrete nature of current flow in the device. Thermal noise arises from the random motion of electrons in a conductor. A third noise source is dark current, which is the current that continues to flow through the bias circuit of a photodiode when no light is incident on the device.

The random arrival rate of signal photons produces shot noise at the photodetector. Because this noise depends on the signal level, it is of particular importance for *pin* receivers that have large optical input levels and for avalanche photodiode receivers. When using an avalanche photodiode, an additional shot noise arises from the statistical nature of the multiplication process. This noise level increases as the avalanche gain M becomes larger. Additional photodetector noise comes from the dark current. This noise is independent of the photodiode illumination and can generally be made very small in relation to other noise currents by a judicious choice of components.

Thermal noises arising from the detector load resistor and from the amplifier electronics tend to dominate in applications with low signal-to-noise ratio when a *pin* photodiode is used. When an avalanche photodiode is used in applications that encounter low optical signal levels, the optimum avalanche gain is determined by a design tradeoff between the thermal noise and the gain-dependent shot noise.

Because the thermal noises are of a Gaussian nature, they can be readily treated by standard techniques. The analyses of the noises and the resulting error probabilities associated with the primary photocurrent generation and the avalanche multiplication are complicated as neither of these processes is Gaussian. The primary photocurrent generated by the photodiode is a time-varying Poisson process resulting from the random arrival of photons at the detector. If the detector is illuminated by an optical signal $P(t)$, then the average number of electron–hole pairs $\overline{N}$ generated in a time τ is

$$\overline{N} = \frac{\eta}{h\nu} \int_0^\tau P(t)\, dt = \frac{\eta E}{h\nu} = \frac{\eta\lambda}{hc} E \tag{7.1}$$

where η is the detector quantum efficiency, $h\nu$ is the photon energy, and E is the energy received in a time interval τ. The actual number of electron–hole pairs n that are generated fluctuates from the average according to the Poisson distribution

$$P_r(n) = \overline{N}^n \frac{e^{-\overline{N}}}{n!} \tag{7.2}$$

where $P_r(n)$ is the probability that n electrons are excited in a time interval τ.

Example 7.1 Using Eq. (7.2) one can find conditions such as the probability that no electrons are excited in a time interval when a pulse of a certain energy E arrives. That is, for example, what is the energy required in a 1 pulse to have a probability of 10^{-9} or smaller that the arriving 1 pulse will not be interpreted as a 0 pulse?

Solution First assume that no electron–hole pairs are created during a 0 pulse, that is, $n = 0$. Then, the energy that is needed to have the condition $P_r(n = 0) < 10^{-9}$ can be calculated from Eq. (7.2). Using the relation $\overline{N} = \eta\lambda E/hc$ from Eq. (7.1), the inequality that needs to be solved is

$$P_r(n = 0) = exp\left(-\frac{\eta\lambda E}{hc}\right) \le 10^{-9} \tag{7.3}$$

Solving this relationship for E yields

$$E \ge (9\ln 10)\frac{hc}{\eta\lambda} = 20.7\frac{hc}{\eta\lambda}$$

Recalling that hc/λ is the energy of a photon and η is the detector quantum efficiency, the above energy inequality means that a 1 pulse must contain at least $20.7/\eta$ photons in order not to be misinterpreted as a 0 pulse with a probability of 10^{-9}. Note that the number of photons must be an integer.

Drill Problem 7.1 Consider a photodetector that has a quantum efficiency η = 0.65.
  (a) Show that an energy

$$E \geq (12 \, ln \, 10) \frac{hc}{\eta\lambda} = 42.5 \frac{hc}{\eta\lambda}$$

is required in a 1 pulse to have a probability of 10^{-12} or smaller in order that the arriving 1 pulse will not be interpreted as a 0 pulse. (*b*) Why must the number of photons in the 1 pulse be greater than or equal to 66?

The fact that it is not possible to predict exactly how many electron–hole pairs are generated by a known optical power incident on the detector is the origin of the excess noise factor F(M) resulting from the random nature of the avalanche multiplication process. Recall from Sect. 6.4 that, for an avalanche detector with a mean gain M, for electron injection F(M) is often approximated by the empirical equation

$$F(M) \approx M^x \tag{7.4}$$

where the factor x ranges between 0 and 1.0 depending on the photodiode material.

A further error source is attributed to *intersymbol interference* (ISI), which results from pulse spreading in the optical fiber. When an optical pulse is transmitted in a given time slot, most of the pulse energy will arrive in the corresponding time slot T_b at the receiver, as shown in Fig. 7.4. However, because of pulse spreading effects induced by the fiber, some of the transmitted energy will progressively spread into neighboring time slots as the pulse propagates along the fiber. The presence of this energy in adjacent time slots results in an interfering signal, hence the term *intersymbol interference*. In Fig. 7.4 the parameter γ designates the fraction of energy remaining in the time slot T_b, so that the fraction of energy that has spread into adjacent time slots is $1 - \gamma$.

Fig. 7.4 Pulse spreading of an optical signal into adjacent bit slots that leads to intersymbol interference

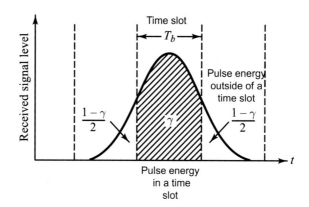

7.1.3   Receiver Front-End Amplifiers

Noise sources at the front end of a receiver dominate the sensitivity and bandwidth, so a major engineering emphasis has been on the design of a low-noise front-end amplifier. The goals generally are to maximize the receiver sensitivity while maintaining a suitable bandwidth. To achieve these goals, a basic concern in front-end design is what load resistor R_L to choose because this parameter affects both the bandwidth and the noise performance. Front-end amplifiers used in optical fiber communication systems can be classified into three broad categories, which are the low-impedance, the high-impedance, and the transimpedance designs. These categories are not actually distinct because a continuum of intermediate configurations is possible, but they serve to illustrate the design approaches.

The *low-impedance (LZ) preamplifier* is the most straightforward configuration, but is not necessarily the optimum preamplifier design. The basic structure is shown in Fig. 7.5. In this design a photodiode operates into a low-impedance amplifier with an effective input resistance R_a (e.g., $R_a = 50\ \Omega$) and a capacitance C. A bias or load resistor R_b in parallel with R_a is used to match the amplifier impedance (i.e., to suppress standing waves in order to achieve uniform frequency response). The total preamplifier load resistance $R_L = R_a R_b/(R_a + R_b)$ is the parallel combination of R_a and R_b. The value of the bias resistor, in conjunction with the amplifier input capacitance C, is such that the preamplifier bandwidth is equal to or greater than the signal bandwidth. As can be seen from Eq. (6.29), a small load resistance yields a large bandwidth. The drawback is that for low load resistances the thermal noise dominates. Thus, although low-impedance preamplifiers can operate over a wide bandwidth, they do not provide high receiver sensitivities because only a small signal voltage can be developed across the total input impedance. This limits the use of these preamplifiers to special short-distance applications in which high receiver sensitivity is not a major concern.

Recall from Eq. (6.15) that the thermal noise is inversely proportional to the load resistance. Thus R_L should be as large as possible to minimize thermal noise. Thus increasing the value of R_b in Fig. 7.5 results in the *high-impedance amplifier* design. Here a tradeoff must be made between noise and receiver bandwidth, because the bandwidth is inversely proportional to the resistance seen by the photodiode. Consequently for a high-impedance front end, a high load resistance results not only in low noise but also gives a low receiver bandwidth. Although equalizers sometimes can be implemented to increase the bandwidth, if the bandwidth is much less than the bit rate, then such a front-end amplifier cannot be used.

Fig. 7.5 Generic structure of low-impedance and high-impedance amplifiers

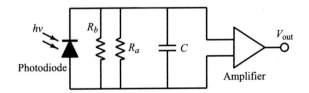

Fig. 7.6 Generic structure of a transimpedance amplifier

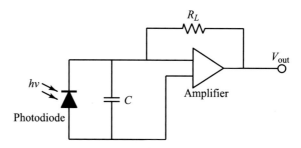

The *transimpedance amplifier* design shown in Fig. 7.6 largely overcomes the drawbacks of the high-impedance amplifier. In this case, R_L is used as a negative feedback resistor around an inverting amplifier. Now R_L can be large because the negative feedback reduces the effective resistance R_P seen by the photodiode by a factor of G, so that $R_P = R_L/(G + 1)$, where G is the gain of the amplifier. This means that compared to the high-impedance design the transimpedance bandwidth increases by a factor of $G + 1$ for the same load resistance. Although this does increase the thermal noise compared to a high-impedance amplifier, the increase usually is less than a factor of 2 and can easily be tolerated. Consequently, the transimpedance front-end design tends to be the amplifier of choice for optical fiber transmission links.

Note that in addition to the thermal noise differences resulting from selection of a particular load resistor, the electronic components in the front-end amplifier that follows the photodetector also add further thermal noise. The magnitude of this additional noise depends on the design of the amplifier, for example, what type of bipolar or field-effect transistors are incorporated in the design. This noise increase can be accounted for by introducing an *amplifier noise figure* F_n into the numerator of Eq. (6.17). This parameter is defined as the ratio of the input SNR to the output SNR of the amplifier. Typical values of the amplifier noise figure range from 3 to 5 dB (a factor of 2 to 3).

Example 7.2 Consider an optical receiver that has a high-impedance amplifier with an input resistance of $R_a = 4$ MΩ. Suppose it is matched to a photodetector bias resistor that has a value $R_b = 4$ MΩ. (*a*) If the total capacitance is $C = 6$ pF, what is the maximum bandwidth achievable without equalization? (*b*) Now consider the case when the high-impedance amplifier is replaced with a transimpedance amplifier that has a 100 kΩ feedback resistor and a gain $G = 350$. What is the maximum achievable bandwidth without equalization in this case?

Solution (*a*) The total preamplifier load resistance R_L is the parallel combination of R_a and R_b, so that $R_L = R_a R_b/(R_a + R_b) = (4 \times 10^6)^2/8 \times 10^6 = 2$ MΩ. Then from Eq. (6.29) the maximum bandwidth is $B = 1/(2\pi R_L C) = 13.3$ kHz.

(*b*) If the total capacitance is again $C = 6$ pF, then for the transimpedance amplifier the bandwidth is given by

$$B = \frac{G}{2\pi R_L C} = \frac{350}{2\pi \left(1 \times 10^5 \Omega\right)\left(6 \times 10^{-12} F\right)} = 92.8 \text{ MHz}$$

7.2 Performance Characteristics of Digital Receivers

Ideally, in a digital receiver the decision-circuit output signal voltage $\upsilon_{out}(t)$ would always exceed the threshold voltage when a 1 is present and would be less than the threshold when no pulse (a 0) was sent. In actual systems, deviations from the average value of $\upsilon_{out}(t)$ are caused by various noises, interference from adjacent pulses, and conditions wherein the light source is not completely extinguished during a zero pulse.

7.2.1 Determining Probability of Error

In practice, there are several ways of measuring the rate of error occurrences in a digital data stream. Chapter 14 describes some of these methods. A simple approach is to divide the number N_e of errors occurring over a certain time interval t by the number N_t of pulses (ones and zeros) transmitted during this interval. This is called either the *error rate* or the *bit-error rate,* which commonly is abbreviated BER. Thus, by definition,

$$BER = \frac{N_e}{N_t} = \frac{N_e}{Bt} \tag{7.5}$$

where $B = 1/T_b$ is the bit rate (i.e., the pulse transmission rate). The error rate is expressed by a number, such as 10^{-9}, for example, which states that, on the average, one error occurs for every billion pulses sent. Typical error rates for optical fiber telecommunication systems range from 10^{-9} to 10^{-12}. This error rate depends on the signal-to-noise ratio at the receiver (the ratio of signal power to noise power). The system error rate requirements and the receiver noise levels thus set a lower limit on the optical signal power level that is required at the photodetector.

Example 7.3 For a given data rate what is the minimum average power required to achieve a certain desired BER? Assume an equal number of 0 and 1 pulses.

Solution Using Eqs. (7.1) and (7.2) the bit-error rate can be written as

$$P_r(n = 0) = exp\left(-\frac{\eta \lambda E}{hc}\right) \leq BER$$

Solving this equation for the energy yields

$$E \geq \frac{hc}{\eta\lambda} \ln\left(\frac{1}{BER}\right)$$

Thus, if E_{min} is the minimum energy needed to achieve a specific BER, then for a bit rate $B = 1/T_b$ the minimum average power is

$$P_{ave} = \frac{E_{min}}{2T_b} = E_{min}B/2$$

Drill Problem 7.2 Consider a photodetector that has a quantum efficiency of 0.65 at a 1310-nm wavelength. Show that the minimum average power needed to achieve a 10^{-12} BER at a 1-Gb/s data rate at 1310 nm is 3.74 nW $= -$ 54.3 dBm.

To compute the bit-error rate at the receiver, it is necessary to know the probability distribution of the signal at the equalizer output [19–21]. Knowing the signal probability distribution at this point is important because it is here that the decision is made as to whether a 0 or a 1 is sent. The shapes of two signal probability distributions are shown in Fig. 7.7. These are

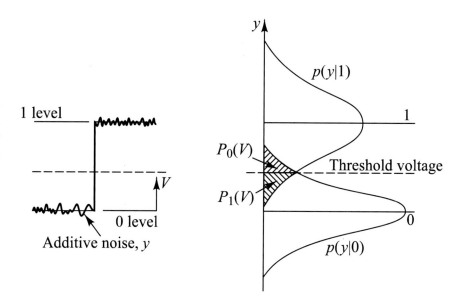

Fig. 7.7 Probability distributions for received logic 0 and 1 signal pulses; various signal distortion effects cause the different widths of the two distributions

$$P_1(v) = \int_{-\infty}^{v} p(y|1)dy \tag{7.6}$$

which is the probability that the equalizer output voltage is less than v when a logical 1 pulse is sent, and

$$P_0(v) = \int_{v}^{\infty} p(y|0)dy \tag{7.7}$$

which is the probability that the output voltage exceeds v when a logical 0 is transmitted. Note that the different shapes of the two probability distributions in Fig. 7.7 indicate that the noise power for a logical 0 is usually not the same as that for a logical 1. This occurs in optical systems because of signal distortion from transmission impairments (e.g., dispersion, optical amplifier noise, and distortion from nonlinear effects) and from noise and ISI contributions at the receiver. The functions $p(y|1)$ and $p(y|0)$ are the conditional probability distribution functions; that is, $p(y|x)$ is the probability that the output voltage is y, given that an x was transmitted.

If the threshold voltage is v_{th} then the error probability P_e is defined as

$$P_e = aP_1(v_{th}) + bP_0(v_{th}) \tag{7.8}$$

The weighting factors a and b are determined by the a priori distribution of the data. That is, a and b are the probabilities that either a 1 or a 0 occurs, respectively. For unbiased data with equal probability of 1 and 0 occurrences, $a = b = 0.5$. The problem to be solved now is to select the decision threshold at that point where P_e is minimum.

To calculate the error probability it is necessary to know the mean-square noise voltage $\langle v_N^2 \rangle$, which is superimposed on the signal voltage at the decision time. The statistics of the output voltage at the sampling time are quite complex, so that an exact calculation is rather tedious to perform. Therefore a number of different approximations have been used to calculate the performance of a binary optical fiber receiver. In applying these approximations, one has to make a tradeoff between computational simplicity and accuracy of the results. The simplest method is based on a Gaussian approximation. In this method, it is assumed that, when the sequence of optical input pulses is known, the equalizer output voltage $v_{out}(t)$ is a Gaussian random variable. Thus, to calculate the error probability, one only needs to know the mean and standard deviation of $v_{out}(t)$.

Thus, assume that a signal s (which can be either a noise disturbance or a desired information-bearing signal) has a Gaussian probability distribution function with a mean value m. If the signal voltage level $s(t)$ is sampled at any arbitrary time t_1, the probability that the measured sample $s(t_1)$ falls in the range s to $s + ds$ is given by

$$f(s)ds = \frac{1}{\sqrt{2\pi}\,\sigma}e^{-(s-m)^2/2\sigma^2}ds \qquad (7.9)$$

where $f(s)$ is the *probability density function*, σ^2 is the noise variance, and its square root σ is the *standard deviation*, which is a measure of the width of the probability distribution. By examining Eq. (7.9) it is observed that the quantity $2\sqrt{2}\,\sigma$ measures the full width of the probability distribution at the point where the amplitude is $1/e$ of the maximum.

The probability density function now can be used to determine the probability of error for a data stream in which the 1 pulses are all of amplitude V. As shown in Fig. 7.8, the mean and variance of the Gaussian output for a 1 pulse are b_{on} and σ_{on}^2, respectively, whereas for a 0 pulse they are b_{off} and σ_{off}^2, respectively. First consider the case of a 0 pulse being sent, so that no pulse is present at the decoding time. The probability of error in this case is the probability that the noise will exceed the threshold voltage v_{th} and be mistaken for a 1 pulse. This probability of error $P_0(v)$ is the chance that the equalizer output voltage $v(t)$ will fall somewhere between v_{th} and ∞. Using Eqs. (7.7) and (7.9), then yields

$$P_0(v_{th}) = \int_{v_{th}}^{\infty} p(y|0)dy = \int_{v_{th}}^{\infty} f_0(v)dv$$

$$= \frac{1}{\sqrt{2\pi}\,\sigma_{off}} \int_{v_{th}}^{\infty} exp\left[-\frac{(v-b_{off})^2}{2\sigma_{off}^2}\right]dv \qquad (7.10)$$

where the subscript 0 denotes the presence of a 0 bit.

Similarly, one can find the probability of error that a transmitted 1 is misinterpreted as a 0 by the decoder electronics following the equalizer. This probability of error is the likelihood that the sampled signal-plus-noise pulse falls below v_{th}. From Eqs. (7.6) and (7.9), this is given by

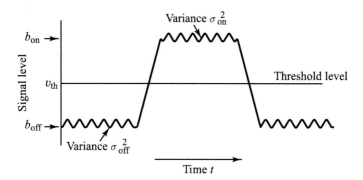

Fig. 7.8 Gaussian noise statistics of a binary signal showing variances around the on and off signal levels

$$P_1(v_{th}) = \int_{-\infty}^{v_{th}} p(y|1)dy = \int_{-\infty}^{v_{th}} f_1(v)dv$$

$$= \frac{1}{\sqrt{2\pi}\,\sigma_{on}} \int_{-\infty}^{v_{th}} exp\left[-\frac{(b_{on}-v)^2}{2\sigma_{on}^2}\right] dv \qquad (7.11)$$

where the subscript 1 denotes the presence of a 1 bit.

If the probabilities of 0 and 1 pulses are equally likely [that is, $a = b = 0.5$ in Eq. (7.8)], then Eqs. (7.6) and (7.7) yield

$$P_0(v_{th}) = P_1(v_{th}) = \frac{1}{2}P_e \qquad (7.12)$$

Thus, using Eqs. (7.10) and (7.11), the bit error rate or the error probability P_e becomes

$$BER = P_e(Q) = \frac{1}{\sqrt{\pi}} \int_{Q/\sqrt{2}}^{\infty} exp(-x^2)\,dx = \frac{1}{2}\left[1 - erf\left(\frac{Q}{\sqrt{2}}\right)\right]$$

$$\approx \frac{1}{\sqrt{2\pi}}\frac{e^{-Q^2/2}}{Q} \qquad (7.13)$$

The approximation on the right-hand side is obtained from the asymptotic expansion of erf (x). Here, the parameter Q is defined as

$$Q = \frac{v_{th} - b_{off}}{\sigma_{off}} = \frac{b_{on} - v_{th}}{\sigma_{on}} = \frac{b_{on} - b_{off}}{\sigma_{on} + \sigma_{off}} \qquad (7.14)$$

and

$$erf(x) = \frac{2}{\sqrt{\pi}} \int_{0}^{x} exp(-y^2)dy \qquad (7.15)$$

is the *error function*, which is tabulated in various mathematical handbooks [22, 23].

The factor Q is widely used to specify receiver performance, because it is related to the signal-to-noise ratio required for achieving a specific bit-error rate. In particular, it takes into account that in optical fiber systems the variances in the noise powers generally are different for received logical 0 and 1 pulses. Figure 7.9 shows how the BER varies with Q. The approximation for P_e given in Eq. (7.13) and shown by the dashed line in Fig. 7.9 is accurate to 1% for $Q \approx 3$ and improves as Q increases. A commonly quoted Q value is 6, because this corresponds to a BER = 10^{-9}.

Example 7.4 When there is little intersymbol interference, $\gamma - 1$ is small, so that $\sigma_{on}^2 = \sigma_{off}^2$. Then, by letting $b_{off} = 0$ in Eq. (7.14) the parameter Q becomes

Fig. 7.9 Plot of the BER (P_e) versus the factor Q where the dashed line represents the approximation from Eq. (7.13)

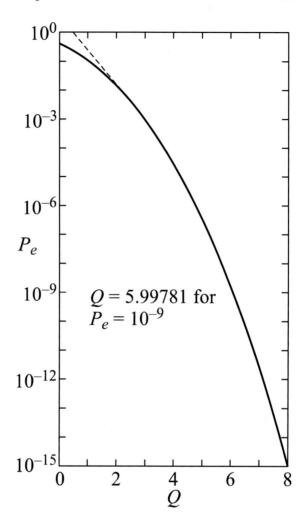

$$Q = \frac{b_{on}}{2\sigma_{on}} = \frac{1}{2}SNR$$

which is one-half the signal-to-noise ratio. In this case, $v_{th} = b_{on}/2$, so that the optimum decision threshold is midway between the 0 and 1 signal levels.

Example 7.5 If the bit error rate is 10^{-9} then from Eq. (7.13)

$$P_e(Q) = 10^{-9} = \frac{1}{2}\left[1 - \text{erf}\left(\frac{Q}{\sqrt{2}}\right)\right]$$

From Fig. 7.9 it follows that $Q = 6$ (an exact evaluation yields $Q = 5.99781$), which gives a signal-to-noise ratio of 12, or 10.8 dB [i.e., 10 log(S/N) = 10 log 12 = 10.8 dB].

Now consider the special case when $\sigma_{off} = \sigma_{on} = \sigma$ and $b_{off} = 0$, so that $b_{on} = V$. Then, from Eq. (7.14) it follows that the threshold voltage $v_{th} = V/2$, so that $Q = V/2\sigma$. Because σ is usually called the *rms noise*, the ratio V/σ is the *peak signal-to-rms-noise ratio*. In this case, Eq. (7.13) becomes

$$P_e(\sigma_{on} = \sigma_{off}) = \frac{1}{2}\left[1 - \text{erf}\left(\frac{V}{2\sqrt{2}\sigma}\right)\right]$$ (7.16)

Equation 7.16 demonstrates the exponential behavior of the probability of error as a function of the signal-to-noise ratio. By increasing V/σ by 2, that is, doubling SNR (a 3-dB power increase), the BER decreases by 10^4. Thus, there exists a narrow range of signal-to-noise ratios above which the error rate is tolerable and below which a highly unacceptable number of errors occur. The SNR at which this transition occurs is called the *threshold level*. In general, a performance safety margin of 3 dB is included in the transmission link design to ensure that this BER threshold is not exceeded when system parameters such as transmitter output, line attenuation, or noise floor vary with time.

Example 7.6 Figure 7.10 shows a plot of the BER expression from Eq. (7.16) as a function of the signal-to- noise ratio. How many bits are misinterpreted in a standard DS1 telephone rate of 1.544 Mb/s for SNR values of 8.5 and 12.0?

Solution If the signal-to-noise ratio is 8.5 (18.6 dB) then $P_e = 10^{-5}$. If this is the received signal level for a standard DS1 telephone rate of 1.544 Mb/s, this BER results in a misinterpreted bit every 0.065 s, which is highly unsatisfactory. However, by increasing the signal strength so that $V/\sigma = 12.0$ (21.6 dB), the BER decreases to $P_e = 10^{-9}$. For the DS1 case, this means that a bit is misinterpreted every 650 s (or 11 min), which, in general, is tolerable.

Drill Problem 7.3 Figure 7.10 shows a plot of the BER expression from Eq. (7.16) as a function of the signal-to-noise ratio. Verify that for a high-speed network that operates at 622 Mb/s, a required BER of 10^{-12} means a value of $V/\sigma = 22.3$ dB is necessary.

7.2.2 Specifying Receiver Sensitivity

Optical communication systems use a BER value to specify the performance requirements for a particular transmission link application. For example, SONET/SDH

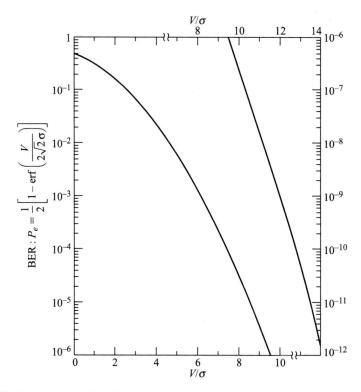

Fig. 7.10 Bit-error rate as a function of signal-to-noise ratio when the standard deviations are equal ($\sigma_{on} = \sigma_{off}$) and when $b_{off} = 0$

networks (see Chap. 13) specify that the BER must be 10^{-10} or lower, whereas Gigabit Ethernet requires no more than a 10^{-12} BER. To achieve a desired BER at a given data rate, a specific minimum average optical power level must arrive at the photodetector. The value of this minimum power level is called the *receiver sensitivity*.

A common method of defining the receiver sensitivity is as an *average optical power* (P_{ave}) in dBm incident on the photodetector. Alternatively it may be defined by the *optical modulation amplitude* (OMA), which is given in terms of a peak-to-peak current at the photodetector output. The receiver sensitivity gives a measure of the minimum average power or OMA needed to maintain a maximum (worst-case) BER at a specific data rate. Many researchers have carried out numerous complex calculations of receiver sensitivity by taking into account different pulse-shape degradation factors. This section presents a simplified analysis to illustrate the basis of receiver sensitivity.

First, expressing Eq. (7.14) in terms of signal currents from 1 and 0 pulses (given by i_1 and i_0, respectively) and their corresponding noise current variations (σ_1 and σ_0, respectively), and assuming there is no optical power in a zero pulse, yields

$$Q = \frac{i_1 - i_0}{\sigma_1 + \sigma_0} = \frac{i_1}{\sigma_1 + \sigma_0} \qquad (7.17)$$

Using Eqs. (6.6), (6.7), and (7.14), the receiver sensitivity $P_{sensitivity}$ is found from the average power contained in a bit period for the specified data rate as

$$P_{sensitivity} = P_1/2 = i_1/(2\mathscr{R}M) = Q(\sigma_1 + \sigma_0)/(2\mathscr{R}M) \qquad (7.18)$$

where $\mathscr{R}$ is the unity-gain responsivity and M is the gain of the photodiode.

If there is no optical amplifier in a fiber transmission link, then thermal noise and shot noise are the dominant noise effects in the receiver. As Sect. 6.2 describes, the thermal noise is independent of the incoming optical signal power, but the shot noise depends on the received power. Therefore, assuming there is no optical power in a received zero pulse, the noise variances for 0 and 1 pulses, respectively, are $\sigma_0^2 = \sigma_{th}^2$ and $\sigma_1^2 = \sigma_{th}^2 + \sigma_{shot}^2$. From Eqs. (6.6) and (6.13), and using the condition from Eq. (7.18), the shot noise variance for a 1 pulse is

$$\sigma_{shot}^2 = 2q\mathscr{R}P_1 M^2 F(M)B_e = 4q\mathscr{R}P_{sensitivity}M^2 F(M)B/2 \qquad (7.19)$$

where $F(M)$ is the photodiode noise figure and the electrical bandwidth B_e of the receiver is assumed to be half the bit rate B (i.e., $B_e = B/2$). In addition to the various photodetector noises, the receiver amplifier also has a noise figure F_n associated with it. Thus, when including this noise in Eq. (6.15), the thermal noise current variance is

$$\sigma_{th}^2 = \frac{4k_B T}{R_L} F_n \frac{B}{2} \qquad (7.20)$$

Substituting $\sigma = \left(\sigma_{shot}^2 + \sigma_{th}^2\right)^{1/2}$ and $\sigma_0 = \sigma_{th}$ into Eq. (7.18) and solving for $P_{sensitivity}$ then gives

$$P_{sensitivity} = (1/\mathscr{R})\frac{Q}{M}\left[\frac{qMF(M)BQ}{2} + \sigma_{th}\right] \qquad (7.21)$$

Example 7.7 To see the behavior of the receiver sensitivity as a function of the BER, first consider the receiver to have a load resistor $R_L = 200\ \Omega$ and let the temperature be $T = 300\ °K$. Letting the amplifier noise figure be $F_n = 3$ dB (a factor of 2), then from Eq. (7.20) the thermal noise current variance is $\sigma_T = 9.10 \times 10^{-12}B^{1/2}$. Next, select an InGaAs photodiode with a unity-gain responsivity $\mathscr{R} = 0.95$ A/W at 1550 nm and assume an operating BER $= 10^{-12}$ so that a value of $Q = 7$ is needed. If the photodiode gain is M, then from Eq. (7.21) the receiver sensitivity is

$$P_{sensitivity} = \frac{7.37}{M}\left[5.6 \times 10^{-19}MF(M)B + 9.10 \times 10^{-12}B^{1/2}\right] \qquad (7.22)$$

Fig. 7.11 Sensitivities as a function of bit rate for generic *pin* and avalanche InGaAs photodiodes at 1550 nm for a 10^{-12} BER

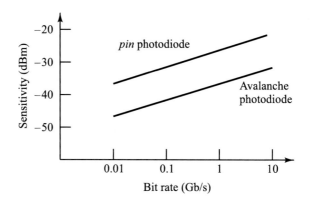

Figure 7.11 shows the receiver sensitivity calculated from Eq. (7.22) as a function of data rate for typical InGaAs *pin* and avalanche photodiodes at 1550 nm for a 10^{-12} BER. In Fig. 7.11 the APD gain was taken to be $M = 10$ and $F(M) = 10^{0.7} = 5$. Note that the curves in Fig. 7.11 are for a BER given by $Q = 7$, a load resistor $R_L = 200\ \Omega$, an amplifier noise figure $F_n = 3$ dB, and a 1550-nm wavelength. The sensitivity curves will change for different values of these parameters.

Example 7.8 Consider an InGaAs *pin* photodiode for which $M = 1$ and $F(M) = 1$. For the conditions in Eq. (7.22), what is the receiver sensitivity at a 1-Gb/s data rate for a 10^{-12} BER requirement?

Solution From Eq. (7.22)

$$P_{sensitivity} = 7.37\left[5.6 \times 10^{-19}\left(1 \times 10^9\right) + 9.10 \times 10^{-12}\left(1 \times 10^9\right)^{1/2}\right]\text{mW}$$
$$= -26.7\ \text{dBm}$$

> **Drill Problem 7.4** Consider an InGaAs avalanche photodiode for which M $= 10$ and $F(M) = 5$. For the conditions in Eq. (7.22), show that the receiver sensitivity at a 1 Gb/s data rate for a 10^{-12} BER requirement is 2.32×10^{-4} mW $= -36.3$ dBm.

7.2.3 The Basic Quantum Limit

In designing an optical system, it is useful to know what the fundamental physical bounds are on the system performance. To see what this bound is for the photodetection process, consider an ideal photodetector that has unity quantum efficiency and produces no dark current; that is, no electron–hole pairs are generated in the absence of an optical pulse. Given this condition, it is possible to find the minimum received optical power required for a specific bit-error rate performance in a digital system. This minimum received power level is known as the *quantum limit*, because all system parameters are assumed ideal and the performance is limited only by the photodetection statistics.

Assume that an optical pulse of energy E falls on the photodetector in a time interval τ. The receiver can only interpret this condition as a 0 pulse if no electron–hole pairs are generated with the pulse present. From Eq. (7.2) the probability that $n = 0$ electrons are excited in a time interval t is

$$P_r(0) = e^{-\overline{N}} \tag{7.23}$$

where the average number of electron–hole pairs $\overline{N}$ is given by Eq. (7.1). Thus, for a given error probability $P_r(0)$, we can find the minimum energy E required at a specific wavelength λ.

Example 7.9 A 10-Mb/s digital fiber optic link operating at 850 nm requires a maximum BER of 10^{-9}.

(*a*) The first step is to find the quantum limit in terms of the quantum efficiency of the detector and the energy of the incident photon. From Eq. (7.23) the probability of error is $P_r(0) = e^{-\overline{N}} = 10^{-9}$. Solving for $\overline{N}$ yields $\overline{N} = 9 \ln 10 = 20.7 \approx 21$. Hence, an average of 21 photons per pulse is required for this BER. Using Eq. (7.1) and solving for E gives $E = 20.7\, h\nu/\eta$.

(*b*) The next step is to find the minimum incident optical power P_i that must fall on the photodetector to achieve a 10^{-9} BER at a data rate of 10 Mb/s for a simple binary-level signaling scheme. If the detector quantum efficiency $\eta = 1$, then

$$E = P_i\tau = 20.7h\nu = 20.7\ h\nu/\lambda$$

where $1/\tau$ is one-half the data rate B; that is, $1/\tau = B/2$. (*Note:* This assumes an equal number of 0 and 1 pulses.) Solving for P_i yields

$$P_i = 20.7\frac{hcB}{2\lambda} = \frac{20.7\left(6.626 \times 10^{-34}\ \text{J s}\right)\left(3.0 \times 10^8\ \text{m/s}\right)\left(10 \times 10^6\ \text{bits/s}\right)}{2\left(0.85 \times 10^{-6}\ \text{m}\right)}$$

$$= 24.2\ \text{pW}$$

or, when the reference power level is 1 mW, then $P_i = -76.2$ dBm.

In practice, the sensitivity of most receivers is around 20 dB higher than the quantum limit because of various nonlinear distortions and noise effects in the transmission link. Furthermore, when specifying the quantum limit, one has to be careful to distinguish between average power and peak power. If one uses average power, the quantum limit given in Example 7.4 would be only 10 photons per bit for a 10^{-9} BER. Sometimes, the literature quotes the quantum limit based on these average powers. However, this can be misleading because the limitation on real components is based on peak and not average power.

7.3 Principles of Eye Diagrams

The eye diagram is a powerful measurement tool for assessing the data-handling ability of a digital transmission system. This method has been used extensively for evaluating the performance of wire line systems and also applies to optical fiber data links. Chapter 14 gives more details on BER test equipment and measurement methods.

7.3.1 Features of Eye Patterns

The eye pattern measurements are made in the time domain and allow the effects of waveform distortion to be shown immediately on the display screen of standard BER test equipment. Figure 7.12 shows a typical display pattern, which is known as an *eye pattern* or an *eye diagram*. The logic 1 and 0 levels, shown by b_{on} and b_{off}, respectively, determine the basic upper and lower bounds.

A great deal of system-performance information can be deduced from the eye pattern display. To interpret the eye pattern, consider Fig. 7.12 and the simplified drawing shown in Fig. 7.13. The following information regarding the signal amplitude distortion, timing jitter, and system rise time can be derived:

- The *width of the eye opening* defines the time interval over which the received signal can be sampled without error due to interference from adjacent pulses (known as *intersymbol interference*).
- The best time to sample the received waveform is when the *height of the eye opening* is largest. This height is reduced as a result of amplitude distortion in the data signal. The vertical distance between the top of the eye opening and the maximum signal level gives the degree of distortion. The more the eye closes, the more difficult it is to distinguish between ones and zeros in the signal.
- The height of the eye opening at the specified sampling time shows the noise margin or immunity to noise. *Noise margin* is the percentage ratio of the peak signal voltage V_1 for an alternating bit sequence (defined by the height of the eye opening) to the maximum signal voltage V_2 as measured from the threshold level, as shown in Fig. 7.13. That is

$$\text{Noise margin } (\%) = \frac{V_1}{V_2} \times 100 \% \qquad (7.24)$$

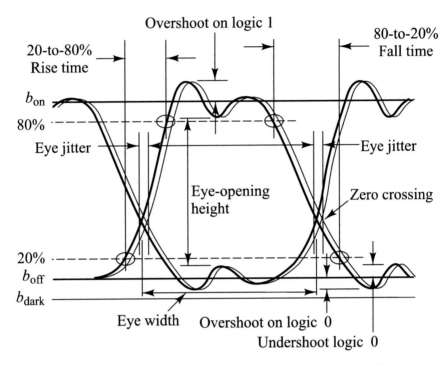

Fig. 7.12 General configuration of an eye diagram showing the definitions of fundamental measurement parameters

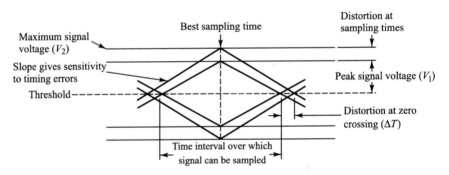

Fig. 7.13 Simplified eye diagram showing the key performance parameters

- The rate at which the eye closes as the sampling time is varied (i.e., the slope of the eye pattern sides) determines the sensitivity of the system to *timing errors*. The possibility of timing errors increases as the slope becomes more horizontal.
- *Timing jitter* (also referred to as *edge jitter* or *phase distortion*) in an optical fiber system arises from noise in the receiver and pulse distortion in the optical fiber [18]. Excessive jitter can result in bit errors, because such jitter can produce

uncertainties in clock timing. This timing uncertainty will cause a receiver to lose synchronization with the incoming bit stream thereby incorrectly interpreting logic 1 and 0 pulses. If the signal is sampled in the middle of the time interval (i.e., midway between the times when the signal crosses the threshold level), then the amount of distortion ΔT at the threshold level indicates the amount of jitter. Timing jitter is thus given by

$$\text{Timing jitter } (\%) = \frac{\Delta T}{T_b} \times 100\% \qquad (7.25)$$

where T_b is one bit interval.

- Traditionally, the *rise time* is defined as the time interval between the points where the rising edge of the signal reaches 10% of its final amplitude to the time where it reaches 90% of its final amplitude. However, when measuring optical signals, these points are often obscured by noise and jitter effects. Thus, the more distinct values at the 20 and 80% threshold points normally are measured. To convert from the 20 to 80% rise time to a 10 to 90% rise time, one can use the approximate relationship

$$T_{10-90} = 1.25 \times T_{20-80} \qquad (7.26)$$

A similar approach is used to determine the fall time.
- Any nonlinear effects in the channel transfer characteristics will create an asymmetry in the eye pattern. If a purely random data stream is passed through a purely linear system, all the eye openings will be identical and symmetrical.

Example 7.10 Consider an eye diagram in which the center opening is about 90% due to intersymbol interference (ISI) degradation. What is the ISI degradation in decibels?

Solution The ISI degradation is given by

$$\text{ISI} = 20 \log \frac{V_1}{V_2} = 20 \log 0.90 = 0.915 \text{ dB}$$

Modern bit-error rate measurement instruments construct and display eye diagrams such as the example shown in Fig. 7.14. Ideally, if the signal impairments are small, the received pattern on the instrument display should exhibit sharp, clearly defined lines. However, time-varying signal impairments in the transmission path can lead to amplitude variations within the signal and to timing skews between the data signal and the associated clock signal. Note that a clock signal, which typically is encoded within a data signal, is used to help the receiver interpret the incoming data correctly. Thus in an actual link the received pattern will become wider or distorted on the sides and on the top and bottom, as shown in Fig. 7.14.

Fig. 7.14 Example of an eye diagram displayed on a bit-error rate measurement instrument showing relatively little signal distortion

7.3.2 BER and Q-Factor Measurements

Because BER is a statistical parameter, its value depends on the measurement time and on the factors that cause the errors. If the errors are due to Gaussian noise in a relatively stable transmission link, then a measurement time in which about 100 errors occur may be needed to ensure a statistically valid BER determination. Longer measurement times may be needed for systems in which bursts of errors can occur. For high-speed communications the required bit error rate typically needs to be 10^{-12} or lower. As an example, for a 10-Gb/s links a 10^{-12} BER means that one bit error occurs every 100 s. Such a level may be unacceptable, so even lower bit-error rates, such as 10^{-15}, may be required to assure customers of a high grade of service. Standards that define acceptable bit-error rates include the ITU-T G.959.1 Recommendation and the IEEE 802.3-2018 Ethernet Standard [20,24–26].

Test times can be quite long. For example, to detect 100 errors for measuring a 10^{-12} BER in a 10-Gb/s link will require 2.8 h. Thus, using such methods test times on installed links could run anywhere from 8 to 72 h. To reduce such costly and time-consuming test periods, a *Q-factor technique* can be used. Although some accuracy is lost in this method, it reduces the test times to minutes instead of hours. In this method the receiver threshold is decreased, which increases the probability of errors and thus decreases test time.

A wide variety of sophisticated bit-error rate test equipment is available for both factory testing and field testing of optical communication equipment and transmission links. In addition to performing tests using standardized patterns or carrying out Q-factor-based measurements, more advanced equipment also measures performance by using a degraded signal that more closely represents what is seen in fielded links. This method is described in the IEEE 802.3ae specification for testing 10-Gigabit Ethernet (10-GbE) devices. This *stressed-eye test* examines the worst-case performance condition by specifying a poor extinction ratio and adding multiple stresses, intersymbol interference (ISI) or vertical eye closure, sinusoidal interference, and

Fig. 7.15 The inclusion of all possible signal distortion effects results in a stressed eye with only a small diamond-shaped opening

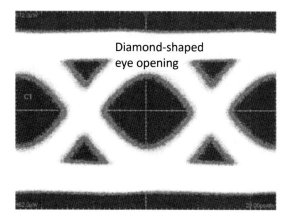

sinusoidal jitter. The concept of this test is to assume that all different possible jitter and intersymbol interference impairments that might occur to a signal in a fielded link will close the eye down to a diamond shape, as shown in Fig. 7.15. If the eye opening of the optical receiver under test is greater than this diamond-shaped area of assured error-free operation, it is expected to operate properly in an actual fielded system. The stressed-eye template height typically is between 0.10 and 0.25 of the full pattern height. Chapter 14 gives more details on this stressed-eye test.

7.4 Burst-Mode Receivers

To address the continuously increasing demands by customers for higher-capacity connections to a central switching facility, network and service providers devised the concept of using a *passive optical network* (PON) [27–30]. Chapter 13 presents more details of passive optical network configurations, which also have become known as *fiber-to-the-premises* (FTTP) networks. In a PON there are no active components between a central switching office and the customer premises (for example, homes, businesses, or government facilities). Instead, only passive optical components are placed in the network transmission path to guide the traffic signals contained within specific optical wavelengths to the user endpoints and back to the central office.

Figure 7.16 illustrates the architecture of a generic PON in which a fiber optic network connects switching equipment in a central office with a number of telecom service subscribers. Examples of equipment in the central office include public telephone switches, video-on-demand servers, Internet protocol (IP) routers, Ethernet switches, and network monitoring and control stations. In the central office, traditionally data and digitized voice are combined and sent *downstream* to customers over an optical link by using a 1490-nm wavelength. The *upstream* (customer to central office) return path for the data and voice uses a 1310-nm wavelength. Video services are sent downstream with a 1550-nm wavelength. The transmission equipment in

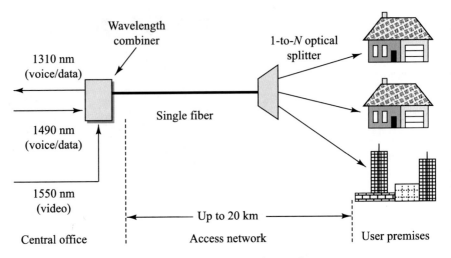

Fig. 7.16 Architecture of a simple generic passive optical network

the network consists of an *optical line termination* (OLT) unit situated at the central office and an *optical network termination* (ONT) unit at each customer premises.

Starting at the central office, one single-mode optical fiber strand runs to a passive *optical power splitter* near a housing complex, an office park, or some other campus environment. At this point a passive splitting device simply divides the optical power into *N* separate paths to the subscribers. The number of splitting directions can vary from 2 to 64, but normally there are 8, 16, or 32 paths. From the optical splitter, individual single-mode fibers then run to each building or serving equipment. The optical fiber transmission span from the central office to the user can be up to 20 km.

For PON applications, at the central office the operational characteristics of an optical receiver in an OLT differ significantly from those of conventional point-to-point links. This arises from the fact that the amplitude and phase of information packets received in successive time slots from different network user (customer) locations can vary widely from packet to packet, as is illustrated in Fig. 7.17. This is due to the possible distance variations of customers from the central office, which can

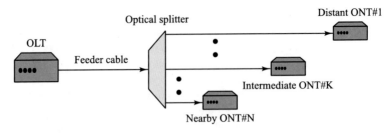

Fig. 7.17 Large distance variations of customers from the central office result in different signal power losses across the PON

range up to 20 km. In one case, suppose the closest and farthest customers attached to a common optical power splitter are 20 km apart and that the fiber attenuation is 0.5 dB/km. Then there is a 10-dB difference in the signal amplitudes that arrive at the OLT from these two users if both have the same upstream laser output level. If there is an additional optical component in the transmission path going to one of the customer sites, then the difference in signal levels arriving at the OLT could vary up to 20 dB.

Figure 7.18 shows the consequence of this effect. The term ONT in this figure refers to the transceiver equipment at the customer location. The top part shows the type of data pattern that would be received in conventional point-to-point links, such as the signal levels arriving at a particular customer site from the central office. Here there is no amplitude variation in the received logic ones. The bottom part illustrates the optical signal pattern levels that might arrive at the OLT from various customers. In this case the signal amplitude changes from one information packet (a formatted unit of data) to another depending on how far away each ONT is from the central office. The *guard time* shown in Fig. 7.18 provides a sufficient delay time to prevent collisions between successive packets that may come from different ONTs.

Because a conventional optical receiver is not capable of instantaneously handling rapidly changing differences in signal amplitude and clock phase alignment, a specially designed *burst-mode receiver* is needed [27–30]. These receivers can

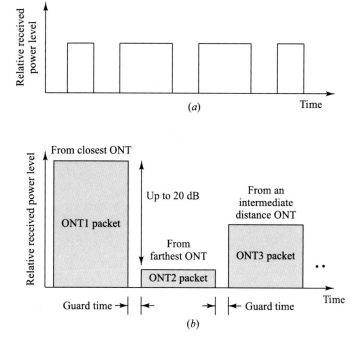

Fig. 7.18 **a** Typical received data pattern in conventional point-to-point links; **b** Optical signal level variations in pulses that may arrive at an OLT

quickly extract the decision threshold and determine the signal phase from a set of overhead bits placed at the beginning of each packet burst. However, this methodology results in a receiver sensitivity power penalty of up to 3 dB.

The key requirements of a burst-mode receiver are high sensitivity, wide dynamic range, and fast response time. The sensitivity is important in relation to the optical power budget, because, for example, a sensitivity improvement of 3 dB can double the size of the power splitter so that more customers can be attached to the PON. A wide dynamic range is essential for achieving a long network reach, that is, to be able to accommodate users located both close and far away from the central office.

The use of a conventional ac-coupling method is not possible in a burst-mode receiver, because the residual charge in a coupling capacitor following any particular data burst cannot dissipate fast enough in order not to affect the initial conditions of the next burst. The burst-mode receiver therefore requires additional circuitry to accommodate dc-coupled operation. Such receivers now are incorporated into standard commercially available OLT equipment.

7.5   Characteristics of Analog Receivers

In addition to the wide usage of fiber optics for the transmission of digital signals, there are many potential applications for analog links. These range from individual 4-kHz voice channels to microwave links operating in the multiple-gigahertz region [31–33]. The previous sections discussed digital receiver performance in terms of error probability. For an analog receiver, the performance fidelity is measured in terms of a *signal-to-noise ratio*. This is defined as the ratio of the mean-square signal current to the mean-square noise current.

The simplest analog technique is to use amplitude modulation of the source. In this scheme, a time-varying electric signal $s(t)$ is used to modulate an optical source directly about some bias point defined by the bias current I_B, as shown in Fig. 7.19. The transmitted optical power $P(t)$ is thus of the form

$$P(t) = P_t[1 + ms(t)] \tag{7.27}$$

where P_t is the average transmitted optical power at a drive current I_B, $s(t)$ is the analog modulation signal, and m is the modulation signal index defined by (see Sect. 4.4)

$$m = \frac{\Delta I}{I'_B} \tag{7.28}$$

Here, $I'_B = I_B$ for LEDs and $I'_B = I_B - I_{\text{th}}$ for laser diodes. The parameter ΔI is the variation in current about the bias point. In order not to introduce distortion into the optical signal, the modulation must be confined to the linear region of the light

Fig. 7.19 Direct analog modulation of an LED source

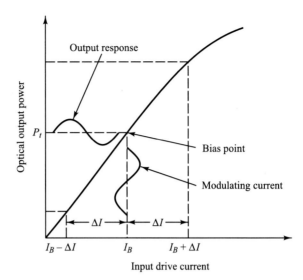

source output curve shown in Fig. 7.19. Also, if $\Delta I > I_B$, severe distortion results because the lower portion of the signal gets cut off.

At the receiver end, the photocurrent generated by the analog optical signal is

$$i_s(t) = \mathscr{R} M P_r [1 + ms(t)] = i_p M[1 + ms(t)] \tag{7.29}$$

where $\mathscr{R}$ is the detector responsivitity, P_r is the average received optical power, $i_P = \mathscr{R} P_r$ is the primary photocurrent, and M is the photodetector gain. If $s(t)$ is a sinusoidally modulated signal, then the mean-square signal current at the photodetector output is (ignoring a dc term)

$$\left\langle i_s^2 \right\rangle = \frac{1}{2} (\mathscr{R} M m P_r)^2 = \frac{1}{2} (M m i_p)^2 \tag{7.30}$$

Recalling from Sect. 6.2.2 that the mean-square noise current for a photodiode receiver is the sum of the mean-square shot noise current, the equivalent-resistance thermal noise current, and the dark noise current, then

$$\left\langle i_N^2 \right\rangle = 2q(i_p + i_D) M^2 F(M) B_e + \frac{4k_B T B_e}{R_{eq}} F_t \tag{7.31}$$

where.
i_P = primary (unmultiplied) photocurrent = $\mathscr{R} P_r$.
i_D = primary bulk dark current.
$F(M)$ = excess photodiode noise factor $\approx M^x$ ($0 < x < 1$).
B_e = effective receiver noise bandwidth.
R_{eq} = equivalent resistance of photodiode load and amplifier.

F_t = noise figure of the baseband amplifier.

The signal-to-noise ratio SNR then is

$$SNR = \frac{\langle i_s^2 \rangle}{\langle i_N^2 \rangle} = \frac{\frac{1}{2}(i_p M m)^2}{2q(i_p + i_D)M^2 F(M)B_e + \frac{4k_B T B_e}{R_{eq}}F_t} \qquad (7.32)$$

For a *pin* photodiode the gain $M = 1$. When the optical power incident on the photodiode is small, the circuit (thermal) noise term dominates the noise current, so that

$$SNR = \frac{\frac{1}{2}(i_p m)^2}{(4k_B T B_e / R_{eq})F_t} \qquad (7.33)$$

Here, the signal-to-noise ratio is directly proportional to the square of the photodiode output current and inversely proportional to the thermal noise of the circuit.

> **Drill Problem 7.5** When thermal noise is the dominant noise mechanism, the signal-to-noise ratio given by Eq. (7.32) is a maximum when
>
> $$M_{opt}^{2+x} = \frac{4k_B T F_t / R_{eq}}{q(i_p + i_D)x}$$
>
> Consider a Si APD detector that has an excess noise factor related parameter $x = 0.3$, a load-resistance/amplifier-noise-figure value of $R_{eq}/F_t = 10^4 \, \Omega$, a dark current of 10 nA, and a responsivity of 0.6 A/W. If at a temperature $T = 300°$ K the photodetector is irradiated with a light power level $P_r = 10$ nW, (*a*) first use Eq. (6.6) to show that the primary photocurrent is 6 nA and (*b*) then use the above relationship to show that the optimum gain is $M_{opt} = 28.1$.

For large optical signals incident on a *pin* photodiode, the shot noise associated with the signal detection process dominates, so that

$$SNR \approx \frac{m^2 i_p}{4q B_e} = (m^2 \mathcal{R} P_r)/(4q B_e) \qquad (7.34)$$

Because the SNR in this case is independent of the circuit noise, it represents the fundamental or quantum limit for analog receiver sensitivity.

When an avalanche photodiode is employed at low signal levels and with low values of gain M, the circuit noise term dominates. At a fixed low signal level, as the gain is increased from a low value, the SNR increases with gain until the shot noise term becomes comparable to the circuit noise term. As the gain is increased further beyond this point, the signal-to-noise ratio *decreases* as $F(M)^{-1}$. Thus for a given set of operating conditions, there exists an optimum value of the avalanche gain for which the signal-to-noise ratio is a maximum. Because an avalanche photodiode increases the SNR for small optical signal levels, it is the preferred photodetector for this situation.

For very large optical signal levels, the shot noise term dominates the receiver noise. In this case, an avalanche photodiode serves no advantage, because the detector noise increases more rapidly with increasing gain M than the signal level. This is shown in Fig. 7.20, which compares the signal-to-noise ratio for a *pin* and an avalanche photodiode receiver as a function of the received optical power. The SNR for the avalanche photodetector is at the optimum gain. The parameter values chosen for Fig. 7.20 are $B_e = 5$ and 25 MHz, $x = 0.5$ for the avalanche photodiode and 0 for the *pin* diode, $m = 80\%$, $\mathscr{R} = 0.5$ A/W, and $R_{eq}/F_t = 10^4\ \Omega$. For low signal levels an avalanche photodiode yields a higher SNR, whereas at large received optical power levels a *pin* photodiode gives equal performance.

Example 7.11 Consider an analog optical fiber system operating at 1550 nm, which has an effective receiver noise bandwidth of 5 MHz. Assuming that the received signal is shot noise limited, what is the incident optical power necessary to have a signal-to-noise ratio of 50 dB at the receiver? Assume the responsivity is 0.9 A/W and that m = 0.5.

Solution First note that a 50-dB SNR means that $SNR = 10^5$. Then, solving Eq. (7.34) for P_r yields

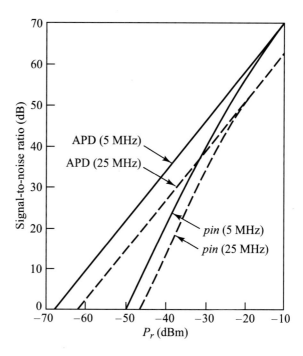

Fig. 7.20 Comparison of the SNR for *pin* and avalanche photodiodes as a function of received optical power for bandwidths of 5 and 25 MHz

$$P_r = (\text{SNR})\, 4qB_e/(m^2\mathscr{R}) = \frac{\left(1 \times 10^5\right)4\left(1.6 \times 10^{-19}\right)\left(5 \times 10^6\right)}{(0.5)^2(0.9)}$$

$$= 1420 \text{ nW} = 1.42 \times 10^{-3} \text{ mW or in dBm}$$

$$P_r(\text{dBm}) = 10 \log P_r = 10 \log 1.42 \times 10^{-3} = -28.5 \text{ dBm}$$

7.6 Summary

The function of an optical receiver is first to convert the optical energy emerging from the end of a fiber into an electrical signal. The receiver then must amplify this signal to a large enough level with an appropriate fidelity so that the information content can be processed by the electronics following the receiver amplifier. In these processes various noise and distortion effects will unavoidably be introduced, which can lead to errors in the interpretation of the received signal. The three basic stages of an optical receiver are a photodetector, an amplifier, and an equalizer. The design of the amplifier that follows the photodiode is of critical importance, because it is in this amplifier where the major noise sources are expected to arise. The equalizer that follows the amplifier is normally a linear frequency-shaping filter, which is used to mitigate the effects of signal distortion and intersymbol interference.

In a digital receiver the amplified and filtered signal emerging from the equalizer is compared with a threshold level once per time slot to determine whether or not a pulse is present at the photodetector in that time slot. Various noises, interference from adjacent pulses, and conditions wherein the light source is not completely extinguished during a 0 pulse can cause errors in the decision-making process. To calculate the error probability, it is necessary to know the mean square noise voltage that is superimposed on the signal voltage during the decision time.

Because the statistics of the output voltage at the sampling time are rather complex, approximations are used to calculate the performance of a binary optical fiber receiver. In applying these approximations, a tradeoff is needed between computational simplicity and accuracy of the results. The simplest method is based on a Gaussian approximation. In this method it is assumed that when the sequence of optical input pulses is known, the equalizer output voltage is a Gaussian random variable. Thus, to calculate the error probability, only the mean and standard deviation of the output voltage need to be known.

The three basic approaches to the design of preamplifiers for optical fiber receivers are the low-impedance, the high-impedance, and the transimpedance configurations. These categories are not actually distinct because a continuum of intermediate configurations is possible, but they illustrate design approaches. The low-impedance preamplifier is the most straightforward, but not necessarily the optimum approach. This design is limited to special short-distance applications in which high receiver sensitivity is not a major concern. The high-impedance design produces the lowest noise, but it has two limitations: (a) equalization methods are required for broadband applications and (b) it has a limited dynamic range. The transimpedance amplifier is less sensitive because of a higher noise penalty, but it has the benefit of a wider dynamic range without equalization.

For passive optical network (PON) applications, a specially designed burst-mode receiver is needed. These receivers can quickly extract the decision threshold and determine the signal phase from a set of overhead bits placed at the beginning of each packet burst. However, this methodology results in a receiver sensitivity power penalty of up to 3 dB. The key requirements of a burst-mode receiver are high sensitivity, wide dynamic range, and fast response time. The sensitivity is important in relation to the optical power budget, because, for example, a 3-dB sensitivity improvement allows more customers to be attached to the PON. A wide dynamic range is essential for achieving a long network reach, that is, to be able to accommodate users located both close and far away from the central office.

The eye diagram is a powerful measurement tool for assessing the data-handling ability of a digital transmission system. This method has been used extensively for evaluating the performance of wire-line systems and also applies to optical fiber data links. To reduce costly and time-consuming test periods, a Q-factor technique can be used. Although some accuracy is lost in this method, it reduces the test times to minutes instead of hours. In this method the receiver threshold is decreased, which increases the probability of errors and thus decreases test time.

Problems

7.1 Consider a photodetector that has a quantum efficiency $\eta = 0.75$.
(a) Show that an energy

$$E \geq 25.3 \frac{hc}{\eta\lambda}$$

is required in a 1 pulse to have a probability of 10^{-11} or smaller in order that the arriving 1 pulse will not be interpreted as a 0 pulse. (b) Why must the number of photons in the 1 pulse be greater than or equal to 34?

7.2 The equalizer in an optical receiver normally is a linear frequency-shaping filter used to mitigate the effects of signal distortion and intersymbol interference. To account for the fact that pulses arrive rounded and distorted at the receiver, the binary digital pulse train incident on the photodetector can be described by

$$P(t) = \sum_{n=-\infty}^{\infty} b_n h_p(t - nT_b)$$

Here $P(t)$ is the received optical power, T_b is the bit period, b_n represents the energy in the nth pulse ($b_n = b_0$ for a 0 pulse and b_1 for a 1 pulse), and $h_p(t)$ is the received pulse shape. Show that the following pulse shapes satisfy the normalization condition

$$\int_{-\infty}^{\infty} h_p(t)dt = 1$$

(a) Rectangular pulse (α = constant)

$$h_p(t) = \begin{cases} \frac{1}{\alpha T_b} & \text{for} \quad -\frac{\alpha T_b}{2} < t < \frac{\alpha T_b}{2} \\ 0 \text{ otherwise} \end{cases}$$

(b) Gaussian pulse

$$h_p(t) = \frac{1}{\sqrt{2\pi}} \frac{1}{\alpha T_b} e^{-t^2/2(\alpha T_b)^2}$$

(c) Exponential pulse

$$h_p(t) = \begin{cases} \frac{1}{\alpha T_b} e^{-t/\alpha T_b} & \text{for} \quad 0 < t < \infty \\ 0 \text{ otherwise} \end{cases}$$

7.3   Use Eq. (7.8) to derive the error probability expression given by Eq. (7.16).

7.4   Consider an optical receiver that has a high-impedance amplifier with an input resistance of $R_a = 3$ MΩ. Suppose it is matched to a photodetector bias resistor that has a value $R_b = 3$ MΩ. (a) If the total capacitance is $C = 6$ pF, show that the maximum bandwidth achievable without equalization is 17.3 kHz. (b) Now consider the case when the high-impedance amplifier is replaced with a transimpedance amplifier that has a 100 $k\Omega$ feedback resistor and a gain $G = 350$. Show that the maximum achievable bandwidth without equalization in this case is 92.8 MHz.

7.5   Consider the probability distributions shown in Fig. 7.7, where the signal voltage for a binary 1 is V_1 and $v_{th} = V_1/2$.

(a)   If $\sigma = 0.20V_1$ for $p(y|0)$ and $\sigma = 0.24V_1$ for $p(y|1)$, use Eqs. (7.10) and (7.11) to show that the error probabilities

$$P_0(v_{th}) = 0.5 [1 - \text{erf} (1.768)] = 0.0065$$
$$\text{and} \quad P_1(v_{th}) = 0.5 [1 - \text{erf} (1.473)] = 0.0185$$

(b) If $a = 0.65$ and $b = 0.35$, show that $P_e = 0.0143$.
(c) If $a = b = 0.5$, show that $P_e = 0.0125$.

7.6   An LED operating at 1300 nm injects 25 μW of optical power into a fiber. Assume that the attenuation between the LED and the photodetector is 40 dB and the photodetector quantum efficiency is 0.65. (a) Show that the power falling on the photodetector is 2.5 nW. (b) Show that the average number of electron–hole pairs generated in a time t = 1 ns is N = 10.6. (c) Show that the probability that fewer than 5 electron − hole pairs will be generated at the detector in the 1-ns interval is P(n = 5) = 0.05 = 5%.

7.7 If a bit-corrupting noise burst lasts for 2 ms, how many bits are affected at data rates of 10 Mb/s, 100 Mb/s, and 2.5 Gb/s? [Answer: (a) 2×10^4; (b) 2×10^5; (c) 5×10^6.]

7.8 Consider a thermal-noise-limited analog optical fiber system that uses a *pin* photodiode with a responsivity of 0.85 A/W at 1310 nm. Assume the system uses a modulation index of 0.5 and operates in a 5-MHz bandwidth. Let the mean-square thermal noise current per unit bandwidth for the receiver be 2 $\times 10^{-23}$ A²/Hz. Show that the peak-to-peak signal power to rms noise ratio at the receiver is 38 dB when the average incident optical power is − 20 dBm (0.010 mW).

7.9 Consider a shot-noise limited analog optical fiber system that uses a *pin* photodiode with a responsivity of 0.85 A/W at 1310 nm. Assume the system uses a modulation index of 0.6 and operates in a 40-MHz bandwidth. Neglecting detector dark current, show that is the signal-to-noise ratio is SNR = 29.1 dB when the incident optical power at the receiver is − 15 dBm (0.032 mW).

7.10 Show that if thermal noise dominates then the signal-to-noise ratio given by Eq. (7.32) is a maximum when the gain is optimized at

$$M_{opt}^{2+x} = \frac{4k_B T F_t / R_{eq}}{q(i_p + i_D)x}$$

7.11 Consider a Si APD detector that has an excess noise factor related parameter $x = 0.3$, a load-resistance/amplifier-noise-figure value of $R_{eq}/F_t = 10^4 \ \Omega$, a dark current of $i_D = 6$ nA, and a responsivity of 0.55 A/W. If at a temperature $T = 300°$K the photodetector is irradiated with a light power level $P_{in} = 15$ nW, (*a*) first use Eq. (6.6) to show that the primary photocurrent is $i_p = 8.25$ nA and (*b*) then use the relationship given in Problem 7.10 to show that the optimum gain is $M_{opt} = 29.6$.

References

1. T. V. Muoi, Receiver design for high-speed optical-fiber systems. J. Lightw. Technol. **LT-2**, 243−267 (1984)
2. M. Brain, T. P. Lee, Optical receivers for lightwave communication systems. J. Lightw. Technol. **LT-3**, 1281−1300 (1985)
3. E. Säckinger, *Broadband Circuits for Optical Fiber Communications* (Wiley, New York, 2005)
4. K. Schneider, H. Zimmermann, *Highly Sensitive Optical Receivers* (Springer, Berlin, 2006)
5. S.D. Personick, Optical detectors and receivers. J. Lightw. Technol. **26**(9), 1005–1020 (2008)

6. S. Bottacchi, *Noise and Signal Interference in Optical Fiber Transmission Systems* (Wiley, New York, 2009)
7. A. Beling, J. C. Campbell, Advances in photodetectors and optical receivers (Chap. 3), in *Optical Fiber Telecommunications VIA*, ed. by A. E. Willner, T. Li, I. Kaminow (Elsevier, Amsterdam, 2010)
8. M. Atef, H. Zimmermann, *Optoelectronic Circuits in Nanometer CMOS Technology* (Springer, Berlin, 2016)
9. B. Razavi, *Design of Integrated Circuits for Optical Communications*, 2nd ed. (Wiley, New York, 2012)
10. F. Aznar, S. Celma, B. Calvo, *CMOS Receiver Front-Ends for Gigabit Short-Range Optical Communications* (Springer, Berlin, 2013)
11. G.-S. Jeong, W. Bae, D.-K. Jeong, Review of CMOS integrated circuit technologies for high-speed photo-detection. Sensors, **17**, 01962 (2017) (40 page review)
12. S. Zohoori, M. Dolatshahi, M. Pourahmadi, & M. Hajisafari, An inverter-based, CMOS, low-power optical receiver front-end. Fiber Integr. Optics, **38**(1), 1–20 (2019)
13. A. B. Carlson, P. Crilly, *Communication Systems* (5th edn.). McGraw-Hill (2010)
14. L. W. Couch II, *Digital and Analog Communication Systems*, 8th edn. (Prentice Hall, 2013)
15. R. E. Ziemer, W. H. Tranter, *Principles of Communications: Systems, Modulation, and Noise*, 7th edn. (Wiley, New York, 2014)
16. M. S. Alencar, V. C. da Rocha, *Communication Systems* (Springer, Berlin, 2020)
17. F. Ling, *Synchronization in Digital Communication Systems* (Cambridge University Press, 2017)
18. T. von Lerber, S. Honkanen, A. Tervonen, H. Ludvigsen, F. Küppers, Optical clock recovery methods: review. Opt. Fiber Technol. **15**(4), 363–372 (2009)
19. Z. Pan, C. Yu, A.E. Willner, Optical performance monitoring for the next generation optical communication networks. Optical Fiber Technol. **16**, 20–45 (2010)
20. ITU-T Recommendation O.201, *Q-factor test equipment to estimate the transmission performance of optical channels*, July 2003
21. J.J. Schiller, R.A. Srinivasan, M.R. Spiegel, *Schaum's Outline of Probability and Statistics*, 4th edn. (McGraw-Hill, 2013)
22. W. Navidi, *Principles of Statistics for Engineers and Scientists*, 2nd edn. (McGraw-Hill, 2021)
23. D. Zwillinger (ed.), *Standard Mathematical Tables and Formulae*, 33rd edn. (CRC Press, 2018)
24. ITU-T G-series Recommendations, Supplement 39, Feb 2016
25. ITU-T Recommendation G.959.1, *Optical transport network physical layer interfaces*, July 2018
26. IEEE 802.3-2018, *IEEE Standard for Ethernet*, June 2018
27. C. Su, L.-K. Chen, K.W. Cheung, Theory of burst-mode receiver and its applications in optical multiaccess networks. J. Lightw. Technol. **15**, 590–606 (1997)
28. G. Keiser, *FTTX Concepts and Applications* (Wiley, New York, 2006)
29. B.C. Thomsen, R. Maher, D.S. Millar, S.J. Savory, Burst mode receiver for 112 Gb/s DP-QPSK with parallel DSP. Opt. Exp. **19**(26), B770–B776 (2011)
30. S.C. Lin, S.L. Lee, C.K. Liu, C.L. Yang, S.C. Ko, T.W. Liaw, G. Keiser, Design and demonstration of REAM-based WDM-PONs with remote amplification and channel fault monitoring. J. Opt. Commun. Netw. **4**(4), 336–343 (2012)
31. A. Brillant, *Digital and Analog Fiber Optic Communications for CATV and FTTx Applications* (Wiley, New York, 2008)
32. C. Lim, A. Nirmalathas, M. Bakaul, P. Gamage, K.L. Lee, Y. Yang, D. Novak, Fiber-wireless networks and subsystem technologies. J. Lightw. Technol. **28**(4), 390–405 (2009)
33. C. Lim, Y. Tian, C. Ranaweera, T.A. Nirmalathas, E. Wong, K.L. Lee, Evolution of radio-over-fiber technology. J. Lightw. Technol. **37**(6), 1647–1656 (2019)

Chapter 8
Digital Optical Fiber Links

Abstract The design of a digital optical link involves many interrelated operating characteristics of the fiber, source, photodetector, and other components in the link. In carrying out an optical fiber link analysis, several iterations with different device characteristics may be required before the analysis is completed satisfactorily. The two basic analyses that usually are carried out to ensure that the desired system performance can be met are the link power budget and the system rise-time analysis. In addition to discussing these two analyses, this chapter addresses procedures for detecting and controlling errors in a digital data stream in order to improve the reliability of a communication link.

The preceding chapters have presented the fundamental characteristics of the individual building blocks of an optical fiber transmission link. These include the optical fiber transmission medium, the optical source, and the photodetector and its associated receiver. Now this chapter will examine how these individual parts can be put together to form a complete optical fiber transmission link. In particular, basic digital links are addressed in this chapter, and analog links in Chap. 9. More complex transmission links and networks are examined in Chap. 13.

The first discussion involves the simplest case of a point-to-point link. This includes considering the transmitted signal format plus examining the components that are available for a particular application and seeing how these components relate to the system performance criteria. From a system performance viewpoint, the key operating criteria include data timing extraction, signal dispersion, power penalties, and bit error rate. For a given set of components and a given set of system requirements, a power budget analysis then is carried out to determine whether the fiber optic link meets the attenuation requirements or if amplifiers are needed within the link to boost the power level at intermediate points. The final step is to perform a system rise-time analysis to verify that the overall system performance requirements are met.

The original version of this chapter was revised: All the incorrect sentences have been corrected. The correction to this chapter can be found at https://doi.org/10.1007/978-981-33-4665-9_15

The analysis in Sect. 8.1 assumes that the optical power falling on the photodetector is a clearly defined function of time within the statistical nature of the quantum detection process. In reality, various signal impairments can degrade the link performance. These impairments can reduce the power of the optical signal arriving at the receiver from the ideal case, which is known as a *power penalty* for that effect. Section 8.2 describes the power penalties associated with some key signal impairments that may be observed in an optical link.

To control errors and to improve the reliability of a communication line, it is necessary to be able to detect the errors and then correct them or retransmit the information. Section 8.3 describes error detection and correction methods that are used in a variety of optical fiber communication links. Basic *coherent detection* schemes are addressed in Sect. 8.4. These schemes are used in place of direct-detection methods to improve receiver sensitivities, especially for high-speed links operating at 40 and 100 Gb/s. Chapter 13 describes additional signal modulation and detection formats, such as *differential phase-shift keying* (DPSK) and *differential quadrature phase-shift keying* (DQPSK).

Versatile and powerful modeling and simulation tools are commercially available for doing many of the tasks described in this chapter. These software-based tools can run on conventional personal computers and include functions such as estimations of the BER and power-penalty effects with different optical receiver models, calculations of link power budgets, and simulation of link performance when using different components.

8.1 Basic Optical Fiber Links

The simplest transmission link is a point-to-point optical fiber line that has a transmitter on one end and a receiver on the other end as shown in Fig. 8.1. At the transmitter end an electrical information signal modulates a light source, which

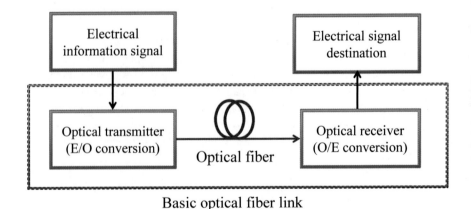

Basic optical fiber link

Fig. 8.1 Basic optical fiber link

performs electrical-to-optical (E/O) conversion to change the electrical signal into an optical format. At the far end of the fiber a photodiode receiver performs optical-to-electrical (O/E) conversion to recover the electrical information signal for processing by the following electronic circuits. This type of link places the least demand on optical fiber technology and thus sets the basis for examining more complex network architectures.

The design of an optical link involves many interrelated variables among the fiber, light source, and photodetector operating characteristics, so that the actual link design and analysis may require several iterations before they are completed satisfactorily. Because performance and cost constraints are very important factors in fiber optic communication links, the designer must carefully choose the components to ensure that the desired performance level can be maintained over the expected system lifetime without over specifying the component characteristics.

The following key system requirements are needed in analyzing a link:

1. The desired (or possible) transmission distance
2. The data rate or channel bandwidth
3. The bit-error rate (BER)

To fulfill these requirements, the designer has a choice of the following components and their associated characteristics:

1. Multimode or single-mode optical fiber

    (a) Core size
    (b) Core refractive-index profile
    (c) Bandwidth or signal dispersion
    (d) Attenuation
    (e) Numerical aperture or mode-field diameter

2. LED or laser diode optical source

    (a) Emission wavelength
    (b) Spectral linewidth
    (c) Optical output power
    (d) Effective radiating area
    (e) Emission pattern
    (f) Number of emitting modes

3. *Pin* or avalanche photodiode

    (a) Responsivity
    (b) Operating wavelength
    (c) Speed
    (d) Sensitivity

Two analyses usually are carried out to ensure that the desired system performance can be met: these are the *link power budget* and the system *rise-time budget* analyses. In the link power budget analysis one first determines the power margin between the optical transmitter output and the minimum receiver sensitivity needed to establish a specified BER. This margin can then be allocated to connector, splice, and fiber losses, plus any additional margins required for other components, possible component degradations, transmission-line impairments, or temperature effects. If the choice of components did not allow the desired transmission distance to be achieved, the components might have to be changed or amplifiers might have to be incorporated into the link. Once the link power budget has been established, the designer can perform a system rise-time analysis to ensure that the desired overall operational performance has been met.

8.1.1 Signal Formats for Transporting Information

In designing a communication link for transporting digitized information, a significant consideration is the format of the transmitted digital signal [1–5]. One important factor concerning the signal format that is sent out from the transmitter is that the receiver must be able to extract precise *timing information* from the incoming signal [6]. The following are the three main purposes of *timing*:

- To allow the signal to be sampled by the receiver at the time the signal-to-noise ratio is a maximum
- To maintain proper spacing between digital pulses
- To indicate the start and end of each timing interval

In addition, it may be desirable for the signal to have an inherent error-detecting capability, as well as an error-correction mechanism, if it is needed or is practical. These timing and error-minimizing features can be incorporated into the data stream by restructuring or encoding the digital signal [7–9]. This process is called *channel coding* or *line coding*. This section examines the basic binary line codes that are used in optical fiber communication systems.

One of the principal functions of a line code is to minimize errors in the bit stream that might arise from noise or other interference effects. Generally one does this by introducing extra redundant bits into the raw data stream at the transmitter, arranging them in a specific pattern, and extracting the redundant bits at the receiver to recover the original signal. Depending on the amount of redundancy that is introduced into the data stream, various degrees of error reduction in the data can be achieved, provided that the data rate is less than the channel capacity.

NRZ and RZ Signal Formats The simplest method for encoding data is the unipolar *nonreturn-to-zero* (NRZ) code. *Unipolar* means that a logic 1 is represented by a voltage or light pulse that fills an entire bit period, whereas for a logic 0 no pulse is transmitted, as shown in Fig. 8.2 for the data sequence 1010110. Because this process turns the light signal on and off, it is known as *amplitude shift keying* (ASK) or *on–off keying* (OOK). If 1 and 0 pulses occur with equal probability, and if the amplitude of the voltage pulse is A, then the average transmitted power for this code is $A^2/2$. In optical systems one typically describes a pulse in terms of its optical power

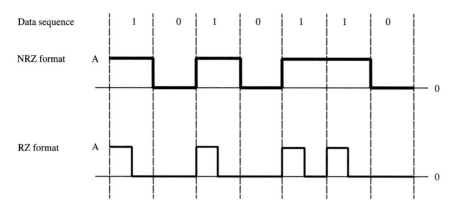

Fig. 8.2 NRZ and RZ code patterns for the data sequence 1010110

level. In this case the average power for an equal number of 1 and 0 pulses is $P/2$, where P is the peak power in a 1 pulse.

The NRZ code needs the minimum bandwidth and is simple to generate and decode. However, the lack of timing capabilities in an NRZ code can lead to misinterpretations of the bit stream at the receiver. For example, because there are no level transitions from which to extract timing information in a long sequence of NRZ ones or zeros, a long string of N identical bits could be interpreted as either $N + 1$ or $N - 1$ bits, unless highly stable (and expensive) timing clocks are used. Two common techniques for restricting the longest time interval in which no level transitions occur are the use of block codes (see below) and scrambling. *Scrambling* produces a random data pattern by modulo 2 addition of a known bit sequence to the data stream. At the receiver the same known bit sequence is again modulo 2 added to the received data, which results in the recovery of the original bit sequence.

If an adequate bandwidth margin exists, the timing problem associated with NRZ encoding can be alleviated with a *return-to-zero* (RZ) code. As shown in the bottom half of Fig. 8.2, the RZ code has an amplitude transition at the beginning of each bit interval when a binary 1 is transmitted and no transition denotes a binary 0. Thus for a RZ pulse a 1 bit occupies only part of the bit interval and returns to zero in the remainder of the bit interval. No pulse is used for a 0 bit.

Although the RZ pulse nominally occupies exactly half a bit period in electronic digital transmission systems, in an optical communication link the RZ pulse might occupy only a fraction of a bit period. A variety of RZ formats are used for links that send data at rates of 10 Gb/s and higher.

Block Codes *Redundant bits* can be introduced into a data stream to provide adequate timing and to allow for error monitoring. A popular and efficient encoding method for this is the class of mBnB *block codes*. In this class of codes, blocks of m binary data bits are converted to longer blocks of $n > m$ binary bits, which include $n - m$ redundant bits. As a result of the additional redundant bits, the required bandwidth increases by the ratio n/m. For example, in an mBnB code with $m = 1$ and $n = 2$, a binary 1 is mapped into the binary pair 10, and a binary 0 becomes 01. The overhead for such a code is 50%.

Suitable mBnB codes for high data rates are the 3B4B, 4B5B, 5B6B, and 8B10B codes. If simplicity of the encoder and decoder circuits is the main criterion, then the 3B4B format is the most convenient code. The 5B6B code is the most advantageous if bandwidth reduction is the major concern. Various versions of Ethernet use the 3B4B, 4B5B, or 8B10B formats. Section 8.3 discusses more advanced codes for error detection and correction.

8.1.2   Considerations for Designing Links

In carrying out a link power budget, one approach is to first decide at which wavelength to transmit and then choose components that operate in this region. If the distance over which the data are to be transmitted is not too far, it may be advantageous to operate in the 770-to-910 nm region or in the T-band. On the other hand, if the transmission distance is relatively long, it is better to take advantage of the lower attenuation and lower dispersion that occur in the O-band through U-band region.

Having decided on a wavelength, the next step is to interrelate the system performances of the three major optical link building blocks; that is, the receiver, transmitter, and optical fiber. Normally, the designer chooses the characteristics of two of these elements and then computes those of the third to see if the system performance requirements are met. If the components have been overspecified or underspecified, then iterations in the design may be needed. The procedure followed here is first to select the photodetector. Then the system designer can choose an optical source and see how far data can be transmitted over a particular fiber before an amplifier is needed in the line to boost up the power level of the optical signal.

In choosing a particular photodetector, a main factor is to determine the minimum optical power that must fall on the photodetector to satisfy the bit-error rate (BER) requirement at the specified data rate. In making this choice, the designer also needs to take into account any design cost and complexity constraints. As noted in Chaps. 6 and 7, a *pin* photodiode receiver is simpler, more stable with changes in temperature, and less expensive than an avalanche photodiode receiver. In addition, *pin* photodiode bias voltages are normally less than 5 V, whereas those of avalanche photodiodes range from 40 V to several hundred volts. However, the advantages of *pin* photodiodes may be overruled by the increased sensitivity of the avalanche photodiode if very low optical power levels are to be detected.

The system parameters involved in deciding between the use of an LED and a laser diode are signal dispersion, data rate, transmission distance, and cost. As shown in Chap. 4, the spectral width of the laser output is much narrower than that of an LED. This is of importance in the 770-to-910-nm region, where the spectral width of an LED and the dispersion characteristics of multimode silica fibers limit the data-rate-distance product to around 150 (Mb/s)·km. For higher values [up to 2500 (Mb/s)·km], a laser must be used at these wavelengths. At wavelengths around

1.3 μm, where signal dispersion is very low, bit-rate-distance products of at least 1500 (Mb/s)·km are achievable with LEDs in multimode fibers. For InGaAsP lasers, distances of 150 m can be achieved at 100-Gb/s rates in OM4 multimode fiber at 1.3 μm (see Sect. 13.4). A single-mode fiber can provide such data rates over much longer distances.

Greater repeaterless transmission distances are possible with a laser, laser diodes typically couple from 10 to 15 dB more optical power into a fiber than an LED. This advantage and the lower dispersion capability of laser diodes may be offset by cost constraints. Not only is a laser diode itself more expensive than an LED, but also the laser transmitter circuitry is much more complex, because the lasing threshold has to be dynamically controlled as a function of temperature and device aging. However, mass production techniques and technology innovations have led to a wide variety of cost-effective laser transmitters being commercially available.

For the optical fiber, there is a choice between single-mode and multimode fiber, either of which could have a step-index or a graded-index core. This choice depends on the type of light source used and on the amount of dispersion that can be tolerated. Light-emitting diodes (LEDs) tend to be used with multimode fibers. The optical power that can be coupled into a fiber from an LED depends on the core-cladding index difference Δ, which, in turn, is related to the numerical aperture of the fiber (for $\Delta = 0.01$, the numerical aperture NA ≈ 0.21). As Δ increases, the fiber-coupled power increases correspondingly. However, because dispersion also becomes greater with increasing Δ, a tradeoff must be made between the optical power that can be launched into the fiber and the maximum tolerable dispersion.

When choosing the attenuation characteristics of a cabled fiber, the excess loss that results from the cabling process must be considered in addition to the attenuation of the fiber itself. This must also include connector and splice losses as well as environmental-induced losses that could arise from temperature variations and dust or moisture on the connector end faces.

8.1.3 Creating a Link Power Budget

An optical power loss model for a point-to-point link is shown in Fig. 8.3. The optical power received at the photodetector depends on the amount of light coupled into the fiber by the transmitter and the losses occurring in the fiber and at the connectors and splices. The link loss budget is derived from the sequential loss contributions of each element in the link. Each of these loss elements is expressed in decibels (dB) as

$$\text{Element loss} = 10 \log \frac{P_{\text{out}}}{P_{\text{in}}} \tag{8.1}$$

where P_{in} and P_{out} are the optical powers entering and leaving the link element, respectively. The loss value corresponding to a particular element generally is called the *insertion loss* for that element.

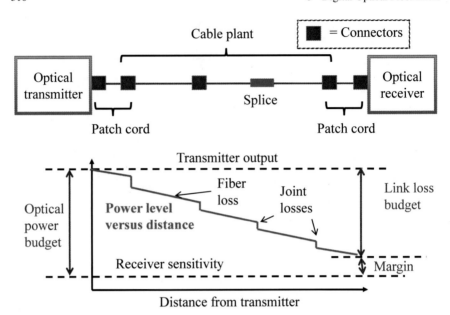

Fig. 8.3 Optical power loss model for a point-to-point link, where the losses occur at connectors, at splices, and in the fiber segments

In addition to the link loss contributors shown in Fig. 8.3, a *link power margin* is normally provided in the analysis to allow for component aging, temperature fluctuations, and losses arising from components that might be added at future dates. A link margin of 3–6 dB is generally used for systems that are not expected to have additional components incorporated into the link in the future.

The link loss budget simply considers the total optical power loss P_T that is allowed between the light source and the photodetector, and allocates this loss to cable attenuation, connector loss, splice loss, and system margin. Thus, if P_S is the optical power emerging from the end of a fiber flylead attached to the light source or from a source-coupled connector, and if P_R is the receiver sensitivity, then for N connectors and splices

$$P_T = P_S - P_R = Nl_c + \alpha L + \text{system margin} \qquad (8.2)$$

Here, l_c is the connector or splice loss, α is the fiber attenuation (given in dB/km), L is the transmission distance, and the system margin is nominally taken as 3 dB.

Example 8.1 To illustrate how a link loss budget is set up, it is helpful to carry out a simple specific design example. The design begins by specifying a data rate of 20 Mb/s and a bit-error rate of 10^{-9} (i.e., at most one error can occur for every 10^9 bits sent). For the receiver, one can choose a silicon *pin* photodiode operating at 850 nm. In Fig. 8.4 the thin dashed line shows that the required receiver input signal is −42 dBm (42 dB below 1 mW). Next one could select a GaAlAs LED

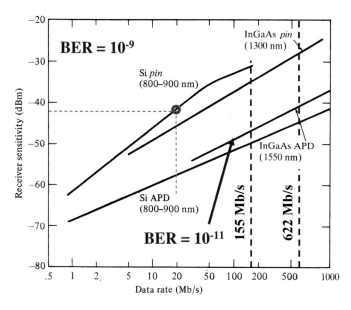

Fig. 8.4 Receiver sensitivities as a function of bit rate: the Si pin, Si APD, and InGaAs pin curves are for a 10^{-9} BER, whereas the InGaAs APD curve is for a 10^{-11} BER

that can couple a 50-μW (-13-dBm) average optical power level into a fiber flylead with a 50-μm core diameter. Thus there is a 29-dB allowable power loss between the light source output and the photodetector sensitivity. Assume further that a 1-dB loss occurs when the source fiber flylead is connected to the cable and another 1-dB connector loss occurs at the cable-photodetector interface. Because this is a short link, there are no other connectors or splices in the transmission path. The losses from the two connectors reduce the link margin to 27 dB. Including a 6-dB system margin, the possible transmission distance for a cable with an attenuation α can be found from Eq. (8.2):

$$P_T = P_S - P_R = 29 \text{ dB} = 2(1 \text{ dB}) + \alpha L + 6 \text{ dB}$$

If $\alpha = 3.5$ dB/km, then a 6.0-km transmission path is possible.

The link power budget can be represented graphically as is shown in Fig. 8.5. The vertical axis represents the optical power loss allowed between the transmitter and the receiver. The horizontal axis gives the transmission distance. The bottom and top lines in Fig. 8.5 show, respectively, a silicon *pin* receiver with a sensitivity of -42 dBm (at 20 Mb/s) and an LED with an output power of -13 dBm coupled into a fiber flylead. Subtracting a 1-dB connector loss at each end leaves a total margin of 27 dB. Subtracting a 6-dB system safety margin leaves a tolerable loss of 21 dB that can be allocated to cable and splice loss. The slope of the line shown in Fig. 8.5 is the 3.5-dB/km cable (and splice, in this case) loss. This line starts at the -14-dBm point (which is the optical power coupled into the cabled fiber) and ends at the $-$

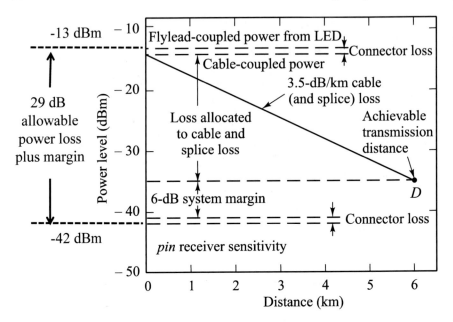

Fig. 8.5 Graphical representation of a link-loss budget for an 850-nm LED/pin system operating at 20 Mb/s

35-dBm level (the receiver sensitivity minus a 1-dB connector loss and a 6-dB system margin). The intersection point D then defines the maximum possible transmission path length.

A convenient procedure for calculating the power budget is to use a tabular or spreadsheet form. This calculation method can be illustrated by way of an example for a 2.5-Gb/s link.

Example 8.2 Consider a 1550-nm laser diode that launches a +3-dBm (2-mW) optical power level into a fiber flylead, an InGaAs APD with a −32-dBm sensitivity at 2.5 Gb/s, and a 60-km long optical cable with a 0.3-dB/km attenuation. Assume that here, because of the way the equipment is arranged, a 5-m optical jumper cable is needed at each end between the end of the transmission cable and the telecom equipment rack as shown in Fig. 8.6. Assume that each jumper cable introduces a loss of 3 dB. In addition, assume a 1-dB connector loss occurs at each fiber joint (two at each end because of the jumper cables).

Table 8.1 lists the components in column 1 and the associated optical output, sensitivity, or loss in column 2. Column 3 gives the power margin available after subtracting the component loss from the total optical power loss that is allowed between the light source and the photodetector, which, in this case, is 35 dB. Adding all the losses results in a final power margin of 7 dB.

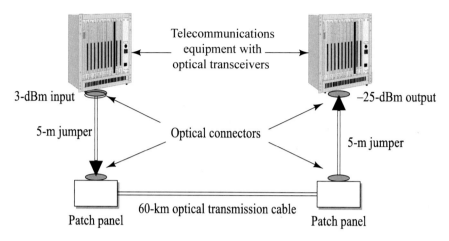

Fig. 8.6 A 2.5-Gb/s 60-km optical fiber link with 5-m optical jumper cables at each end

Table 8.1 Example of a spreadsheet for calculating an optical link power budget

Component/loss parameter	Output/sensitivity/loss	Power margin (dB)
Laser output	3 dBm	
APD sensitivity at 2.5 Gb/s	−32 dBm	
Allowed loss [3 − (−32)]		35
Source connector loss	1 dB	34
Jumper + connector loss	3 + 1 dB	30
Cable attenuation (60 km)	18 dB	12
Jumper + connector loss	3 + 1 dB	8
Receiver connector loss	1 dB	7 (final margin)

Drill Problem 8.1 Suppose that the components of an optical link operating at 1310 nm have the following parameter values:

(a) A laser diode that emits 0 dBm of optical power from an attached fiber flylead

(b) A pin photodiode with a −20-dBm sensitivity at 2.5 Gb/s

(c) A 20-km optical fiber with an attenuation of 0.4-dB/km at 1310 nm

(d) A 1-dB connector loss at each end of the link

Show that the power margin is 2 dB.

Drill Problem 8.2 A single-mode optical fiber link operating at a 1310-nm wavelength is intended for a 1-Gb/s metro network. Suppose that the components of the link have the following parameter values:

(a) A laser diode that emits 0 dBm of optical power from an attached fiber flylead
(b) A pin photodiode with a -22-dBm sensitivity at 1.0 Gb/s
(c) An optical fiber with an attenuation of 0.4-dB/km at 1310 nm
(d) A 1-dB connector loss at each end of the link
(e) A required power margin of 8 dB

   Show that the maximum link length is 30 km.

8.1.4 Formulating a Rise-Time Budget

A *rise-time budget analysis* is a convenient method for determining the dispersion limitation of an optical fiber link. This is particularly useful for digital systems. In this approach, the total rise time t_{sys} of the link is the root sum square of the rise times from each contributor t_i to the pulse rise-time degradation:

$$t_{sys} = \left(\sum_{i=1}^{N} t_i^2 \right)^{1/2} \tag{8.3}$$

The four basic elements that may significantly limit system speed are the transmitter rise time t_{tx}, the group-velocity dispersion (GVD) rise time t_{GVD} of the fiber, the modal dispersion rise time t_{mod} of the fiber, and the receiver rise time t_{rx}. Single-mode fibers do not experience modal dispersion, so in these fibers the rise time is related only to GVD. Generally, the total transition-time degradation of a digital link should not exceed 70% of an NRZ (non-return-to-zero) bit period or 35% of a bit period for RZ (return-to-zero) data, where one bit period is defined as the reciprocal of the data rate.

The rise times of transmitters and receivers are generally known from data sheets. The transmitter rise time is attributable primarily to the light source and its drive circuitry. The receiver rise time results from the photodetector response and the 3-dB electrical bandwidth of the receiver front end. The response of the receiver front end can be modeled by a first-order lowpass filter that has a step response [10]

$$g(t) = \left[1 - \exp(-2\pi B_e t) \right] u(t) \tag{8.4}$$

where B_e is the 3-dB electrical bandwidth of the receiver and $u(t)$ is the unit step function which is 1 for $t \geq 0$ and 0 for $t < 0$. The rise time t_{rx} of the receiver is usually defined as the time interval between $g(t) = 0.1$ and $g(t) = 0.9$, that is, $t_{rx} = t_{10\%} - t_{90\%}$. This is known as the *10-to-90 percent rise time*. Thus, if B_e is given in megahertz, then by solving Eq. (8.4) the receiver front-end rise time t_{rx} in nanoseconds is (see Problem 8.5)

$$t_{rx} = \frac{350}{B_e} \tag{8.5}$$

In practice, an optical fiber link seldom consists of a uniform, continuous, jointless fiber. Instead, a transmission link nominally is formed from several concatenated (joined in tandem) fibers that may have different dispersion characteristics. This is especially true for dispersion-compensated links operating at 10 Gb/s and higher (see Chap. 13). In addition, multimode fibers experience modal distributions at fiber-to-fiber joints owing to misaligned joints, different core index profiles in each fiber, and/or different degrees of mode mixing in individual fibers. Determining the fiber rise times resulting from GVD and modal dispersion then becomes more complex than for the case of a single uniform fiber.

By using Eq. (3.49) the fiber rise time t_{GVD} resulting from GVD over a length L can be approximated as

$$t_{GVD} = |D|L\sigma_\lambda \tag{8.6}$$

where σ_λ is the half-power spectral width of the source. The dispersion D is given by Eq. (3.52) for a non-dispersion-shifted fiber and by Eq. (3.54) for a dispersion-shifted fiber. Because the dispersion value generally changes from fiber section to section in a long link, an average value should be used for D in Eq. (8.6).

The difficulty in predicting the bandwidth (and hence the modal rise time) of a series of concatenated multimode fibers arises from the observation that the total route bandwidth can be a function of the order in which fibers are joined. For example, instead of randomly joining together arbitrary (but very similar) fibers, an improved total link bandwidth can be obtained by selecting adjoining fibers with alternating overcompensated and undercompensated refractive-index profiles to provide some modal delay equalization. Although the ultimate concatenated fiber bandwidth can be obtained by judiciously selecting adjoining fibers for optimum modal delay equalization, in practice this is unwieldy and time-consuming, particularly because the initial fiber in the link appears to control the final link characteristics.

A variety of empirical expressions for modal dispersion have thus been developed [11–13]. From practical field experience, it has been found that for modal dispersion the bandwidth B_M in a link of length L can be expressed to a reasonable approximation by the empirical relation

$$B_M(L) = \frac{B_0}{L^q} \tag{8.7}$$

where the parameter q ranges between 0.5 and 1, and B_0 is the bandwidth of a 1-km length of cable. A value of $q = 0.5$ indicates that a steady-state modal equilibrium has been reached, whereas $q = 1$ indicates little mode mixing. Based on field experience, a reasonable estimate is $q = 0.7$.

Another expression for N concatenated fibers that has been proposed for B_M based on curve fitting of experimental data, is

$$\frac{1}{B_M} = \left[\sum_{n=1}^{N} \left(\frac{1}{B_n} \right)^{1/q} \right]^q \tag{8.8}$$

where the parameter q ranges between 0.5 (quadrature addition) and 1.0 (linear addition), and B_n is the bandwidth of the nth fiber section.

The next step is to find the relation between the fiber rise time and the 3-dB bandwidth. First assume that the optical power emerging from the fiber has a Gaussian temporal response described by

$$g(t) = \frac{1}{\sqrt{2\pi}\sigma} e^{-t^2/2\sigma^2} \tag{8.9}$$

where σ is the rms pulse width. The Fourier transform of this function is

$$G(\omega) = \frac{1}{\sqrt{2\pi}} e^{-\omega^2\sigma^2/2} \tag{8.10}$$

From Eq. (8.9) the time $t_{1/2}$ required for the pulse to reach its half-maximum value, that is, the time required to have

$$g(t_{1/2}) = 0.5g(0) \tag{8.11}$$

is given by

$$t_{1/2} = (2 \ln 2)^{1/2}\sigma \tag{8.12}$$

Defining the time t_{FWHM} as the full width of the pulse at its half-maximum (FWHM) value, then yields

$$t_{FWHM} = 2t_{1/2} = 2\sigma (2 \ln 2)^{1/2} \tag{8.13}$$

The 3-dB optical bandwidth B_{3dB} is defined as the modulation frequency f_{3dB} at which the received optical power has fallen to 0.5 of the zero frequency value. Thus, setting Eq. (8.10) equal to $0.5G(0)$ to find the 3-dB frequency and using Eq. (8.13), the relation between the FWHM rise time t_{FWHM} and the 3-dB optical bandwidth is

$$f_{3dB} = B_{3dB} = \frac{0.44}{t_{FWHM}} \tag{8.14}$$

Using Eq. (8.6) for the 3-dB optical bandwidth of the fiber link and letting t_{FWHM} be the rise time resulting from modal dispersion, then, from Eq. (8.14),

$$t_{mod} = \frac{0.44}{B_M} = \frac{0.44L^q}{B_0} \tag{8.15}$$

If t_{mod} is expressed in nanoseconds and B_M is given in megahertz, then

$$t_{mod} = \frac{440}{B_M} = \frac{440L^q}{B_0} \tag{8.16}$$

Substituting Eqs. (3.27), (8.5), and (8.16) into Eq. (8.3) gives a total system rise time of

$$t_{sys} = \left[t_{tx}^2 + t_{mod}^2 + t_{GVD}^2 + t_{rx}^2\right]^{1/2} = \left[t_{tx}^2 + \left(\frac{440L^q}{B_0}\right)^2 + D^2\sigma_\lambda^2 L^2 + \left(\frac{350}{B_e}\right)^2\right]^{1/2} \tag{8.17}$$

where all the times are given in nanoseconds, σ_λ is the half-power spectral width of the source, and the dispersion D [expressed in ns/(nm km)] is given by Eq. (3.52) for a non-dispersion-shifted fiber and by Eq. (3.54) for a dispersion-shifted fiber. As indicated by the curves in Fig. 3.18 for G.652 single-mode fiber, the dispersion D is less than $+3.5$ ps/(nm km) in the O-band and about $+17$ ps/(nm km) at 1550 nm. For G.655 fiber the dispersion values range from -10 to -3 ps/(nm km) across the O-band and from $+5$ to $+10$ ps/(nm km) in the C-band.

Example 8.3 As an example of a rise-time budget for a multimode link, consider the continuation of the analysis that was started in Sect. 8.1.3. First assume that the LED together with its drive circuit has a rise time of 15 ns. Taking a typical LED spectral width of 40 nm yields a material-dispersion-related rise-time degradation of 21 ns over the 6-km link. Assuming the receiver has a 25-MHz bandwidth, then from Eq. (8.5) the contribution to the rise-time degradation from the receiver is 14 ns. If the selected fiber has a 400-MHz km bandwidth-distance product and with $q = 0.7$ in Eq. (8.7), then from Eq. (8.15) the modal-dispersion-induced fiber rise time is 3.9 ns. Substituting all these values back into Eq. (8.17) results in a link rise time of

$$t_{sys} = \left[t_{tx}^2 + t_{mod}^2 + t_{GVD}^2 + t_{rx}^2\right]^{1/2} = \left[(15 \text{ ns})^2 + (21 \text{ ns})^2 + (3.9 \text{ ns})^2 + (14 \text{ ns})^2\right]^{1/2}$$
$$= 30 \text{ ns}$$

This value falls below the maximum allowable 35-ns rise-time degradation for a 20-Mb/s NRZ data stream (0.70/bit rate). The choice of components was thus adequate to meet the system design criteria.

Analogous to power budget calculations, a convenient procedure for keeping track of the various rise-time values in the rise-time budget is to use a tabular or spreadsheet

Table 8.2 Example for a tabular form for keeping track of component contributions to an optical-link rise-time budget

Component	Rise time	Rise-time budget
Allowed rise-time budget		$t_{sys} = 0.7/B_{NRZ} = 0.28$ ns
Laser transmitter	25 ps	
GVD in fiber	12 ps	
Receiver rise time	0.14 ns	
System rise time (Eq. 8.17)		0.14 ns

form. This can be illustrated by way of an example for the SONET OC-48 (2.5 Gb/s) link that was looked at in Example 8.2.

Example 8.4 Assume that the laser diode together with its drive circuit has a rise time of 0.025 ns (25 ps). Take a 1550-nm laser diode spectral width of 0.1 nm and an average dispersion of 2 ps/(nm km) for the fiber. These parameter values then yield a GVD-related rise-time degradation of 12 ps (0.012 ns) over a 60-km long optical cable. Assuming the InGaAs-APD-based receiver has a 2.5-GHz bandwidth, then from Eq. (8.5) the receiver rise time is 0.14 ns. Using Eq. (8.17) to add up the various contributions results in a total rise time of 0.14 ns.

Table 8.2 lists the components in column 1 and the associated rise times in column 2. Column 3 gives the allowed system rise-time budget of 0.28 ns for a 2.5-Gb/s NRZ data stream at the top. This is found from the expression $0.7/B_{NRZ}$ where B_{NRZ} is the bit rate for the NRZ signal. The calculated system rise time of 0.14 ns is shown at the bottom. The system rise time, in this case, is dominated by the receiver and is well within the required limits.

Drill Problem 8.3 A single-mode optical fiber link that uses a 1310-nm Fabry–Perot laser diode is designed to operate over a 6-km distance in a campus network. Suppose that the components of the link operating over this distance have the following rise time values:

1. The laser transmitter rise time = 0.30 ns.
2. A *pin* photodiode receiver rise time is 0.14 ns.
3. The dispersion in the optical fiber is 2.0 ps/(nm km).
4. The laser diode has a 2.0-nm spectral width.

   (a) Show that the system rise time is 0.33 ns.
   (b) Show that the maximum bit rate for an NRZ signal is $B_{NRZ} = 2.1$ Gb/s.

8.1.5 Transmission at Short Wavelengths

Figure 8.7 shows the attenuation and dispersion limitation on the repeaterless transmission distance as a function of data rate for the short-wavelength (770 − 910-nm)
LED/*pin* combination. The BER was taken as 10^{-9} for all data rates. The fiber-
coupled LED output power was assumed to be a constant −13 dBm for all data
rates up to 200 Mb/s. The attenuation limit curve then results by using a fiber loss
of 3.5 dB/km and the receiver sensitivities shown in Fig. 8.4. Because the minimum
optical power required at the receiver for a given BER becomes higher for increasing
data rates, the attenuation limit curve slopes downward to the right. The analysis
includes a 1-dB connector-coupling loss at each end and a 6-dB system-operating
margin.

The dispersion limit depends on material and modal dispersion. Material dispersion at 800 nm is taken as 0.07 ns/(nm km) or 3.5 ns/km for an LED with a 50-nm
spectral width. The curve shown is the material dispersion limit in the absence of
modal dispersion. This limit was taken to be the distance at which t_{mat} is 70% of a
bit period. The modal dispersion was derived from Eq. (8.15) for a fiber with an 800-
MHz·km bandwidth–distance product and with $q = 0.7$. The modal dispersion limit
was then taken to be the distance at which t_{mod} is 70% of a bit period. The achievable transmission distances are those that fall below the attenuation limit curve and
to the left of the dispersion line, as indicated by the hatched area. The transmission distance is attenuation-limited up to about 40 Mb/s, after which it becomes
material-dispersion-limited.

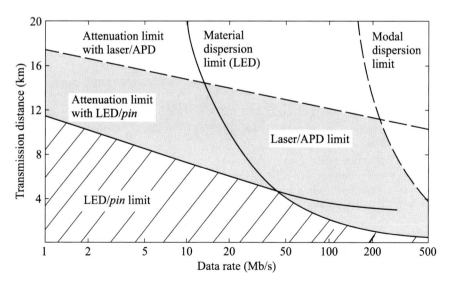

Fig. 8.7 Transmission-distance limits as a function of data rate for an 800-MHz·km fiber, a combination of an 800-nm LED source with a Si pin photodiode, and an 850-nm laser diode with a Si
APD

Greater transmission distances are possible when a laser diode is used in conjunction with an avalanche photodiode. Consider an AlGaAs laser emitting at 850 nm with a spectral width of 1 nm that couples 0 dBm (1 mW) into a fiber flylead. The receiver uses an APD with a sensitivity depicted in Fig. 8.4. The fiber is the same as described above in this section. In this case, the material-dispersion-limit curve lies off the graph to the right of the modal-dispersion-limit curve, and the attenuation limit (with an 8-dB system margin) is as shown in Fig. 8.7. The achievable transmission distances now include those indicated by the shaded area.

8.1.6 Attenuation Limits for SMF Links

For single-mode links there is no modal dispersion. In this case, in addition to attenuation factors, the repeaterless transmission distance is limited by dispersion arising from the source spectral width, from polarization mode dispersion, and from nonlinear effects in the fiber. This section examines the limits that signal attenuation imposes on repeaterless transmission distances. The transmission limits due to chromatic dispersion and polarization-mode dispersion are described in Sect. 8.2. Here it is assumed that the optical power launched into the fiber is no more than 0 dBm (1 mW), so that there are negligible nonlinear effects on the optical signals. Discussions on signal distortion arising from the nonlinear effects resulting from high optical powers in fibers are given in Chap. 12.

Example 8.5 To illustrate the attenuation-limited repeaterless transmission limits, consider two single-mode links operating at 1550 nm based on using the *pin* and APD receivers described for Fig. 7.11. The component and performance characteristics for the two links are taken to be as follows:

1. The optical source is a DFB laser that has a fiber-coupled output of 0 dBm at 1550 nm;
2. At 1550 nm the single-mode fiber has a 0.20-dB/km attenuation;
3. Consider the receiver to have a load resistor $R_L = 200\ \Omega$ and let the temperature be 300 °K;
4. The performances of the two links are measured at a 10^{-12} BER, so that a value of $Q = 7$ is needed;
5. The InGaAs *pin* and APD photodiodes each have a responsivity of 0.95 A/W. The gain of the APD is $M = 10$ and the noise figure $F(M) = 5$ dB.

What are the attenuation-limited repeaterless transmission distances?

Solution (a) From the receiver sensitivity curves shown in Fig. 7.11, it can be deduced that for an InGaAs *pin* photodiode operating at 1550 nm with a 10^{-12} BER, the receiver sensitivity can be approximated by the straight-line equation $P_R = 8 \log B - 28$ dBm, where B is the data rate in Gb/s. To find the attenuation-limited repeaterless transmission distance L_{pin}, use Eq. (8.2) with a combined connector loss plus system margin of 3 dB, so that

$$L_{pin} = (P_S - P_R - 3 \text{ dB})/\alpha = (0 \text{ dBm} - 8 \log B + 28 \text{ dB} - 3 \text{ dB})/\alpha$$
$$= (-8 \log B + 25)/0.2$$
$$= -40 \log B + 125$$

(b) Similarly, from the receiver sensitivity curves shown in Fig. 7.11, for the InGaAs APD the receiver sensitivity can be approximated by the straight-line equation $P_R = 5 \log B - 38$ dBm, where B is the data rate in Gb/s. Again, use Eq. (8.2) with a combined connector loss plus system margin of 3 dB, so that the attenuation-limited repeaterless transmission distance L_{APD} when using an APD is

$$L_{APD} = (P_S - P_R - 3 \text{ dB})/\alpha = (0 \text{ dBm} - 5 \log B + 38 \text{ dB} - 3 \text{ dB})/\alpha$$
$$= (-5 \log B + 35)/0.2 = -25 \log B + 175$$

The results for the attenuation-limited repeaterless transmission distances L_{pin} and L_{APD} are plotted in Fig. 8.8.

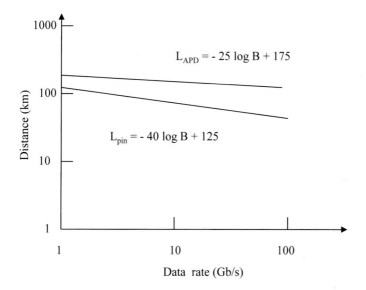

Fig. 8.8 Transmission-distance limits as a function of data rate for 1550-nm laser diodes with 0 dBm fiber-coupled power, InGaAs pin and avalanche photodiodes, and a single-mode fiber with a 0.2-dB/km attenuation

8.2   Concepts of Link Power Penalties

The analysis in Sect. 8.1 assumed that the optical power falling on the photodetector is a clearly defined function of time within the statistical nature of the quantum detection process. In reality, a number of signal impairments that are inherent in optical fiber transmission systems can degrade the link performance.

When any signal impairments are present in a link, a lower optical power level arrives at the receiver compared to the ideal reception case. This lower power results in a reduced signal-to-noise ratio of the link compared to the case when there are no impairments. Because a reduced SNR leads to a higher BER, a higher signal power is required at the receiver in order to maintain the same BER as in the ideal case. The ratio of the reduced received signal power to the ideal received power is known as the *power penalty* for that effect and generally is expressed in decibels. If P_{ideal} and P_{impair} are the received optical powers for the ideal and impaired cases, respectively, then the power penalty PP_x in decibels for impairment condition x is given by

$$PP_x = -10 \log \frac{P_{impair}}{P_{ideal}} \tag{8.18}$$

In some cases one can increase the optical power level at the receiver to reduce the power penalty. For other situations, for example for some nonlinear effects described in Chap. 12, increasing the power level will have no effect on the power penalty. The main power penalties are due to chromatic and polarization-mode dispersions, modal or speckle noise, mode-partition noise, the extinction ratio, wavelength chirp, timing jitter, optical reflection noise, and nonlinear effects that arise when there is a high optical power level in a fiber link. Modal noise is present only in multimode links, but all the other effects can be serious in single-mode links. This section addresses these performance impairments except nonlinear effects, which Chap. 12 describes. Additional power penalties due to optical amplifiers and WDM-channel crosstalk are given in Chaps. 11 and 12, respectively.

8.2.1   Power Penalties from Chromatic Dispersion

Chromatic dispersion originates from the fact that each unique wavelength travels at a slightly different velocity in a fiber. Thus distinct simultaneously launched wavelengths arrive at different times at the fiber end. Therefore, the range of arrival times at the fiber end of a spectrum of wavelengths will lead to temporal spreading of a pulse. As noted in Sect. 3.3, chromatic dispersion is a fixed quantity at a specific wavelength and is measured in units of ps/(nm km). Figure 3.18 shows the chromatic dispersion behavior as a function of wavelength for several different standard single-mode fiber types. For example, a G.652 fiber typically has a chromatic dispersion value of $D_{CD} = 18$ ps/(nm km) at 1550 nm, whereas a G.655 fiber has a chromatic dispersion value of about $D_{CD} = 4$ ps/(nm km) at 1550 nm.

The accumulated chromatic dispersion increases with distance along a link. Therefore, either a transmission system has to be designed to tolerate the total dispersion, or some type of optical or electronic dispersion compensation method has to be employed [14–17]. A basic estimate of what limitation chromatic dispersion imposes on link performance can be made by specifying that the accumulated dispersion should be less than a fraction ε of the bit period $T_b = 1/B$, where B is the bit rate. This gives the relationship $|D_{CD}|L\ \sigma_\lambda < \varepsilon T_b$, or equivalently,

$$|D_{CD}|L\ B\sigma_\lambda < \varepsilon \tag{8.19}$$

The ITU-T Recommendation G.957 for SDH and the *Telcordia Generic Requirement* GR-253 for SONET specify that for a 1-dB power penalty the accumulated dispersion should be less than 0.306 of a bit period [18, 19]. For a 2-dB power penalty the requirement is $\varepsilon = 0.491$.

Shifting the operating wavelength to 1310 nm for a G.652 fiber where $D_{CD} \approx 6$ ps/(nm km) will increase the maximum transmission distance for a 10-Gb/s data rate to about 100 km. However, because fiber attenuation is larger at 1310 nm than at 1550 nm, operation at 1310 nm may become attenuation limited.

Several methods have been examined to mitigate the effects of chromatic dispersion-induced intersymbol interference. Historically, first G.653 dispersion-shifted fibers were developed to reduce the value of D_{CD} at 1550 nm. Although this is useful for links carrying a single wavelength, these fibers are not suitable for WDM systems (see Chaps. 10 and 13) owing to nonlinear crosstalk between different wavelength channels. A more successful method to overcome the dispersion limit is by means of dispersion compensation. This is done via dispersion-compensating modules (DCM) in which the dispersion has the opposite sign of that in the transmission fiber. Through a proper design of such modules, the overall accumulated dispersion in a lightwave transmission system can be reduced to an acceptable level.

Example 8.6 What are the dispersion-limited repeaterless transmission distances L_{CD} at 1550 nm as a function of the bit rate in a G.652 single-mode fiber for the following three cases? Let the chromatic dispersion be $D_{CD} = 18$ ps/(nm km) at 1550 nm.

(a)   A directly modulated laser source with a $\sigma_\lambda = 1.0$-nm spectral width
(b)   A directly modulated laser source with a $\sigma_\lambda = 0.2$-nm spectral width
(c)   An externally modulated single-longitudinal-mode (SLM) DFB laser source with a spectral width that corresponds to the modulation bandwidth

Solution For this case, select an NRZ data format and choose the criterion that for the maximum allowed pulse dispersion with a 2-dB penalty the product $D_{CD}L_{CD}\sigma_\lambda$ is less than or equal to 0.491 of the bit period $1/B$. Thus it is necessary to have the condition $D_{CD}BL_{CD}\sigma_\lambda \leq 0.491$.

(a)   Solving for the bit-rate distance product yields, where B is the data rate in Gb/s,

$$BL_{CD} \leq \frac{0.491}{D_{CD}\sigma_\lambda} = \frac{0.491}{[(18\,\text{ps/nm km}) \times 1\,\text{nm}} = 27\text{Gb/s km}$$

This spectral width imposes a severe limitation on the transmission distance. The chromatic dispersion-limited repeaterless transmission distances L_{CD} are plotted as a function of the bit rate in Fig. 8.9.

(b)   Solving for the bit-rate distance product yields, where B is the data rate in Gb/s,

$$BL_{CD} \leq \frac{0.491}{D_{CD}\sigma_\lambda} = \frac{0.491}{[(18\,\text{ps/nm km}) \times 0.2\,\text{nm}} \right] = 135\,\text{Gb/s km}$$

Narrowing the spectral width to 0.2 nm shows some improvement in the transmission distance, but it is still not adequate for high-speed long-distance optical communication systems. The chromatic dispersion-limited repeaterless transmission distances L_{CD} for this case also are plotted as a function of the bit rate in Fig. 8.9.

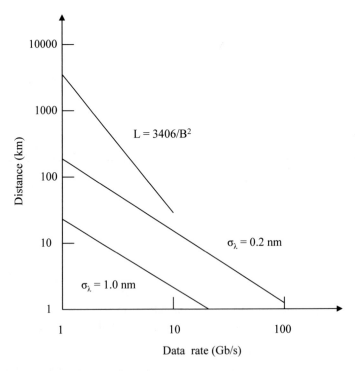

Fig. 8.9 Chromatic dispersion limits for two different chromatic dispersion values and two different source spectral widths

(c) When external modulation is used, the spectral width of the signal is proportional to the bit rate. For example, using a factor of $\Delta f = B$, a 10-Gb/s externally modulated signal would have a spectral width of $\Delta f = 10$ GHz. To view this particular spectral width in terms of wavelength, differentiate the basic equation $c = f\lambda$ to get $\Delta\lambda = (c/f^2)\Delta f = (\lambda^2/c)\Delta f$. Substituting $\sigma_\lambda = \Delta\lambda = (\lambda^2/c)B$ into Eq. (8.19) yields (for a 2-dB power penalty)

$D_{CD}B^2L_{CD}\lambda^2/c \leq 0.491$. Using the parameter values $D_{CD} = 18$ ps/(nm km) at $\lambda = 1550$ nm yields $B^2L_{CD} \leq 3406(\text{Gb/s})^2$ km

Thus, at $B = 2.5$ Gb/s the transmission distance limit is 545 km, whereas at $B = 10$ Gb/s the transmission distance limit is 34 km when using a G.652 fiber at 1550 nm. This condition for the chromatic dispersion-limited repeaterless transmission distances L_{CD} is plotted as a function of the bit rate in the top curve in Fig. 8.9.

Drill Problem 8.4 Repeat the exercise given in Example 8.6 for the case when the maximum allowed pulse dispersion with a 1-dB penalty requires that the product $D_{CD}L_{CD}\sigma_\lambda$ is less than or equal to 0.306 of the bit period $1/B$.
  [Answers: (a) $B L_{CD} \leq 17$ Gb/s km; (b) 85 Gb/s km; (c) $B^2 L_{CD} \leq 2123$ (Gb/s)^2 km.]

8.2.2 Power Penalties Arising from PMD

As Sect. 3.2 describes, *polarization-mode dispersion* (PMD) results from the fact that light-signal energy at a given wavelength in a single-mode fiber actually occupies two orthogonal polarization states or modes. Figure 3.12 shows this condition. PMD arises because the two fundamental orthogonal polarization modes travel at slightly different speeds owing to fiber birefringence. That is, each polarization mode encounters a slightly different refractive index in the fiber. The resulting difference in propagation times between the two orthogonal polarization modes will result in pulse spreading. This PMD effect cannot be mitigated easily and can be a very serious impediment for links operating at 10 Gb/s and higher.

PMD is not a fixed quantity but fluctuates with time due to factors such as temperature variations and stress changes on the fiber [20–22]. Because these external stresses vary slowly with time, the resulting PMD also fluctuates slowly. PMD varies as the square root of distance and thus is specified as a maximum value in units of ps/√km. A typical PMD value for a fiber is $D_{PMD} = 0.05$ ps/√km, but the cabling process can increase this value. The PMD value does not fluctuate widely for cables that are enclosed in underground ducts or in buildings. However it can increase periodically to over 1 ps/$\sqrt{km}$ for outside cables that are suspended on poles, because such cables are subject to wide variations in temperature, wind-induced stresses, and elongations caused by ice loading.

To have a power penalty of less than 1.0 dB, the pulse spreading $\Delta\tau_{PMD}$ resulting from polarization mode dispersion must on the average be less than 10% of a bit period T_b. Using Eq. (3.40) this condition is given by

$$\Delta\tau_{PMD} = D_{PMD}\sqrt{L} < 0.1T_b \tag{8.20}$$

Example 8.7 Consider a 100-km long fiber for which $D_{PMD} = 0.5$ ps/$\sqrt{km}$. What is the maximum possible data rate for an NRZ-encoded signal if the pulse spread can be no more than 10% of a pulse width?

Solution From Eq. (8.20) the pulse spread over the 100-km distance is $\Delta\tau_{PMD} = 5.0$ ps. Because this pulse spread can be no more than 10% of a pulse width, it follows that

$$\Delta\tau_{PMD} = 5.0 \text{ ps} \le 0.1\ T_b$$

Therefore the maximum NRZ bit rate is $1/T_b = 0.1/(5 \text{ ps}) = 20$ Gb/s.

8.2.3   Extinction Ratio Power Penalties

The *extinction ratio* r_e in a laser is defined as the ratio of the optical power level P_1 for a logic 1 to the power level P_0 for a logic 0, that is, $r_e = P_1/P_0$. Ideally one would like the extinction ratio to be infinite, so that there would be no power penalty from this condition. In this case, if P_{ave} is the average power, then $P_0 = 0$ and $P_1 = 2P_{ave} = P_{ideal}$. However, the extinction ratio must be finite in an actual system in order to reduce the rise time of laser pulses. That is, the laser must be slightly on during a zero pulse.

Letting P_{1-ER} and P_{0-ER} be the 1 and 0 power levels, respectively, with a finite extinction ratio, and defining $r_e = P_{1-ER}/P_{0-ER}$, the average power is

$$P_{ave} = \frac{P_{1-ER} + P_{0-ER}}{2} = P_{0-ER}\frac{r_e + 1}{2} = P_{1-ER}\frac{r_e + 1}{2r_e} \tag{8.21}$$

When receiver thermal noise dominates, then the 1 and 0 noise powers are equal and independent of the signal level. In this case, letting $P_0 = 0$ and $P_1 = 2P_{ave}$, the power penalty given by Eq. (8.18) becomes

$$PP_{ER} = -10\log\frac{P_{1-ER} - P_{0-ER}}{P_1} = -10\log\frac{r_e - 1}{r_e + 1} \tag{8.22}$$

In practice, optical transmitters have minimum extinction ratios ranging from 7 to 10 (8.5–10 dB), for which the power penalties range from 1.25 to 0.87 dB. A minimum extinction ratio of 18 is needed in order to have a power penalty of less than 0.5 dB. Note that the power penalty increases significantly for lower extinction ratios.

Drill Problem 8.5 (a) Show that for an extinction ratio $r_e = 8$ the power penalty is 1.09 dB. (b) Show that a minimum extinction ratio of 18 is needed to have a power penalty of less than 0.5 dB.

8.2.4 *Modal Noise Power Penalties*

When light from a coherent laser is launched into a multimode fiber, normally a number of propagating modes of the fiber are excited [23, 24]. As long as these modes retain their relative phase coherence, the radiation pattern seen at the end of the fiber (or at any point along the fiber) takes on the form of a speckle pattern. This is the result of constructive and destructive interference between propagating modes at any given plane. An example of this is shown in Fig. 8.10. The number of speckles in the pattern approximates the number of propagating modes. As the light travels along the fiber, a combination of mode-dependent losses, changes in phase between modes, and fluctuations in the distribution of energy among the various fiber modes will change the modal interference and result in a different speckle pattern. *Modal* or *speckle noise* occurs when any losses that are speckle-pattern dependent are present in a link. Examples of such losses are splices, connectors, microbends, and

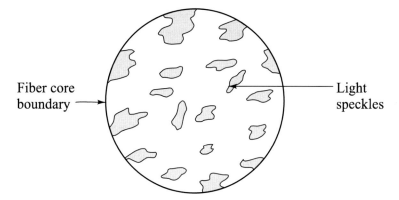

Fig. 8.10 Example of a speckle pattern that is produced when coherent laser light is launched into a multimode fiber

photodetectors with nonuniform responsivity across the photosensitive area. Noise is generated when the speckle pattern *changes in time* so as to vary the optical power transmitted through the particular loss element. The continually changing speckle pattern that falls on the photodetector thus produces a time-varying noise in the received signal, which degrades receiver performance.

The modal distortion resulting from interference between a single pair of modes will appear as a sinusoidal ripple of frequency

$$v = \delta T \frac{d v_{source}}{dt} \tag{8.23}$$

where $d v_{source}/dt$ is the rate of change of optical frequency.

The performance of a high-speed, laser-based multimode fiber link is difficult to predict because the degree of modal noise that can appear depends greatly on the particular installation. Thus the best policy is to take steps to avoid it. This can be done by the following measures:

1. Use LEDs (which are incoherent sources). This totally avoids modal noise.
2. Use a laser that has a large number of longitudinal modes (10 or more). This increases the graininess of the speckle pattern, thus reducing intensity fluctuations at mechanical disruptions in the link.
3. Use a fiber with a large numerical aperture because it supports a large number of modes and hence gives a greater number of speckles.
4. Use a single-mode fiber because it supports only one mode and thus has no modal interference.

8.2.5 Power Penalties Due to Mode-Partition Noise

Mode-partition noise is associated with intensity fluctuations in the longitudinal modes of a multimode laser diode [25]; that is, the side modes are not sufficiently suppressed. This is the dominant noise in single-mode fibers when using multimode devices, such as Fabry–Perot lasers. Intensity fluctuations can occur among the various modes in a multimode laser even when the total optical output is constant, as exhibited in Fig. 8.11. This power distribution can vary significantly both within a pulse and from pulse to pulse.

Because the output pattern of a laser diode is highly directional, the light from these fluctuating modes can be coupled into a single-mode fiber with high efficiency. Each longitudinal mode that is coupled into the fiber has a different attenuation and time delay because each mode is associated with a slightly different wavelength. Because the power fluctuations among the dominant modes can be quite large, significant variations in signal levels can occur at the receiver in systems with high fiber dispersion.

The signal-to-noise ratio due to mode-partition noise is independent of signal power, so that the overall system error rate cannot be improved beyond the limit

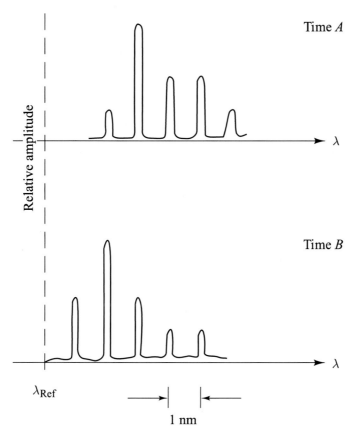

Fig. 8.11 Time-resolved dynamic spectra of a laser diode showing 1-nm-spaced modes or groups of modes dominating the optical output at different times

set by this noise. This is an important difference from the degradation of receiver sensitivity normally associated with chromatic dispersion, which one can compensate for by increasing the signal power.

The power penalty in decibels caused by laser mode-partition noise can be approximated by [26]

$$PP_{mpn} = -5\frac{x+2}{x+1} \log\left[1 - \frac{k^2 Q^2}{2}(\pi B L D_{CD}\sigma_\lambda)^4\right] \tag{8.24}$$

where x is the excess noise factor of an APD, Q is the signal-to-noise factor (see Fig. 7.9), B is the bit rate in Gb/s, L is the fiber length in km, D_{CD} is the fiber chromatic dispersion in ps/(nm km), σ_λ is the rms spectral width of the source in nm, and k is the mode-partition noise factor. The parameter k is difficult to quantify because it can vary from 0 to 1 depending on the laser. However, experimental values of k

range from 0.6 to 0.8. To keep the power penalty less than 0.5 dB, a well-designed system should have the quantity $BL\ D_{CD}\ \sigma_\lambda < 0.1$.

Mode-partition noise becomes more pronounced for higher bit rates. The errors due to mode-partition noise can be reduced and sometimes eliminated by setting the bias point of the laser above threshold. However, raising the bias power level reduces the available signal-pulse power, thereby reducing the achievable signal-to-thermal-noise ratio.

8.2.6   Chirping-Induced Power Penalties

A laser that oscillates in a single longitudinal mode under CW operation may experience dynamic line broadening when the injection current is directly modulated above about 2.5 Gb/s [27–29]. This line broadening is a frequency "chirp" associated with modulation-induced changes in the carrier density. Laser chirping can lead to significant dispersion effects for intensity-modulated pulses when the laser emission wavelength is displaced from the zero-dispersion wavelength of the fiber. This is particularly true in systems operating at 1550 nm, where dispersion in G.652 non-dispersion-shifted fibers is much greater than at 1300 nm.

To a good approximation, the time-dependent frequency change $\Delta v(t)$ of the laser can be given in terms of the output optical power $P(t)$ as [27]

$$\Delta v(t) = \frac{-\alpha}{4\pi}\left[\frac{d}{dt}\ln P(t) + \kappa P(t)\right]\qquad(8.25)$$

where α is the *linewidth enhancement factor* and κ is a frequency-independent factor that depends on the laser structure. The factor α ranges from -3.5 to -5.5 for AlGaAs lasers and from -6 to -8 for InGaAsP lasers.

When the effect of laser chirp is small, the eye closure Δ can be approximated by

$$\Delta = \left(\frac{4}{3}\pi^2 - 8\right)t_{chirp}DLB^2\delta\lambda\left[1 + \frac{2}{3}\left(DL\delta\lambda - t_{chirp}\right)\right]\qquad(8.26)$$

where t_{chirp} is the chirp duration, B is the bit rate, D is the fiber chromatic dispersion, L is the fiber length, and $\delta\lambda$ is the chirp-induced wavelength excursion.

The power penalty for an APD system can be estimated from the signal-to-noise ratio degradation (in dB) due to the signal amplitude decrease as

$$PP_{chirp} = -10\frac{x+2}{x+1}log(1-\Delta)\qquad(8.27)$$

where x is the excess noise factor of an APD.

One approach to minimize chirp is to increase the bias level of the laser so that the modulation current does not drive it below threshold where $\ln P$ and P change rapidly. However, this results in a lower extinction ratio (the ratio of on-state power to off-state power), which leads to an extinction-ratio power penalty at the receiver because of a reduced signal-to-background noise ratio.

However, current systems operating above about 2.5 GB/s now use the external modulation techniques described in Sect. 4.3.9. These external modulators are available in standard miniaturized electronic packages in which they are integrated together with the laser diode.

8.2.7 Link Instabilities from Reflection Noise

When light travels through a fiber link, some optical power gets reflected at refractive-index discontinuities such as in splices, couplers, and filters, or at air–glass interfaces in connectors. The reflected signals can degrade transmitter and receiver performance. In high-speed systems, this reflected power causes optical feedback, which can induce laser instabilities. These instabilities can show up as intensity noise (output power fluctuations), jitter (pulse distortion), or phase noise in the laser, and they can change its wavelength, linewidth, and threshold current. Because they reduce the signal-to-noise ratio, these effects cause two types of power penalties in receiver sensitivities. First, as shown in Fig. 8.12a, multiple reflection points set up an interferometric cavity that feeds power back into the laser cavity, thereby converting phase noise into intensity noise. A second effect created by multiple optical paths is the appearance of spurious signals arriving at the receiver with variable delays, thereby causing intersymbol interference. Figure 8.12b illustrates this.

Unfortunately, these effects are signal-dependent, so that increasing the transmitted or received optical power does not improve the bit-error rate performance. Thus one has to find ways to eliminate reflections. The first step is to look at their magnitudes. As shown by Eq. (5.10), a cleaved silica-fiber end face in air typically will reflect about

$$R = \left(\frac{1.47 - 1.00}{1.47 + 1.00} \right)^2 = 3.6\%$$

This corresponds to an optical return loss of 14.4 dB down from the incident signal. Polishing the fiber ends can create a thin surface layer with an increased refractive index of about 1.6. This increases the reflectance to 5.3% (a 12.7-dB optical return loss). A further increase in the optical feedback level occurs when the distance between multiple reflection points equals an integral number of half-wavelengths of the transmitted wavelength. In this case, all roundtrip distances equal an integral number of in-phase wavelengths, so that constructive interference arises. This quadruples the reflection to 14% or 8.5 dB for unpolished end faces and to over 22% (a 6.6-dB optical return loss) for polished end faces.

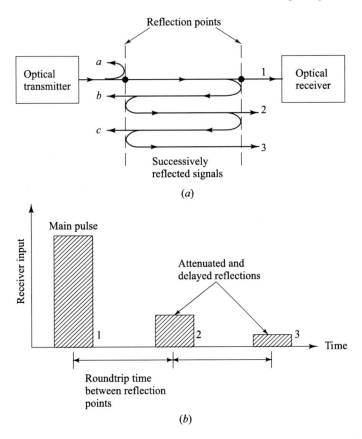

Fig. 8.12 **a** Refractive-index discontinuities can set up multiple reflections in a fiber link; **b** each round trip of a light pulse reflection creates another attenuated and delayed pulse, which can cause intersymbol interference

The power penalties can be reduced to a few tenths of a decibel by keeping the return losses below values ranging from −15 to −32 dB. Techniques for reducing optical feedback include the following:

1. Prepare fiber end faces with a curved surface or an angle relative to the emitting facet of the laser. This directs reflected light away from the fiber axis, so it does not reenter the waveguide. Return losses of 45 dB or higher can be achieved with end-face angles of 5–15°. However, this increases both the insertion loss and the complexity of the connector.
2. Use index-matching oil or gel at air–glass interfaces. The return loss with this technique is usually greater than 30 dB. However, this may not be practical or recommended if connectors need to be disconnected and rejoined often, because contaminants can collect on the interface.

3. Use connectors in which the end faces make physical contact (the so-called *PC connectors*). Return losses of 25–40 dB have been measured with these connectors.
4. Use optical isolators within the laser transmitter module. These devices easily achieve 25-dB return losses, but they also can introduce up to 1 dB of forward loss in the link.

> **Drill Problem 8.6** A GaAs optical source with a refractive index of 3.60 is coupled to an optical fiber that has a core refractive index of 1.48. Show that the optical return loss is 7.59 dB down from the signal power level incident on the material interface (see Sect. 5.1).

8.3 Detection and Control of Errors

In any digital transmission system, errors are likely to occur even when there is a sufficient SNR to provide a low bit-error rate. The acceptance of a certain level of errors depends on the network user. For example, digitized speech or video can tolerate occasional high error rates. However, applications such as financial transactions require almost completely error-free transmission. In this case, the transport protocol of the network must compensate the difference between the desired and the actual bit-error rates.

To control errors and to improve the reliability of a communication line, first it is necessary to be able to detect the errors and then either to correct them or retransmit the information. Error detection methods encode the information stream to have a specific pattern. If segments in the received data stream violate this pattern, then errors have occurred. Sections 8.3.1–8.3.3 discuss the concept and several popular methods of error detection.

The two basic schemes for error correction are *automatic repeat request* (ARQ) and *forward error correction* (FEC) [7, 30–32]. ARQ schemes have been used for many years in applications such as computer communication links that use telephone lines and for data transmission over the Internet. As shown in Fig. 8.13, the ARQ technique uses a feedback channel between the receiver and the transmitter to request

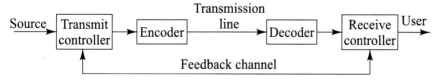

Fig. 8.13 Basic setup for an automatic-repeat-request (ARQ) error-correction scheme

message retransmission in case errors are detected at the receiver. Because each such retransmission adds at least one roundtrip time of latency, ARQ may not be feasible for applications in which data must arrive within a certain time in order to be useful. Forward error correction avoids the shortcomings of ARQ for high-bandwidth optical networks requiring low delays. In FEC techniques, redundant information is transmitted along with the original information. If some of the original data is lost or received in error, the redundant information is used to reconstruct it. Section 8.3.4 gives an overview of popular Reed-Solomon codes used in FEC techniques.

8.3.1 Concept of Error Detection

An error in a data stream can be categorized as a single-bit error or a burst error. As its name implies, a *single-bit error* means that only one bit within a data unit (e.g., a byte, code word, a packet, or a frame) is changed from a 1 to a 0, or vice versa. Single-bit errors are not very common in a typical transmission system because most bit-corrupting noise effects last longer than a bit period.

A *burst error* refers to the fact that more than a single bit within a data unit has changed. This type of error happens most often in a typical transmission system because the duration of a noise burst lasts over several bit periods. A burst error does not necessarily change every bit in a data segment that contains errors. As shown in Fig. 8.14, the length of an error burst is measured from the first corrupted bit to the last corrupted bit. Not all the bits in this particular segment were damaged.

The basic concept of error detection is straightforward. Prior to being inserted into a transmission channel, the information bit stream coming from a communication device is encoded so that it satisfies a certain *pattern* or a *specific set of code words*. At the destination the receiver checks the arriving information stream to verify that the pattern is satisfied. If the data stream contains segments (that is, invalid code words) that do not conform to valid code words, then an error has occurred in that segment.

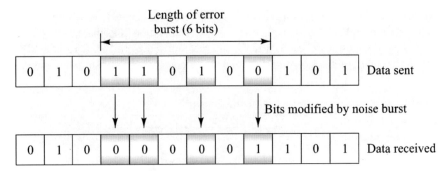

Fig. 8.14 The length of an error burst is measured from the first to the last corrupted bit

Example 8.8 The number of bits affected by a burst error depends on the data rate and the duration of the noise burst. If a bit-corrupting burst noise lasts for 1 ms, then 10 bits are affected for a 10-kb/s data rate, whereas a 10,000-bit segment is damaged for a 10-Mb/s rate.

8.3.2 Codes Used for Linear Error Detection

The *single parity check code* is one of the simplest error detection methods. This code forms a *code word* from the combination of k information bits and a single added *check bit*. If the k information bits contain an odd number of 1 bits, then the check bit is set to 1; otherwise it is set to 0. This procedure ensures that the code word has an even number of ones, which is called having an *even parity*. Hence the check bit is called a *parity bit*. The single parity check code thus can detect when an odd number of errors has occurred in a code word. However, if the received code word contains an even number of errors, this method will fail to detect the errors. The single parity check code is called a *linear code* because the parity bit b_{k+1} is calculated as the modulo 2 sum of the k information bits, that is,

$$b_{k+1} = b_1 + b_2 + \cdots + b_k \text{ modulo } 2 \tag{8.28}$$

where $b_1, b_2, \ldots, b_k$ are the information bits.

A more general linear code with stronger error detection capabilities is called a *binary linear code*. This linear code adds $n - k$ check bits to a group of k information bits, thereby forming a code word consisting of n bits. Such a code is designated by the notation (n, k). One example is the (7, 4) linear Hamming code in which the first four bits of a code word are the information bits b_1, b_2, b_3, b_4 and the next three bits b_5, b_6, b_7 are check bits. Among the wide variety of *Hamming codes*, this particular one can detect all single and double bit errors, but fails to detect some triple errors.

8.3.3 Error Detection with Polynomial Codes

Polynomial codes are used widely for error detection because these codes are easy to implement using shift-register circuits. The term *polynomial code* comes from the fact that the information symbols, the code words, and the error vector are represented by polynomials with binary coefficients. Here, if a transmitted code word has n bits, then the error vector is defined by $(e_1, e_2, \ldots, e_n)$, where, $e_j = 1$ if an error has occurred in the jth transmitted bit and $e_j = 0$ otherwise. Because the encoding process generates check bits by means of a process called a *cyclic redundancy check* (CRC), a polynomial code also is known as a CRC code.

The *cyclic redundancy check* technique is based on a binary division process involving the data portion of a packet and a sequence of redundant bits. Figure 8.15 outlines the following basic CRC procedure:

Step 1. At the sender end a string of n zeros is added to the data unit on which error detection will be performed. This data unit may be a packet (data plus routing and control bits). The characteristic of the redundant bits is such that the result (packet plus redundant bits) is exactly divisible by a second predetermined binary number.

Step 2. The new enlarged data unit is divided by the predetermined divisor using binary division. If the number of bits added to the data unit is n, then the number of bits in the predetermined divisor is $n + 1$. The remainder that results from this division is called the *CRC remainder* or simply the CRC. The number of digits in this remainder is equal to n. For example, if $n = 3$ it may be the binary number 101. Note that the remainder also might be 000, if the two numbers are exactly divisible.

Step 3. The n zeros that were added to the data unit in step 1 are replaced by the n-bit CRC. The composite data unit then is sent through the transmission channel.

Step 4. When the data unit plus the appended CRC arrives at the destination, the receiver divides this incoming composite unit by the same divisor that was used to generate the CRC.

Step 5. If there is no remainder after this division occurs, then the assumption is that there are no errors in the data unit and the receiver accepts the data unit. A remainder indicates that some bits became corrupted during the transmission process and therefore the data unit is rejected.

Instead of using a string of 1 and 0 bits, the CRC generator normally is represented by an algebraic polynomial with binary coefficients. The advantage of using a

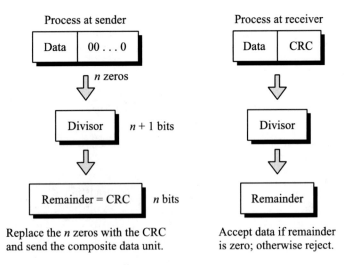

Fig. 8.15 The basic procedure for the cyclic redundancy check (CRC) technique

polynomial is that it is simple to visualize and perform the division mathematically. Numerous varieties of cyclic redundancy check polynomials have been incorporated into technical standards. This can be confusing to developers because they need to select a polynomial according to the application requirements. Table 8.3 shows examples of several commonly used polynomials and their binary equivalents for the CRC generation [31, 32]. The most commonly used polynomial lengths are designated as CRC-8, CRC-16, CRC-32, and CRC-64. The numbers 8, 16, 32, and 64, respectively, refer to the size of the CRC remainder. Thus the CRC divisors for these polynomials are 9, 17, 33, and 65 bits, respectively. Various CRC-8 polynomials are used in applications such as Bluetooth and mobile wireless networks, whereas a usage example of CRC-32 is in IEEE-802 LANs. CRC-16 is used in bit-oriented protocols, such as the High-Level Data Link Control (HDLC) Standard, where frames are viewed as a collection of bits. A polynomial needs to have the following properties:

- It should not be divisible by x. This condition guarantees that the CRC can detect all burst errors that have a length less than or equal to the degree of the polynomial.
- It should be divisible by $x + 1$. This allows the CRC to detect all bursts that affect an odd number of bits.

Given these two rules, the CRC also can find with an error-detection probability.

$$P_{ed} = 1 - 1/2^N \tag{8.29}$$

any burst errors that have a length greater than the degree N of the generator polynomial.

Example 8.9 The generator polynomial $x^7 + x^5 + x^2 + x + 1$ can be written as

$$1 \times x^7 + 0 \times x^6 + 1 \times x^5 + 0 \times x^4 + 0 \times x^3 + 1 \times x^2 + 1 \times x^1 + 1 \times x^0$$

where the exponents on the variable x represent bit positions in a binary number and the coefficients correspond to the binary digits at these positions. Thus the generator polynomial given here corresponds to the 8-bit binary representation 10100111.

Example 8.10 The generator polynomial $x^3 + x + 1$ can be written in binary form as 1011. For the example information unit 11110 the CRC can be found through either binary or algebraic division using steps 1 through 3 outlined earlier. Because

Table 8.3 Commonly used polynomials and their binary equivalents for CRC generation

CRC type	Generator polynomial	Binary equivalent
CRC-8	$x^8 + x^2 + x + 1$	100000111
CRC-16	$x^{16} + x^{15} + x^2 + 1$	11000000000000101
CRC-32	$x^{32} + x^{26} + x^{23} + x^{22} + x^{16} + x^{12} + x^{11}$ $+ x^{10} + x^8 + x^7 + x^5 + x^4 + x^2 + x + 1$	100000100110000010001110110110111

there are 4 bits in the divisor, three 0 s are added to the data for the binary arithmetic operation. Figure 8.16 shows the two different procedures using polynomial and binary arithmetic division. For the polynomial division process the remainder is $x^2 + 1$, which is equivalent to the remainder 101 found by the binary division method. The resulting composite data unit plus CRC that get transmitted is 11110101. Note that when following the binary division method, if the leftmost bit of a remainder is zero, one must use 0000 as the divisor instead of the original 1011 divisor.

Drill Problem 8.7 (a) Verify that the binomial equivalent of the polynomial $x^8 + x^6 + x^4 + x + 1$ is 101010011. (b) Verify the polynomial equivalent of the binomial sequence 10101011110101101 is $x^{16} + x^{14} + x^{12} + x^{10} + x^9 + x^8 + x^7 + x^5 + x^3 + x^2 + 1$.

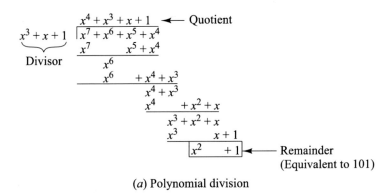

(a) Polynomial division

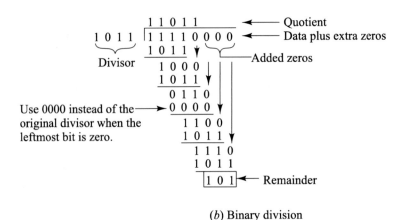

(b) Binary division

Fig. 8.16 Two different procedures for finding the CRC using polynomial and binary arithmetic divisions

Example 8.11 The CRC-32 given in Table 8.3 has a degree of 32. Thus it will detect all burst errors affecting an odd number of bits, all burst errors with a length less than or equal to 32, and from Eq. (8.29) more than 99.99% of burst errors with a length of 32 or more.

8.3.4 Using Redundant Bits for Error Correction

Error correction may be done by the use of *redundancy* in the data stream. With this method, extra bits are introduced into the raw data stream at the transmitter on a regular and logical basis and are extracted at the receiver. These digits themselves convey no information but allow the receiver to detect and correct a certain percentage of errors in the information-bearing bits. The degree of error-free transmission that can be achieved depends on the amount of redundancy introduced. Note: The data rate that includes this redundancy must be less than or equal to the channel capacity.

The method of introducing redundant bits into the information stream at the transmitter for error-reducing purposes is called *forward error correction* (FEC). Typically the amount of added redundancy is small, so the FEC scheme does not use up much additional bandwidth and thus remains efficient. The most popular error-correcting codes are *cyclic codes*, such as *Reed-Solomon* (RS) codes. These codes add a redundant set of r symbols to blocks of k data symbols, with each symbol being s bits long, for example, $s = 8$. The codes are designated by the notation (n, k) where n equals the number of original information symbols k plus the number of redundant symbols r. For a given symbol size s, the maximum length of a Reed-Solomon code word is $n = 2^s - 1$.

Example 8.12 The (255,239) Reed-Solomon code with $s = 8$ (one byte) is used in high-speed undersea optical fiber links. This means that $r = n - k = 255 - 239 = 16$ redundant bytes are sent for every block of 239 information bytes. The code is quite efficient, because the 16 redundant bytes add less than 7 percent of overhead to the information stream.

A Reed-Solomon decoder can correct up to t symbol errors, where $2t = n - k$. For example, the (255,239) RS code can correct up to 8 errors in a block of 239 bytes. One symbol error occurs when one or more bits in a symbol are wrong. Thus the number of bits that are corrected depends on the distribution of the errors. If each incorrect byte contains only one bit error, then the (255,239) RS code will correct 8 bit errors. At the other extreme, if all the bits in each of the 8 incorrect bytes are corrupted, then the (255,239) code will correct $8 \times 8 = 64$ bit errors. Thus, a key feature of RS codes is their ability to correct burst errors, where a sequence of bytes is received incorrectly.

Another advantage of a Reed-Solomon code is that it allows transmission at a lower power level to achieve the same BER that would result without encoding. The resulting power saving is called the *coding gain*. The (255,239) RS code provides about a 6-dB coding gain. Concatenated Reed-Solomon codes (several codes used sequentially) can provide even higher coding gains.

Current terrestrial and undersea high-speed optical communication systems use a number of different FEC codes. For example, as part of the G.709 Digital Wrapper Recommendation, the ITU-T has selected the (255,239) and (255,223) Reed-Solomon codes.[33-35] The (255,223) code has a higher overhead (15%) compared to the (255,239) code, but is somewhat stronger because it is able to correct 16 errors in a block of 223 bits. The Digital Wrapper uses the same error-monitoring techniques as is employed in the earlier SDH and SONET standards. The performance metrics that are calculated include code violations in the incoming bit stream, the number of seconds in which at least one error occurs, the number of seconds in which multiple errors occur (called *severely errored seconds*), and the total number of seconds in which service is not available.

8.4 Coherent Detection Schemes

The basic receiver analysis in Chap. 7 considered a simple and cost-effective light-wave transmission scheme in which the light intensity of the optical source is modulated linearly with respect to the input electrical signal voltage. This scheme pays no attention to the frequency or phase of the optical carrier, because a photodetector at the receiving end only responds to changes in the power level (intensity) that falls directly on it. The photodetector then transforms the optical power level variations back to the original electrical signal format. This method is known as *intensity modulation with direct detection* (IM/DD). Although these IM/DD methods offer system simplicity and relatively low cost, their sensitivities are limited by noises generated in the photodetector and the receiver preamplifier. These noises degrade the receiver sensitivities of square-law IM/DD transmission systems by 10 to 20 dB from the fundamental quantum noise limit.

Around 1978 component researchers had improved the spectral purity and frequency stability of semiconductor lasers to the point where alternative techniques using homodyne or heterodyne detection of the optical signal appeared to be feasible. Optical communication systems that use homodyne or heterodyne detection are called *coherent optical communication systems*, because their implementation depends on *phase coherence* of the optical carrier. In coherent detection techniques the light is treated as a carrier medium that can be amplitude-, frequency-, or phase-modulated similar to the methods used in microwave radio systems.[36-39].

Coherent systems were examined extensively during the 1980s and early 1990s as a method for increasing the transmission spans for long-haul links. However, interest in these methods declined when optical amplifiers were introduced because these amplification devices offered dramatic increases in the transmission distances of multi-wavelength OOK systems. Fortunately research on coherent techniques continued, because a decade later there was renewed interest as data transmission speeds moved to 10 Gb/s and beyond. This interest was spurred by the fact that coherent detection techniques enable a higher spectral efficiency and greater tolerance to chromatic dispersion and polarization-mode dispersion than direct detection methods.

8.4.1 Fundamental Concepts

Figure 8.17 illustrates the fundamental concept in coherent lightwave systems. The key principle of the coherent detection technique is to provide gain to the incoming optical signal by combining or mixing it with a locally generated *continuous-wave* (CW) optical field. The term *mixing* means that when two waves with frequencies ω_1 and ω_2 are combined, the result will be other waves with frequencies equal to $2\omega_1$, $2\omega_2$, and $\omega_1 \pm \omega_2$. For coherent lightwave systems, all frequency components except $\omega_1 - \omega_2$ are filtered out at the receiver. The device used for creating the CW signal is a narrow-linewidth laser called a *local oscillator* (LO). The result of this mixing procedure is that the dominant noise in the receiver is the shot noise coming from the local oscillator. This means the receiver can achieve a sensitivity limited by shot noise.

For simplicity, to understand how this mixing can increase the coherent receiver performance, consider the electric field of the transmitted optical signal to be a plane wave having the form

$$E_s = A_s \cos\lfloor\omega_s t + \varphi_s(t)\rfloor \tag{8.30}$$

where A_s is the amplitude of the optical signal field, ω_s is the optical signal carrier frequency, and $\varphi_s(t)$ is the phase of the optical signal. To send information, one can modulate the amplitude, frequency, or phase of the optical carrier. Thus one of the following three modulation techniques can be implemented:

1. *Amplitude shift keying (ASK) or on–off keying (OOK)*. In this case φ_s is constant and the signal amplitude A_s takes one of two values during each bit period, depending on whether a 0 or a 1 is being transmitted.

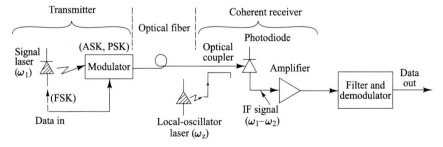

Fig. 8.17 The fundamental concept in coherent lightwave system

2. *Frequency shift keying (FSK)*. For FSK modulation the amplitude A_s is constant and $\varphi_s(t)$ is either $\omega_1 t$ or $\omega_2 t$, where the frequencies ω_1 and ω_2 represent binary signal values.
3. *Phase shift keying (PSK)*. In the PSK method, information is conveyed by varying the phase with a sine wave $\varphi_s(t) = \beta \sin \omega_m t$, where β is the modulation index and ω_m is the modulation frequency.

In a *direct-detection* system the electrical signal coming into the transmitter *amplitude modulates* the optical power level of the light source. Thus the optical power is proportional to the signal current level. At the receiver the incoming optical signal is converted directly into a demodulated electrical output. This directly detected current is proportional to the intensity I_{DD} (the square of the electric field) of the optical signal, yielding

$$I_{DD} = E_s E_s^* = \frac{1}{2} A_s^2 [1 + \cos(2\omega_s t + 2\varphi_s)] \tag{8.31}$$

The term involving $\cos(2\omega_s t + 2\varphi_s)$ gets eliminated from the receiver because its frequency is twice the optical carrier frequency, which is beyond the response capability of the detector. Thus for direct detection Eq. (8.31) becomes

$$I_{DD} = E_s E_s^* = \frac{1}{2} A_s^2 \tag{8.32}$$

At the receiving end in coherent lightwave systems, the receiver first adds a locally generated optical wave to the incoming information-bearing signal and then detects the combination. There are four basic demodulation formats, depending on how the optical signal is mixed with the local oscillator (which gives heterodyne or homodyne detection) and how the electrical signal is detected (either synchronously or asynchronously). As described in this section, for a given modulation format homodyne receivers are more sensitive than heterodyne receivers, and synchronous detection is more sensitive than asynchronous detection.

The mixing of the information-bearing and local-oscillator signals is done on the surface of the photodetector (before photodetection takes place). If the local-oscillator (LO) field has the form

$$E_{LO} = A_{LO} \cos \lfloor \omega_{LO} t + \varphi_{LO}(t) \rfloor \tag{8.33}$$

where A_{LO} is the amplitude of the local oscillator field, and ω_{LO} and $\varphi_{LO}(t)$ are the local-oscillator frequency and phase, respectively, then the detected current is proportional to the square of the total electric field of the signal falling on the photodetector. That is, the intensity $I_{coh}(t)$ is

$$I_{coh}(t) = (E_s + E_{LO})^2 = \frac{1}{2} A_s^2 + \frac{1}{2} A_{LO}^2 + A_s A_{LO} \cos[(\omega_s - \omega_{LO})t + \varphi(t)] \cos \theta(t) \tag{8.34}$$

where $\varphi(t) = \varphi_s(t) - \varphi_{LO}(t)$ is the relative phase difference between the incoming information-bearing signal and the local-oscillator signal, and

$$\cos\theta(t) = \frac{E_s \cdot E_{LO}}{|E_s||E_{LO}|} \tag{8.35}$$

represents the polarization misalignment between the signal wave and the local-oscillator wave. Here again the analysis used the condition that the photodetector does not respond to higher-frequency terms oscillating near the frequency $2\omega_s$ and the frequency $2\omega_{LO}$.

Because the optical power $P(t)$ is proportional to the intensity, then at the photodetector

$$P(t) = P_s + P_{LO} + 2\sqrt{P_s P_{LO}} \cos[(\omega_s - \omega_{LO})t + \varphi(t)]\cos\theta(t) \tag{8.36}$$

where P_s and P_{LO} are the signal and local-oscillator optical powers, respectively, with $P_{LO} \gg P_s$. Thus it can be seen that the angular-frequency difference $\omega_{IF} = \omega_s - \omega_{LO}$ is an intermediate frequency, and the phase angle $\varphi(t)$ gives the time-varying phase difference between the signal and local-oscillator levels. The frequency ω_{IF} is normally in the radio-frequency range of a few tens or hundreds of megahertz.

8.4.2 Homodyne Detection

When the frequencies of the signal carrier and the local oscillator are equal, that is, when $\omega_{IF} = 0$, we have the special case of *homodyne detection*. Equation (8.36) then becomes

$$P(t) = P_s + P_{LO} + 2\sqrt{P_s P_{LO}} \cos\varphi(t)\cos\theta(t) \tag{8.37}$$

Thus one can use either OOK [varying the signal level P_s while keeping $\varphi(t)$ constant] or PSK [varying the phase $\varphi_s(t)$ of the signal and keeping P_s constant] modulation schemes to transmit information. Note that because $P_{LO} \gg P_s$ and P_{LO} is constant, the last term on the right-hand side of Eq. (8.37) contains the transmitted information. Because this term increases with increasing laser power, the local oscillator effectively acts as a signal amplifier, thereby giving greater receiver sensitivity than direct detection.

As can be seen from Eq. (8.37), homodyne detection brings the signal directly to the baseband frequency, so that no further electrical demodulation is required. Homodyne receivers yield the most sensitive coherent systems. However, they are also the most difficult to build, because the local oscillator must be controlled by an optical phase-locked loop. In addition, the need for the signal and the local-oscillator lasers to have the same frequencies puts very stringent requirements on these two optical sources. These criteria include an extremely narrow spectral width (linewidth) and a high degree of wavelength tunability.

8.4.3 Heterodyne Detection

In *heterodyne detection* the intermediate frequency ω_{IF} is nonzero and an optical phase-locked loop is not needed. Consequently heterodyne receivers are much easier to implement than homodyne receivers. However, the price for this simplification is a 3-dB degradation in sensitivity compared to homodyne detection.

Any of the OOK, FSK, or PSK modulation techniques can be used. For analyzing the output current at the receiver, the condition $P_s \ll P_{LO}$ implies that the first term on the right-hand side of Eq. (8.36) can be ignored. The receiver output current then contains a dc term given by

$$i_{dc} = \frac{\eta q}{hv} P_{LO} \tag{8.38}$$

and a time-varying IF term given by

$$i_{IF}(t) = \frac{2\eta q}{hv} \sqrt{P_s P_{LO}} \cos[\omega_{IF}t + \varphi(t)] \cos\theta(t) \tag{8.39}$$

The dc-current is normally filtered out in the receiver, and the IF current gets amplified. One then recovers the information from the amplified current using conventional RF demodulation techniques.

8.4.4 SNR in Coherent Detection

In an optical coherent detection receiver the SNR is mainly determined by the shot noise, because the local oscillator power is generally much stronger than the received optical signal. Thus considering only shot noise and thermal noise, the SNR of the receiver is

$$SNR = \frac{\langle i_s^2(t) \rangle}{\langle i_{th}^2(t) \rangle + \langle i_{shot}^2(t) \rangle} \tag{8.40}$$

For homodyne detection with optical phase locking between the received optical signal and the local oscillator, the signal power is

$$\langle i_s^2(t) \rangle_{\text{homodyne}} = \mathcal{R}^2 P_s(t) P_{LO} \tag{8.41}$$

For heterodyne detection with $\langle \cos^2 \varphi(t) \rangle = 1/2$, the signal power is

$$\langle i_s^2(t) \rangle_{\text{heterodyne}} = \mathscr{R}^2 P_s(t) P_{LO}/2 \tag{8.42}$$

The thermal noise power and the shot noise power in Eq. (8.40) are, respectively,

$$\langle i_{th}^2 \rangle = \frac{4k_B T B_e}{R_L}$$

$$\langle i_{th}^2 \rangle = \frac{4k_B T B_e}{R_L} \tag{8.43}$$

$$\langle i_{\text{shot}}^2 \rangle = 2q\mathscr{R}[P_s(t)/2 + P_{LO}]B_e \tag{8.44}$$

where B_e is the bandwidth of the electrical signal and R_L is the load resistor. The factor $P_s(t)/2$ is due to a 3-dB splitting loss that arises from the coupler used for combining the optical information signal and the local oscillator signal. Using the above expressions then yields for homodyne detection

$$SNR_{\text{homodyne}} = \frac{\mathscr{R}^2 P_s(t) P_{LO}}{4k_B T/R_L + 2q\mathscr{R}[P_s(t)/2 + P_{LO}]} \frac{1}{B_e} \tag{8.45}$$

and for heterodyne detection

$$SNR_{\text{heterodyne}} = \frac{\mathscr{R}^2 P_s(t) P_{LO}/2}{4k_B T/R_L + 2q\mathscr{R}[P_s(t)/2 + P_{LO}]} \frac{1}{2B_e} \tag{8.46}$$

where $2B_e$ is for the double bandwidth needed for heterodyne detection. If the LO power is strong compared to the signal power, then the shot noise created by the local oscillator dominates the thermal noise. In this case, Eq. (8.45) and Eq. (8.46) can be simplified as

$$SNR_{\text{homodyne}} \approx \frac{\mathscr{R}}{2q B_e} P_s(t) \tag{8.47}$$

and

$$SNR_{\text{heterodyne}} \approx \frac{\mathscr{R}}{8q B_e} P_s(t) \tag{8.48}$$

8.4.5 BER Comparisons in Coherent Detection

For making a comparison between the various coherent detection techniques, generally one characterizes the performance of a digital communication system in terms of the bit-error rate. The BER depends on the signal-to-noise ratio (SNR) and the probability density function (PDF) at the receiver output (at the input to the comparator). Because for high local oscillator powers the PDF is Gaussian for both homodyne and heterodyne techniques, the BER depends only on the SNR. Thus one can describe receiver sensitivity in terms of the SNR available at the receiver output, which is directly proportional to the received optical signal power. Traditionally the receiver sensitivity for coherent detection techniques has been described in terms of the average number of photons required to achieve a 10^{-9} BER. That criterion will be used here.

Direct-Detection OOK

Consider an OOK system in which 1 and 0 pulses occur with equal probability. Because the OOK data stream is in an on state only half of the time, the average number of photons required per bit is half the number required per 1 pulse. Thus if N and 0 electron–hole pairs are created during 1 and 0 pulses, respectively, then the average number of photons per bit $\overline{N}_p$ for unity quantum efficiency ($\eta = 1$) is

$$\overline{N}_p = \frac{1}{2}\overline{N} + \frac{1}{2}(0) \tag{8.49}$$

so that the number of electron-hole pairs created in a 1 pulse is $\overline{N} = 2\overline{N}_p$. From Eq. (7.23) the chance of making an error is

$$\frac{1}{2}P_r(0) = \frac{1}{2}e^{-2\overline{N}_p} \tag{8.50}$$

Equation (8.50) implies that about 10 photons per bit are required to get a BER of 10^{-9} for a direct-detection OOK system.

In practice this fundamental quantum limit is very difficult to achieve for direct-detection receivers. The amplification electronics that follow the photodetector add both thermal noise and shot noise, so that the required received power level lies between 13 and 20 dB above the quantum limit.

OOK Homodyne System

As noted in Sect. 8.4.1, either homodyne or heterodyne type receivers can be used with OOK modulation. First consider the homodyne case. When a 0 pulse of duration T_b is received, the average number $\overline{N}_0$ of electron–hole pairs created is simply the number generated by the local oscillator; that is,

$$\overline{N}_0 = A_{LO}^2 T_b \tag{8.51}$$

For a 1 pulse the average number of electron–hole pairs $\overline{N}_1$ is

$$\overline{N}_1 = (A_{LO} + A_s)^2 T_b \approx \left(A_{LO}^2 + 2A_{LO}A_s \right) T_b \tag{8.52}$$

where the approximation arises from the condition $A_{LO}^2 \gg A_s^2$. Because the local-oscillator output power is much higher than the received signal level, the voltage V seen by the decoder in the receiver during a 1 pulse is

$$V = \overline{N}_1 - \overline{N}_0 = 2A_{LO}A_s T_b \tag{8.53}$$

and the associated rms noise σ is

$$\sigma \approx \sqrt{\overline{N}_1} \approx \sqrt{\overline{N}_0} \tag{8.54}$$

Thus from Eq. (7.16) the probability of error or BER is

$$P_e = \text{BER} = \frac{1}{2}\left[1 - \text{erf}\left(\frac{V}{2\sqrt{2}\sigma} \right) \right] = \frac{1}{2}\text{erfc}\left(\frac{V}{2\sqrt{2}\sigma} \right) = \frac{1}{2}\text{erfc}\left(\frac{A_s T_b^{1/2}}{\sqrt{2}} \right) \tag{8.55}$$

where $\text{erfc}(x) = 1 - \text{erf}(x)$ is the complementary error function.

Recall from Example 7.8 that to achieve a BER of 10^{-9} one needs to have $V/\sigma = 12$. Using Eqs. (8.53) and (8.54), this implies

$$A_s^2 T_b = 36 \tag{8.56}$$

which is the expected number of signal photons created per pulse. Thus for OOK homodyne detection, the average energy of each pulse must produce 36 electron–hole pairs. In the ideal case when the quantum efficiency is unity, a 10^{-9} BER is achieved with an average received optical energy of 36 photons per pulse. Assuming an OOK sequence of 1 and 0 pulses, which occur with equal probability, then the average number of received photons per bit of information, $\overline{N}_p$, is 18 (half the number required per pulse). Thus for OOK homodyne detection the BER is given by

$$\text{BER} = \frac{1}{2}\text{erfc}\left(\sqrt{\eta \overline{N}_p} \right) \tag{8.57}$$

To simplify this, note that a useful approximation to $\text{erfc}(\sqrt{x})$ for x $\geq$ 5 is

$$\text{erfc}(\sqrt{x}) \approx \frac{e^{-x}}{\sqrt{\pi x}} \tag{8.58}$$

so that

$$\text{BER} = \frac{e^{-\eta \overline{N}_p}}{\left(\pi \eta \overline{N}_p\right)^{1/2}} \tag{8.59}$$

for $\eta \overline{N}_p \geq 5$ in OOK homodyne detection.

> **Drill Problem 8.8** (a) Verify that 10 photons per bit are required to get a
> bit-error rate of 10^{-9} for an ideal direct-detection OOK system.
>   (b) Show that for an ideal OOK homodyne system, one needs 36 photons
> per pulse to achieve a 10^{-9} BER.

PSK Homodyne System

Homodyne detection of PSK modulation gives the best theoretical receiver sensitivity, but it is also the most difficult method to implement. Figure 8.18 shows the
fundamental setup for a homodyne receiver. The incoming optical signal is first
combined with a strong optical wave being emitted from the local oscillator. This is
done using either a fiber directional coupler (see Chap. 11) or a partially reflecting
plate called a *beam splitter*. When a beam splitter is used, it is made almost completely
transparent, because the incoming signal is much weaker than the local-oscillator
output.

As Eq. (8.37) shows, the information is sent by changing the phase of the transmitted wave. For a 0 pulse the signal and local oscillator are out of phase, so that the
resultant number of electron–hole pairs generated is

$$\overline{N}_0 = (A_{LO} - A_s)^2 T_b \tag{8.60}$$

Similarly, for a 1 pulse the signals are in phase, so that

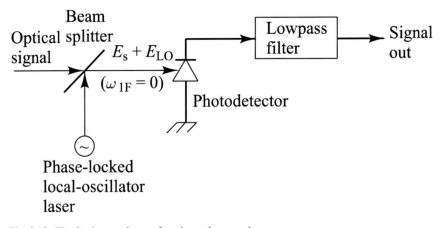

Fig. 8.18 The fundamental setup for a homodyne receiver

$$\overline{N}_1 = (A_{LO} + A_s)^2 T_b \tag{8.61}$$

Consequently, the voltage seen by the decoder in the receiver is

$$V = \overline{N}_1 - \overline{N}_0 = (A_{LO} + A_s)^2 T_b - (A_{LO} - A_s)^2 T_b = 4 A_{LO} A_s T_b \tag{8.62}$$

and the associated rms noise is

$$\sigma = \sqrt{A_{LO}^2 T_b} \tag{8.63}$$

Again, as in the case of homodyne OOK detection, the condition $V/\sigma = 12$ for a BER of 10^{-9} yields

$$A_{LO}^2 T_b = 9 \tag{8.64}$$

This says that for ideal PSK homodyne detection ($\eta = 1$), an average of 9 photons per bit is required to achieve a 10^{-9} BER. Note that here it is not necessary to consider the difference between photons per pulse and photons per bit as in the OOK case, because a PSK optical signal is on all the time.

Again using Eq. (7.16),

$$BER = \frac{1}{2} \text{erfc}\left(\sqrt{2\eta \overline{N}_p}\right) \tag{8.65}$$

for PSK homodyne detection.

Heterodyne Detection Schemes

The analysis for heterodyne receivers is more complicated than in the homodyne case because the photodetector output appears at an intermediate frequency ω_{IF}. The detailed derivations of the BER for various modulation schemes are given in the literature [36–39] so only the results are given here.

An attractive feature of heterodyne receivers is that they can employ either synchronous or asynchronous detection. Figure 8.19 shows the general receiver configuration. In synchronous PSK detection (Fig. 8.19a) one uses a carrier-recovery circuit, which is usually a microwave phase-locked loop (PLL), to generate a local phase reference. The intermediate-frequency carrier is recovered by mixing the output of the PLL with the intermediate-frequency signal. One then uses a low-pass filter to recover the baseband signal. The BER for synchronous heterodyne PSK is given by

$$BER = \frac{1}{2} \text{erfc}\left(\sqrt{\eta \overline{N}_p}\right) \tag{8.66}$$

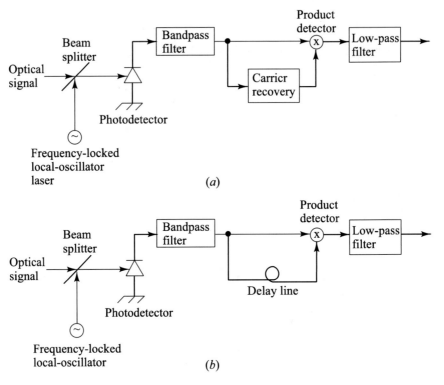

Fig. 8.19 General heterodyne receiver configurations for **a** synchronous PSK detection and **b** asynchronous detection

In this case the ideal PSK receiver requires 18 photons per bit for a 10^{-9} BER. Note that this is the same as for OOK homodyne detection.

A simpler but robust technique that does not use a PLL is *asynchronous detection,* as illustrated in Fig. 8.19b. This technique is called *differential PSK* or DPSK. Here a simple one-bit delay line replaces the carrier-recovery circuit. Because with a PSK method information is encoded by means of changes in the optical phase, the mixer will produce a positive or negative output depending on whether the phase of the received signal has changed from the previous bit. The transmitted information is thus recovered from this output. This sensitivity of the DPSK technique is close to that of synchronous heterodyne detection of PSK, with a bit-error rate of

$$\text{BER} = \frac{1}{2}\text{erfc}\left(\sqrt{\eta \overline{N}_P}\right) \tag{8.67}$$

A BER of 10^{-9} thus requires 20 photons per bit, which is a 0.5-dB penalty with respect to synchronous heterodyne detection of PSK.

Analogous to the PSK case, synchronous heterodyne OOK detection is 3 dB less sensitive than homodyne OOK. Thus the BER is given by

$$BER = \frac{1}{2}\text{erf}\sqrt{\frac{1}{2}\eta\overline{N}_p} \qquad (8.68)$$

Here one needs a minimum of 36 photons per bit for a 10^{-9} BER. In the case of asynchronous heterodyne OOK detection, the bit error rate is given by

$$BER = \frac{1}{2}\exp\left(-\frac{1}{2}\eta\overline{N}_p\right) \qquad (8.69)$$

Thus asynchronous heterodyne OOK detection requires 40 photons per bit for a 10^{-9} BER, which is 3 dB less sensitive than DPSK.

Tables 8.4 and 8.5 summarize the receiver sensitivities for the various modulation techniques. Table 8.4 gives the probability of error as a function of the average number of received photons per bit, N_p, and Table 8.5 shows the number of photons required for a 10^{-9} BER by an ideal receiver having a photodetector with a quantum efficiency of $\eta = 1$.

Table 8.4 Summary of the probability of error as a function of the number of the received photons per bit for coherent optical fiber systems

Probability of error

| Modulation | Homodyne | Heterodyne | | Direct detection |
		Synchronous detection	Asynchronous detection	
On–off keying (OOK)	$\frac{1}{2}\text{erfc}\left(\sqrt{\eta\overline{N}_p}\right)$	$\frac{1}{2}\text{erfc}\left(\sqrt{\frac{1}{2}\eta\overline{N}_p}\right)$	$\frac{1}{2}\exp\left(-\frac{1}{2}\eta\overline{N}_p\right)$	$\frac{1}{2}\exp\left(-2\eta\overline{N}_p\right)$
Phase-shift keying (PSK)	$\frac{1}{2}\text{erfc}\left(\sqrt{2\eta\overline{N}_p}\right)$	$\frac{1}{2}\text{erfc}\left(\sqrt{\eta\overline{N}_p}\right)$	$\frac{1}{2}\exp\left(-\eta\overline{N}_p\right)$	–
Frequency-shift keying (FSK)	–	$\frac{1}{2}\text{erfc}\left(\sqrt{\frac{1}{2}\eta\overline{N}_p}\right)$	$\frac{1}{2}\exp\left(-\frac{1}{2}\eta\overline{N}_p\right)$	–

Table 8.5 Summary of the number of photons required for a 10^{-9} BER by an ideal receiver having a photodetector with unity quantum efficiency

Number of photons

| Modulation | Homodyne | Heterodyne | | Direct detection |
		Synchronous detection	Asynchronous detection	
On–off keying (OOK)	18	36	40	10
Phase-shift keying (PSK)	9	18	20	–
Frequency-shift keying (FSK)	–	36	40	–

8.5   Higher-Order Signal Modulation Formats

The individual channel capacity of fiber communication links using on–off keying (OOK) has steadily increased to 10-Gb/s rates per wavelength. Combined with dense wavelength division multiplexing or DWDM technology (see Chap. 10), this has resulted in transmission rates of 1 Tb/s and beyond per fiber. To increase the single-fiber capacity even more, different modulation formats with higher spectral efficiencies are being tested and deployed [40]. This section discusses the concept of spectral efficiency and describes some multilevel modulation being used in high-speed high-capacity telecom links.

8.5.1   Concept of Spectral Efficiency

Spectral efficiency is a measure of how efficient an optical modulation scheme is at using the available fiber frequency spectrum. This efficiency parameter is given as the number of bits transmitted per second per hertz of optical frequency, that is, it is measured in units of bits/s/Hz.

Transceivers operating at 10 Gb/s per channel with simple OOK modulation occupy only a portion of the standard 50-GHz DWDM channel grid spacing. Thus, much of the 50-GHz channel is unused, which results in a relatively low spectral efficiency of 0.2 bits/s/Hz. For transceivers employing the 100-Gb/s modulation techniques described in Sect. 8.5.4, ten times the capacity is transmitted in the same 50-GHz channel spacing, resulting in a spectral efficiency of 2 bits/s/Hz. Modulation schemes for 400-Gb/s transmission can achieve spectral efficiencies of up to 8 bits/s/Hz.

8.5.2   Phase Shift Keying or IQ Modulation

For high-speed data rates, information can be transmitted using a *quadrature modulation* technique. In this technique the information signal is separated into two parts, one part is called the *in-phase component* (designated by I) and the other is an *out-of-phase component* (designated by Q). This means that the in-phase component (I) will be the real value of the signal and the out-of-phase component (Q) will be a phase-shifted version of the signal. This method is referred to as *phase-shift keying* (PSK) or *IQ modulation*. Therefore rather than turning the optical power on and off as is done in OOK modulation, PSK techniques use changes of the phase of the optical carrier to encode the data.

Differential phase shift keying (DPSK) is the simplest PSK format. DPSK carries the information in the optical phase. Optical power appears in each DPSK bit slot and can occupy the entire slot for NRZ-DPSK, or it can occupy part of the slot in the form of a pulse in an RZ-DPSK format. Binary data is encoded as either an optical phase shift of 0 or π between adjacent slots. For example, the information bit 1 may be transmitted by a 180° carrier phase shift relative to the carrier phase in the previous slot, whereas the information bit 0 is transmitted by no phase shift relative to the carrier phase in the previous signaling interval. There are many other varieties of PSK, such as DQPSK described in Sect. 8.5.3.

8.5.3 Differential Quadrature Phase-Shift Keying

Until about 2002, traffic was transmitted over most optical communication systems at data rates up to 2.5 Gb/s per wavelength using OOK signals in either NRZ or RZ formats. As the desire grew to transmit data at higher speeds, such as 10 and 40 Gb/s, the idea of using a multilevel modulation format received much attention. Of particular interest for high-speed transmission is the use of the *differential quadrature phase-shift keying* (DQPSK) method. In a multilevel modulation format, more than one bit per symbol is transmitted. In the DQPSK method, information is encoded by means of the four phase shifts $\{\pi/4, +3\pi/4, -\pi/4, -3\pi/4\}$. The set of bit pairs $\{00, 10, 01, 11\}$ can be assigned to each of the four phase shifts, respectively. The data points on the IQ diagram are shown in Fig. 8.20. For example, a phase shift of $\pi/4$ means that the bit pair 11 was sent. Thus DQPSK transmits at a *symbol rate of half the aggregate bit rate*.

Because for a given data rate the symbol rate in DQPSK is reduced by a factor of 2 compared to a modulation scheme such as OOK, the spectral occupancy is reduced and the transmitter and receiver requirements are lowered. In addition, the chromatic dispersion and polarization-mode dispersion limits are extended. However, compared to DPSK the SNR needed to achieve a specific BER is increased by a factor of 1 to 2 dB. Also, the design of the receiver becomes more complex because the tolerance to frequency drifts between the transmit laser and the delay interferometers is six times lower than for DPSK.

Fig. 8.20 The data points on an IQ diagram for DQPSK modulation

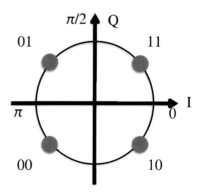

8.5.4   Quadrature Amplitude Modulation (QAM)

The PSK concept can be extended to higher-order modulation formats by encoding
$m = \log_2 M$ data bits on M states per symbol. This method results in a reduction
of the spectral width and allows upgrading to higher data rates with lower-speed
components. Figure 8.21 illustrates the data constellations for 8PSK where every
phase shift of 45° represents a different block of three data bits. Figure 8.22 shows
two higher-order modulation formats of quadrature amplitude modulation (QAM) for
up to 16 states. The format in Fig. 8.22a uses 3 amplitudes and 12 phases, whereas the
format in Fig. 8.22b uses 4 amplitudes and 8 phases This modulation format is known
as 16QAM and is a leading candidate for data transmission at 400 Gb/s per channel.
Figure 8.23 illustrates the four-bit data points for a square 16QAM format. Other

Fig. 8.21   The data
constellation for 8PSK

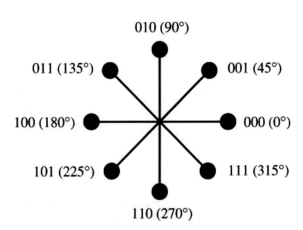

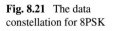

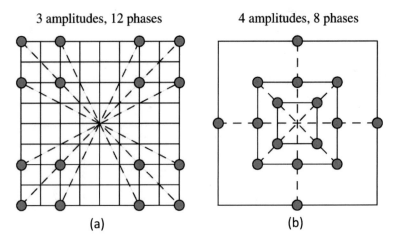

(a)   (b)

Fig. 8.22   Two possible modulation formats for 16QAM

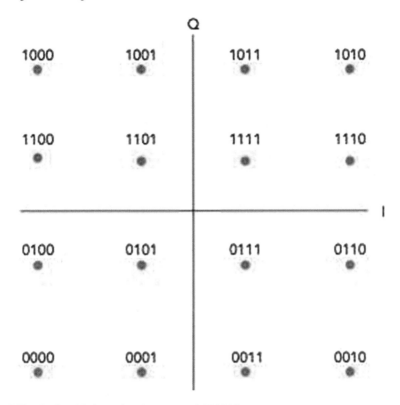

Fig. 8.23 The four-bit data points for a square 16QAM format

QAM formats include 32QAM, 64QAM, 128QAM, and 256QAM. The 256QAM format can achieve a high spectral efficiency of up to 8 b/s/Hz.

8.6 Summary

The design of an optical link involves many interrelated operating characteristics of the fiber, source, photodetector, and other components in the link. Two design analyses usually are carried out to ensure that the desired system performance can be met. These are the link power budget and the system rise-time analysis. In the link power budget analysis one first determines the power margin between the optical transmitter output and the minimum receiver sensitivity needed to establish a specified BER. This margin then can be allocated to link component losses and any additional power margins needed to compensate for unexpected component degradations.

Then the designer makes a system rise-time analysis to ensure that the dispersion limit of the link has not been exceeded. The four basic factors that may limit the system response speed significantly are the transmitter rise time, the material dispersion of the fiber, the modal dispersion of the fiber, and the receiver rise time.

The basic link power budget and the system rise-time analysis assume that the optical power falling on the photodetector is a clearly defined function of time within the statistical nature of the quantum detection process. However, in an actual link various signal impairments can reduce the power of the optical signal arriving at the receiver from the ideal case, which is known as a power penalty for that effect. The main power penalties are due to dispersions, modal or speckle noise, mode-partition noise, the extinction ratio, wavelength chirp, timing jitter, optical reflection noise, and nonlinear effects. Modal noise is present only in multimode links, but all the other effects can be serious in single- mode links.

To control errors and to improve the reliability of a communication line, first it is necessary to be able to detect the errors and then either to correct them or retransmit the information. Error detection methods encode the information stream to have a specific pattern. Error correction may be done by the use of redundancy in the data stream. With this method, extra bits are introduced into the raw data stream at the transmitter on a regular and logical basis and are extracted at the receiver. These digits themselves convey no information but allow the receiver to detect and correct a certain percentage of errors in the information-bearing bits. This method is called forward error correction (FEC).

In a basic optical receiver the light intensity of the optical source is modulated linearly with respect to the input electrical signal voltage. This method is known as intensity modulation with direct detection (IM/DD). Optical communication systems that use homodyne or heterodyne detection instead of IMDD are called coherent optical communication systems, because their implementation depends on phase coherence of the optical carrier. In coherent detection techniques the light is treated as a carrier medium that can be amplitude-, frequency-, or phase-modulated similar to the methods used in microwave radio systems. Coherent detection techniques enable a higher spectral efficiency and greater tolerance to dispersion effects than direct detection methods.

Problems

8.1   A 3B4B code converts blocks of 3 bits to blocks of 4 bits according to the rules given in Table 8.6. When there are two or more consecutive blocks of three zeros, the coded binary blocks 0010 and 1101 are used alternately. Similarly, the coded blocks 1011 and 0100 are used alternately for consecutive blocks of three ones.

(a)   Using these translation rules, find the coded bit stream for the data input 010001111111101000000001111110.

(b)   What is the maximum number of consecutive identical bits in the coded pattern?

8.2   A 4B5B code has $2^4 = 16$ four-bit data characters. The code maps these characters into the 5-bit sequences listed in Table 8.7. Using this information,

Table 8.6 Conversion rules for a 3B4B code

Original code word	3B4B code word	
	Mode 1	Mode 2
000	0010	1101
001	0011	
010	0101	
011	0110	
100	1001	
101	1010	
110	1100	
111	1011	0100

Table 8.7 Data sequence used in 4B5B code conversion

Data sequence	Encoded sequence
0000	11110
0001	01001
0010	10100
0011	10101
0100	01010
0101	01011
0110	01110
0111	01111
1000	10010
1001	10011
1010	10110
1011	10111
1100	11010
1101	11011
1110	11100
1111	11101

encode the following bit stream: 010111010010111010100111.

8.3 Find the maximum attenuation-limited transmission distance of the following two systems operating at 100 Mb/s:

System 1 operating at 850 nm

(a) GaAlAs laser diode: 0-dBm (1-mW) fiber-coupled power
(b) Silicon avalanche photodiode: −50-dBm sensitivity
(c) Graded-index fiber: 3.5-dB/km attenuation at 850 nm
(d) Connector loss: 1 dB/connector

System 2 operating at 1300 nm

(a) InGaAsP LED: −13-dBm fiber-coupled power
(b) InGaAspinphotodiode: −38-dBm sensitivity
(c) Graded-index fiber: 1.5-dB/km attenuation at 1300 nm
(d) Connector loss: 1 dB/connector

Allow a 6-dB system operating margin in each case.

8.4   An engineer has the following components available:

(a) GaAlAs laser diode operating at 850 nm and capable of coupling 1 mW
     (0 dBm) into a fiber
(b) Ten sections of cable each of which is 500 m long, has a 4-dB/km
     attenuation, and has connectors on both ends
(c) Connector loss of 2 dB/connector
(d) A *pin* photodiode receiver
(e) An avalanche photodiode receiver

Using these components, the engineer wishes to construct a 5-km link operating at 20 Mb/s. If the sensitivities of the *pin* and APD receivers are −45
and −56 dBm, respectively, show that the APD receiver should be used if a
6-dB operating margin is required for the system.

8.5   Using the step response $g(t) = [1 − \exp(−2\pi B_e t)]u(t)$ from Eq. (8.4), show
that the 10-to-90% receiver rise time is given by Eq. (8.5). [Hint: First find
the values of $t_{10\%}$ and $t_{90\%}$ from $g(t_{10\%}) = 0.1$ and $g(t_{90\%}) = 0.9$. Then take
the natural log (ln) of each expression to find $t_{rx} = t_{90\%} − t_{10\%}$ which yields
Eq. (8.5).]

8.6   (a) Use Eq. (8.9) to verify the steps leading from Eqs. (8.11) to (8.12).
(*b*) As described in the text, show that Eq. (8.14) follows from Eqs. (8.10)
and (8.13).

8.7   A 90-Mb/s NRZ data transmission system that sends two DS3 (45-Mb/s)
channels uses a GaAlAs laser diode that has a 1-nm spectral width. The
rise time of the laser transmitter output is 2 ns. The transmission distance is
7 km over a graded-index fiber that has an 800-MHz·km bandwidth-distance
product.

(a) If the receiver bandwidth is 90 MHz and the mode-mixing factor $q =$
     0.7, what is the system rise time? Does this rise time meet the NRZ
     data requirement of being less than 70% of a pulse width?
(b) What is the system rise time if there is no mode mixing in the 7-km
     link; that is, $q = 1.0$?

8.8   Verify the plot in Fig. 8.7 of the transmission distance versus data rate of
the following system. The transmitter is a GaAlAs laser diode operating at
850 nm. The laser power coupled into a fiber flylead is 0 dBm (1 mW), and
the source spectral width is 1 nm. The fiber has a 3.5-dB/km attenuation at
850 nm and a bandwidth of 800 MHz km. The receiver uses a silicon avalanche
photodiode that has the sensitivity versus data rate shown in Fig. 8.4. For
simplicity, the receiver sensitivity (in dBm) can be approximated from curve

fitting by $P_R = 9\log B - 68.5$ where B is the data rate in Mb/s. To find the attenuation-limited transmission distance, include a 1-dB connector loss at each end and a 6-dB system margin.

8.9 A 1550-nm single-mode digital fiber optic link needs to operate at 622 Mb/s over 80 km without amplifiers. A single-mode InGaAsP laser launches an average optical power of 3.0 dBm into the fiber. The fiber has a loss of 0.25 dB/km, and there is a splice with a loss of 0.1 dB every kilometer. The coupling loss at the receiver is 0.5 dB, and the receiver uses an InGaAs APD with a sensitivity of -39 dBm. Excess-noise penalties are predicted to be 1.5 dB. (a) Set up an optical power budget for this link and find the system margin. (b) What is the system margin at 2.5 Gb/s with an APD sensitivity of -31 dBm?

8.10 In the (7,4) linear Hamming code the first four bits of a code word are the information bits b_1, b_2, b_3, b_4 and the next three bits b_5, b_6, b_7 are the check bits, which are given by $b_5 = b_1 + b_3 + b_4, b_6 = b_1 + b_2 + b_4$, and $b_7 = b_2 + b_3 + b_4$ Make a table listing the sixteen possible 4-bit information words, that is, 0000 through 1111, and the corresponding 7-bit code words.

8.11 (a) Find the binary equivalent of the polynomial $x^8 + x^7 + x^3 + x + 1$. (b) Find the polynomial equivalent of 10011011110110101.

8.12 Consider the 10-bit data unit 1010011110 and the divisor 1011. Use both binary and algebraic division to show that the CRC remainder is 001.

8.13 Consider the generator polynomial $x^3 + x + 1$.

(a) Show that the CRC for the data unit 1001 is given by 110.

(b) If the resulting code word has an error in the first bit when it arrives at the destination, show that the CRC calculated by the receiver is 101.

8.14 Why can the (255, 223) Reed-Solomon code correct up to 16 bytes, whereas the (255, 239) is limited to correcting 8 byte errors? What is the overhead for each of these two codes?

8.15 Verify the resulting expression in Eq. (8.34) for the intensity due to the combined signal and local oscillator fields.

8.16 A homodyne ASK receiver has a 100-MHz bandwidth and contains a 1310-nm *pin* photodiode with a responsivity of 0.6 A/W. It is shot noise limited and needs a signal-to- noise ratio of 12 to achieve a 10^{-9} BER. Find the photocurrent that is generated if the local oscillator power is -3 dBm and the phase error is $10°$. Assume that both the signal and the local oscillator have the same polarization.

8.17 For a Bit error rate of 10^{-9} assume that the combined spectral width of the signal carrier wave and the local oscillator should be 1 percent of the transmitted bit rate.

(a) What spectral width is needed at 1310 nm for a 100-Mb/s data rate?

(b) What is the maximum allowed spectral width at 2.5 Gb/s?

8.18 (a) Verify that 10 photons per bit are required to get a bit error rate of 10^{-9} for a direct-detection OOK system.

(b) Show that for an ideal OOK homodyne system, one needs 36 photons per pulse to achieve a 10^{-9} BER.

Answers to Selected Problems

8.1   The spaces in the following answer are inserted for clarity purposes only:
       (a) Original code: 010 001 111 111 101 000 000 001 111 110
       3B4B encoded: 0101 0011 1011 0100 1010 0010 1101 0011 1011 1100
       (b) The maximum number of consecutive identical bits is three.

8.2   (a) Original code: 0101 1101 0010 1110 1010 0111
       4B5B encoded: 01011 11011 10100 11100 10110 01111

8.3   For system1, $L = 12$ km; for system 2, $L = 11.3$ km

8.4   (a) For the *pin* receiver $L = 4.25$ km; (b) For the APD receiver $L = 7.0$ km

8.7   (a) $t_{sys} = 4.90$ ns; $t_{sys} < 0.7\, Tb = 7.78$ ns (b) $t_{sys} = 5.85$ ns

8.9   12.1 dB at 622 Mb/s; 4.1 dB at 2.5 Gb/s

8.10

Information word				Code word						
b1	b2	b3	b4	b1	b2	b3	b4	b5	b6	b7
0	0	0	0	0	0	0	0	0	0	0
0	0	0	1	0	0	0	1	1	1	1
0	0	1	0	0	0	1	0	1	0	1
0	0	1	1	0	0	1	1	0	1	0
0	1	0	0	0	1	0	0	0	1	1
0	1	0	1	0	1	0	1	1	0	0
0	1	1	0	0	1	1	0	1	1	0
0	1	1	1	0	1	1	1	0	0	1
1	0	0	0	1	0	0	0	1	1	0
1	0	0	1	1	0	0	1	0	0	1
1	0	1	0	1	0	1	0	0	1	1
1	0	1	1	1	0	1	1	1	0	0
1	1	0	0	1	1	0	0	1	0	1
1	1	0	1	1	1	0	1	0	1	0
1	1	1	0	1	1	1	0	0	0	0
1	1	1	1	1	1	1	1	1	1	1

8.11   (a) 110001011; (b) $x^{16} + x^{13} + x^{12} + x^{10} + x^9 + x^8 + x^7 + x^5 + x^4 + x^2 + 1$

8.16 $i_{IF}(t) = 0.67\ \mu A$

8.17   (a) $\Delta\lambda = 5.6 \times 10^{-6}$ nm; (b) $\Delta\lambda = 1.0 \times 10^{-4}$ nm.

References

1. A. B. Carlson, P. Crilly, *Communication Systems*, 5th edn. (McGraw-Hill, 2010)
2. L.W. Couch II, *Digital and Analog Communication Systems*, 8th edn. (Prentice Hall, 2013)
3. R.E. Ziemer, W.H. Tranter, *Principles of Communications: Systems, Modulation, and Noise*, 7th edn. (Wiley, 2014)
4. J.M. Giron-Sierra, *Digital Signal Processing with Matlab Examples, Volume 1: Signals and Data, Filtering, Non-stationary Signals, Modulation* (Springer, 2017)
5. M.S. Alencar, V.C. da Rocha, *Communication Systems* (Springer, 2020)
6. T. von Lerber, S. Honkanen, A. Tervonen, H. Ludvigsen, F. Küppers, Optical clock recovery methods: Review. Opt. Fiber Technol. **15**(4), 363–372 (2009)
7. S. Benedetto, G. Bosco, Channel coding for optical communications, in *Optical Communication Theory and Techniques*, ed. by E. Forestieri, pp. 63–78 (Springer, 2005)
8. P.J. Winzer, R.J. Essiambre, Advanced optical modulation formats. Proc. IEEE **94**, 952–985 (2006)
9. P.V. Kumar, M.Z. Win, H.-F. Lu, C.N. Georghiades, Error-control coding techniques and applications (Chap. 17), in *Optical Fiber Telecommunications-V*, ed. by I.P. Kaminov, T. Li, A.E. Willner, vol. B. Academic Press (2008)
10. R. Keim, What is a low pass filter? A tutorial on the basics of passive RC filters. *All About Circuits*, 12 May 2019
11. T. Kanada, Evaluation of modal noise in multimode fiber-optic systems. J. Lightw. Technol. **2**, 11–18 (1984)
12. R.D. de la Iglesia, E.T. Azpitarte, Dispersion statistics in concatenated single-mode fibers. J. Lightw. Technol. **5**, 1768–1772 (1987)
13. M. Suzuki, N. Edagawa, Dispersion-managed high-capacity ultra-long-haul transmission. J. Lightw. Technol. **21**, 916–929 (2003)
14. L. Grüner-Nielsen, M. Wandel, P. Kristensen, C. Jørgensen, L.V. Jørgensen, B. Edvold, B. Pálsdóttir, D. Jakobsen, Dispersion-compensating fibers. J. Lightw. Technol. **23**, 3566–3579 (2005)
15. A.B. Dar, R.K. Jha, Chromatic dispersion compensation techniques and characterization of fiber Bragg grating for dispersion compensation. Opt Quant Electron **49**, 108 (2017)
16. T. Xu, G. Jacobsen, S. Popov, J. Li, S. Sergeyev, A.T. Friberg, T. Liu, Y. Zhang, Analysis of chromatic dispersion compensation and carrier phase recovery in long-haul optical transmission system influenced by equalization enhanced phase noise. Optik **138**, 494–508 (2017)
17. H. Bülow, F. Buchali, A. Klekamp, Electronic dispersion compensation. J. Lightw. Technol. **26**, 158–167 (2008)
18. ITU-T Recommendation G.957, *Optical Interfaces for Equipments and Systems Relating to the Synchronous Digital Hierarchy*, Mar 2006
19. Telcordia, *SONET Transport Systems—Common Generic Criteria GR-253,* Issue 5, Dec 2009
20. M. Brodsky, N.J. Frigo, M. Tur, Polarization mode dispersion (Chap. 16), in *Optical Fiber Telecommunications-V*, ed. by I.P. Kaminov, T. Li, A.E. Willner, vol. A, pp. 593–603 (Academic Press, 2008)
21. D.S. Waddy, L. Chen, X. Bao, Polarization effects in aerial fibers. Optical Fiber Tech. **11**(1), 1–19 (2005)
22. M. Zamani, Polarization-induced distortion effects on the information rate in single-mode fibers. Appl. Opt. **57**, 6572–6581 (2018)
23. A.M.J. Koonen, Bit-error-rate degradation in a multimode fiber optic transmission link due to modal noise. IEEE J. Sel. Areas. Commun. **SAC-4**, 1515–1522 (1986)
24. P.M. Shankar, Bit-error-rate degradation due to modal noise in single-mode fiber optic communication systems. J. Opt. Commun. **10**, 19–23 (1989)
25. P. Pepeljugoski, Dynamic behavior of mode partition noise in multimode fiber links. J. Lightw. Technol. **30**(15), 2514–2519 (2012)
26. K. Ogawa, Analysis of mode partition noise in laser transmission systems. IEEE J. Quantum Electron. **QE-18**, 849–855 (1982)

27. R. Heidemann, Investigation on the dominant dispersion penalties occurring in multigigabit direct detection systems. J. Lightw. Technol. **6**, 1693–1697 (1988)
28. Y. Matsui, D. Mahgerefteh, X. Zheng, C. Liao, Z.F. Fan, K. McCallion, P. Tayebati, Chirp-managed directly modulated laser (CML). IEEE Photonics Tech. Lett. **18**, 385–387 (2006)
29. S.L. Chuang, G. Liu, P.K. Kondratko, High-speed low-chirp semiconductor lasers (Chap. 3), in *Optical Fiber Telecommunications-V*, ed. by I.P. Kaminov, T. Li, A.E. Willnerc, vol. A, pp. 53–80 (Academic Press, 2008)
30. S. Lin, D.J. Costello, *Error control coding*, 2nd edn. (Prentice-Hall, 2005)
31. B. A. Forouzan, *Data Communication and Networking*, McGraw-Hill, 5th ed., 2013.
32. F. Kienle, Channel coding basics, in *Architectures for Baseband Signal Processing* (Springer, 2014)
33. ITU-T Recommendation G.709, *Interfaces for the Optical Transport Network* (OTN), June 2016
34. ITU-T Recommendation G.975, *Forward Error Correction for Submarine Systems*, Oct 2000
35. ITU-T Recommendation G.975.1, *Forward Error Correction for High Bit Rate DWDM Submarine Systems*, Feb 2004
36. G. Li, Recent advances in coherent optical communication. Adv. Opt. Photon. **1**(2), 279–307 (2009)
37. S.J. Savory, Digital coherent optical receivers: Algorithms and subsystems. IEEE J. Sel. Topics Quantum. Electron. **16**(5), 1164–1179 (2010)
38. K. Kikuchi, Fundamentals of coherent optical fiber communications (tutorial review). J. Lightw. Technol. **34**(1), 157–179 (2016)
39. Z. Jia, L.A. Campos, *Coherent Optics for Access Networks* (CRC Press, 2019)
40. K. Kitayama, Optical Code Division Multiple Access (Cambridge University Press, 2014)

Chapter 9
Analog Optical Fiber Channels

Abstract Although usually there is a major emphasis on digital transmission links, in many instances it is more advantageous to transmit information in its original analog form instead of first converting it to a digital format. For example, as a result of the increasing use of broadband wireless communication devices, schemes have been investigated and implemented for using analog optical fiber links for distributing broadband microwave-frequency signals in a variety of applications. This chapter addresses these methods, which have become known as RF-over-fiber techniques.

Historically in optical fiber networks the trend has been to use digital transmission schemes. A major reason for this was that digital integrated-circuit technology offered a reliable and economic method of transmitting both voice and data signals. However, in many instances it is more advantageous to transmit information in its original analog form instead of first converting the information to a digital format. Key applications are in microwave photonics and radio-over-fiber channels [1–9]. Most analog applications use laser diode transmitters, so the discussion here will concentrate on this optical source.

When implementing an analog fiber optic system, the main parameters to consider are the carrier-to-noise ratio, bandwidth, spurious-free dynamic range, and signal distortion resulting from nonlinearities in the transmission system. Section 9.1 describes the general operational aspects and components of an analog fiber optic link. Traditionally, in an analog system a carrier-to-noise ratio analysis is used instead of a signal-to-noise ratio analysis, because the information signal is normally superimposed on a radio-frequency (RF) carrier. Thus in Sect. 9.2 carrier-to-noise ratio requirements are examined. This is first done for a single channel under the assumption that the information signal is directly modulated onto an optical carrier (e.g., a constant laser output).

For transmitting multiple signals over the same channel, a multichannel modulation technique can be used. In this method, which is described in Sect. 9.3, the information signals are first superimposed on underlying RF subcarriers. These carriers then are combined and the resulting electrical signal is used to modulate the optical carrier. A limiting factor in these systems is the signal impairment arising from harmonic and intermodulation distortions.

© The Author(s), under exclusive license to Springer Nature Singapore Pte Ltd. 2021   363
G. Keiser, *Fiber Optic Communications*,
https://doi.org/10.1007/978-981-33-4665-9_9

As a result of the increasing use of broadband wireless communication devices, schemes have been investigated and implemented for using analog optical fiber links for distributing broadband microwave-frequency signals in a variety of applications. The methods for transmitting microwave analog signals in the 0.3–300-GHz range over an optical fiber link have become known as *RF-over-fiber* techniques. Section 9.4 examines the basics of these techniques. Section 9.5 gives an example of radio-over-fiber links used for in-building distributed antenna systems to provide wireless LAN and mobile telephony services over a single fiber.

To enable the efficient application of RF-over-fiber techniques, the field of *microwave photonics* came into existence. Research in this field encompasses the study and applications of photonic devices operating at microwave frequencies. In addition to device developments, microwave photonics also addresses optical signal processing at microwave speeds and the design and implementation of RF photonic transmission systems. Section 9.6 gives a brief overview of microwave-photonic components and their uses.

9.1   Basic Elements of Analog Links

Figure 9.1 shows the basic elements of an analog link. The transmitter contains either an LED or a laser diode optical source. As noted in Sect. 4.5 and shown in Fig. 4.34, in analog applications one first sets a bias point on the source approximately at the midpoint of the linear optical output region. The analog signal can then be sent using one of several modulation techniques. The simplest form for optical fiber links is direct intensity modulation. This method uses an analog electrical signal to vary the drive current of a laser diode current around the bias point. Thereby the amplitude of the optical output from the source is in exact proportion to the message electrical signal level. Thus the information signal is transmitted directly in the baseband.

A somewhat more complex but often more efficient method is to superimpose the baseband signal onto an electrical subcarrier prior to intensity modulation of the optical source. This is done using standard amplitude modulation (AM), frequency modulation (FM), or phase modulation (PM) techniques [10]. No matter which

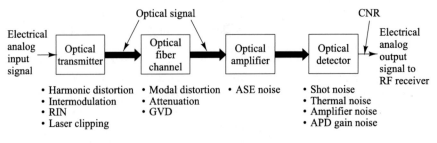

Fig. 9.1   Basic elements of an analog link and the major noise contributors

method is implemented, one must pay careful attention to signal impairments in the optical source. These include harmonic distortions, intermodulation products, relative intensity noise (RIN) in the laser, and laser clipping.

In addition, the frequency dependence of the amplitude, phase, and group delay in the optical fiber must be taken into account. Thus the fiber should have a flat amplitude and group-delay response within the passband required to send the signal free of linear distortion. Because modal-distortion-limited bandwidth is difficult to equalize, it is best to choose a single-mode fiber. The fiber attenuation is also important, because the carrier-to-noise performance of the system will change as a function of the received optical power.

The use of an optical amplifier in the link leads to additional noise, known as amplified spontaneous emission (ASE), which is described in Chap. 11. In the optical receiver, the principal impairments are shot noise, APD gain noise, and thermal noise.

9.2 Concept of Carrier-to-Noise Ratio

In analyzing the performance of analog systems, one usually calculates the ratio of root-mean-square (rms) carrier power to rms noise power at the input of the RF receiver following the photodetection process. This is known as the *carrier-to-noise ratio* (CNR). For digital data, consider the use of frequency-shift keying (FSK). In this modulation scheme, the amplitude of a sinusoidal carrier remains constant, but the phase shifts from one frequency to another to represent binary signals. For FSK, BERs of 10^{-9} and 10^{-15} translate into CNR values of 36 (15.6 dB) and 64 (18.0 dB), respectively. The analysis for analog signal quality is more complex compared to digital signals, because sometimes the signal quality depends on user perception of images, such as in viewing a television picture. A widely used analog signal is a 525-line studio-quality television signal. Using amplitude modulation (AM) for such a signal requires a CNR of 56 dB, because the need for bandwidth efficiency leads to a high signal-to-noise ratio. Frequency modulation (FM), on the other hand, only needs CNR values of 15–18 dB.

If CNRi represents the carrier-to-noise ratio related to a particular signal contaminant (e.g., shot noise), then for N signal-impairment factors the total CNR is given by Eq. (9.1).

$$\left(\frac{1}{CNR_{total}}\right)^{-1} = \sum_{i=1}^{N}\left(\frac{1}{CNR_i}\right)^{-1} \tag{9.1}$$

For links in which only a single information channel is transmitted, the important signal impairments include laser intensity noise fluctuations, laser clipping, photodetector noise, and optical amplifier noise. When multiple message channels operating at different carrier frequencies are sent simultaneously over the same fiber, then harmonic and intermodulation distortions arise. Furthermore, the inclusion of an optical amplifier gives rise to ASE noise. In principle, the three dominant factors that cause signal impairments in a fiber link are shot noise, optical amplifier noise, and laser clipping.

Most other degradation effects can be sufficiently reduced or eliminated.

This section examines a simple single-channel amplitude-modulated signal sent at baseband frequencies. Section 9.3 addresses multichannel systems in which intermodulation noise becomes important.

9.2.1 Carrier Power

To find the carrier power, first consider the signal generated at the transmitter. As shown in Fig. 9.2, the drive current through the optical source is the sum of the fixed bias current and a time-varying sinusoid. The source acts as a square-law device, so that the envelope of the output optical power $P(t)$ has the same form as the input drive current. If the time-varying analog drive signal is $s(t)$, then

$$P(t) = P_t[1 + ms(t)] \qquad (9.2)$$

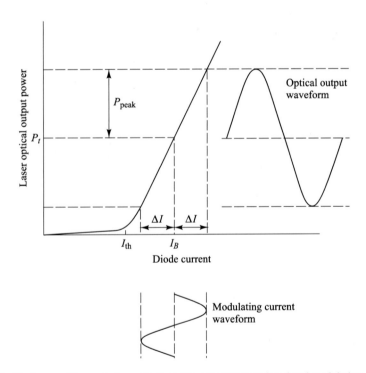

Fig. 9.2 Biasing conditions of a laser diode and its response to analog signal modulation

where P_t is the optical output power at the bias current level and the modulation index m is defined by Eq. (4.56). In terms of optical power, the modulation index is given by

$$m = \frac{P_{peak}}{P_t} \tag{9.3}$$

where P_{peak} and P_t are defined in Fig. 9.2. Typical values of m for analog applications range from 0.25 to 0.50.

For a sinusoidal received signal, the carrier power C at the output of the receiver (in units of A^2) is

$$C = \frac{1}{2}\left(m\,\mathscr{R}M\overline{P}\right)^2 \tag{9.4}$$

where $\mathscr{R}$ is the unity gain responsivity of the photodetector, M is the photodetector gain ($M = 1$ for *pin* photodiodes), and $\overline{P}$ is the average received optical power.

9.2.2 Photodetector and Preamplifier Noises

The expressions for the photodiode and preamplifier noises are given by Eqs. (6.14) and (6.15), respectively. These expressions then yield the following total mean-square photodetector noise current

$$\langle i_N^2 \rangle = \sigma_N^2 = 2q\left(i_p + i_D\right)M^2 F(M)B_e \tag{9.5}$$

Here, as defined in Chap. 6, $i_p = \mathscr{R}\overline{P}$ is the primary photocurrent, i_D is the detector dark current, M is the photodiode gain with $F(M)$ being its associated noise figure, and B_e is the receiver bandwidth. Then, the CNR for the photodetector only is $\mathrm{CNR}_{det} = C/\sigma_N^2$

Generalizing Eq. (6.15) for the preamplifier noise gives

$$\langle i_{th}^2 \rangle = \sigma_{th}^2 = \frac{4k_B T}{R_{eq}} B_e F_t \tag{9.6}$$

Here, R_{eq} is the equivalent resistance of the photodetector load and the preamplifier, and F_t is the noise factor of the preamplifier. Then, the CNR for the preamplifier only is $\mathrm{CNR}_{preamp} = C/\sigma_{th}^2$.

Drill Problem 9.1 A 20-km analog optical fiber link has a fiber attenuation of 1.0 dB/km. Assume the transmitting laser diode injects an average power of

0.1 mW into the fiber and the receiving *pin* photodiode receiver has a 0.6-A/W responsivity. (*a*) Show that for a modulation index $m = 0.5$, the carrier power at the receiver output is 1.12×10^{-10} A^2. (*b*) If the receiver is thermal noise limited, show that the CNR value is 2.40×10^3 if the amplifier equivalent load resistance is 50 Ω with a 1.5-dB noise figure at $T = 300°$K and the signal is measured in a 100 MHz bandwidth.

9.2.3   Effects of Relative Intensity Noise (RIN)

Within a semiconductor laser, fluctuations in the amplitude or intensity of the output produce *optical intensity noise*. These fluctuations could arise from temperature variations or from spontaneous emission contained in the laser output. The noise resulting from the random intensity fluctuations is called *relative intensity noise* (RIN), which may be defined in terms of the mean-square intensity variations. The resultant mean-square noise current is given by

$$\langle i_{RIN}^2 \rangle = \sigma_{RIN}^2 = \mathrm{RIN}(\mathscr{R}M\bar{P})^2 B_e \tag{9.7}$$

Then, the CNR due to laser amplitude fluctuations only is $\mathrm{CNR_{RIN}} = C/\sigma_{RIN}^2$. Here, the RIN, which is measured in dB/Hz, is defined by the noise-to-signal power ratio

$$\mathrm{RIN} = \frac{\langle (\Delta P_L)^2 \rangle}{\bar{P}_L^2} \tag{9.8}$$

where $\langle (\Delta P_L)^2 \rangle$ is the mean-square intensity fluctuation of the laser output and $\bar{P}_L$ is the average laser light output power. This noise decreases as the injection current level increases according to the relationship

$$\mathrm{RIN} \propto \left(\frac{I_B}{I_{th}} - 1 \right)^{-3} \tag{9.9}$$

where I_B is the bias current and I_{th} is the threshold current as shown in Fig. 9.2. Vendor data sheets for 1550-nm DFB lasers typically quote RIN values of -152 to -158 dB/Hz. Substituting the CNRs resulting from Eq. (9.4) through Eq. (9.7) into Eq. (9.1) yields the following carrier-to-noise ratio for a single-channel AM system:

$$\frac{C}{N} = \frac{\frac{1}{2}(m\mathscr{R}M\bar{P})^2}{RIN(\mathscr{R}M\bar{P})^2 B_e + 2q(i_p + i_D)M^2 F(M)B_e + (4k_B T/R_{eq})B_e F_t} \tag{9.10}$$

Drill Problem 9.2 Consider a single-mode InGaAsP laser that has RIN $= -155$ dB/Hz $= 3.16 \times 10^{-16}$ Hz^{-1}. Suppose the laser has an average optical output power of 2 mW. If the laser output is incident directly on a *pin* photodetector receiver that has a responsivity of 0.6 A/W and a 100-MHz bandwidth, (*a*) show that the rms value of the power fluctuation in a 100-MHz bandwidth is 3.16×10^{-7} W and (*b*) that the rms noise current due to the laser RIN is 2.13×10^{-7} A.

9.2.4 Limiting C/N Conditions

This section looks at some limiting conditions on C/N, which are illustrated in Fig. 9.3. When the optical power level at the receiver is low, the preamplifier circuit noise dominates the system noise. For this condition

$$\left(\frac{C}{N}\right)_{\text{limit 1}} = \frac{\frac{1}{2}(m\mathscr{R}M\bar{P})^2}{(4k_BT/R_{eq})B_eF_t} \tag{9.11}$$

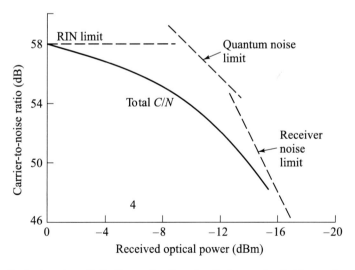

Fig. 9.3 Carrier-to-noise ratio limits as a function of optical power level at the receiver, showing that RIN dominates at high powers, quantum noise gives a 1-dB drop in C/N for each 1-dB power decrease at intermediate levels, and receiver thermal noise yields a 2-dB C/N roll-off per 1-dB drop in received power at low light levels

In this case, the carrier-to-noise ratio is directly proportional to the square of the received optical power, so that for each 1-dB variation in received optical power, C/N will change by 2 dB.

For well-designed photodiodes, the dark current is small compared with the shot (quantum) noise for intermediate optical signal levels at the receiver. Thus, at intermediate power levels the shot noise term of the photodiode will dominate the system noise. In this case,

$$\left(\frac{C}{N}\right)_{\text{limit 2}} = \frac{\frac{1}{2}m^2\mathscr{R}\bar{P}}{2qF(M)B_e} \tag{9.12}$$

so that the carrier-to-noise ratio will vary by 1-dB for every 1-dB change in the received optical power.

If the laser has a high RIN value so that the reflection noise dominates over other noise terms, then the carrier-to-noise ratio becomes

$$\left(\frac{C}{N}\right)_{\text{limit 3}} = \frac{\frac{1}{2}m^2}{\text{RIN}B_e} \tag{9.13}$$

which is a constant. In this case, the performance cannot be improved unless the modulation index is increased.

Example 9.1 As an example of the limiting conditions, consider a link with a laser transmitter and a *pin* photodiode receiver having the following characteristics:

Transmitter	Receiver
$m = 0.25$	$\mathscr{R} = 0.6$ A/W
RIN $= -143$ dB/Hz	$B_e = 10$ MHz
$P_c = 0$ dBm	$i_D = 10$ nA
	$R_{\text{eq}} = 750\ \Omega$
	$F_t = 3$ dB

where P_c is the optical power coupled into the fiber. To see the effects of the different noise terms on the carrier-to-noise ratio, Fig. 9.3 shows a plot of C/N as a function of the optical power level at the receiver. For high levels of received power the source noise dominates to give a constant C/N. At intermediate levels, the quantum noise (shot noise) is the main contributor, with a 1-dB drop in C/N for every 1-dB decrease in received optical power. For low light levels, the thermal noise of the receiver is the limiting noise term, yielding a 2-dB rolloff in C/N for each 1-dB drop in received optical power. It is important to note that the limiting factors can vary significantly depending on the transmitter and receiver characteristics. For example, for low-impedance amplifiers the thermal noise of the receiver can be the dominating performance limiter for all practical link lengths (see Problem 9.1).

9.3 Multichannel Amplitude Modulation

The initial widespread application of analog fiber optic links, which started in the late 1980s, was to CATV networks [11]. These coax-based television networks operate in a frequency range from 50 to 88 MHz and from 120 to 550 MHz. The band from 88 to 120 MHz is not used, because it is reserved for FM radio broadcast. The CATV networks can deliver over 80 amplitude-modulated vestigial-sideband (AM-VSB) video channels, each having a noise bandwidth of 4 MHz within a channel bandwidth of 6 MHz with signal-to-noise ratios exceeding 47 dB. To remain compatible with existing coax-based networks, a multichannel AM-VSB format is chosen for the fiber optic system.

Figure 9.4 depicts the technique for combining N independent messages. An information bearing signal on channel i amplitude modulates a carrier wave that has a frequency f_i, where $i = 1, 2, ..., N$. An RF power combiner then sums these N amplitude-modulated carriers to yield a composite frequency-division-multiplexed (FDM) signal that intensity modulates a laser diode. Following the optical receiver, a bank of parallel bandpass filters separates the combined carriers back into individual channels. The individual message signals are recovered from the carriers by standard RF techniques.

For a large number of FDM carriers with random phases, the carriers add on a power basis. Thus, for N channels the optical modulation index m is related to the per channel modulation index m_i by

$$m = \left(\sum_{i=1}^{N} m_i^2 \right)^{1/2} \tag{9.14}$$

If each channel modulation index m_i has the same value m_c, then

$$m = m_c N^{0.5} \tag{9.15}$$

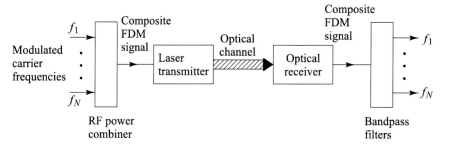

Fig. 9.4 Standard technique for frequency-division multiplexing of N independent information-bearing signals

As a result, when N signals are frequency multiplexed and used to modulate a single optical source, the carrier-to-noise ratio of a single channel is degraded by 10 log N. If only a few channels are combined, the signals will add in voltage rather than power, so that the degradation will have a 20 log N characteristic.

When multiple carrier frequencies pass through a nonlinear device such as a laser diode, signal products other than the original frequencies can be produced. These undesirable signals are called *intermodulation products*, and they can cause serious interference in both in-band and out-of-band channels. The result is a degradation of the transmitted signal. Among the intermodulation products, generally only the second-order terms and third-order terms are considered because higher-order products tend to be significantly smaller.

Third-order intermodulation (IM) distortion products at frequencies $f_i + f_j - f_k$ (which are known as *triple-beat IM products*) and $2f_i - f_j$ (which are known as *two-tone third-order IM products*) are the most dominant, because many of these fall within the bandwidth of a multichannel system. For example, a 50-channel CATV network operating over a standard frequency range of 55.25–373.25 MHz has 39 second-order IM products at 54.0 MHz and 786 third-order IM tones at 229.25 MHz. The amplitudes of the triple-beat products are 3-dB higher than the two-tone third-order IM products. In addition, because there are $N(N - 1)(N - 2)/2$ triple-beat terms compared with $N(N - 1)$ two-tone third-order terms, the triple- beat products tend to be the major source of IM noise.

If a signal passband contains a large number of equally spaced carriers, several IM terms will exist at or near the same frequency. This so-called *beat stacking* is additive on a power basis. For example, for N equally spaced equal-amplitude carriers, the number of third-order IM products that fall right on the rth carrier is given by [12]

$$D_{1,2} = \frac{1}{2}\left\{N - 2 - \frac{1}{2}[1 - (-1)^N](-1)^r\right\} \tag{9.16}$$

for two-tone terms of the type $2f_i - f_j$, and by

$$D_{1,1,1} = \frac{r}{2}(N - r + 1) + \frac{1}{4}\left\{(N - 3)^2 - 5 - \frac{1}{2}[1 - (-1)^N](-1)^{N+r}\right\} \tag{9.17}$$

for triple-beat terms of the type $f_i + f_j - f_k$.

Whereas the two-tone third-order terms are fairly evenly spread through the operating passband, the triple-beat products tend to be concentrated in the middle of the channel passband, so that the center carriers receive the most intermodulation interference. Tables 9.1 and 9.2 show the distributions of the third-order triple-beat and two-tone IM products for the number of channels N ranging from 1 to 8.

The results of beat stacking are referred to as *composite second order* (CSO) and *composite triple beat* (CTB) and they describe the performance of multichannel AM links. The word *composite* means that the overall distortion is due to a collection of discrete distortions. CSO and CTB are defined as [1, 12, 13]

Table 9.1 Distribution of the number of third-order triple-beat intermodulation products for the number of channels N ranging from 1 to 8

N	r							
	1	2	3	4	5	6	7	8
1	0							
2	0	0						
3	0	1	0					
4	1	2	2	1				
5	2	4	4	4	2			
6	4	6	7	7	6	4		
7	6	9	10	11	10	9	6	
8	9	12	14	15	15	14	12	9

Table 9.2 Distribution of the number of third-order two-tone intermodulation products for the number of channels N ranging from 1 to 8

N	r							
	1	2	3	4	5	6	7	8
1	0							
2	0	0						
3	1	0	1					
4	1	1	1	1				
5	2	1	2	1	2			
6	2	2	2	2	2	2		
7	3	2	3	2	3	2	3	
8	3	3	3	3	3	3	3	3

$$CSO = \frac{\text{peak carrier power}}{\text{peak power in composite 2nd-order IM tone}} \qquad (9.18)$$

and

$$CTB = \frac{\text{peak carrier power}}{\text{peak power in composite 3rd-order IM tone}} \qquad (9.19)$$

9.4 Spurious-Free Dynamic Range

The dynamic range of an analog link is defined in relation to two-tone third-order intermodulation frequencies. First consider two large equal-power signals at fundamental frequencies f_1 and f_2, as shown in Fig. 9.5. These two signals will produce second-order modulation products at $2f_1$, $2f_2$, and $f_1 \pm f_2$, and third-order intermodulation products at frequencies $2f_1 \pm f_2$ and $2f_2 \pm f_1$. The second-order terms normally fall outside of the passband of a system, so they can be ignored. However, the third-order products are of concern because they could fall on a signal frequency within the system bandwidth and cannot be removed by a simple filtering technique. To determine the system operating requirements, consider the case shown in Fig. 9.5. Here a third-order intermodulation product (dashed line) resulting from the two strongest fundamental carriers falls at the frequency where the weakest channel operates. The parameter ΔP is the power difference between the strongest and weakest channels, and CNR_{min} is the minimum required carrier-to-noise ratio for the weakest signal. For the case shown in Fig. 9.5, the intermodulation products resulting from the strongest equal-power fundamental carriers are equal to the noise floor.

For standard analog links the third-order intermodulation distortion (designated by IMD3) varies as the cube of the RF input power. Figure 9.6 shows this relationship and also shows the linear relationship of the output power of the fundamental carriers as a function of the RF input power. The *spurious-free dynamic range* (SFDR) is

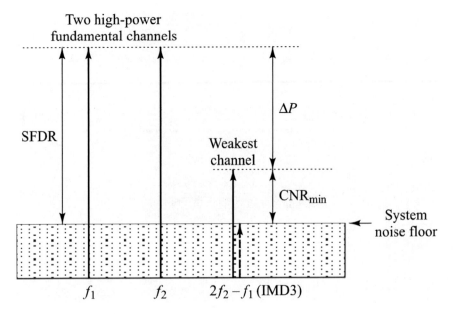

Fig. 9.5 The relationship of third-order intermodulation products (dashed line) to system operating requirements

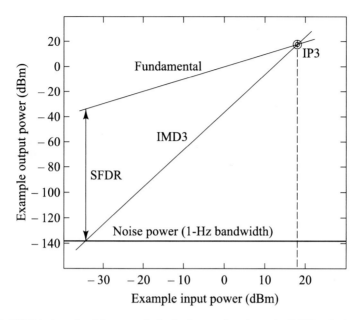

Fig. 9.6 SFDR is the ratio of the power in the fundamental carrier to the IMD3 at the input power level where the IMD3 where the IMD3 power equals the noise level

defined as the ratio between the powers in the fundamental carrier and the third-order intermodulation at that power level where the IMD3 is equal to the noise floor. This means that the SFDR is the usable dynamic range before spurious noise interferes with or distorts the fundamental signal. Thus, referring back to Fig. 9.5, the SFDR must be larger than $CNR_{min} + \Delta P$.

In Fig. 9.6 the point IP3 designates the input power at which the IMD3 is equal to the output carrier power. From the curves in Fig. 9.6 the SFDR is given by

$$SFDR = \frac{2}{3} 10 \log \frac{IP3}{N_{out} R_{load}} \qquad (9.20)$$

Here N_{out} is the total output noise power and R_{load} is the detector load resistance. The SFDR is measured in units of $dB \cdot Hz^{2/3}$. Many different measurements of SFDR that have been reported in the literature are summarized in Ref. [14]. The general trends of these measurement values as a function of frequency are that directly modulated microwave links can have a large SFDR (up to 125 $dB \cdot Hz^{2/3}$ at 1 GHz), but the SFDR decreases significantly as the frequency increases beyond about 1 GHz. This is due to inherent distortion effects in the laser, which get worse as the operating frequency gets closer to the relaxation-oscillation peak (see Fig. 4.27). The SFDR for externally modulated links is not as high below 1 GHz as for direct modulation setups, but it remains at the upper level out to higher frequencies. For example,

links using a Mach–Zehnder interferometer-based modulator are able to maintain a
112-dB · Hz$^{2/3}$ SFDR out to about 17 GHz.

Drill Problem 9.4 Consider an analog link that has equal signal powers at
frequencies 3.000 and 3.001 GHz. Show that (*a*) second-order harmonics occur
at 6.000 and 6.002 GHz; (*b*) second-order intermodulation spurs occur at 0.001
and 6.001 GHz; and (*c*) third-order intermodulation spurs occur at 2.999 and
3.002 GHz.

9.5   Radio-Over-Fiber Links

The transition in the use of wireless devices from pure voice communications to a
wide selection of broadband services created much interest in developing radio-over-
fiber (ROF) links [2–9]. The convergence of optical and wireless access networks is
driven by the need to have seamless connectivity between an access network and both
stationary and fast-moving mobile users at data rates of at least 2.5 Gb/s. The implementations of ROF links include the interconnection of antenna base stations with a
central controlling office in a wireless access network, access to wireless services for
indoor environments (e.g., large office buildings, airport departure lounges, hospitals, conference centers, hospitals, and hotel rooms), and connections to personal
area networks in homes. Services of interest to mobile users include broadband
Internet access, fast peer-to-peer file transfers, high-definition video, and online
multiparty gaming. Depending on the network type, the optical links can use single-
mode fibers, 50- or 62.5-μm core-diameter multimode glass fibers, or large-core
multimode polymer optical fibers.

One application of radio-over-fiber technology is in broadband wireless access
networks for interconnecting antenna base stations (BSs) with the central controlling
office. Figure 9.7 shows the basic network architecture for such a scheme. Here a
number of antenna base stations provide wireless connectivity to subscribers by
means of millimeter-wave frequencies. Subscribers are located up to 1 km from a
local base station. The transmission range around a BS is called a *microcell* (diameter
less than 1 km) or a *picocell* or *hotspot* (radii ranging from 5 to 50 m). The BSs
are connected to a microcell control station (CS) in the central office, which is
responsible for functions such as RF modulation and demodulation, channel control,
and switching and routing of customer calls.

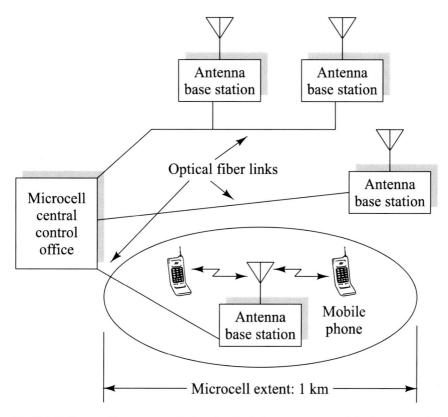

Fig. 9.7 Radio-over-fiber concept of a broadband wireless access network for interconnecting antenna base stations with the central controlling office

9.6 Microwave Photonics

The field of microwave photonics encompasses the study and applications of photonic devices operating at microwave frequencies. The key components being developed and applied include the following:

- High-frequency low-loss external optical modulators that have linear transfer functions and can withstand continuous-wave optical powers up to 60 mW
- Optical sources with high slope efficiencies and low RIN that can be modulated at tens of GHz
- High-speed photodiodes and optical receivers that can respond to signal frequencies of 20–60 GHz
- Microwave photonic filters that perform the same tasks as standard RF filters.

In addition to device developments, microwave photonics also addresses optical signal processing at microwave speeds and the design and implementation of RF photonic transmission systems. For example, applications of photonic signal processing at multiple-gigahertz sampling frequencies include signal filtering, analog-to-digital conversion, frequency conversion and mixings, signal correlation, generation of arbitrary waveforms, and beam-forming methodologies for phased array radars.

9.7 Summary

Although digital transmission techniques have been used most widely in optical fiber communication links, sometimes it is more advantageous to transmit information in its original analog form instead of first converting it to a digital format. In fact, as a result of the emerging use of broadband wireless communication devices, schemes have been investigated and implemented for using analog optical fiber links for distributing broadband microwave-frequency signals in a variety of applications.

The transition in the use of wireless devices from pure voice communications to a wide selection of broadband services created much interest in developing radio-over-fiber (ROF) links. The convergence of optical and wireless access networks is driven by the need to have seamless connectivity between an access network and both stationary and fast-moving mobile users at data rates of at least 2.5 Gb/s. The implementations of ROF links include the interconnection of antenna base stations with a central controlling office in a wireless access network, access to wireless services for indoor environments (e.g., large office buildings, airport departure lounges, hospitals, conference centers, hospitals, and hotel rooms), and connections to personal area networks in homes. Services of interest to mobile users include broadband Internet access, fast peer-to-peer file transfers, high-definition video, and online multiparty gaming. Depending on the network type, the optical links can use single-mode fibers, 50- or 62.5-μm core-diameter multimode glass fibers, or large-core multimode polymer optical fibers.

Problems

9.1 Commercially available wideband receivers have equivalent resistances $R_{eq} = 75\ \Omega$. With this value of R_{eq} and letting the remaining transmitter and receiver parameters be the same as in Example 9.1, plot the total carrier- to-noise ratio and its limiting expressions, as given by Eq. (9.10) through Eq. (9.13), for received power levels ranging from 0 to -16 dBm. Show that the thermal noise of the receiver dominates over the quantum noise at all power levels when $R_{eq} = 75\ \Omega$.

9.2 Consider a five-channel frequency-division-multiplexed (FDM) system having carriers at $f_1, f_2 = f_1 + \Delta, f_3 = f_1 + 2\Delta, f_4 = f_1 + 3\Delta$, and $f_5 = f_1 + 4\Delta$, where Δ is the spacing between carriers. On a frequency plot, show the number and location of the triple-beat and two-tone third-order intermodulation products.

9.3 Suppose an engineer wants to frequency-division multiplex
60 FM signals. If 30 of these signals have a per-channel modulation index m_i

$= 3\%$ and the other 30 signals have $m_i = 4\%$, find the optical modulation index of the laser.

9.4 Consider an analog system having 120 channels, each modulated at 2.3%. The link consists of 12 km of single-mode fiber having a loss of 1 dB/km, plus a connector having a 0.5-dB loss on each end. The laser source couples 2 mW of optical power into the fiber and has RIN $= -135$ dB/Hz. The *pin* photodiode receiver has a responsivity of 0.6 A/W, $B_e = 5$ GHz, $i_D = 10$ nA, $R_{eq} = 50$ Ω, and $F_t = 3$ dB. Find the carrier-to-noise ratio for this system.

9.5 What is the carrier-to-noise ratio for the system described in Prob. 9.4 if the *pin* photodiode is replaced with an InGaAs avalanche photodiode having $M = 10$ and $F(M) = M^{0.7}$?

9.6 Consider a 32-channel FDM system with a 4.4% modulation index per channel. Let RIN $= -135$ dB/Hz, and assume the *pin* photodiode receiver has a responsivity of 0.6A/W, $B_e = 5$ GHz, $i_D = 10$ nA, $R_{eq} = 50$ Ω, and $F_t = 3$ dB.

(a) Find the carrier-to-noise ratio for this link if the received optical power is -10 dBm.

(b) Find the carrier-to-noise ratio if the modulation index is increased to 7% per channel and the received optical power is decreased to -13 dBm.

Answers

9.2

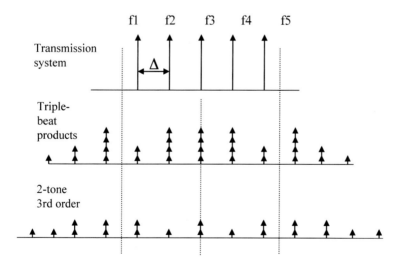

9.3 From Eq. (9.14), the total optical modulation index is $0.274 = 27.4\%$.

9.4 From Eq. (9.14), the total optical modulation index is $0.25 = 25.0\%$.
The received power is -10 dBm $= 100$ μW.
The carrier power is $0.5(15 \times 10^{-6}$ A$)^2$.

The source noise is, with RIN $= -135$ dB/Hz $= 3.162 \times 10^{-14}$/Hz,

$$\langle i_s^2 \rangle = \text{RIN}(\mathscr{R}\text{P})^2 B_e = 5.69 \times 10^{-13} \text{A}^2$$

The shot noise is

$$\langle i_{\text{shot}}^2 \rangle = 2q(\mathscr{R}\text{P} + i_\text{D})B_e = 9.5 \times 10^{-14} \text{A}^2$$

The thermal noise is

$$\langle i_{th}^2 \rangle = \frac{4K_\text{B}\text{T}}{R_{\text{eq}}} F_e = 8.25 \times 10^{-13} \text{A}^2$$

Thus the carrier-to-noise ratio is C/N $= 75.6$
or in dB: C/N $= 10 \log 75.6 = 18.8$ dB.

9.5   When an APD is used, the carrier power and the shot noise change.
The carrier power is $0.5 (15 \times 10^{-5} \text{ A}^2)$.
The shot noise is

$$\langle i_{\text{shot}}^2 \rangle = 2q(\mathscr{R}P + i_\text{D})M^2 F(M)B_e = 4.76 \times 10^{-10} \text{ A}^2$$

Thus the carrier-to-noise ratio is C/N $= 236.3$ or in dB, C/N $= 23.7$ dB.

9.6   (a) The modulation index is $m = \left[\displaystyle\sum_{i=1}^{32} (0.044)^2 \right]^{1/2} = 0.25$.

The received power is -10 dBm $= 100$ μW.
The rest of the answers for part (a) are the same as in Problem 9.4.

(b) The modulation index is $m = \left[\displaystyle\sum_{i=1}^{32} (0.07)^2 \right]^{1/2} = 0.396$.

The received power is -13 dBm $= 50$ μW.
The carrier power is $0.5(1.19 \times 10^{-5} \text{ A})^2 = 7.06 \times 10^{-11} \text{ A}^2$.
The source noise is, with RIN $= -135$ dB/Hz $= 3.162 \times 10^{-14}$/Hz.

$$\langle i_s^2 \rangle = \text{RIN}(\mathscr{R}\text{P})^2 B_e = 1.42 \times 10^{-13} \text{A}^2$$

The shot noise is

$$\langle i_{\text{shot}}^2 \rangle = 2q(\mathscr{R}\text{P} + i_\text{D})B_e = 4.8 \times 10^{-14} \text{A}^2$$

The thermal noise is

$$\langle i_{th}^2 \rangle = \frac{4\text{kBT}}{R_{\text{eq}}} F_e = 8.25 \times 10^{-13} \text{ A}^2$$

Thus the carrier-to-noise ratio is C/N $= 69.6$
or in dB: C/N $= 10 \log 69.6 = 18.4$ dB.

References

1. A. Brillant, *Digital and Analog Fiber Optic Communication for CATV and FTTx Applications* (Wiley, Hoboken, 2008)
2. T. Nagatsuma, G. Ducoumau, C.C. Renaud, Advances in terahertz communications accelerated by photonics. Nat. Photonics **10**(6), 371–379 (2016)
3. J. Yao, Microwave Photonics (Tutorial). J. Lightwave Technol. **27**(3), 314–335 (2009)
4. K.Y. Lau, RF transport over optical fiber in urban wireless infrastructures. IEEE J. Opt. Commun. Netw. **4**(4), 326–335 (2012)
5. J. Capmany, G. Li, C. Lim, and J. Yao, Microwave photonics: current challenges towards widespread application. Opt. Expr. **21**(19), 22862–22867 (2013)
6. D. Novak, R.B. Waterhouse, A. Nirmalathus, C. Lim, P.A. Gamage, T.R. Clark, Jr., M.L. Dennis, J.A. Nanzer, Radio-over-fiber technologies for emerging wireless systems. IEEE J. Quantum Electr. **52** (2016)
7. W. Zhang, J. Yao, Silicon-based integrated microwave photonics (review). IEEE J. Quantum Electr. **52** (2016)
8. C. Ranaweera, E. Wong, A. Nirmalathas, C. Jayasundara, C. Lim, 5G C-RAN with optical fronthaul: an analysis from a deployment perspective. J. Lightwave Technol. **36**(11), 2059–2068 (2017)
9. R. Katti, S. Prince, A survey on role of photonic technologies in 5G communication systems. Photon Netw. Commun. **38**, 185–205 (2019)
10. R.E. Ziemer, W.H. Tranter, *Principles of Communications: Systems, Modulation, and Noise*, 7th edn.(Wiley, Hoboken, 2014)
11. H.H. Lu, W.S. Tsai, A hybrid CATV/256-QAM/OC-48 DWDM system over an 80-km LEAF transport. IEEE Trans. Broadcasting **49**(1), 97–102 (2003)
12. C. Cox, *Analog Optical Links* (Cambridge University Press, 2004 (print version), 2009 (online version))
13. D. Large, J. Farmer, *Modern Cable Television Technology* (Elsevier, Amsterdam, 2004)
14. C.H. Cox, E.I. Ackerman, G.E. Betts, J.L. Prince, Limits on the performance of RF-over-fiber links and their impact on device design. IEEE Microwave Theory Tech. **54**, 906–920 (2006)

Chapter 10
Wavelength Division Multiplexing (WDM)

Abstract Wavelength division multiplexing or WDM allows the combining of a number of independent information-carrying wavelengths onto the same fiber, because of the wide spectral region in which optical signals can be transmitted efficiently. This chapter addresses the operating principles of WDM, describes a wide array of optical sources and passive components needed to implement WDM links, examines the functions of a generic WDM link, and discusses the internationally standardized spectral grids that designate independent channels for wavelength multiplexing schemes.

A distinctive operational feature of an optical fiber is that it has a wide spectral region in which optical signals can be transmitted efficiently. For full-spectrum fibers used over long distances this region includes an over 400-nm spectrum in the O-band through the L-band, whereas the 260-nm spectrum in the T-band can be employed for shorter links such as in data centers. The light sources used in high-capacity optical fiber communication systems emit in a narrow wavelength band of less than 1 nm, so many different independent optical channels can be used simultaneously in different segments of a desired wavelength range. The technology of combining a number of such independent information-carrying wavelengths onto the same fiber is known as *wavelength division multiplexing* or WDM [1–6].

Section 10.1 addresses the operating principles of WDM, examines the functions of a generic WDM link, and discusses the internationally standardized spectral grids that designate independent channels for wavelength multiplexing schemes. Sections 10.2 through 10.6 describe various categories of passive optical components that are needed to insert separate wavelengths into a fiber at the transmitting end and separate them into individual channels at the destination. An important factor in deploying these components in a WDM system is to avoid inter-channel interference by ensuring that optical signal power from one channel does not drift into the spectral territory occupied by adjacent channels.

The original version of this chapter was revised: All the incorrect sentences have been corrected. The correction to this chapter can be found at https://doi.org/10.1007/978-981-33-4665-9_15

G. Keiser, *Fiber Optic Communications*,
https://doi.org/10.1007/978-981-33-4665-9_10

Applications of WDM techniques are found in all levels of communication links including long-distance terrestrial and undersea transmission systems, metro networks, data center links, and fiber-to-the-premises (FTTP) networks. Chapter 13 shows how WDM methodologies apply to several categories of communication networks.

10.1   Concepts of WDM

The original optical fiber links that were deployed around 1980 consisted of simple point-to-point connections. These links contained a single fiber with one light source at the transmitting end and one photodetector at the receiving end. In these early systems, signals from different light sources used separate and uniquely assigned optical fibers. Because the spectral width of a typical laser source occupies only a narrow slice of optical bandwidth, these simplex systems greatly underutilize the large bandwidth capacity of a fiber. The first use of WDM was to upgrade the capacity of installed point-to-point transmission links. This was achieved with wavelengths that were separated from several tens up to 200 nm in order not to impose strict wavelength-tolerance requirements on the different laser sources and the wavelength-separating components at the receiving end.

With the advent of high-quality light sources with extremely narrow spectral emission widths (less than 1 nm), many independent wavelength channels spaced less than a nanometer apart could be placed simultaneously on the same fiber. Thus the use of WDM allows a dramatic increase in the capacity of an optical fiber compared to the original simple point-to-point link that carried only a single wavelength. For example, if each wavelength supports an independent transmission rate of 10 Gb/s, then each additional channel provides the fiber with significantly more capacity. Another advantage of WDM is that the various optical channels can support different transmission formats. Thus, by using separate wavelengths, differently formatted signals at any data rate can be sent simultaneously and independently over the same fiber without the need for a common signal structure.

10.1.1   WDM Operational Principles

A characteristic of WDM is that the discrete wavelengths form an *orthogonal set* of carriers that can be separated, routed, and switched without interfering with each other. This isolation between channels holds as long as the total optical power intensity is kept sufficiently low to prevent nonlinear effects such as stimulated Brillouin scattering and four-wave mixing processes from degrading the link performance (see Chap. 12).

The implementation of sophisticated WDM networks requires a variety of *passive* and *active* devices to combine, distribute, isolate, and amplify optical power at different wavelengths. *Passive devices* require no external control for their operation,

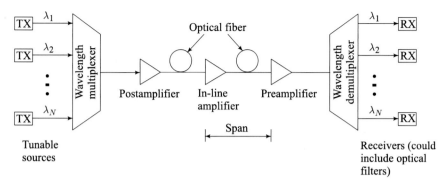

Fig. 10.1 Implementation of a typical WDM network containing various types of optical amplifiers

so they are somewhat limited in their application flexibility. These components are mainly used to split and combine or tap off optical signals. The wavelength-dependent performance of *active devices* can be controlled electronically or optically, thereby providing a large degree of network flexibility. As later chapters discuss, active WDM components include optical amplifiers, wavelength switches, and optical wavelength converters.

Figure 10.1 shows the implementation of passive and active components in a typical WDM link containing various types of optical amplifiers (see Chap. 11). At the transmitting end there are several independently modulated light sources, each emitting signals at a unique wavelength. Here a *wavelength multiplexer* is needed to combine these optical outputs into a continuous spectrum of signals and couple them onto a single fiber. At the receiving end a *wavelength demultiplexer* is required to separate the optical signals into appropriate detection channels at different wavelengths for signal processing. Examples of active devices within the link are various types of optical amplifiers used to compensate for power losses along the transmission path (see Chap. 11).

As Fig. 10.2 shows, there are many independent operating regions across the spectrum ranging from the O-band through the L-band in which narrow-linewidth optical sources can be used simultaneously. These regions can be viewed either in terms of *spectral width* (the wavelength band occupied by the light signal) or by means of *optical bandwidth* (the frequency band occupied by the light signal). To find the optical bandwidth corresponding to a particular spectral width in these regions, consider the fundamental relationship $c = \lambda \nu$, which relates the wavelength λ to the carrier frequency ν, where c is the speed of light. Differentiating this equation yields for $\Delta \lambda \ll \lambda^2$

$$|\Delta \nu| = \frac{c}{\lambda^2} |\Delta \lambda| \qquad (10.1)$$

where the frequency deviation $\Delta \nu$ corresponds to the wavelength deviation $\Delta \lambda$ around λ.

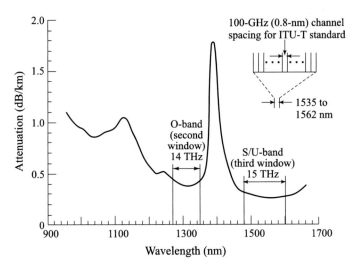

Fig. 10.2 The transmission-band widths in the O-band and C-band (the 1310-nm and 1550-nm windows) allow the use of many simultaneous channels for sources with narrow spectral widths, such as the ITU-T standard 100-GHz channel spacing for WDM

Example 10.1 Consider a fiber that has the attenuation characteristic shown in Fig. 10.2. What are the usable spectral bands (*a*) in the O-band centered at 1420 nm; (*b*) in the combined S-band and C-band with a 1520-nm center wavelength?

Solution: (*a*) From Eq. (10.1) the optical bandwidth is $\Delta v = 14$ THz for a usable spectral band $\Delta \lambda = 100$ nm covering the O-band with a 1420-nm center wavelength.

(*b*) Similarly, $\Delta v = 14$ THz for a usable spectral band $\Delta \lambda = 105$ nm in the low-loss region covering the S-band and C-band with a 1520-nm center wavelength.

The operational frequency band allocated to a particular light source normally ranges from 25 to 100 GHz (or equivalently, a spectral band of 0.25–0.8 nm at a 1550-nm wavelength). The exact width of the frequency or spectral band that is selected needs to take into account possible drifts in the peak wavelength emitted by the laser and temporal variations in the wavelength response of other link components. These parameter changes can result from effects such as *component aging* or *temperature variations.*

Depending on the frequency bands chosen for the optical transmission link, many operational regions are available in the various spectral bands. The engineering challenge for using such a large number of light sources, each of which is emitting at a different wavelength, is to ensure that each source is spaced sufficiently far from its neighbors so as not to create interference between them. This means that highly stabilized optical transmitters are needed so that the integrities of the independent message streams from each source are maintained for subsequent conversion back to electrical signals at the receiving end. For example, a designated source might be specified to have a wavelength of 1557.363 ± 0.005 nm (or equivalently, 192.50 THz).

Example 10.2 Consider a spectral band of 0.8 nm (or equivalently, a mean frequency spacing of 100 GHz at a 1550-nm wavelength) within which lasers with narrow linewidths are transmitting. How many of such signal channels fit into (a) the C band, and (b) the combined S-band and C-band?

Solution: (a) Because the C-band ranges from 1530 to 1565 nm, one can have $N = (35 \text{ nm})/(0.8 \text{ nm per channel}) = 43$ independent signal channels.

(b) Because the combined S-band and C-band cover the 1460- to-1565 nm range, one can have $N = (105 \text{ nm})/(0.8 \text{ nm per channel}) = 131$ independent signal channels.

Example 10.3 Assume that a 16-channel WDM system has a uniform channel spacing $\Delta v = 200$ GHz and let the frequency v_n correspond to the wavelength λ_n. Let the wavelength $\lambda_1 = 1550$ nm. Calculate the wavelength spacing between the first two channels (channels 1 and 2) and between the last two channels (channels 15 and 16). From the result, what can be concluded about using an equal-wavelength spacing definition in this wavelength band instead of the standard equal-frequency channel spacing specification?

Solution: To find the wavelength spacing between the first two channels (channels 1 and 2), first look at the frequency difference between these channels. The frequency for channel N is given by $v_N = v_1 + (N - 1)\Delta v$, where $\Delta v = 200$ GHz. Therefore,
$v_1 = c/\lambda_1 = (3 \times 10^8 \text{ m/s})/(1550 \text{ nm}) = 193.5483$ THz yields.
$\lambda_2 = c/v_2 = c/[v_1 + (2 - 1)\Delta v] = (3 \times 10^8 \text{ m/s})/[193.5483 + 0.2 \text{ THz}] = 1548.40$ nm. Thus between λ_1 and λ_2 the channel spacing is $\Delta \lambda = 1.60$ nm. Similarly.
$\lambda_{15} = c/v_{15} = c/[v_1 + (15 - 1)\Delta v] = (3 \times 10^8 \text{ m/s})/[193.5483 + 2.8] = 1527.90$ nm. and
$\lambda_{16} = c/v_{16} = c/[v_1 + (16 - 1)\Delta v] = (3 \times 10^8 \text{ m/s})/[193.5483 + 3.0] = 1526.34$ nm. Thus between λ_{15} and λ_{16} the channel spacing is $\Delta \lambda = 1.55$ nm.

This shows that the spacing between different adjacent wavelengths is not uniform. Because the channel spacing difference at the wavelength extremes in this case is small (about 3%), an equal-wavelength approximation is close to an equal-frequency spacing definition.

10.1.2   Standards for WDM

Because WDM is essentially frequency division multiplexing at optical carrier frequencies, the WDM standards developed by the International Telecommunication Union (ITU) specify channel spacing in terms of frequency [7–10]. A key reason for selecting a frequency spacing that is fixed, rather than a wavelength spacing that is constant, is that when locking a laser to a particular operating mode, it is the frequency of the laser that is fixed. Recommendation G.692 was the first ITU-T specification for WDM. This document specifies selecting the channels from a grid of frequencies referenced to 193.100 THz (1552.524 nm) and spacing them 100 GHz (about 0.8 nm at 1550 nm) apart. Suggested alternative spacings in G.692 include 50 GHz and 200 GHz, which correspond to spectral widths of 0.4 and 1.6 nm, respectively, at 1550 nm.

Historically the term *dense WDM* (DWDM) generally referred to small wavelength separations. The ITU-T Recommendation G.694.1 is aimed specifically at DWDM. This document specifies WDM operation in the S-, C-, and L-bands for high-quality, high-rate metro area network (MAN) and wide area network (WAN) services. It calls for narrow frequency spacings of 100–12.5 GHz (or, equivalently, 0.8 to 0.1 nm at 1550 nm). This implementation requires the use of stable, high-quality, temperature-controlled and wavelength-controlled (frequency-locked) laser diode light sources. For example, the wavelength-drift tolerances for 25-GHz channels are ±0.02 nm.

Table 10.1 lists part of the ITU-T G.694.1 dense WDM frequency grid for 100-GHz and 50-GHz spacings in the L- and C-bands. The column labeled "50-GHz offset" means that for the 50-GHz grid, one interleaves these 50-GHz values with the 100-GHz spacings. For example, the 50-GHz channels in the L-band would be at 186.00, 186.05, 186.10 THz, and so on. Note that when the frequency spacing is uniform, the wavelengths are not spaced uniformly because of the relationship given in Eq. (10.1).

To designate which C-band channel is under consideration in 100-GHz applications, the ITU-T uses a *channel numbering convention*. For this, the frequency 19 N.M THz is designated as ITU channel number NM. For example, the frequency 194.3 THz is ITU channel 43.

Table 10.1 Portion of the ITU-T G.694.1 dense WDM grid for 100- and 50-GHz spacings in the L- and C-bands

L-band				C-band			
100-GHz		50-GHz offset		100-GHz		50-GHz offset	
THz	nm	THz	nm	THz	nm	THz	nm
186.00	1611.79	186.05	1611.35	191.00	1569.59	191.05	1569.18
186.10	1610.92	186.15	1610.49	191.10	1568.77	191.15	1568.36
186.20	1610.06	186.25	1609.62	191.20	1576.95	191.25	1567.54
186.30	1609.19	186.35	1608.76	191.30	1567.13	191.35	1566.72
186.40	1608.33	186.45	1607.90	191.40	1566.31	191.45	1565.90
186.50	1607.47	186.55	1607.04	191.50	1565.50	191.55	1565.09
186.60	1606.60	186.65	1606.17	191.60	1564.68	191.65	1564.27
186.70	1605.74	186.75	1605.31	191.70	1563.86	191.75	1563.45
186.80	1604.88	186.85	1604.46	191.80	1563.05	191.85	1562.64
186.90	1604.03	186.95	1603.60	191.90	1562.23	191.95	1561.83

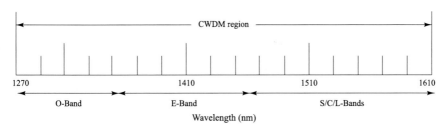

Fig. 10.3 Spectral grid for coarse wavelength-division multiplexing (CWDM)

The concept of *coarse WDM* (CWDM) emerged from the combination of the production of full-spectrum (low-water-content) G.652C and G.652D fibers, the development of relatively inexpensive optical sources, and the desire to have low-cost optical links operating in access networks and local area networks. The ITU-T Recommendation G.694.2 defines the spectral grid for CWDM. As shown in Fig. 10.3, the CWDM grid is made up of 18 wavelengths defined within the range 1270–1610 nm (O-band through L-band) spaced by 20 nm with wavelength-drift tolerances of ±2 nm. This can be achieved with inexpensive light sources that are not temperature controlled.

The ITU-T Recommendation G.695 outlines optical interface specifications for multiple-channel CWDM over distances of 40 and 80 km. Both unidirectional and bidirectional systems (such as used in passive optical network applications) are included in the recommendation. The applications for G.695 cover all or part of the 1270-to-1610-nm range. The main deployments are for single-mode fibers, such as those specified in ITU-T Recommendations G.652 and G.655.

10.2 Passive Optical Couplers

Passive devices operate completely in the optical domain to split and combine light streams. They include $N \times N$ couplers (with $N \geq 2$), power splitters, power taps, and star couplers. These components can be fabricated either from optical fibers or by means of planar optical waveguides using material such as lithium niobate (LiNbO$_3$), InP, silica, silicon oxynitride, or various polymers.

Most passive WDM devices are variations of a star-coupler concept. Figure 10.4 shows a generic star coupler, which can perform both power combining and power splitting. In the broadest application, star couplers combine the light streams from two or more input fibers at the transmitting end and divide them among several output fibers at the destination. In the general case the splitting is done uniformly for all wavelengths, so that each of the N outputs receives $1/N$ of the power entering the device, as indicated in Fig. 10.4. A common fabrication method for an $N \times N$ splitter is to fuse together N single-mode fibers that have thinned core-cladding regions over a length W of a few millimeters. The optical power inserted through one of the N fiber

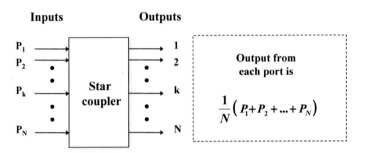

Fig. 10.4 Basic star coupler concept for combining or splitting optical powers

entrance ports gets divided uniformly into the cores of the N output fibers by means of evanescent power coupling through the thinned claddings in the fused region (see Sect. 10.2.1).

In principle, any size star coupler can be made, provided that all fibers can be heated uniformly during the coupler fabrication process. Couplers with 64 inputs and outputs are possible, although more commonly the size tends to be less than 10. One simple device is a power tap. Taps are nonuniform 2×2 couplers used to extract a small portion of optical power from a fiber line for monitoring signal quality.

The three fundamental technologies for making passive components are based on optical fibers, integrated optical waveguides, and bulk micro-optics. The next sections describe the physical principles of several simple examples of fiber-based and integrated-optic devices to illustrate the fundamental operating principles. Couplers using micro-optic designs are not widely used because the strict tolerances required in the fabrication and alignment processes affect their cost, performance, and robustness.

10.2.1   The 2 × 2 Fiber Coupler

When discussing couplers and splitters, it is customary to refer to them in terms of the number of input and output ports on the device. For example, a device with two inputs and two outputs would be called a "2×2 coupler." In general, an $N \times M$ coupler has N inputs and M outputs.

The 2×2 coupler [11–13] is a simple fundamental device that can be used to demonstrate the operational principles. A common construction is the fused-fiber coupler. The coupler is fabricated by twisting together, melting, and pulling two single-mode fibers so they get fused together over a uniform section of length W, as shown in Fig. 10.5. Each input and output fiber has a long tapered section of length L, because the transverse dimensions are gradually reduced down to that of the coupling region when the fibers are pulled during the fusion process. The total draw length is $L_{draw} = 2L + W$. This device is known as a *fused biconical tapered*

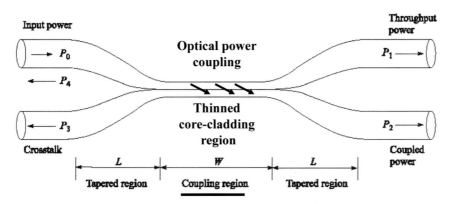

Fig. 10.5 Cross-sectional view of a fused-fiber coupler having a coupling region W and two tapered regions of length L, which gives a 2L + W coupler draw length

coupler. In Fig. 10.5, the parameter P_0 is the input power, P_1 is the throughput power, and P_2 is the power coupled into the second fiber. The parameters P_3 and P_4 are extremely low signal levels (-50 to -70 dB below the input level) that result from backward reflections and from scattering due to bending in and packaging of the device, respectively.

As the input light P_0 propagates along the taper in fiber 1 and into the coupling region W, there is a significant decrease in the V number owing to the reduction in the ratio r/λ [see Eq. (2.27)]. Here r is the reduced fiber radius and λ is the wavelength of the input optical power. Consequently, as the optical signal enters the coupling region, an increasingly larger portion of the input field now propagates outside the core of the fiber. This portion of the field thus penetrates through the thinned cladding regions and gets coupled into the adjacent fiber, as indicated by the heavy arrows. Depending on the dimensioning of the coupling region, any desired fraction of this decoupled field can be transferred into the other fiber. By making the tapers very gradual, only a negligible fraction of the incoming optical power is reflected back into either of the input ports. Thus these devices are also known as *directional couplers.*

The optical power coupled from one fiber to another can be varied through three parameters: the axial length of the coupling region over which the fields from the two fibers interact; the size of the reduced radius r in the coupling region; and the spacing between the axes of the two coupled fibers. In making a fused fiber coupler, the coupling length W is normally fixed by the width of the heating flame, so that only L and r change as the coupler is elongated. Typical values for W and L are a few millimeters, the exact values depending on the coupling ratios desired for a specific wavelength. During fabrication the input and output powers from different ports can be monitored in real time until the desired coupling ratio is reached. Assuming that the coupler is lossless, the expression for the power P_2 coupled from one fiber to another over an axial distance z is

$$P_2 = P_0 \sin^2(\kappa z) \tag{10.2}$$

where κ is the *coupling coefficient* describing the interaction between the fields in the two fibers. By conservation of power, for identical-core fibers it follows that

$$P_1 = P_0 - P_2 = P_0[1 - \sin^2(\kappa z)] = P_0 \cos^2(\kappa z) \tag{10.3}$$

This shows that the phase of the driven fiber always lags 90° behind the phase of the driving fiber, as Fig. 10.6a illustrates. Thus, when power is launched into fiber 1, at $z = 0$ the phase in fiber 2 lags 90° behind that in fiber 1. This lagging phase relationship continues for increasing path length z, until at a distance that satisfies $\kappa z = \pi/2$, all of the power has been transferred from fiber 1 to fiber 2. Now fiber 2 becomes the driving fiber, so that for $\pi/2 \leq \kappa z \leq \pi$ the phase in fiber 1 lags behind that in fiber 2, and so on. As a result of this phase relationship, the 2×2 coupler is a *directional coupler*. That is, no energy can be coupled into a wave traveling backward in the negative-z direction in the driven waveguide.

Fig. 10.6 a Normalized coupled powers P_2/P_0 and P_1/P_0 as functions of the coupler draw length for a 1300-nm power level P_0 launched into fiber 1; **b** dependence on wavelength of the coupled powers in the completed 15-mm-long coupler

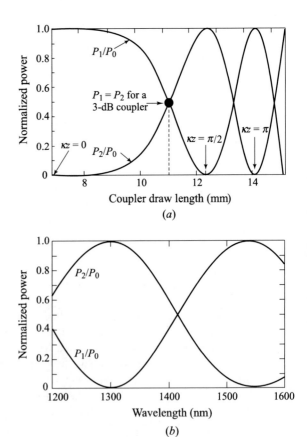

Example 10.4 The coupling coefficient κ is a complex parameter that depends on a variety of factors, such as the wavelength, the core and cladding refractive indices of the fibers, the fiber radius a, and the spacing d between the axes of the two coupled fibers. A simplified and accurate empirical expression for κ in a directional coupler made from two identical step-index fibers is given by [14].

$$\kappa = \frac{\pi}{2} \frac{\sqrt{\delta}}{a} \exp\left[-\left(A + Bx + Cx^2\right)\right]$$

where $x = d/a$

$$\delta = \frac{n_1^2 - n_2^2}{n_1^2}$$

$$A = 5.2789 - 3.663V + 0.3841V^2$$
$$B = -0.7769 + 1.2252V - 0.0152V^2$$
$$C = -0.0175 - 0.0064V - 0.0009V^2$$

with V defined by Eq. (2.27). Consider two fibers for which $n_1 = 1.4532$, $n_2 = 1.4500$, and $a = 5.0$ μm. If the spacing between the centers of the fibers is $d = 12$ μm, what is the coupling coefficient κ at a 1300-nm wavelength?

Solution: Using Eq. (2.27) it follows that $V = 2.329$. From the above equation the coupling coefficient is

$$\kappa = 20.8 \exp\left[-\left(-1.1693 + 1.9945x - 0.0373x^2\right)\right] = 0.694 \text{ mm}^{-1}$$

for $x = 12/5 = 2.4$.

Figure 10.6b shows how the normalized coupled power ratios P_2/P_0 and P_1/P_0 vary with wavelength for a 15-mm long coupler. Couplers with different performances result by varying the parameters W, L, and r for a specific wavelength.

In specifying the performance of an optical coupler, one usually indicates the percentage division of optical power between the output ports by means of the *splitting ratio* or *coupling ratio*. Referring to Fig. 10.5, with P_0 being the input power and P_1 and P_2 the output powers, then

$$\text{Coupling ratio} = \frac{P_2}{P_1 + P_2} \times 100\% \tag{10.4}$$

By adjusting the coupler parameters so that power is divided evenly, with half of the input power going to each output, one creates a *3-dB coupler*. A coupler could also be made in which almost all the optical power at 1500 nm goes to one port and almost all the energy around 1300 nm goes to the other port (see Prob. 10.5).

In the above analysis, for simplicity it was assumed that the device is lossless. However, in any practical coupler some light is always lost when a signal goes through it. The two basic losses are excess loss and insertion loss. The *excess loss* is defined as the ratio of the input power to the total output power. Thus in decibels, the excess loss for a 2 × 2 coupler is

$$\text{Excess loss} = 10\log\left(\frac{P_0}{P_1 + P_2}\right) \tag{10.5}$$

The *insertion loss* refers to the loss for a particular port-to-port path. For example, for the path from input port i to output port j, the insertion loss in decibels is

$$\text{Insertion loss} = P_{ij} = 10\log\left(\frac{P_i}{P_j}\right) \tag{10.6}$$

Another performance parameter is *crosstalk* or *return loss*, which measures the degree of isolation between the input at one port and the optical power scattered or reflected back to the other input port. That is, it is a measure of the optical power level P_3 shown in Fig. 10.5 and is given by

$$\text{Return loss} = 10\log\left(\frac{P_3}{P_0}\right) \tag{10.7}$$

Example 10.5 A 2 × 2 biconical tapered fiber coupler has an input optical power level of $P_0 = 200\ \mu\text{W}$. The output powers at the other three ports are $P_1 = 90\ \mu\text{W}$, $P_2 = 85\ \mu\text{W}$, and $P_3 = 6.3\ \text{nW}$. What are the coupling ratio, excess loss, insertion losses, and return loss for this coupler?

Solution: From Eq. (10.4), the coupling ratio is

$$\text{Coupling ratio} = \frac{85}{90 + 85} \times 100\% = 48.6\%$$

From Eq. (10.5), the excess loss is

$$\text{Excess loss} = 10\log\left(\frac{200}{90 + 85}\right) = 0.58\ \text{dB}$$

Using Eq. (10.6), the insertion losses are

$$\text{Insertion loss(port 0 to port 1)} = 10\log\left(\frac{200}{90}\right) = 3.47\ \text{dB}$$

$$\text{Insertion loss(port 0 to port 2)} = 10\log\left(\frac{200}{85}\right) = 3.72\ \text{dB}$$

From Eq. (10.7) the return loss is

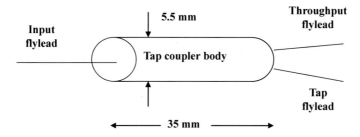

Fig. 10.7 Typical configuration and package dimensions for a tap coupler

Table 10.2 Representative specifications for a 2 × 2 tap coupler

Parameter	Unit	Specification
Tap ratio	Percent	1–5
Insertion loss (throughput)	dB	0.5
Return loss	dB	55
Power handling	mW	1000
Flylead length	m	1
Size (diameter × length)	mm	5.5 × 35

$$\text{Return loss} = 10 \log\left(\frac{6.3 \times 10^{-3}}{200}\right) = -45 \text{ dB}$$

Example 10.6 To monitor the light signal level or quality in a link, one can use a 2 × 2 device that has a coupling fraction of around 1 to 5%, which is selected and fixed during fabrication. This is known as a *tap coupler*. Nominally the tap coupler is packaged as a three-port device with one arm of the 2 × 2 coupler being terminated inside the package. Figure 10.7 shows a typical package for such a tap coupler and Table 10.2 lists some representative specifications.

Drill Problem 10.1 A 2 × 2 biconical tapered fiber coupler has an input optical power level of $P_0 = 400 \, \mu$W. The output powers at the other three ports are $P_1 = 180 \, \mu$W, $P_2 = 170 \, \mu$W, and $P_3 = 12.6$ nW. Show that the coupling ratio $= 48.6\%$, the excess loss $= 0.58$ dB, the insertion losses are $P_{01} = 3.47$ dB and $P_{02} = 3.72$ dB, and the return loss $= -45$ dB.

10.2.2   Scattering Matrix Analyses of Couplers

One also can analyze a 2×2 guided-wave coupler as a four-terminal device that has two inputs and two outputs, as shown in Fig. 10.8. Either all-fiber or integrated-optics devices can be analyzed in terms of the *scattering matrix* (also called the *propagation matrix*) **S**, which defines the relationship between the two input field strengths a_1 and a_2, and the two output field strengths b_1 and b_2. By definition [15]

$$\mathbf{b} = \mathbf{Sa}, \text{ where } \mathbf{b} = \begin{bmatrix} b_1 \\ b_2 \end{bmatrix}$$

$$\mathbf{a} = \begin{bmatrix} a_1 \\ a_2 \end{bmatrix}$$

$$\text{and} \quad \mathbf{S} = \begin{bmatrix} s_{11} & s_{21} \\ s_{12} & s_{22} \end{bmatrix} \tag{10.8}$$

Here, $s_{ij} = |s_{ij}| \exp(j\varphi_{ij})$ represents the *coupling coefficient* of optical power transfer from input port i to output port j, with $|s_{ij}|$ being the magnitude of s_{ij} and φ_{ij} being its phase at port j relative to port i.

For an actual physical device, two restrictions apply to the scattering matrix **S**. One is a result of the reciprocity condition arising from the fact that Maxwell's equations are invariant for time inversion; that is, they have two solutions in opposite propagating directions through the device, assuming single-mode operation. The other restriction arises from energy-conservation principles under the assumption that the device is lossless. From the first condition, it follows that

$$s_{12} = s_{21} \tag{10.9}$$

From the second restriction, if the device is lossless, the sum of the output intensities I_o must equal the sum of the input intensities I_i:

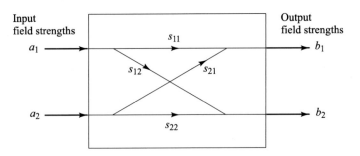

Fig. 10.8 Generic 2×2 guided-wave coupler, where a_i and b_j represent the field strengths of input port i and output port j, respectively, and the s_{ij} are the scattering matrix parameters

$$I_0 = b_1^* b_1 + b_2^* b_2 = I_i = a_1^* a_1 + a_2^* a_2$$

$$\text{or } b^+ b = a^+ a \tag{10.10}$$

where the superscript * means the complex conjugate and the superscript + indicates the transpose conjugate. Substituting Eqs. (10.8) and (10.9) into Eq. (10.10) yields the following set of three equations:

$$s_{11}^* s_{11} + s_{12}^* s_{12} = 1 \tag{10.11}$$

$$s_{11}^* s_{12} + s_{12}^* s_{22} = 0 \tag{10.12}$$

$$s_{22}^* s_{22} + s_{12}^* s_{12} = 1 \tag{10.13}$$

Now assume that the coupler has been constructed so that the fraction $(1 - \varepsilon)$ of the optical power from input 1 appears at output port 1, with the remainder ε going to port 2. Then it follows that $s_{11} = \sqrt{1 - \varepsilon}$, which is a real number between 0 and 1. Here, without loss of generality, it is assumed that the electric field at output 1 has a zero phase shift relative to the input at port 1; that is $\varphi_{11} = 0$. Because the parameter of interest is the phase change that occurs when the coupled optical power from input 1 emerges from port 2, the simplifying assumption is made that the coupler is symmetric. Then, analogous to the effect at port 1, it follows that $s_{22} = \sqrt{1 - \varepsilon}$, with $\varphi_{22} = 0$. Using these expressions, it is possible to determine the phase φ_{12} of the coupled outputs relative to the input signals and to find the constraints on the composite outputs when both input ports are receiving signals.

Inserting the expressions for s_{11} and s_{22} into Eq. (10.12) and letting $s_{12} = |s_{12}| \exp(j\varphi_{12})$, where $|s_{12}|$ is the magnitude of s_{12} and φ_{12} is its phase, then

$$\exp(j2\varphi_{12}) = -1 \tag{10.14}$$

which holds when

$$\varphi_{12} = (2n + 1)\frac{\pi}{2} \quad \text{where } n = 0, 1, 2, \ldots \tag{10.15}$$

so that the scattering matrix from Eq. (10.8) becomes

$$S = \begin{bmatrix} \sqrt{1 - \varepsilon} & j\sqrt{\varepsilon} \\ j\sqrt{\varepsilon} & \sqrt{1 - \varepsilon} \end{bmatrix} \tag{10.16}$$

Example 10.7 Assume we have a 3-dB coupler, so that half of the input power gets coupled to the second fiber. What are the output powers $P_{out,1}$ and $P_{out,2}$?

Solution: Because the input power is divided evenly, $\varepsilon = 0.5$ and the output field intensities $E_{out,1}$ and $E_{out,2}$ can be found from the input intensities $E_{in,1}$ and $E_{in,2}$ and the scattering matrix in Eq. (10.16):

$$\begin{bmatrix} E_{out,1} \\ E_{out,2} \end{bmatrix} = \frac{1}{\sqrt{2}} \begin{bmatrix} 1 & j \\ j & 1 \end{bmatrix} \begin{bmatrix} E_{in,1} \\ E_{in,2} \end{bmatrix}$$

Letting $E_{in,2} = 0$, then $E_{out,1} = \left(1/\sqrt{2}\right) E_{in,1}$ and $E_{out,2} = \left(j/\sqrt{2}\right) E_{in,1}$. The output powers are then given by

$$P_{out,1} = E_{out,1} E^*_{out,1} = \frac{1}{2} E^2_{in,1} = \frac{1}{2} P_0$$

Similarly,

$$P_{out,2} = E_{out,2} E^*_{out,2} = \frac{1}{2} E^2_{in,1} = \frac{1}{2} P_0$$

so that half the input power appears at each output of the coupler.

It also is important to note that when it is desired to have a large portion of the input power from, say, port 1 to emerge from output 1, then it is necessary for ε to be small. However, this, in turn, means that the amount of power at the same wavelength coupled to output 1 from input 2 is small. Consequently, if one is using the same wavelength, it is not possible, in a passive 2×2 coupler, to have all the power from both inputs coupled simultaneously to the same output. The best that can be done is to have half of the power from each input appear at the same output. However, if the wavelengths are different at each input, it is possible to couple a large portion of both input power levels onto the same output fiber [11].

10.2.3   Basis of the 2 × 2 Waveguide Coupler

More versatile 2×2 couplers are possible with waveguide-type devices [12, 13, 16]. Fig. 10.9 shows two types of 2×2 waveguide couplers. The uniformly symmetric device in Fig. 10.9a has two identical parallel guides in the coupling region, whereas the uniformly asymmetric coupler in Fig. 10.9b has one guide wider than the other. Analogous to fused-fiber couplers, waveguide devices have an intrinsic wavelength dependence in the coupling region. The degree of interaction between the guides varies when changing the guide width w, the gap s between the guides, and the refractive index n_1 between the guides. In Fig. 10.9, the z direction lies along the coupler length and the y axis lies in the coupler plane transverse to the two waveguides.

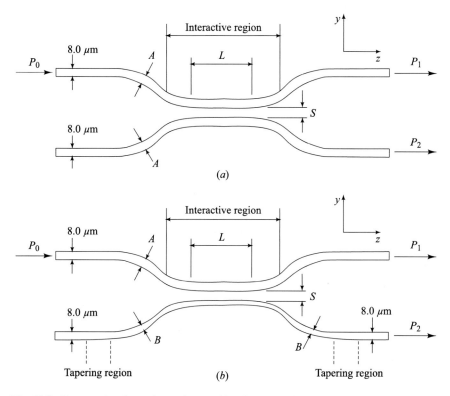

Fig. 10.9 Cross-sectional top views of **a** a uniformly symmetric directional waveguide coupler with both guides having a width A = 8 μm, **b** a uniformly asymmetric directional coupler in which one guide has a narrower width B in the coupling region

For the analysis, first consider the symmetric coupler. In real waveguides, with absorption and scattering losses, the propagation constant β_z is a complex number given by

$$\beta_z = \beta_r + j\frac{\alpha}{2} \tag{10.17}$$

where β_r *is* the real part of the propagation constant and α is the optical loss coefficient in the guide. Hence, the total power contained in both guides decreases by a factor $\exp(-\alpha z)$ along their length. For example, losses in semiconductor and silicon oxynitride waveguide devices fall in the $0.05 < \alpha < 0.35$ cm^{-1} range (or, equivalently, about $0.2 < \alpha < 1.5$ dB/cm). Losses in silica waveguides are less than 0.1 dB/cm. Recall from Eq. (3.1) the relationship α(dB/cm) $= 4.343\alpha$(cm^{-1}).

The transmission characteristics of the symmetric coupler can be expressed through the coupled-mode theory approach to yield [16]

$$P_2 = P_0 \sin^2(\kappa z)e^{-\alpha z} \tag{10.18}$$

where the coupling coefficient is

$$\kappa = \frac{2\beta_y^2 q e^{-qs}}{\beta_z w (q^2 + \beta_y^2)} \tag{10.19}$$

This is a function of the waveguide propagation constants β_y and β_z (in the y and z directions, respectively), the gap width d and separation s, and the extinction coefficient q in the y direction (i.e., the exponential falloff in the y direction) outside the waveguide, which is

$$q^2 = \beta_z^2 - k_1^2 \tag{10.20}$$

The theoretical power distribution as a function of the guide length is shown in Fig. 10.10 where the parameter values were chosen to be $\kappa = 0.6$ mm^{-1} and $\alpha = 0.02$ mm^{-1}. Analogous to the fused-fiber coupler, complete power transfer to the second guide occurs when the guide length L is

$$L = \frac{\pi}{2\kappa}(m + 1) \text{ with } m = 0, 1, 2, \ldots \tag{10.21}$$

Because κ is found to be almost monotonically proportional to wavelength, the coupling ratio P_2/P_0 rises and falls sinusoidally from 0 to 100% as a function of wavelength, as Fig. 10.11 illustrates generically (assuming here, for simplicity, that the guide loss is negligible).

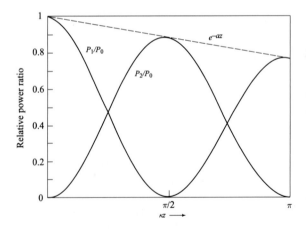

Fig. 10.10 Theoretical through-path and coupled power distributions as a function of the guide length in a symmetric 2×2 guided-wave coupler with $\kappa = 0.6$ mm^{-1} and $\alpha = 0.02$ mm^{-1}

Fig. 10.11 Wavelength response of the coupled power P_2/P_0 in the symmetric 2×2 guided-wave coupler shown in Fig. 10.9a

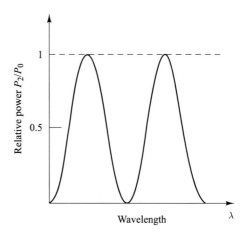

Drill Problem 10.2 Consider a symmetric waveguide coupler for which the fraction of *coupled* power $P_2/P_0 = 0.48$ at $\kappa z = \pi/4$. Show that the length of the coupler is 2.04 mm if the optical loss coefficient $\alpha = 0.02$/mm.

Example 10.8 A symmetric waveguide coupler has a coupling coefficient $\kappa = 0.6 \, \text{mm}^{-1}$. What is the coupling length for complete power transfer?

Solution: Using Eq. (10.21), the coupling length for $m = 1$ is found to be $L = 5.24$ mm.

When the two guides do not have the same widths, as shown in Fig. 10.9b, the amplitude of the coupled power is dependent on wavelength, and the coupling ratio becomes

$$P_2/P_0 = \frac{\kappa^2}{g^2} \sin^2(gz) \, e^{-\alpha z} \qquad (10.22)$$

where

$$g^2 = \kappa^2 + \left(\frac{\Delta\beta}{2}\right)^2 \qquad (10.23)$$

with $\Delta\beta$ being the phase difference between the two guides in the z direction. With this type of configuration, one can fabricate devices that have a flattened response in which the coupling ratio is less than 100% in a specific desired wavelength range, as shown in Fig. 10.12. The main cause of the wave-flattened response at the lower wavelength results from suppression by the amplitude term κ^2/g^2. This asymmetric characteristic can be used in a device where only a fraction of power from a specific

Fig. 10.12 Wavelength
response of the coupled
power P_2/P_0 in the
asymmetric 2×2
guided-wave coupler shown
in Fig. 10.9b

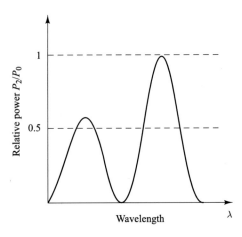

wavelength should be tapped off. Note also that when $\Delta\beta = 0$, Eq. (10.22) reduces
to the symmetric case given by Eq. (10.18).

More complex structures are readily fabricated in which the widths of the guides
are tapered. These nonsymmetric structures can be used to flatten the wavelength
response over a particular spectral range. It is also important to note that the above
analysis based on the coupled-mode theory holds when the indices of the two waveguides are identical, but a more complex analytical treatment is needed for different
refractive indices [12].

10.2.4 Principal Role of Star Couplers

The principal role of star couplers is to combine the powers from N inputs and
divide them equally (usually) among M output ports. Techniques for creating star
couplers include fused fibers, gratings, micro-optic technologies, and integrated-
optics schemes. The fiber-fusion technique has been a popular construction method
for $N \times N$ star couplers. For example, 7×7 devices and 1×19 splitters or combiners
with excess losses at 1300 nm of 0.4 dB and 0.85 dB, respectively, have been demonstrated [17]. However, large-scale fabrication of these devices for $N > 2$ is limited
because of the difficulty in controlling the coupling response between the numerous
fibers during the heating and pulling process. Figure 10.13 shows a generic 4×4
fused-fiber star coupler.

In an ideal star coupler, the optical power from any input is evenly divided among
the output ports. The total loss of the device consists of its splitting loss plus the
excess loss in each path through the star. The *splitting loss* is given in decibels by

$$\text{Splitting loss} = -10 \log\left(\frac{1}{N}\right) = 10 \log N \qquad (10.24)$$

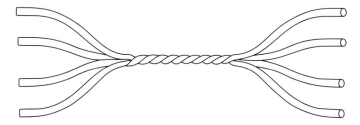

Fig. 10.13 Generic 4 × 4 fused-fiber star coupler fabricated by twisting, heating, and pulling on four fibers to fuse them together

Similar to Eq. (10.5), for a single input power P_{in} and N output powers, the excess loss in decibels is given by

$$\text{Fiber star excess loss} = 10 \log \left(\frac{P_{in}}{\sum_{i=1}^{N} P_{out,i}} \right) \tag{10.25}$$

The insertion loss and return loss can be found from Eqs. (10.6) and (10.7), respectively.

> **Drill Problem 10.3** Commonly used optical power dividing components for fiber-to-the-premises (FTTP) networks include 8 × 8, 16 × 16, and 32 × 32 star couplers. Show that the splitting losses for these devices are 9, 12, and 15 dB, respectively.

> **Drill Problem 10.4** If an optical signal that has a 20-μW power level enters a 16 × 16 star coupler, show that if the device has no excess loss, then the coupler splitting loss is 12 dB and the power level emerging from each branch of the coupler is −29 dBm.

An alternative is to construct star couplers by cascading a number of 3-dB couplers. Figure 10.14 shows an example for an 8 × 8 device formed by using twelve 2 × 2 couplers. This device could be made from either fused-fiber or integrated-optic components. As can be seen from this figure, a fraction $1/N$ of the launched power from each input port appears at all output ports. A limitation to the flexibility or modularity of this technique is that N is a multiple of 2; that is, $N = 2^n$ with the integer $n \geq 1$. The consequence is that if an extra node needs to be added to a fully connected $N \times N$ network, the $N \times N$ star needs to be replaced by a $2N \times 2N$ star, thereby leaving $2(N - 1)$ new ports being unused. Alternatively, one extra 2 × 2

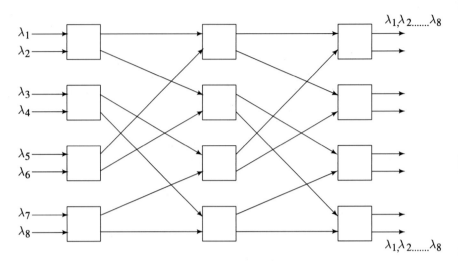

Fig. 10.14 Example of an 8 × 8 star coupler formed by interconnecting twelve 2 × 2 couplers

coupler can be used at one port to get $N + 1$ outputs. However, these two new ports each have an additional 3-dB loss.

As can be deduced from Fig. 10.14, the number of 3-dB couplers N_c needed to construct an $N \times N$ star is

$$N_c = \frac{N}{2} \log_2 N = \frac{N}{2} \frac{\log N}{\log 2} \tag{10.26}$$

because there are $N/2$ elements in the vertical direction and $\log_2 N = \log N/\log 2$ elements horizontally. (Reminder: The term "$\log x$" designates the base-10 logarithm of x.)

Example 10.9 A device engineer wants to construct a 32 × 32 coupler from a cascade of 2 × 2 single-mode 3-dB fiber couplers. How many 2 × 2 elements are needed for this?

Solution: In this case there will be 16 coupler elements in the vertical direction. From Eq. (10.26), the required number of 2 × 2 elements is

$$N_c = \frac{32}{2} \frac{\log 32}{\log 2} = 80$$

If the fraction of power traversing each 3-dB coupler element is F_T, with $0 \le F_T \le 1$ (i.e., a fraction $1 - F_T$ of power is lost in each 2 × 2 element), then the *excess loss* in decibels is

$$\text{Excess loss} = -10 \log\left(F_T^{\log_2 N}\right) \tag{10.27}$$

The splitting loss for this star is, again, given by Eq. (10.24). Thus the total loss experienced by a signal as it passes through the $\log_2 N$ stages of the $N \times N$ star and gets divided into N outputs is (in decibels)

$$\text{Total loss} = \text{splitting loss} + \text{excess loss}$$

$$= -10 \log \left(\frac{F_T^{\log_2 N}}{N} \right)$$

$$= -10 \left(\frac{\log N \log F_T}{\log 2} - \log N \right)$$

$$= 10(1 - 3.322 \log F_T) \log N \qquad (10.28)$$

This shows that the loss increases logarithmically with N.

Example 10.10 Consider a commercially available 32×32 single-mode coupler made from a cascade of 3-dB fused-fiber 2×2 couplers, where 5% of the power is lost in each element. What are the excess and splitting losses for this coupler?

Solution: From Eq. (10.27), the excess loss is

$$\text{Excess loss} = -10 \log \left(0.95^{\log 32 / \log 2} \right) = 1.1 \text{ dB}$$

and, from Eq. (10.24), the splitting loss is

$$\text{Splitting loss} = -10 \log 32 = 15 \text{ dB}$$

Hence, the total loss is 16.1 dB.

10.2.5 Mach–Zehnder Interferometry Techniques

Wavelength-dependent multiplexers can also be made using Mach–Zehnder interferometry techniques [13]. These devices can be either active or passive. This section describes passive multiplexers. Figure 10.15 illustrates the constituents of an individual Mach–Zehnder interferometer (MZI). This 2×2 MZI consists of three stages: an initial 3-dB directional coupler that splits the input signals, a central section where one of the waveguides is longer by ΔL to give a wavelength-dependent phase shift between the two arms, and another 3-dB coupler that recombines the signals at the output. As shown in the following derivation, the function of this arrangement is that, by splitting the input beam and introducing a phase shift in one of the paths, the recombined signals will interfere constructively at one output and destructively at the other. The signals then finally emerge from only one output port. For simplicity, the following analysis does not take into account waveguide material losses or bend losses.

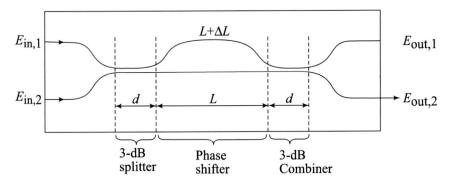

Fig. 10.15 Layout of a basic 2 × 2 Mach–Zehnder interferometer

The propagation matrix $\mathbf{M}_{coupler}$ for a coupler of length d is

$$\mathbf{M}_{coupler} = \begin{bmatrix} \cos\kappa d & j\sin\kappa d \\ j\sin\kappa d & \cos\kappa d \end{bmatrix} \tag{10.29}$$

where κ is the coupling coefficient. Because the devices of interest here are 3-dB couplers that divide the power equally, then $2\kappa d = \pi/2$, so that

$$\mathbf{M}_{couplex} = \frac{1}{\sqrt{2}} \begin{bmatrix} 1 & j \\ j & 1 \end{bmatrix} \tag{10.30}$$

In the central region, when the signals in the two arms come from the same light source, the outputs from these two guides have a phase difference $\Delta\varphi$ given by

$$\Delta\varphi = \frac{2\pi n_1}{\lambda}L - \frac{2\pi n_2}{\lambda}(L + \Delta L) \tag{10.31}$$

Note that this phase difference can arise either from a different path length (given by ΔL) or through a refractive index difference if $n_1 \neq n_2$. Here, assume both arms to have the same index and let $n_1 = n_2 = n_{eff}$ (the effective refractive index in the waveguide; see Sect. 2.5.4). Then Eq. (10.31) can be rewritten as

$$\Delta\varphi = k\Delta L \tag{10.32}$$

where $k = 2\pi n_{eff}/\lambda$.

For a given phase difference $\Delta\varphi$, the propagation matrix $\mathbf{M}_{\Delta\varphi}$ for the phase shifter is

$$\mathbf{M}_{\Delta\varphi} = \begin{bmatrix} \exp(jk\Delta L/2) & 0 \\ 0 & \exp(-jk\Delta L/2) \end{bmatrix} \tag{10.33}$$

The optical output fields $E_{out,1}$ and $E_{out,2}$ from the two central arms can be related to the input fields $E_{in,1}$ and $E_{in,2}$ by

$$\begin{bmatrix} E_{out,1} \\ E_{out,2} \end{bmatrix} = M \begin{bmatrix} E_{in,1} \\ E_{in,2} \end{bmatrix} \tag{10.34}$$

where M is the product of the matrices $\mathbf{M}_{coupler}$, $\mathbf{M}_{\Delta\varphi}$, and $\mathbf{M}_{coupler}$,

$$M = M_{coupler} \cdot M_{\Delta\varphi} \cdot M_{coupler} = \begin{bmatrix} M_{11} & M_{21} \\ M_{12} & M_{22} \end{bmatrix}$$

$$= j \begin{bmatrix} \sin(k\Delta L/2) & \cos(k\Delta L/2) \\ \cos(k\Delta L/2) & -\sin(k\Delta L/2) \end{bmatrix} \tag{10.35}$$

Because the device of interest is a multiplexer, it is necessary to have the inputs to the MZI be at different wavelengths; that is, $E_{in,1}$ is at λ_1 and $E_{in,2}$ is at λ_2. Then, from Eq. (10.34), the output field $E_{out,1}$ and $E_{out,2}$ are each the sum of the individual contributions from the two input fields:

$$E_{out,1} = j\left[E_{in,1}(\lambda_1)\sin(k_1\Delta L/2) + E_{in,2}(\lambda_2)\cos(k_2\Delta L/2) \right] \tag{10.36}$$

$$E_{out,2} = j\left[E_{in,1}(\lambda_1)\cos(k_1\Delta L/2) - E_{in,2}(\lambda_2)\sin(k_2\Delta L/2) \right] \tag{10.37}$$

where $k_j = 2\pi n_{eff}/\lambda_j$. The output powers are then found from the light intensity, which is the square of the field strengths. Thus,

$$P_{out,1} = E_{out,1}E_{out,1}^* = \sin^2(k_1\Delta L/2)P_{in,1} + \cos^2(k_2\Delta L/2)P_{in,2} \tag{10.38}$$

$$P_{out,2} = E_{out,2}E_{out,2}^* = \cos^2(k_1\Delta L/2)P_{in,1} + \sin^2(k_2\Delta L/2)P_{in,2} \tag{10.39}$$

where $P_{in,j} = |E_{in,j}|^2 = E_{in,j} \cdot E_{in,j}^*$. In Eqs. (10.38) and (10.39), the cross terms are dropped because their frequency is twice the optical carrier frequency, which is beyond the response capability of the photodetector.

From Eqs. (10.38) and (10.39), it can be seen that if one wants all the power from both inputs to leave the same output port (e.g., port 2), then it is necessary to have $k_1\Delta L/2 = \pi$ and $k_2\Delta L/2 = \pi/2$, or

$$(k_1 - k_2)\Delta L = 2\pi n_{eff}\left(\frac{1}{\lambda_1} - \frac{1}{\lambda_2} \right)\Delta L = \pi \tag{10.40}$$

Hence, the length difference in the interferometer arms should be

$$\Delta L = \left[2n_{eff} \left(\frac{1}{\lambda_1} - \frac{1}{\lambda_2} \right) \right]^{-1} = \frac{c}{2n_{eff} \Delta v} \tag{10.41}$$

where Δv is the frequency separation of the two wavelengths.

Example 10.11

(a)   Assume that the input wavelengths of a 2×2 silicon MZI are separated by 10 GHz (i.e., $\Delta \lambda = 0.08$ nm at 1550 nm). With $n_{\text{eff}} = 1.5$ in a silicon waveguide, it can be seen from (Eq. 10.41) that the waveguide length difference must be.

$$\Delta L = \frac{3 \times 10^8 \text{m/s}}{2(1.5)10^{10}/\text{s}} = 10 \text{ mm}$$

(b)   If the frequency separation is 130 GHz (i.e., $\Delta \lambda = 1$ nm), then $\Delta L = 0.77$ mm.

Using basic 2×2 MZIs, any size $N \times N$ multiplexer (with $N = 2^n$) can be constructed. Figure 10.16 gives an example for a 4×4 multiplexer. Here the inputs to MZI$_1$ are v and $v + 2\Delta v$ (which here are called λ_1 and λ_3, respectively), and the inputs to MZI$_2$ are $v + \Delta v$ and $v + 3\Delta v$ (λ_2 and λ_4, respectively). Because the signals in both interferometers of the first stage are separated by $2\Delta v$, the path differences satisfy the condition

$$\Delta L_1 = \Delta L_2 = \frac{c}{2n_{eff}(2\Delta v)} \tag{10.42}$$

In the next stage, the inputs are separated by Δv. Consequently, it is necessary to have

$$\Delta L_3 = \frac{c}{2n_{eff} \Delta v} = 2\Delta L_1 \tag{10.43}$$

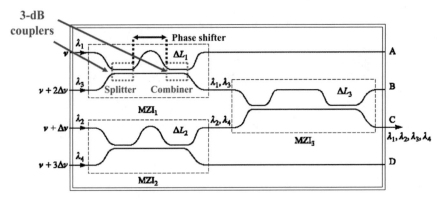

Fig. 10.16 Example of a 4-channel wavelength multiplexer using three 2×2 MZI elements

When these conditions are satisfied, all four input powers will emerge from port C.

From this design example, it can be deduced that for an N-to-1 MZI multiplexer, where $N = 2^n$ with the integer $n \geq 1$, the number of multiplexer stages is n and the number of MZIs in stage j is 2^{n-j}. The path difference in an interferometer element of stage j is thus

$$\Delta L_{stage\,j} = \frac{c}{2^{n-j} n_{eff} \Delta v} \tag{10.44}$$

The N-to-1 MZI multiplexer can also be used as a 1-to-N demultiplexer by reversing the light-propagation direction. For a real MZI, the ideal case given in these examples needs to be modified to have slightly different values of ΔL_1 and ΔL_2.

10.3 Nonreciprocal Isolators and Circulators

In a number of applications it is desirable to have a passive optical device that is nonreciprocal; that is, it works differently when its inputs and outputs are reversed. Two examples of such a device are isolators and circulators. To understand the operation of these devices, first it is necessary to recall some facts about polarization and polarization-sensitive components from Chap. 2:

- Light can be represented as a combination of a parallel vibration and a perpendicular vibration, which are called the two *orthogonal plane polarization states* of a lightwave.
- A *polarizer* is a material or device that transmits only one polarization component and blocks the other.
- A *Faraday rotator* is a device that rotates the state of polarization (SOP) of light passing through it by a specific angular amount.
- A device made from birefringent materials (called a *walk-off polarizer*) splits the light signal entering it into two orthogonally (perpendicularly) polarized beams, which then follow different paths through the material.
- A *half-wave plate* rotates the SOP clockwise by 45° for signals going from left to right, and counterclockwise by 45° for signals propagating in the other direction.

10.3.1 Functions of Optical Isolators

Optical isolators are devices that allow light to pass through them in only one direction. This is important in a number of instances to prevent scattered or reflected light from traveling in the reverse direction. One common application of an optical isolator is to keep such backward-traveling light from entering a laser diode and possibly causing instabilities in the optical output.

Many design configurations of varying complexity exist for optical isolators. The simple ones depend on the state of polarization of the input light. However, such a design results in a 3-dB loss (half the power) when unpolarized light is passed through the device because it blocks half of the input signal. In practice the optical isolator should be independent of the SOP because light in an optical link normally is not polarized.

Figure 10.17 shows a design for a *polarization-independent isolator* that is made of three miniature optical components. The core of the device consists of a 45° Faraday rotator that is placed between two wedge-shaped birefringent plates or walk-off polarizers. These plates consist of a material such as YVO$_4$ or TiO$_2$, as described in Chap. 2. Light traveling in the forward direction (left to right in the top of Fig. 10.17) is separated into ordinary and extraordinary rays by the first birefringent plate. The Faraday rotator then rotates the polarization plane of each ray by 45°. After exiting the Faraday rotator, the two rays pass through the second birefringent plate. The axis of this polarizer plate is oriented in such a way that the relationship between the two types of rays is maintained. Thus, when they exit the polarizer, they both are refracted in an identical parallel direction. Going in the reverse direction (right to left as shown in the bottom diagram), the relationship of the ordinary and extraordinary rays is

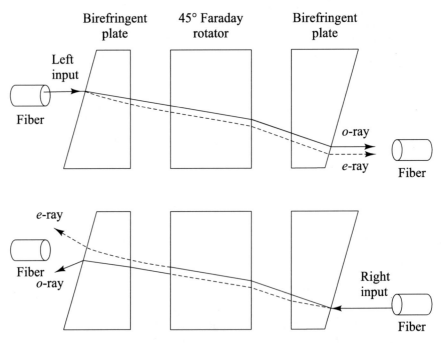

Fig. 10.17 Design and operation of a polarization-independent isolator made of three miniature optical components

Table 10.3 Typical parameter values of commercially available optical isolators

Parameter	Unit	Value
Central wavelength λ_c	nm	1310, 1550
Peak isolation	dB	40
Isolation at $\lambda_c \pm 20$ nm	dB	30
Insertion loss	dB	<0.5
Polarization-dependent loss	dB	<0.1
Polarization-mode dispersion	ps	<0.25
Size (diameter × length)	mm	6 × 35

reversed when exiting the Faraday rotator due to the nonreciprocity of the Faraday rotation. Consequently, the rays diverge when they exit the left-hand birefringent plate and are not coupled to the left-hand fiber.

Table 10.3 lists some operational characteristics of commercially available isolators. The packages have similar configurations as the tap coupler shown in Fig. 10.7.

10.3.2 Characteristics of Optical Circulators

An *optical circulator* is a nonreciprocal multiport passive device that directs light sequentially from port to port in only one direction. This device is used in optical amplifiers, add/drop multiplexers, and dispersion compensation modules. The operation of a circulator is similar to that of an isolator except that its construction is more complex. Typically it consists of a number of walk-off polarizers, half-wave plates, and Faraday rotators and has three or four input /output ports. To see how it functions, consider the three-port circulator, as shown in Fig. 10.18. Here an input on port 1 is sent out on port 2, an input on port 2 is sent out on port 3, and an input on port 3 is sent out on port 1.

Similarly, in a four-port device ideally one could have four inputs and four outputs if the circulator is perfectly symmetrical. However, in actual applications it usually is not necessary to have four inputs and four outputs. Furthermore, such a perfectly symmetrical circulator is rather tedious to fabricate. Therefore in a four-port circulator it is common to have three input ports and three output ports, making port 1 be an input-only port, 2 and 3 being input and output ports, and port 4 be an output-only port.

A variety of circulators are available commercially. These devices have low insertion loss, high isolation over a wide wavelength range, minimal polarization-dependent loss (PDL), and low polarization-mode dispersion (PMD). Table 10.4 lists some operational characteristics of commercially available circulators.

Fig. 10.18 Operational concept of a three-port circulator

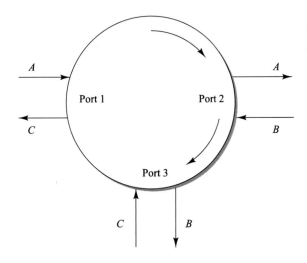

Table 10.4 Typical parameter values of commercially available optical circulators

Parameter	Unit	Value
Wavelength band	nm	C-band: 1525–1565 L-band: 1570–1610
Insertion loss	dB	<0.6
Channel isolation	dB	>40
Optical return loss	dB	>50
Operating power	mW	<500
Polarization-dependent loss	dB	<0.1
Polarization-mode dispersion	ps	<0.1
Size (diameter × length)	mm	5.5 × 50

10.4   WDM Devices Based on Grating Principles

A grating is an important element in WDM systems for combining and separating individual wavelengths. Basically, a grating is a periodic structure or perturbation in a material. This variation in the material has the property of reflecting or transmitting light in a certain direction depending on the wavelength. Thus, gratings can be categorized as either reflecting or transmitting gratings.

10.4.1   Grating Basics

Figure 10.19 defines various parameters for a reflection grating. Here, θ_i is the incident angle of the light, θ_d is the diffracted angle, and Λ is the *period of the grating* (the periodicity of the structural variation in the material). In a transmission grating

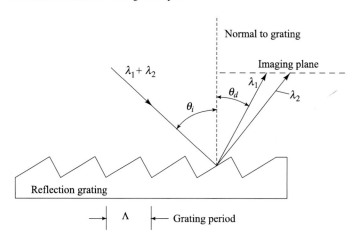

Fig. 10.19 Basic parameters in a reflection grating

consisting of a series of equally spaced slits, the spacing between two adjacent slits is called the *pitch* of the grating. Constructive interference at a wavelength λ occurs in the imaging plane when the rays diffracted at the angle θ_d satisfy the *grating equation* given by

$$\Lambda(\sin\theta_i - \sin\theta_d) = m\lambda \qquad (10.45)$$

Here, m is called the *order* of the grating. In general, only the first-order diffraction condition $m = 1$ is considered. (Note that in some texts the incidence and refraction angles are defined as being measured from the same side of the normal to the grating. In this case, the sign in front of the term $\sin\theta_d$ changes.) A grating can separate individual wavelengths because the grating equation is satisfied at different points in the imaging plane for different wavelengths.

10.4.2 Optical Fiber Bragg Grating (FBG)

A Bragg grating constructed within an optical fiber constitutes a high-performance device for accessing individual wavelengths in the closely spaced spectrum of dense WDM systems [18]. Because this fiber Bragg grating (FBG) is an all-fiber device, its main advantages are simple packaging, low cost, low insertion loss (around 0.3 dB), ease of coupling with other fibers, polarization insensitivity, and a low temperature coefficient of <0.7 pm/°C for an athermal device, that is, one made insensitive to temperature changes. A fiber Bragg grating is a narrowband reflection filter that is fabricated through a photo-imprinting process. The technique is based on the observation that germanium-doped silica fiber exhibits high photosensitivity to ultraviolet light. This means that one can induce a change in the refractive index of the core by exposing it to ultraviolet radiation such as 244 nm.

Several methods can be used to create a fiber phase grating. Figure 10.20 demonstrates the so-called *external-writing technique*. The grating fabrication is accomplished by means of two ultraviolet beams transversely irradiating the fiber to produce an interference pattern in the core. Here, the regions of high intensity (denoted by the shaded ovals) cause an increase in the local refractive index of the photosensitive core, whereas it remains unaffected in the zero-intensity regions. A permanent reflective Bragg grating is thus written into the core. When a multi-wavelength signal encounters the grating, those wavelengths that are phase-matched to the Bragg reflection condition [given in Eq. (10.47) below] are reflected and all others are transmitted.

Using the standard grating equation given by Eq. (10.45), with λ being the wavelength of the ultraviolet light λ_{uv}, the period Λ of the interference pattern (and hence the period of the grating) can be calculated from the angle θ between the two interfering beams of free-space wavelength λ_{uv}. Note from Fig. 10.20 that θ is measured outside of the fiber.

The imprinted grating can be represented as a uniform sinusoidal modulation of the refractive index along the core:

$$n(z) = n_{core} + \delta n \left[1 + \cos\left(\frac{2\pi z}{\Lambda} \right) \right]$$

(10.46)

where n_{core} is the unexposed core refractive index and δn is the photoinduced change in the index.

The maximum reflectivity R of the grating occurs when the *Bragg condition* holds, that is, at a reflection wavelength λ_{Bragg} where

$$\lambda_{Bragg} = 2\Lambda n_{eff}$$

(10.47)

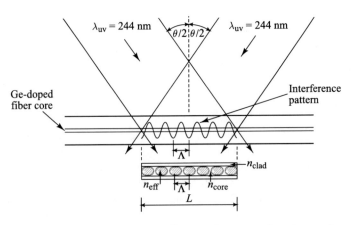

Fig. 10.20 Formation of a Bragg grating in a fiber core by means of two intersecting ultraviolet light beams

and n_{eff} is the effective index of the core as described in Eq. (10.32). At this wavelength, the peak reflectivity R_{max} for the grating of length L and coupling coefficient κ is given by

$$R_{max} = \tanh^2(\kappa L) \qquad (10.48)$$

The full bandwidth $\Delta\lambda$ over which the maximum reflectivity holds is [18]

$$\Delta\lambda = \frac{\lambda_{Bragg}^2}{\pi n_{eff} L}\left[(\kappa L)^2 + \pi^2\right]^{1/2} \qquad (10.49)$$

An approximation for the full-width half-maximum (FWHM) bandwidth is

$$\Delta\lambda_{FWHM} = \lambda_{Bragg} s \left[\left(\frac{\delta n}{2n_{core}}\right)^2 + \left(\frac{\Lambda}{L}\right)^2\right]^{1/2} \qquad (10.50)$$

where $s \approx 1$ for strong gratings with near 100% reflectivity, and $s \approx 0.5$ for weak gratings.

For a uniform sinusoidal modulation of the index throughout the core, the coupling coefficient κ is given by

$$\kappa = \frac{\pi \delta n \eta}{\lambda_{Bragg}} \qquad (10.51)$$

with η being the fraction of optical power contained in the fiber core. Under the assumption that the grating is uniform in the core, η can be approximated by

$$\eta \approx 1 - V^{-2} \qquad (10.52)$$

where V is the V number of the fiber.

Example 10.12

(a) The table below shows the values of R_{max} as given by Eq. (10.48) for different values of κL:

κL	R_{max} (%)
1	58
2	93
3	98

Table 10.5 Typical parameter values of commercially available fiber Bragg gratings

Parameter	Typical values for three channel spacings		
	25 GHz	50 GHz	100 GHz
Reflection bandwidth	>0.08 nm @ −0.5 dB <0.2 nm @ −3 dB <0.25 nm @ −25 dB	>0.15 nm @ −0.5 dB <0.4 nm @ −3 dB <0.5 nm @ −25 dB	>0.3 nm @ −0.5 dB <0.75 nm @ −3 dB <1 nm @ −25 dB
Transmission bandwidth	>0.05 nm @ −25 dB	>0.1 nm @ −25 dB	>0.2 nm @ −25 dB
Adjacent channel isolation	>30 dB		
Insertion loss	<0.25 dB		
Central λ tolerance	< ± 0.05 nm @ 25 °C		
Thermal λ drift	<1 pm/°C (for an athermal design)		
Package size	5 mm (diameter) × 80 mm (length)		

(b)   Consider a fiber grating with the following parameters: $L = 0.5$ cm, λ_{Bragg} = 1530 nm, $n_{\text{eff}} = 1.48$, $\delta n = 2.5 \times 10^{-4}$, and $\eta = 82\%$. From Eq. (10.51) the coupling coefficient $\kappa = 4.2$ cm^{-1}. Substituting this into Eq. (10.49) then yields $\Delta\lambda = 0.38$ nm.

Fiber Bragg gratings are available in a wide range of reflection bandwidths varying from 25 GHz and higher. Table 10.5 lists some operational characteristics of commercially available 25-, 50-, and 100-GHz fiber Bragg gratings for use in optical communication systems.

In the fiber Bragg grating (FBG) illustrated in Fig. 10.20, the grating spacing is uniform along its length. It is also possible to have the spacing vary along the length of the fiber, which means that a range of different wavelengths will be reflected by the FBG. This is the basis of what is known as a *chirped grating*.

10.4.3 WDM FBG Applications

Figure 10.21 shows a simple concept of a demultiplexing function of four wavelengths using a FBG. To extract the desired wavelength, a circulator is used in conjunction with the grating. Here the four wavelengths enter through port 1 of the circulator and leave from port 2. All wavelengths except λ_2 pass through the grating. Because λ_2 satisfies the Bragg condition of the grating, it gets reflected, enters port 2 of the circulator, and exits at port 3.

To create a device for combining or separating N wavelengths, one needs to cascade $N - 1$ FBGs and $N - 1$ circulators. Figure 10.22 illustrates a multiplexing function for the four wavelengths λ_1, λ_2, λ_3, and λ_4 using three FBGs and three circulators (labeled C_2, C_3, and C_4). The fiber grating filters labeled FBG$_2$, FBG$_3$,

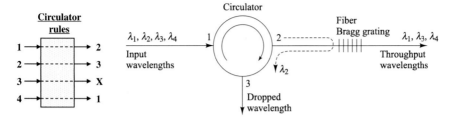

Fig. 10.21 Simple concept of a demultiplexing function using a fiber grating and an optical circulator

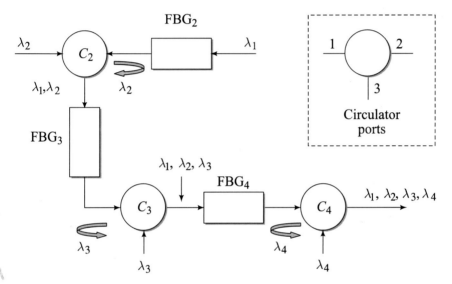

Fig. 10.22 Multiplexing of four wavelengths using three FBG devices and three circulators

and FBG$_4$ are constructed to reflect wavelengths λ_2, λ_3, and λ_4, respectively, and to pass all others.

The following steps show the multiplexer functions:

(a)   First consider the combination of circulator C_2 and fiber Bragg filter FBG$_2$. Here filter FBG$_2$ is designed to reflect wavelength λ_2 and allow wavelength λ_1 to pass through.

(b)   After the incoming wavelength λ_1 passes through FBG$_2$ it enters port 2 of circulator C_2 and exits from port 3. The incoming wavelength λ_2 enters port 1 of circulator C_2 and exits from port 2. After being reflected from FBG$_2$ it enters port 2 of circulator C_2 and exits from port 3 together with wavelength λ_1. These two wavelengths then continue passing through the next four elements to emerge from port 2 of circulator C_4.

(c) Next at circulator C_3 the incoming wavelength λ_3 enters port 3 of circulator C_3, exits from port 1, and then travels toward FBG$_3$. After being reflected from FBG$_3$ it enters port 1 of circulator C_3 and exits from port 2 together with wavelengths λ_1 and λ_2.

(d) After a similar process takes place at circulator C_4 and filter FBG$_4$ to insert wavelength λ_4, the four wavelengths all exit together from port 2 of circulator C_4 and can be coupled easily into a fiber.

The coupler size limitation when using FBGs is that one filter is needed for each wavelength and normally the operation is sequential with wavelengths being transmitted by one filter after another. Therefore the losses are not uniform from channel to channel, because each wavelength goes through a different number of circulators and fiber gratings, each of which adds loss to that channel. This may be acceptable for a small number of channels, but the loss differential between the first and last inserted wavelengths is a restriction for large channel counts.

10.5   Dielectric Thin-Film Filter (TFF)

A dielectric *thin-film filter* (TFF) is used as an *optical bandpass filter* [19–21]. This means that it allows a particular very narrow wavelength band to pass straight through it and reflects all others. The basis of these devices is a classical Fabry–Perot filter structure, which is a cavity formed by two parallel highly reflective mirror surfaces, as shown in Fig. 10.23. This structure is called a *Fabry–Perot interferometer* or an *etalon*. It also is known as a *thin-film resonant cavity filter*.

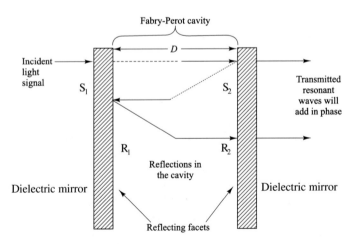

Fig. 10.23 Two parallel light-reflecting mirrored surfaces define a Fabry–Perot resonator cavity or an etalon

To see how it works, consider a light signal that is incident on the left surface S_1 of the etalon. After the light passes through the cavity and hits the inside surface S_2 on the right, some of the light leaves the cavity and some is reflected. The amount of light that is reflected depends on the reflectivity R of the surface S_2. If the roundtrip distance between the two mirrors is an integral multiple of a wavelength λ (i.e., λ, 2λ, 3λ, etc.), then all the light at those wavelengths that pass through the right facet *add in phase*. This means that these wavelengths *interfere constructively* in the device output beam so they add in intensity. These wavelengths are called the *resonant wavelengths* of the cavity. The etalon rejects all other wavelengths.

10.5.1 Applications of Etalon Theory

The transmission T of an ideal etalon in which there is no light absorption by the dielectric mirrors is an *Airy function* given by

$$T = \left[1 + \frac{4R}{(1-R)^2} \sin^2\left(\frac{\varphi}{2}\right)\right]^{-1} \qquad (10.53)$$

where R is the *reflectivity* of the mirrors (the fraction of light reflected by the mirror) and φ is the roundtrip phase change of the light beam. If one ignores any phase change at the mirror surface, then the phase change for a wavelength λ is

$$\varphi = \frac{2\pi}{\lambda} 2nD \cos\theta \qquad (10.54)$$

where n is the refractive index of the dielectric layer that forms the mirror, D is the distance between the mirrors, and θ is the angle to the normal of the incoming light beam.

Figure 10.24 gives a generalized plot of Eq. (10.53) over the range $-3\pi \leq \varphi \leq 3\pi$ for three different reflectivities ($R = 0.4, 0.7$, and 0.9). Because φ is proportional to the optical frequency $f = 2\pi/\lambda$, Fig. 10.24 shows that the power transfer function T is periodic in f (or λ). The wavelength range that can pass through the filter is called the *passband*. The peaks of the spacings occur at those wavelengths that satisfy the condition $N\lambda = 2nD$, where N is an integer. Thus in order for a single wavelength or a narrow spectral segment to be selected by the etalon filter from a wider spectral range, all the wanted wavelengths must lie within a passband of the filter transfer function. The distance between adjacent peaks (shown in Fig. 10.24) is called the *free spectral range* or FSR. The FSR can be expressed in terms of either the frequency or the wavelength. Respectively, these expressions (and two common notations) are

$$\text{FSR}_v = \Delta v_{FSR} = \frac{c}{2nD} \qquad (10.55)$$

Fig. 10.24 The behavior of
the resonant wavelengths in a
Fabry–Perot cavity for three
values of the mirror
reflectivity based on the Airy
function

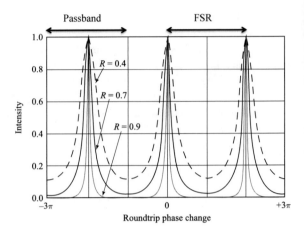

$$\text{FSR}_\lambda = \Delta\lambda_{FSR} = \frac{\lambda^2}{2nD} \qquad (10.56)$$

Here λ is the wavelength at the transmission peak.

Drill Problem 10.5 (*a*) If an engineer wants to use a material with a refractive index $n = 1.50$ to make an etalon with an FSR of $\Delta\nu_{FSR} = 50$ GHz, show that the etalon thickness should be 2.0 mm. (*b*) If this etalon is to be used at 1550 nm, show that the corresponding FSR in terms of wavelength is $\Delta\lambda_{FSR} = 0.40$ nm.

Another important parameter of an etalon is the measure of the full width of the transmission passband at its half-maximum value, which is designated by $\delta\lambda$ and is called the *full-width half-maximum* (FWHM). The FSR is related to the FWHM through the ratio

$$F = \frac{\Delta\lambda_{FSR}}{\delta\lambda} \qquad (10.57)$$

The parameter F is known as the *finesse* of the filter. For mirror reflectivities greater than 0.5, F can be approximated by the expression

$$F \approx \frac{\pi (R_1 R_2)^{1/4}}{1 - \sqrt{R_1 R_2}} \qquad (10.58)$$

Here R_1 and R_2 are the reflectivities of the etalon mirrors. This shows that the major factor influencing the finesse is the reflectivity of the mirrors; mirrors that have greater reflectivities yield a higher finesse. Absorption within the device, especially within the mirrors, reduces the sharpness of the filter peaks. Etalons with high finesse show sharper transmission peaks with lower minimum transmission coefficients. That is, the higher the finesse the narrower the passband and the sharper the boundaries.

> **Drill Problem 10.6** Suppose the end mirrors in an etalon have the same reflectivities $R_1 = R_2 = R$. Show that for devices with $R = 0.5$, 0.7, and 0.9, the corresponding values of the finesse are 4.4, 8.8, and 29.8. This shows that the finesse value is larger for sharper peaked Airy functions, that is, very high finesse values require highly reflective mirrors.

A typical TFF consists of multilayer thin-film coatings of alternating low-index and high-index materials, such as SiO_2 and Ta_2O_5, as shown in Fig. 10.25. The layers usually are deposited on a glass substrate. Each dielectric layer acts as a non-absorbing reflecting surface, so that the structure is a series of resonance cavities each of which is surrounded by mirrors. As shown in Fig. 10.26, the passband of the filter sharpens up to create a flat top for the filter when the number of cavities increases, which is a desirable characteristic for a practical filter. In Fig. 10.25 the

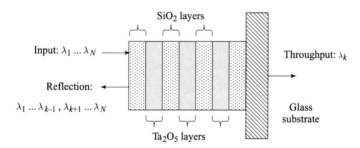

Fig. 10.25 A multilayer optical thin film filter consists of a stack of several dielectric thin films separated by resonance cavities

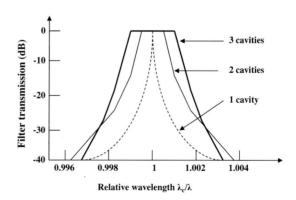

Fig. 10.26 Passband sharpening of a TFF when the number of resonance cavities increases

Table 10.6 Typical
parameter values of
commercial available 50-GHz
thin-film filters

Parameter	Unit	Value
Channel passband	GHz	$>\pm 10$ at 0.5 dB
Insertion loss at f_c ± 10 GHz	dB	<3.5
Polarization-dependent loss	dB	<0.20
Isolation (adjacent channels)	dB	>25
Isolation (nonadjacent channels)	dB	>40
Optical return loss	dB	>45
Polarization-mode dispersion	ps	<0.2
Chromatic dispersion	ps/nm	<50

filter is made such that if the input spectrum contains wavelengths λ_1 through λ_N, then only λ_k passes through the device. All the other wavelengths are reflected.

Thin-film filters are available in a wide range of passbands varying from 50 to 800 GHz and higher for widely spaced channels. Table 10.6 lists some operational characteristics of commercially available 50-GHz multilayer dielectric thin-film filters for use in fiber optic communication systems.

10.5.2   TFF Applications to WDM Links

To create a multiplexing device for combining or separating N wavelength channels, one needs to cascade $N - 1$ thin-film filters. Figure 10.27 illustrates a multiplexing function for the four wavelengths λ_1, λ_2, λ_3, and λ_4. Here the filters labeled TFF$_2$, TFF$_3$, and TFF$_4$ pass wavelengths λ_2, λ_3, and λ_4, respectively, and reflect all others. The filters are set at a slight angle in order to direct light from one TFF to another. First filter TFF$_2$ reflects wavelength λ_1 and allows wavelength λ_2 to pass through. These

Fig. 10.27 Multiplexing of
four wavelengths using
thin-film filters

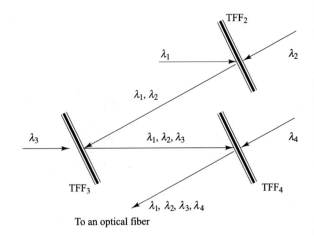

To an optical fiber

Table 10.7 Typical performance parameters for 8-channel DWDM and CWDM multiplexers based on thin-film-filter technology

Parameter	50-GHz DWDM	100-GHz DWDM	20-nm CWDM
Center wavelength accuracy	±0.1 nm	±0.1 nm	±0.3 nm
Channel passband @ 0.5-dB bandwidth	±0.20 nm	±0.11 nm	±6.5 nm
Insertion loss (dB)	≤1.0	≤1.0	≤2.0
Ripple in passband (dB)	≤0.5	≤0.5	≤0.5
Adjacent channel isolation (dB)	≥23	≥20	≥15
Directivity (dB)	≥50	≥55	≥50
Optical return loss (dB)	≥40	≥50	≥45
Polarization dependent loss (dB)	≤0.1	≤ 0.1	≤0.1
Thermal wavelength drift (nm/°C)	<0.001	<0.001	<0.003
Optical power capability (mW)	500	500	500

two signals then are reflected from filter TFF$_3$ where they are joined by wavelength λ_3, which has passed through TFF$_3$. After a similar process at filter TFF$_4$, the four wavelengths can be coupled into a fiber by means of a lens mechanism.

To separate the four wavelengths from one fiber into four physically independent channels, the directions of the arrows in Fig. 10.27 are reversed. Because a light beam loses some of its power at each TFF because the filters are not perfect, this multiplexing architecture works for only a limited number of channels. This usually is specified as being 16 channels or less.

Table 10.7 lists typical performance parameters for commercially available wavelength multiplexers based on thin-film filter technology. The parameters address 8-channel DWDM devices with 50- and 100-GHz channel spacings and an 8-channel 20-nm CWDM module.

10.6 Arrayed Waveguide Devices

A versatile WDM device is based on using an arrayed waveguide grating (AWG) [22, 23]. This device can function as a multiplexer, a demultiplexer, a drop-and-insert element, or a wavelength router. The arrayed waveguide grating is a generalization of the 2×2 Mach–Zehnder interferometer multiplexer. As Fig. 10.28 shows, the design consists of M_{in} input and M_{out} output slab waveguides designated as regions 1 and 6, respectively. The slab waveguides interface to two identical focusing planar star couplers located in regions 2 and 5. An array of N uncoupled waveguides that have a propagation constant β connect the star couplers together. In the grating array region, the path length of each waveguide differs by a very precise amount ΔL from the lengths in adjacent arms, so that the array forms a Mach–Zehnder type grating. For a pure multiplexer, $M_{in} = N$ and $M_{out} = 1$. The reverse holds for a demultiplexer,

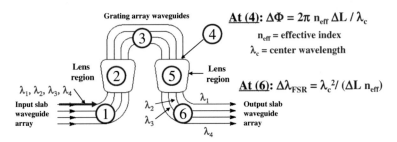

Fig. 10.28 Top view of a typical arrayed waveguide grating and designation of its various operating regions

that is $M_{in} = 1$ and $M_{out} = N$. In the case of a network routing application, it is common to have $M_{in} = M_{out} = N$.

Figure 10.29 depicts the geometry of the star coupler. The coupler acts as a lens of focal length L_f so that the object and image planes are located at a distance L_f from the transmitter and receiver slab waveguides, respectively. Both the input and output waveguides are positioned on the focal lines, which are circles of radius L_f /2. In Fig. 10.29, the parameter x is the center-to-center spacing between the input waveguides and the output waveguides at the star coupler interfaces, d is the spacing between the grating array waveguides, and θ is the diffraction angle in the input or output slab waveguide. The refractive indices of the star coupler and the grating array waveguides are n_s and n_c, respectively.

From a top-level viewpoint of Fig. 10.28, the AWG functions are as follows:

- Starting from the left, the input slab waveguides in region 1 are connected to a planar star coupler (region 2) that acts as a lens.

Fig. 10.29 Geometry of the star coupler used in the arrayed waveguide grating WDM device

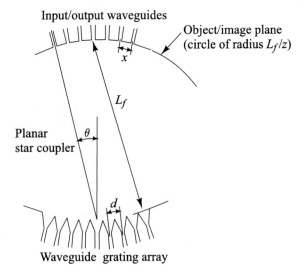

- The lens distributes the entering optical power among the different waveguides in the grating array in region 3.
- Adjacent waveguides of the grating array in region 3 differ in path length by a precise length ΔL. The path length differences ΔL can be chosen such that all input wavelengths emerge at point 4 with different phase delays $\Delta \Phi = 2\pi n_c \Delta L / \lambda$.
- The second lens in region 5 refocuses the light from all the grating array waveguides onto the output slab waveguide array in region 6.
- Thus each wavelength is focused into a different output waveguide in region

From the phase-matching condition, the light emitted from the output channel waveguides must satisfy the grating equation

$$n_s d \sin \theta + n_c \Delta L = m\lambda \tag{10.59}$$

where the integer m is the diffraction order of the grating.

Focusing is achieved by making the path length difference ΔL between adjacent array waveguides, measured inside the array, to be an integer multiple of the central design wavelength of the demultiplexer

$$\Delta L = m \frac{\lambda_c}{n_c} \tag{10.60}$$

where λ_c is the central wavelength in vacuum; that is, it is defined as the pass wavelength for the path from the center input waveguide to the center output waveguide.

To determine the channel spacing, it is necessary to find the angular dispersion. This is defined as the incremental lateral displacement of the focal spot along the image plane per unit frequency change, and is found by differentiating Eq. (10.59) with respect to frequency. Doing so, and considering the result in the vicinity of $\theta = 0$, yields

$$\frac{d\theta}{d\upsilon} = -\frac{m\lambda^2}{n_s c d} \frac{n_g}{n_c} \tag{10.61}$$

where the group index of the grating array waveguide is defined as

$$n_g = n_c - \lambda \frac{dn_c}{d\lambda} \tag{10.62}$$

In terms of frequency, the channel spacing $\Delta \upsilon$ is

$$\Delta \upsilon = \frac{x}{L_f} \left(\frac{d\theta}{d\upsilon} \right)^{-1} = \frac{x}{L_f} \frac{n_s c d}{m\lambda^2} \frac{n_c}{n_g} \tag{10.63}$$

or, in terms of wavelength,

$$\Delta\lambda = \frac{x}{L_f}\frac{n_s d}{m}\frac{n_c}{n_g} = \frac{x}{L_f}\frac{\lambda_c d}{\Delta L}\frac{n_s}{n_g} \tag{10.64}$$

Equations (10.63) and (10.64) thus define the pass frequencies or wavelengths for which the multiplexer operates, given that it is designed for a central wavelength λ_c. Note that by making ΔL large, the device can multiplex and demultiplex optical signals with very small wavelength spacings.

Example 10.13 Consider an $N \times N$ waveguide grating multiplexer having $L_f = 10$ mm, $x = d = 5$ μm, $n_c = 1.45$, and a central design wavelength $\lambda_c = 1550$ nm. What is the waveguide length difference ΔL and what is the channel spacing $\Delta\lambda$ for $m = 1$?

Solution: For $m = 1$, the waveguide length difference from Eq. (10.60) is

$$\Delta L = (1)\frac{1.550}{1.45} = 1.069 \ \mu m$$

If $n_s = 1.45$ and $n_g = 1.47$, then from Eq. (10.64) the channel spacing is

$$\Delta\lambda = \frac{x}{L_f}\frac{n_s d}{m}\frac{n_c}{n_g} = \frac{5}{10^4}\frac{(1.45)(5)}{1}\frac{1.45}{1.47} = 3.58 \text{ nm}$$

Equation (10.59) shows that the phased array is periodic for each path through the device, so that after every change of 2π in θ between adjacent waveguides the field will again be imaged at the same spot. The period between two successive field maxima in the frequency domain is called the *free spectral range* (FSR) $\Delta\nu_{FSR}$ and can be represented by the relationship

$$\Delta\nu_{FSR} = \frac{c}{n_g(\Delta L + d\sin\theta_i + d\sin\theta_o)} \tag{10.65}$$

where θ_i and θ_o are the diffraction angles in the input and output waveguides, respectively. These angles generally are measured from the center of the array, so that $\theta_i = jx/L_f$ and $\theta_o = kx/L_f$ for the jth input port and the kth output port, respectively, on either side of the central port. This shows that the FSR depends on which input and output ports the optical signal utilizes. When the ports are across from each other, so that $\theta_i = \theta_o = 0$, then

$$\Delta\nu_{FSR} = \frac{c}{n_g \Delta L} \tag{10.66}$$

Alternatively, the FSR can be expressed in terms of a wavelength separation $\Delta\lambda_{FSR}$ as

$$FSR = \Delta\lambda_{FSR} = \frac{\lambda_c^2}{\Delta L n_c} \tag{10.67}$$

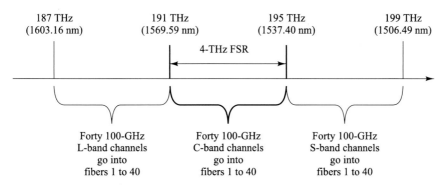

Fig. 10.30 The FSR specifies the spectral width that will be separated across the output waveguides of an AWG

Example 10.14 As shown in Fig. 10.30, suppose an AWG is designed to separate light in the 4-THz-wide frequency range in the C-band running from 195.00 THz (1537.40 nm) to 191.00 THz (1569.59 nm) into forty 100-GHz channels. Then it also will separate the next higher-frequency 4-THz spectral segment in the S-band and lower-frequency 4-THz spectral segment in the L-band into the same forty output fibers. The free spectral range $\Delta\lambda_{FSR}$ can be determined from Eq. (10.67). For the 4-THz frequency range denoted here, the center wavelength λ_c is 1550.5 nm, the free spectral range $\Delta\lambda_{FSR}$ should be at least 32.2 nm in order to separate all the wavelengths into distinct fibers, and the effective refractive index n_c is nominally 1.45 in silica. Then the length difference between adjacent array waveguides is $\Delta L = 51.49 \ \mu m$.

The passband shape of the AWG filter versus wavelength can be altered by the design of the input and output slab waveguides. Two common passband shapes are shown in Fig. 10.31. On the left is the *normal* or *Gaussian* passband. This passband shape exhibits the lowest loss at the peak, but the fact that it rolls off quickly on either side of the peak means that it requires a high stabilization of the laser wavelength. Furthermore, for applications where the light passes through several AWGs, the accumulative effect of the filtering function reduces the passband to an extremely small value. An alternative to the Gaussian passband shape is the *flattop* or *wideband* shape, as shown on the right in Fig. 10.31. This wideband device has a uniform insertion loss across the passband, and is therefore not as sensitive to laser drift or the sensitivity of a cascade of filters as is the Gaussian passband. However, the loss in a flattop device is usually 2–3 dB higher than that in a Gaussian AWG. Table 10.8 compares the main operating characteristics of these two designs for a typical 40-channel AWG.

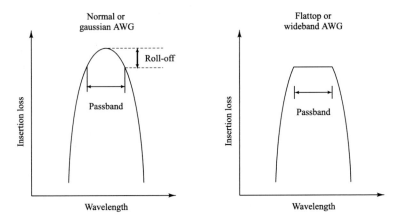

Fig. 10.31 Two common optical-filter passband shapes: normal or Gaussian and flat-top or wideband

Table 10.8 Performance characteristics of typical 40-channel arrayed waveguide gratings (AWGs)

Parameter	Gaussian	Wideband
Channel spacing	100 GHz	100 GHz
1-dB bandwidth	>0.2 nm	>0.4 nm
3-dB bandwidth	>0.4 nm	>0.6 nm
Insertion loss	<5 dB	<7 dB
Polarization dependent loss	<0.25 dB	<0.15 dB
Adjacent channel crosstalk	30 dB	30 dB
Passband ripple	1.5 dB	0.5 dB
Optical return loss	45 dB	45 dB
Size ($L \times W \times H$)	130 × 65 × 15 (mm)	130 × 65 × 15 (mm)

10.7   WDM Applications of Diffraction Gratings

Another DWDM technology is based on diffraction gratings [24, 25]. A *diffraction grating* is a conventional optical device that spatially separates the different wavelengths contained in a beam of light. The device consists of a set of diffracting elements, such as narrow parallel slits or grooves, separated by a distance comparable to the wavelength of light. These diffracting elements can be either reflective or transmitting, thereby forming a reflection grating or a transmission grating, respectively. Separating and combining wavelengths with diffraction gratings is a parallel process, as opposed to the serial process that is used with the fiber-based Bragg gratings.

Reflection gratings are fine ruled or etched parallel lines on some type of reflective surface. With these gratings, light will bounce off the grating at an angle. The angle at which the light leaves the grating depends on its wavelength, so the reflected light fans out in a spectrum. For DWDM applications, the lines are spaced equally and each individual wavelength will be reflected at a slightly different angle, as shown in Fig. 10.32. There can be a reception fiber at each of the positions where the reflected light gets focused. Thus individual wavelengths will be directed to separate fibers. The reflective diffraction grating works reciprocally; that is, if different wavelengths come into the device on the individual input fibers, all of the wavelengths will be focused back into one fiber after traveling through the device. One also could have a photodiode array in place of the receiving fibers for functions such as monitoring the power in each wavelength.

One type of *transmission grating*, which is known as a *phase grating,* consists of a periodic variation of the refractive index of the grating. These may be characterized by a Q-parameter, which is defined as

$$Q = \frac{2\pi \lambda d}{n_g \Lambda^2 \cos \alpha} \tag{10.68}$$

where λ is wavelength, d is the thickness of the grating, n_g is the refractive index of the material, Λ is the grating period, and α is the incident angle, as shown in Fig. 10.33. The phase grating is called *thin* for $Q < 1$ and *thick* for $Q > 10$. After a spectrum of wavelength channels passes through the grating, each wavelength emerges at a slightly different angle and can be focused into a receiving fiber.

Fig. 10.32 The angle at which reflected light leaves a reflection grating depends on its wavelength

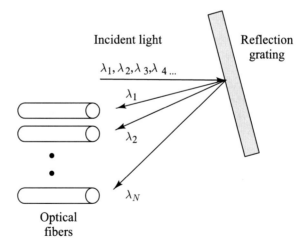

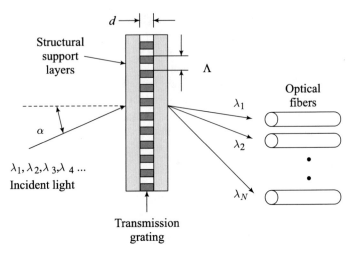

Fig. 10.33 Each wavelength emerges at a slightly different angle after passing through a transmission grating

10.8   Summary

Wavelength division multiplexing or WDM allows the combining of a number of independent information-carrying wavelengths onto the same fiber. This multiplexing can be accomplished because an optical fiber has a wide spectral region in which optical signals can be transmitted efficiently. For full-spectrum fibers this region includes the O-band through the L-band, which ranges from about 1260–1675 nm. The light sources used in high-capacity optical fiber communication systems emit in a narrow wavelength band of less than 1 nm, so many different independent optical channels can be used simultaneously in different segments of this wavelength range.

At the transmitting end in a WDM system there are several independently modulated light sources, each emitting signals at a unique wavelength. Here a multiplexer is needed to combine these optical outputs into a continuous spectrum of signals and couple them onto a single fiber. At the receiving end a demultiplexer is required to separate the optical signals into appropriate detection channels for signal processing.

The implementation of sophisticated WDM networks requires a variety of optical devices to combine, distribute, isolate, and separate optical power at different wavelengths. Passive devices require no external control for their operation. These components are mainly used to split and combine or tap off optical signals. The wavelength-dependent performance of active devices can be controlled electronically or optically, thereby providing a large degree of network flexibility. As later chapters discuss, active WDM components include optical amplifiers, wavelength switches, and optical wavelength converters.

Problems

10.1 A DWDM optical transmission system is designed to have 100-GHz channel spacings. How many wavelength channels can be utilized in the 1536-to-1556-nm spectral band?

10.2 Assume that a 32-channel DWDM system has uniform channel spacings of $\Delta\nu = 100$ GHz and let the frequency ν_n correspond to the wavelength λ_n. Using this correspondence, let the wavelength $\lambda_1 = 1550$ nm. Calculate the wavelength spacing between the first two channels (between channels 1 and 2) and between the last two channels (between channels 31 and 32). From the result, what can be concluded about using an equal wavelength spacing definition in this wavelength band instead of the standard equal frequency channel spacing specification?

10.3 Assume that for a given tap coupler the throughput and coupled powers are 230 μW and 5 μW, respectively, for an input power of 250 μW.

    (a) What is the coupling ratio?
    (b) What are the insertion losses?
    (c) Find the excess loss of the coupler.

10.4 A product sheet for a 2 × 2 single-mode biconical tapered coupler with a 40/60 splitting ratio states that the insertion losses are 2.7 dB for the 60-percent channel and 4.7 dB for the 40-percent channel.

    (a) If the input power $P_0 = 200$ μW, find the output levels P_1 and P_2.
    (b) Find the excess loss of the coupler.
    (c) From the calculated values of P_1 and P_2, verify that the splitting ratio is 40/60.

10.5 Consider the coupling ratios as a function of pull lengths shown in Fig. 10.34 for a fused biconical tapered coupler. The performances are given for 1310-nm and 1540-nm operation. Discuss the behavior of the coupler for each wavelength if its pull length is stopped at the following points: A, B, C, D, E, and each F.

10.6 Suppose an engineer has two 2 × 2 waveguide couplers (couplers A and B) that have identical channel geometries and spacings, and are formed on the same substrate material. If the index of refraction of coupler A is larger than that of coupler B, which device has a larger coupling coefficient κ? What does this imply about the device lengths needed in each case to form a 3-dB coupler?

10.7 Consider an optical fiber transmission star coupler that has seven inputs and seven outputs. Suppose the coupler is constructed by arranging the seven fibers in a circular pattern (a ring of six with one in the center). This fiber bundle then is put against the end a glass rod that serves as the mixing element. Let the rod diameter be 300 μm.

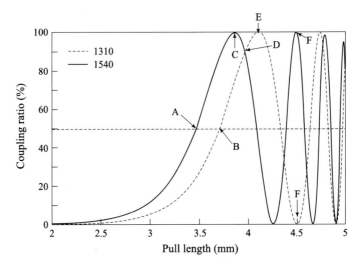

Fig. 10.34 Pull-length-dependent coupling ratios (for Prob. 10.5)

(a)   If the fibers have 50-μm core diameters and 125-μm outer cladding diameters, what is the coupling loss resulting from light escaping between the output fiber cores? Assume the fiber cladding is not removed.If the fibers have 50-μm core diameters and 125-μm outer cladding diameters, what is the coupling loss resulting from light escaping between the output fiber cores? Assume the fiber cladding is not removed.

(b)   What is the coupling loss if the fiber ends are arranged in a row and a 50-μm × 800-μm glass plate is used as the star coupler?

10.8   Repeat Problem 10.7 for seven fibers that have 200-μm core diameters and 400-μm outer cladding diameters. What should the sizes of the glass rod and the glass plate be in this case?

10.9   Suppose an $N \times N$ star coupler is constructed of n 3-dB 2 × 2 couplers, each of which has a 0.1-dB excess loss. Find the maximum value of n and the maximum size N if the power budget for the star coupler is 30 dB.

10.10   Consider the 4 × 4 multiplexer shown in Fig. 10.16.

(a)   If $\lambda_1 = 1548$ nm and $\Delta v = 125$ GHz, what are the four input wavelengths?

(b)   If $n_{eff} = 1.5$, what are the values of ΔL_1 and $\Delta L3$?

10.11   Following the same line of analysis as in Example 10.11, use 2 × 2 Mach–Zehnder interferometers to design an 8-to-1 multiplexer that can handle a channel separation of 25 GHz. Let the shortest wavelength be 1550 nm. Specify the value of ΔL for the 2 × 2 MZIs in each stage.

10.12 Consider a wavelength multiplexer made from circulators and fiber Bragg gratings as shown in Fig. 10.22. Assume the input optical power of each wavelength λ_1 through λ_4 is 1 mW. Let both the insertion loss and the throughput loss of each FBG be 0.25 dB, and let the insertion loss of a circulator be 0.6 dB. What is the power level of each wavelength λ_1 through λ_4 emerging from the final circulator C_4?

10.13 A 0.5-cm-long fiber Bragg grating is constructed by irradiating a single-mode fiber with a pair of 244-nm ultraviolet light beams. The fiber has V $= 2.405$ and $n_{eff} = 1.48$. The half- angle between the two beams is $\theta/2 =$ 13.5°. If the photo-induced index change is 2.5×10^{-4}, find the following:

    (a)   The grating period,
    (b)   The Bragg wavelength,
    (c)   The coupling coefficient,
    (d)   The full bandwidth $\Delta\lambda$ measured between the zeros on either side of R_{max},
    (e)   The maximum reflectivity.

10.14 Consider a wavelength multiplexer made from a series of thin-film filters as shown in Fig. 10.27. Assume the input optical power of each wavelength λ_1 through λ_4 is 1 mW. For each TFF let the throughput loss equal 1.0 dB and let the reflection loss be 0.4 dB. What is the power level of each wavelength λ_1 through λ_4 emerging from the final thin-film filter TFF_4?

Answers to Selected Problems.

10.1 Channel spacing is 4 nm; 5 channels can be utilized

10.2 $\lambda_1 - \lambda_2 = 1550.00 - 1549.20$ nm $= 0.80$ nm;

$$\lambda_{31} - \lambda_{32} = 1526.34 - 1525.56 \, \text{nm} = 0.78 \, \text{nm}$$

10.3 (a) 2%; (b) $IL_{0\to1} = 0.36$ dB, $IL_{0\to2} = 17.0$ (c) 0.27 dB

10.4 (a) $P_1 = 107.4 \, \mu W$, $P_2 = 67.8 \, \mu W$; (b) 0.58 dB; (c) $P_1/P_0 = 61\%$, $P_2/P_0 = 39$

10.5 The following coupling percentages are realized when the pull length is stopped at the designated points:

Coupling percentages from input fiber to output 2.

Points	A	B	C	D	E	F
1310 nm	25	50	75	90	100	0
1540 nm	50	88	100	90	50	100

10.6 Because $\beta_z \propto n$, then for $n_A > n_B$ we have $\kappa_A < \kappa_B$. Thus, because one needs to have $\kappa_A L_A = \kappa_B L_B$, it is necessary to have $L_A > L_B$.

10.7   (a) The coupling loss is found from the area mismatch between the fiber-core end-face areas and the coupling-rod cross-sectional area. If a is the fiber-core radius and R is the coupling-rod radius, then the coupling loss is

$$L_{coupling} = 10 \log \frac{P_{out}}{P_{in}} = 10 \log \frac{7\pi a^2}{\pi R^2} = 10 \log \frac{7(25)^2}{(150)^2} = -7.11 \text{ dB}$$

(b)   Similarly, for the linear-plate coupler

$$L_{coupling} = 10 \log 7\pi a^2 / (\text{length} \times \text{width}) = 10 \log \frac{7\pi(25)^2}{800(50)} = -4.64 \text{ dB}$$

10.8   (a) The diameter of the circular coupling rod must be 1000 μm. The coupling loss is

$$L_{coupling} = 10 \log \frac{7\pi a^2}{\pi R^2} = 10 \log \frac{7(100)^2}{(500)^2} = -5.53 \text{ dB}$$

(b)   The size of the plate coupler must be 200 μm by 2600 μm.

$$\text{The coupling loss} = 10 \log \frac{7\pi(100)^2}{200(2600)} = -3.74 \text{ dB}$$

10.9 $n = 9$ and $N = 2^n = 512$
10.10   (a) 1549, 1550, and 1551 nm;
(b) $\Delta L_1 = 0.4$ mm, $\Delta L_3 = 0.8$ mm
10.11 $\Delta L_1 = 0.75$ mm, $\Delta L_2 = 1.5$ mm, $\Delta L_3 = 3.0$ mm
10.12 λ_1 loss: 2.55 dB, λ_2 loss: 3.15 dB, λ_3 loss: 2.30 dB, λ_4 loss: 1.45 dB
10.13   (a) 523 nm; (b) 1547 nm; (c) 4.2 cm^{-1}; (d) 3.9 nm; (e) 94%
10.14 λ_1 loss: 1.2 dB, λ_2 loss: 1.8 dB, λ_3 loss: 1.4 dB, λ_4 loss: 1.0 dB

References

1. C.A. Brackett, Dense wavelength division multiplexing networks: principles and applications. IEEE J. Sel. Areas Commun. **8**, 948–964 (1990)
2. G. Keiser, *FTTX Concepts and Applications* (Wiley, New York, 2006)
3. N. Antoniades, G. Ellias, I. Roudas, (eds.), *WDM Systems and Networks* (Springer, 2012)
4. C. Kachris, K. Bergman, I. Tomkos, *Optical Interconnects for Future Data Center Networks* (Springer, Berlin, 2013)

5. D. Chadha, *Optical WDM Networks: From Static to Elastic Networks* (Wiley-IEEE, 2019)
6. A. Paradisi, R. Carvalho Figueirdo, A. Chiuchiarelli, E. de Sousa Rosa, (eds.) *Optical Communications* (Springer, Berlin, 2019)
7. ITU-T Recommendation G.692, *Optical Interfaces for Multichannel Systems with Optical Amplifiers*, Oct. 1998; Amendment 1 (2005)
8. ITU-T Recommendation G.694.1, *Dense Wavelength Division Multiplexing (DWDM)* (2012)
9. ITU-T Recommendation G.694.2, *Coarse Wavelength Division Multiplexing (CWDM)* (2003)
10. ITU-T Recommendation G.695, *Optical Interfaces for Coarse Wavelength Division Multiplexing Applications* (2018)
11. V.J. Tekippe, Passive fiber optic components made by the fused biconical taper process. Fiber Integr. Opt. **9**(2), 97–123 (1990)
12. C.-L. Chen, *Foundations for Guided Wave Optics* (Wiley, New York, 2007)
13. B.E.A. Saleh, M. Teich, *Fundamentals of Photonics*, 3rd edn. (Wiley, New York, 2019)
14. R. Tewari, K. Thyagarajan, Analysis of tunable single-mode fiber directional couplers using simple and accurate relations. J. Lightw. Technol. **4**, 386–390 (1986)
15. J. Pietzsch, Scattering matrix analysis of 3 × 3 fiber couplers. J. Lightw. Technol. **7**, 303–307 (1989)
16. R.G. Hunsperger, *Integrated Optics: Theory and Technology*, 6th edn. (Springer, Berlin, 2009)
17. J.W. Arkwright, D.B. Mortimore, R.M. Adams, Monolithic 1 × 19 single-mode fused fiber couplers. Electron. Lett. **27**, 737–738 (1991)
18. R. Kashyap, *Fiber Bragg Gratings*, 2nd edn. (Academic Press, 2010)
19. J. Jiang, J.J. Pan, Y.H. Guo, G. Keiser, Model for analyzing manufacturing-induced internal stresses in 50-GHz DWDM multilayer thin film filters and evaluation of their effects on optical performances. J. Lightw. Technol. **23**, 495–503 (2005)
20. V. Kochergin, *Omnidirectional Optical Filters* (Springer, Berlin, 2003)
21. H.A. Macleod, *Thin-Film Optical Filters*, 5th edn. (CRC Press, 2018)
22. M.K. Smit, C. van Dam, PHASAR-based WDM devices: principles, design and applications. IEEE J. Sel. Top. Quantum Electron. **2**, 236–250 (1996)
23. H. Uetsuka, AWG technologies for dense WDM applications. IEEE J. Sel. Top. Quantum Electron. **10**, 393–402 (2004)
24. C.F. Lin, *Optical Components for Communications: Principles and Applications*, (Springer, Berlin, 2004)
25. H. Venghaus, *Wavelength Filters in Fibre Optics* (Springer, Berlin, 2006)

Chapter 11
Basics of Optical Amplifiers

Abstract The creation and development of optical amplifiers has provided significant increases in information capacity in applications ranging from ultra-long undersea links to short links in access networks. This chapter describes the three main optical amplifier types, which are semiconductor optical amplifiers, active fiber or doped-fiber amplifiers, and Raman amplifiers. The topics include noise effects generated in the amplification process, the concept of optical signal-to-noise ratio and its relation to bit-error rate, the operation and use of optical amplifiers based on the Raman scattering mechanism, and wideband optical amplifiers that operate over several wavelength bands simultaneously.

Traditionally, when setting up an optical link, one formulates a power budget and adds repeaters when the path loss exceeds the available power margin. To amplify an optical signal with a conventional repeater, it is necessary to perform photon-to-electron conversion, electrical amplification, retiming, pulse shaping, and then electron-to-photon conversion. Although this process works well for moderate-speed single-wavelength operation, it can present a data transmission bottleneck for high-speed multiple-wavelength systems. Thus, to eliminate the transmission delay problem a great deal of effort has been expended to develop all-optical amplifiers. These devices operate completely in the optical domain to boost the power levels of multiple lightwave signals over spectral bands of 30 nm and more [1–5].

The three fundamental amplifier types are semiconductor optical amplifiers (SOAs), doped-fiber amplifiers (DFAs), and Raman amplifiers. In addition, the literature uses the term hybrid optical amplifier (HOA), which is the combination of more than one optical amplifier in any type of configuration. This chapter first classifies the three basic types and usage of optical amplifiers in Sect. 11.1. Then Sect. 11.2 discusses SOAs, which are based on the same operating principles as laser diodes. This discussion includes external pumping principles and signal gain mechanisms. Next, Sect. 11.3 gives details on erbium-doped fiber amplifiers (EDFAs), which are widely used in the C-band (1530–1565 nm) for optical communication networks. Noise effects generated in the amplification process are discussed in Sect. 11.4. The concept of optical signal-to-noise ratio (OSNR) and its relation to bit-error rate is discussed in Sect. 11.5. The topic of Sect. 11.6 concerns system applications of

© The Author(s), under exclusive license to Springer Nature Singapore Pte Ltd. 2021   437
G. Keiser, *Fiber Optic Communications*,
https://doi.org/10.1007/978-981-33-4665-9_11

EDFAs when they are used in three basic locations. Section 11.7 addresses the operation and use of optical amplifiers based on the Raman scattering mechanism. Finally, Sect. 11.8 describes wideband optical amplifiers that operate over several wavelength bands simultaneously.

11.1   Fundamental Optical Amplifier Types

Optical amplifiers have found widespread use in diverse applications ranging from ultra-long undersea links to short links in access networks. In long-distance undersea and terrestrial point-to-point links, the traffic patterns are relatively stable, so that the input power levels to an optical amplifier do not vary significantly. However, because many closely spaced wavelength channels are being transported over these links, the amplifier must have a wide spectral response range and be highly reliable. Usually fewer wavelengths are carried on metro and access network links, but the traffic patterns can come in bursts and wavelengths often can be added or dropped randomly depending on customers' demand for service. Optical amplifiers for these applications thus need to be able to recover quickly from rapid input power variations when the number of amplified channels suddenly changes. Although these diverse applications offer different optical amplifier design challenges, all devices share some basic operational requirements and performance characteristics, which are given in this section.

11.1.1   General Applications of Optical Amplifiers

Figure 11.1 shows general applications of the following three classes of optical amplifiers:

In-line Optical Amplifiers In a single-mode link, the effects of fiber dispersion may be small so that the main limitation to repeater spacing is fiber attenuation. Because such a link does not necessarily require a complete regeneration of the signal, simple amplification of the optical signal is sufficient. Thus an in-line optical amplifier can be used to compensate for transmission loss and increase the distance between regenerative repeaters, as Fig. 11.1a illustrates.

Preamplifier Fig. 11.1b shows an optical amplifier being used as a frontend preamplifier for an optical receiver. Thereby, a weak received optical signal is amplified before photodetection so that the signal-to-noise ratio degradation caused by thermal noise in the receiver electronics can be suppressed. Compared with other frontend devices such as avalanche photodiodes or optical heterodyne detectors, an optical preamplifier provides a larger gain factor and a broader bandwidth.

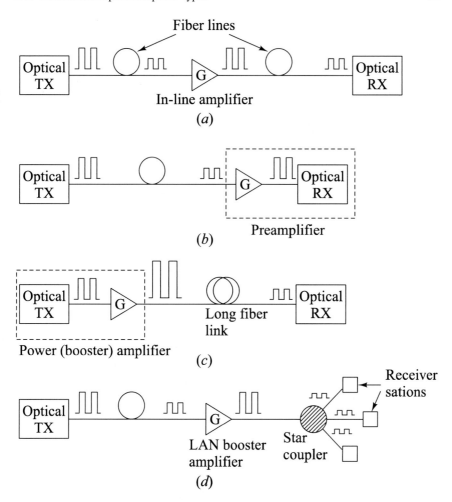

Fig. 11.1 Four possible applications of optical amplifiers: **a** in-line amplifier to increase transmission distance, **b** preamplifier to improve receiver sensitivity, **c** booster of transmitted power, **d** booster of signal level in a local area network

Power Amplifier Power or booster amplifier applications include placing the device immediately after an optical transmitter to boost the transmitted power, as Fig. 11.1c shows. This serves to increase the transmission distance by 10–100 km depending on the amplifier gain and fiber loss. As an example, using this technique together with an optical preamplifier at the receiving end can enable repeaterless undersea transmission distances of 200–250 km. One can also employ an optical amplifier in a local area network to compensate for coupler-insertion loss and power-splitting loss. Figure 11.1d shows an example for boosting the optical signal in front of a passive star coupler so sufficient power arrives at each receiver.

11.1.2 Amplifier Classifications

The three main optical amplifier types can be classified as semiconductor optical amplifiers (SOAs), active-fiber or doped-fiber amplifiers (DFAs), and Raman amplifiers. This section first gives a concise overview of these amplifier types; details are in the following sections. All optical amplifiers increase the power level of incident light through a stimulated emission or an optical power transfer process. In SOAs and DFAs the mechanism for creating the population inversion that is needed for stimulated emission to occur is the same as the process that is used in laser diodes. Although the structure of such an optical amplifier is similar to that of a laser, it does not have the optical feedback mechanism that is necessary for lasing to take place. Thus an optical amplifier can boost incoming signal levels, but it cannot generate a coherent optical output by itself.

The basic operation of a generic optical amplifier is shown in Fig. 11.2. Here, the device absorbs energy supplied from an external source called the *pump*. The pump supplies energy to electrons in an active medium, which raises them to higher energy levels to produce a population inversion. An incoming signal photon will trigger these excited electrons to drop to lower levels through a stimulated-emission process. Because one incoming trigger photon stimulates a cascade effect in which many excited electrons emit photons of equal energy as they drop to the ground state, the result is an amplified optical signal.

As Sect. 11.7 describes, in contrast to the amplification mechanisms used in an SOA or DFA, in Raman amplification there is a transfer of optical power from a high-power pump wavelength (e.g., 500 mW at 1480 nm) to lightwave signals at longer wavelengths (e.g., a -25-dBm signal around 1550 nm). This Raman amplification mechanism is done without the need for a population-inversion process.

Alloys of semiconductor materials from groups III and V (e.g., phosphorous, gallium, indium, and arsenic) make up the active medium in SOAs. The attractiveness of SOAs is that these devices can be made to work in the O-band (around 1310 nm) as well as in the C-band. They can be integrated easily on the same substrate as other optical devices and circuits (e.g., couplers, optical isolators, and receiver circuits), and compared with DFAs they consume less electrical power, have fewer components, and are more compact. SOAs have a more rapid gain response, which is on the

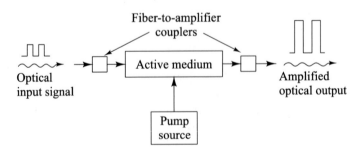

Fig. 11.2 Basic operation of a generic optical amplifier

order of 1–100 ps. This rapid response results in both advantages and limitations. The advantage is that SOAs can be implemented when both switching and signal-processing functions are called for in optical networks. The limitation is that the rapid carrier response causes the gain at a particular wavelength to fluctuate with the signal rate for speeds up to several Gb/s. Because this fluctuation affects the overall gain, the signal gain at other wavelengths also fluctuates. Thus the rapid gain response gives rise to crosstalk effects when a broad spectrum of wavelengths must be amplified simultaneously.

In DFAs, the active medium for operation in the S-, C-, and L-bands is created by lightly doping a silica (silicon dioxide) or tellurite (tellurium oxide) fiber core with rare-earth elements such as thulium (Tm), erbium (Er), or ytterbium (Yb). The DFAs for the O-band are achieved through doping fluoride-based fibers (rather than silica fibers) with elements such as neodymium (Nd) and praseodymium (Pr). The important features of DFAs include the ability to pump the devices simultaneously at several different wavelengths, low coupling loss to the compatible-sized fiber transmission medium, and very low dependence of gain on light polarization. In addition, DFAs are highly transparent to signal format and bit rate, because they exhibit slow gain dynamics, with carrier lifetimes on the order of 0.1–10 ms. The result is that, in contrast to SOAs, the gain responses of DFAs are basically constant for signal modulations greater than a few kilohertz. Consequently, they are immune from interference effects (such as crosstalk and intermodulation distortion) between different optical channels when wavelength channels in a broad spectrum (e.g., in a 30-nm spectral band ranging from 1530 to 1560 nm) are injected simultaneously into the amplifier.

A Raman optical amplifier is based on a nonlinear effect called stimulated Raman scattering (SRS), which occurs in fibers at high optical powers (see Chap. 12). Whereas a DFA requires a specially constructed optical fiber for its operation, Raman amplification takes place within a standard transmission fiber. The Raman gain mechanism can be achieved through either a lumped (or discrete) amplifier or a distributed amplifier. In the lumped Raman amplifier configuration, a spool of about 80 m of small-core fiber along with appropriate pump lasers is inserted into the transmission path as a distinct packaged unit. For the distributed Raman amplifier application, one or more Raman pump lasers convert the final 20–40 km of the transmission fiber into a preamplifier. Because the Raman gain in a particular spectral range is derived from the SRS-induced transfer of optical power from shorter pump wavelengths to longer signal wavelengths, these amplifiers can be designed for use in any wavelength band.

Table 11.1 lists some possible optical amplifier structures and their operating ranges. The following sections give details on their characteristics.

11.2 Semiconductor Optical Amplifiers

A *semiconductor optical amplifier* is essentially an InGaAsP laser that is operating below its lasing threshold point [6, 7]. Analogous to the construction of a laser diode,

Table 11.1 Various optical amplifier structures and their operating regions

Acronym	Structure	Operating band
GC-SOA	Gain-clamped semiconductor optical amplifier	O- or C-band
PDFFA	Praseodynium-doped fluoride fiber amplifier	O-band
TDFA	Thulium-doped fiber amplifier	S-band
EDFA	Erbium-doped fiber amplifier	C-band
GS-EDFA	Gain-shifted EDFA	L-band
ETDFA	Er/Tm-doped tellurite (tellurium oxide) glass fiber	C- and L-bands
RFA	Raman fiber amplifier	1260–1650 nm

the gain peak of an SOA can be selected in any narrow wavelength band extending from 1280 nm in the O-band to 1650 nm in the U-band by varying the composition of the active InGaAsP material. Most SOAs belong to the *traveling-wave* (TW) *amplifier* category. This means that in contrast to the laser feedback mechanism where the optical signal makes many back and forth passes through the lasing cavity in the SOA, in a TW amplifier the optical signal travels through the device only once. During this single passage the optical signal gains energy and emerges intensified at the other end of the amplifier.

The SOA construction is similar to a resonator cavity structure of a laser diode. The SOA has an active region of length L, width w, and height d. The end facets have reflectivities R_1 and R_2. However, in contrast to a semiconductor laser diode in which the reflectivities are around 0.3, the parameters R_1 and R_2 for an SOA are dramatically lower in order for the optical signal to pass through the amplification cavity only once. Low reflectivities of about 10^{-4} are achieved by depositing thin layers of silicon oxide, silicon nitride, or titanium oxide on the SOA end facets.

11.2.1 External Pumping of Active Medium

External current injection is the pumping method used to create the population inversion needed for having a gain mechanism in SOAs. This is similar to the operation of laser diodes (see Sect. 4.3). Thus, from Eq. (4.31), the sum of the injection, stimulated emission, and spontaneous recombination rates gives the rate equation that governs the carrier density $n(t)$ in the excited state

$$\frac{\partial n(t)}{\partial t} = R_p(t) - R_{st}(t) - \frac{n(t)}{\tau_r} \tag{11.1}$$

where

$$R_p(t) = \frac{J(t)}{qd} \tag{11.2}$$

is the external pumping rate from the injection current density $J(t)$ into an active layer of thickness d, τ_r is the combined time constant coming from spontaneous emission and carrier-recombination mechanisms, and

$$R_{st}(t) = \Gamma a V_g (n - n_{th}) N_{ph} \equiv g V_g N_{ph} \tag{11.3}$$

is the net stimulated emission rate. Here, V_g is the group velocity of the incident light, Γ is the optical confinement factor, a is a gain constant (which depends on the optical frequency v), n_{th} is the threshold carrier density, N_{ph} is the photon density, and g is the overall gain per unit length. Given that the active area of the optical amplifier has a width w and a thickness d, then for an optical signal of power P_s with photons of energy hv and group velocity V_g, the photon density is

$$N_{ph} = \frac{P_s}{V_g \, hv \, (wd)} \tag{11.4}$$

In the steady state, $\partial n(t)/\partial t = 0$, so that Eq. (11.1) becomes

$$R_p = R_{st} + \frac{n}{\tau_r} \tag{11.5}$$

Now substitute Eq. (11.2) for R_p, the second equality in Eq. (11.3) for R_{st}, and the first equality in Eq. (11.3) solved for n into Eq. (11.5). Solving for g then yields the *steady-state gain per unit length*

$$g = \frac{\frac{J}{qd} - \frac{n_{th}}{\tau_r}}{V_g N_{ph} + 1/\Gamma a \tau_r} = \frac{g_0}{1 + N_{ph}/N_{ph;sat}} \tag{11.6}$$

where

$$N_{ph;sat} = \frac{1}{\Gamma a V_g \tau_r} \tag{11.7a}$$

is defined as the *saturation photon density*, and

$$g_0 = \Gamma a \tau_r \left(\frac{J}{qd} - \frac{n_{th}}{\tau_r} \right) \tag{11.7b}$$

is the medium gain per unit length in the absence of signal input (when the photon density is zero), which is called the *zero-signal* or *small-signal gain per unit length*.

Example 11.1 Consider an InGaAsP SOA with $w = 5$ μm and $d = 0.5$ μm. Given that $V_g = 2 \times 10^8$ m/s, if a 1.0-μW optical signal at 1550 nm enters the device, what is the photon density?

Solution: From Eq. (11.4) the photon density is

$$N_{ph} = \frac{1 \times 10^{-6} \text{W}}{\left(2 \times 10^8 \text{m/s}\right) \frac{\left(6.626 \times 10^{-34} \text{J·s}\right)\left(3 \times 10^8 \text{m/s}\right)}{\left(1.55 \times 10^{-6} \text{m}\right)} (5\,\mu\text{m})(0.5\,\mu\text{m})}$$

$$= 1.56 \times 10^6 \text{ photons/m}^3$$

Example 11.2 Consider the following parameters for a 1300-nm InGaAsP SOA:

Symbol	Parameter	Value
w	Active area width	$3\,\mu\text{m}$
d	Active area thickness	$0.3\,\mu\text{m}$
L	Amplifier length	$500\,\mu\text{m}$
Γ	Confinement factor	0.3
τ_r	Time constant	1 ns
a	Gain coefficient	$2 \times 10^{-20} \text{ m}^2$
n_{th}	Threshold density	$1 \times 10^{24} \text{ m}^{-3}$

(a) What is the pumping rate for the SOA?
(b) What is the zero-signal gain?

Solution:

(a) If a 100-mA bias current is applied to the device, then from Eq. (11.2) the pumping rate is

$$R_p = \frac{J}{qd} = \frac{i}{qdwL} = \frac{0.1\,\text{A}}{\left(1.6 \times 10^{-19}\text{C}\right)(0.3\,\mu m)(3\,\mu m)(500\,\mu m)}$$

$$= 1.39 \times 10^{33} \text{ (electrons/m}^3)/\text{s}$$

(b) From Eq. (11.7b), the zero-signal gain is

$$g_0 = 0.3\left(2.0 \times 10^{-20}\text{m}^2\right)(1\text{ ns})\left(1.39 \times 10^{33}\text{m}^{-3}\text{s}^{-1} - \frac{1.39 \times 10^{24}\text{m}^{-3}}{1.0\,\text{ns}}\right)$$

$$= 2340\,\text{m}^{-1} = 23.4\,\text{cm}^{-1}$$

Drill Problem 11.1 Consider an InGaAsP SOA that has cavity dimensions $w = 5\,\mu\text{m}$, $d = 0.5\,\mu\text{m}$, and $L = 200\,\mu\text{m}$. Assume the SOA has a gain coefficient $a = 1.0 \times 10^{-20} \text{ m}^2$, a confinement factor $\Gamma = 0.3$, a 1-ns time constant, and a threshold carrier density $n_{\text{th}} = 1.0 \times 10^{24} \text{ m}^{-3}$. (a) If a 100-mA bias current is applied to the device, show that the pumping rate $R_p = 1.25 \times 10^{33}$ (electrons/m^3)/s. (b) Show that the zero-signal gain is $g_0 = 750$ m^{-1}.

11.2.2 Amplifier Signal Gain

One of the most obviously important parameters of an optical amplifier is the *signal gain* or *amplifier gain G*, which is defined as

$$G = \frac{P_{s,out}}{P_{s,in}} \tag{11.8}$$

where $P_{s,in}$ and $P_{s,out}$ are the input and output powers, respectively, of the optical signal being amplified. As noted in Chap. 4, the radiation intensity at a photon energy $h\nu$ varies exponentially with the distance traversed in a lasing cavity. Hence, using Eq. (4.23), the single-pass gain in the active medium of the SOA is

$$G = \exp[\Gamma(g_m - \alpha_{mat})L] \equiv \exp[g(z)L] \tag{11.9}$$

where Γ is the optical confinement factor in the cavity, g_m is the material gain coefficient, α_{mat} is the effective absorption coefficient of the material in the optical path, L is the amplifier length, and $g(z)$ is the overall gain per unit length.

Equation (11.9) shows that the gain increases with device length. However, the internal gain is limited by gain saturation. This occurs because the carrier density in the gain region of the amplifier depends on the optical input intensity. As the input signal level is increased, excited carriers (electron–hole pairs) are depleted from the active region. When there is a sufficiently large optical input power, further increases in the input signal level no longer yield an appreciable change in the output level because there are not enough excited carriers to provide an appropriate level of stimulated emission. Note here that the carrier density at any point z in the amplifying cavity depends on the signal level $P_s(z)$ at that point. In particular, near the input where z is small, incremental portions of the device may not have reached saturation at the same time as the sections farther down the device, where incremental portions may be saturated because of higher values of $P_s(z)$.

An expression for the gain G as a function of the input power can be derived by examining the gain parameter $g(z)$ in Eq. (11.9). This parameter depends on the carrier density and the signal wavelength. Using Eqs. (11.4) and (11.6), then at a distance z from the input end, $g(z)$ is given by

$$g(z) = \frac{g_0}{1 + \dfrac{P_s(z)}{P_{amp,sat}}} \tag{11.10}$$

where g_0 is the unsaturated medium gain per unit length in the absence of signal input, $P_s(z)$ is the internal signal power at point z, and $P_{amp,sat}$ is the *amplifier saturation power*, which is defined as the internal power level at which the gain per unit length has been halved. Thus, from Eq. (11.9) the gain decreases with increasing signal power. In particular, the gain parameter in Eq. (11.10) is reduced by a factor of 2 when the internal signal power is equal to the amplifier saturation power.

Given that $g(z)$ is the gain per unit length, in an incremental length dz the light power increases by

$$dP = g(z)P_s(z)dz \tag{11.11}$$

Substituting Eq. (11.10) into Eq. (11.11) and rearranging terms gives

$$g_0(z)dz = \left[\frac{1}{P_s(z)} + \frac{1}{P_{amp,sat}} \right] dP \tag{11.12a}$$

Integrating this equation from $z = 0$ to $z = L$ yields

$$\int_0^L g_0 dz = \int_{P_{s,in}}^{P_{s,out}} \left[\frac{1}{P_s(z)} + \frac{1}{P_{amp,sat}} \right] dP \tag{11.12b}$$

Defining the single-pass gain in the absence of light to be $G_0 = \exp(g_0 L)$, and using Eq. (11.8), then yields

$$G = 1 + \frac{P_{amp,sat}}{P_{s,in}} ln\left(\frac{G_0}{G} \right) = G_0 exp\left(-\frac{G-1}{G} \frac{P_{s,out}}{P_{amp,sat}} \right) \tag{11.13}$$

Figure 11.3 illustrates the dependence of the gain on the input power. Here, the zero-signal gain (or small-signal gain) is $G_0 = 30$ dB, which is a gain factor of 1000. The curve shows that as the input signal power is increased, the gain first stays near the small-signal level and then starts to decrease. After decreasing linearly in the gain saturation region, it finally approaches an asymptotic value of 0 dB (a unity gain) for high input powers. Also shown is the *output saturation power* $P_{amp,sat}$, which is the point at which the gain is reduced by 3 dB.

The wavelength at which the SOA has a maximum gain can be tailored to occur anywhere between about 1200 and 1700 nm by changing the composition of the active InGaAsP material. As an example, Fig. 11.4 shows a typical gain versus wavelength characteristic for a device with a peak gain of 25 dB at 1530 nm. The wavelength span over which the gain decreases by less than 3 dB with respect to the maximum gain is known as the *gain bandwidth* or the *3-dB optical bandwidth*. In the example shown in Fig. 11.4 the 3-dB optical bandwidth is 85 nm. Values of up to 100 nm can be achieved.

Fig. 11.3 Typical dependence of the single-pass gain on optical input power for a small-signal gain of $G_0 = 30$ dB (a gain of 1000)

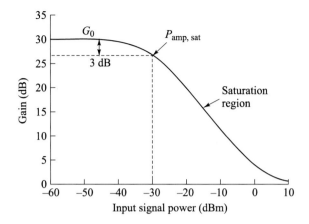

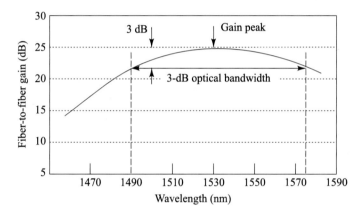

Fig. 11.4 Example of 3-dB optical bandwidth for an SOA and the gain-versus-wavelength characteristic with a peak gain of 25 dB at 1530 nm

11.2.3 SOA Bandwidth

A general expression for the cavity gain G_c as a function of signal frequency f is given by

$$G_c(f) = \frac{(1 - R_1)(1 - R_2)G}{\left(1 - \sqrt{R_1 R_2}G\right)^2 + 4\sqrt{R_1 R_2}G \sin^2 \varphi} \quad (11.14)$$

where G is the single-pass gain, R_1 and R_2 are the input and output facet reflectivities, respectively, and φ is the single-pass phase shift through the amplifier. The phase can be expressed as $\varphi = \pi(f - f_0)/\Delta f_{FSR}$, where f_0 is the cavity resonance frequency and Δf_{FSR} is the free spectral range of the SOA (see Sect. 10.5.1).

From Eq. (11.14) the 3-dB spectral bandwidth B_{SOA} of an SOA can be expressed by [8]

$$B_{SOA} = 2(f - f_0) = \frac{2\Delta f_{FSR}}{\pi} \sin^{-1}\left[\frac{1 - \sqrt{R_1 R_2}G}{2\left(\sqrt{R_1 R_2}G\right)^{1/2}}\right]$$

$$= \frac{c}{\pi n L} \sin^{-1}\left[\frac{1 - \sqrt{R_1 R_2}G}{2\left(\sqrt{R_1 R_2}G\right)^{1/2}}\right] \tag{11.15}$$

Here L is the length of the amplifier and n is its refractive index.

Drill Problem 11.2 Consider an InGaAsP SOA that has a refractive index n = 3.25, a cavity length $L = 300$ μm, and a 25-dB gain at 1530 nm. Suppose the facet reflectivity values are $R_1 = R_2 = 0.001$. Show that the 3-dB spectral bandwidth B_{SOA} is 64.1 GHz.

11.3 Erbium-Doped Fiber Amplifiers

The active medium in an optical fiber amplifier consists of a nominally 10 to 30-m length of optical fiber that has been lightly doped (e.g., 1000 parts per million weight) with a rare-earth element, such as erbium (Er), ytterbium (Yb), thulium (Tm), or praseodymium (Pr). The host fiber material can be standard silica, a fluoride-based glass, or a tellurite glass.

The operating regions of these devices depend on the host material and the doping elements. A popular material for long-haul telecommunication applications is a silica fiber doped with erbium, which is called an *erbium-doped fiber amplifier* or EDFA [8–12]. In some cases Yb is added to increase the pumping efficiency and the amplifier gain [13]. The operation of a standard EDFA normally is limited to the 1530–1565 nm region. Actually the fact that an EDFA operates in this spectral band is the origin of the term *C-band* or *conventional band* (see Chap. 1). However, various techniques have been proposed and used to extend the operation to the S-band and the L-band. Section 11.8 describes some of these techniques for creating wideband optical amplifiers.

11.3.1 Basics of Fiber Amplifier Pumping

Whereas semiconductor optical amplifiers use external current injection to excite electrons to higher energy levels, optical fiber amplifiers use *optical pumping* [14].

In this process, one uses photons to directly raise electrons into excited states. The optical pumping process requires three or more energy levels. The top energy level to which the electron is elevated initially must lie energetically above the desired final emission level. After reaching its initial excited state, the electron must quickly release some of its energy and drop to a slightly lower energy level. A signal photon can then trigger the excited electron sitting in this new lower level into stimulated emission, whereby the electron releases its remaining energy in the form of a new photon with a wavelength identical to that of the signal photon. Because the pump photon must have a higher energy than the signal photon, the pump wavelength is shorter than the signal wavelength.

To get a phenomenological understanding of how an EDFA works, it is helpful to look at the energy-level structure of erbium. The erbium atoms in silica are Er^{3+} ions, which are erbium atoms that have lost three of their outer electrons. In describing the transitions of the outer electrons in these ions to higher energy states, it is common to refer to the process as "raising the ions to higher energy levels." Fig. 11.5 shows a simplified energy-level diagram and various energy-level transition processes of these Er^{3+} ions in silica glass. The two principal levels for telecommunication applications are a *metastable level* (the so-called $^4I_{13/2}$ level) and the $^4I_{11/2}$ *pump level*. The term "metastable" means that the lifetimes for transitions from this state to the ground state are very long compared with the lifetimes of the states that led to this level. (Note that, by convention, the possible states of a multielectron atom are referred to by the symbol $^{2S+1}L_J$, where $2S + 1$ is the spin multiplicity, L is the orbital angular momentum, and J is the total angular momentum.) The metastable, the pump, and the ground-state levels are actually bands of closely spaced energy levels that form a

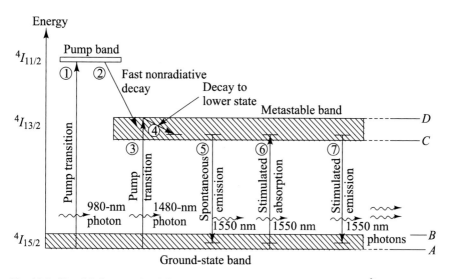

Fig. 11.5 Simplified energy-level diagrams and various transition processes of Er^{3+} ions in silica, with an example for 1550-nm signal photons

manifold (a set of many levels) due to the effect known as *Stark splitting*. Furthermore, thermal effects broaden each Stark level into an almost continuous band.

To understand the various energy transitions and photon emission ranges, consider the following conditions:

- The pump band shown in the top left of Fig. 11.5 exists at a 1.27-eV separation from the bottom of the $^4I_{15/2}$ ground state. This energy corresponds to a 980-nm wavelength.
- The top of the $^4I_{13/2}$ metastable band (level D in Fig. 11.5) is separated from the bottom of the $^4I_{15/2}$ ground state band (level A in Fig. 11.5) by 0.841 eV. This energy corresponds to a 1480-nm wavelength.
- The bottom of the $^4I_{13/2}$ metastable band (level C in Fig. 11.5) is separated from the bottom of the $^4I_{15/2}$ ground state band (level A in Fig. 11.5) by 0.814 eV. This energy corresponds to a 1530-nm wavelength.
- The bottom of the $^4I_{13/2}$ metastable band (level C in Fig. 11.5) is separated from the top of the $^4I_{15/2}$ ground state band (level B in Fig. 11.5) by about 0.775 eV. This energy corresponds to a 1600-nm wavelength.

This means that possible pump wavelengths are 980 and 1480 nm. The photons emitted during transitions of electrons between possible energy levels in the metastable band and ground-state band can range from 1530 to 1600 nm.

In normal operation, a pump laser emitting 980-nm photons is used to excite ions from the ground state to the pump level, as shown by transition process 1 in Fig. 11.5. These excited ions decay (relax) very quickly (in about 1 μs) from the pump band to the metastable band, shown as transition process 2. During this decay, the excess energy is released as phonons or, equivalently, mechanical vibrations in the fiber. Within the metastable band, the electrons of the excited ions tend to populate the lower end of the band. Here, they are characterized by a very long fluorescence time of about 10 ms.

Another possible pump wavelength is 1480 nm. The energy of these pump photons is very similar to the signal-photon energy, but slightly higher. The absorption of a 1480-nm pump photon excites an electron from the ground state directly to the lightly populated top of the metastable level, as indicated by transition process 3 in Fig. 11.5. These electrons then tend to move down to the more populated lower end of the metastable level (transition 4). Some of the ions sitting at the metastable level can decay back to the ground state in the absence of an externally stimulating photon flux, as shown by transition process 5. This decay phenomenon is known as *spontaneous emission* and adds to the amplifier noise because it emits a background noise photon.

Two more types of transitions occur when a flux of signal photons with energies corresponding to the bandgap energy between the ground state and the metastable level passes through the device. First, a small portion of the external photons will be absorbed by ions in the ground state, which raises these ions to the metastable level, as shown by transition process 6. Second, in the stimulated emission process (transition process 7) a signal photon triggers an excited ion to drop to the ground state, thereby emitting a new photon of the same energy, wave vector, and polarization as the incoming signal photon. The widths of the metastable level and the ground-state level allow high degrees of stimulated emissions to occur in the 1530-to-1560-nm range.

11.3.2 Construction of an EDFA

An optical fiber amplifier consists of a doped fiber, one or more pump lasers, a passive wavelength coupler, optical isolators, and tap couplers, as shown in Fig. 11.6. The dichroic (two-wavelength) coupler handles either 980/1550-nm or 1480/1550-nm wavelength combinations to couple both the pump and signal optical powers efficiently into the fiber amplifier. The tap couplers are wavelength-insensitive with typical splitting ratios ranging from 99:1 to 95:5. They generally are used on both sides of the amplifier to compare the incoming signal with the amplified output. The optical isolators prevent the amplified signal from reflecting back into the device, where it could increase the amplifier noise and decrease the amplifier efficiency.

The pump light is usually injected from the same direction as the signal flow. This is known as *codirectional pumping*. It is also possible to inject the pump power in the

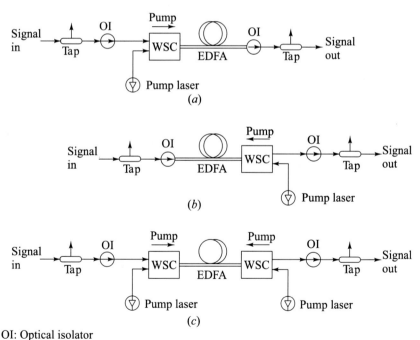

OI: Optical isolator
WSC: Wavelength-selective coupler

Fig. 11.6 Three possible configurations of an EDFA: **a** co-directional pumping, **b** counter-directional pumping, and **c** dual pumping

opposite direction to the signal flow, which is known as *counterdirectional pumping*. As shown in Fig. 11.6, one can employ either a single pump source or use *dual-pump schemes*, with the resultant gains typically being +17 dB and +35 dB, respectively. Counter-directional pumping allows higher gains, but codirectional pumping gives better noise performance. In addition, pumping at 980 nm is preferred, because it produces less noise and achieves larger population inversions than pumping at 1480 nm.

11.3.3 EDFA Power-Conversion Efficiency and Gain

As is the case with any amplifier, as the magnitude of the output signal from an EDFA increases, the amplifier gain eventually starts to saturate. The reduction of gain in an EDFA occurs when the population inversion is reduced significantly by a large signal, thereby yielding the typical gain-versus-power performance curve shown in Fig. 11.3.

The input and output signal powers of an EDFA can be expressed in terms of the principle of energy conservation:

$$P_{s,out} \leq P_{s,in} + \frac{\lambda_p}{\lambda_s} P_{p,in} \tag{11.16}$$

where $P_{p,in}$ is the input pump power, and λ_p and λ_s are the pump and signal wavelengths, respectively. The fundamental physical principle here is that the amount of signal energy that can be extracted from an EDFA cannot exceed the pump energy that is stored in the device. The inequality in Eq. (11.16) reflects the possibility of effects such as pump photons being lost due to various causes (such as interactions with impurities) or pump energy lost due to spontaneous emission.

From Eq. (11.16), it can be seen that the maximum output signal power depends on the ratio λ_p/λ_s. For the pumping scheme to work, it is necessary to have $\lambda_p < \lambda_s$, and, to have an appropriate gain, it is necessary that $P_{s,in} \leq P_{p,in}$. Thus, the *power conversion efficiency* (PCE), defined as

$$PCE = \frac{P_{s,out} - P_{s,in}}{P_{p,in}} \approx \frac{P_{s,out}}{P_{p,in}} \leq \frac{\lambda_p}{\lambda_s} \leq 1 \tag{11.17}$$

is less than unity. The maximum theoretical value of the PCE is λ_p/λ_s. For absolute reference purposes, it is helpful to use the *quantum conversion efficiency* (QCE), which is wavelength-independent and is defined by

$$QCE = \frac{\lambda_s}{\lambda_p} PCE \tag{11.18}$$

The maximum value of QCE is unity, in which case all the pump photons are converted to signal photons.

One also can rewrite Eq. (11.16) in terms of the amplifier gain G. Assuming there is no spontaneous emission, then

$$G = \frac{P_{s,out}}{P_{s,in}} \leq 1 + \frac{\lambda_p}{\lambda_s} \frac{P_{p,in}}{P_{s,in}} \tag{11.19}$$

This shows an important relationship between signal input power and gain. When the input signal power is very large so that $P_{s,in} \gg (\lambda_p/\lambda_s)P_{p,in}$, then the maximum amplifier gain is unity. This means that the device is transparent to the signal. From Eq. (11.19), it can be seen that in order to achieve a specific maximum gain G, the input signal power cannot exceed a value given by

$$P_{s,in} \leq \frac{\lambda_p}{\lambda_s} \frac{P_{p,in}}{(G-1)} \tag{11.20}$$

Example 11.3 Consider an EDFA being pumped at 980 nm with a 30-mW pump power. If the gain at 1550 nm is 20 dB, what are the maximum input and output powers?

Solution: From Eq. (11.20), the maximum input power is

$$P_{s,in} \leq \frac{980}{1550} \frac{30\,\text{mW}}{(100-1)} = 190\,\mu\text{W}$$

From Eq. (11.16), the maximum output power is

$$P_{s,out}(\text{max}) = P_{s,in}(\text{max}) + \frac{\lambda_p}{\lambda_s} P_{p,in}$$

$$= 190\,\mu\text{W} + 0.63(30\,\text{mW})$$

$$= 19.1\,\text{mW} = 12.8\,\text{dBm}$$

Drill Problem 11.3 Consider an EDFA being pumped with a 36-mW pump power at 1480 nm. (*a*) If the gain is 23 dB at 1550 nm, show that the maximum input power is 0.172 mW = −7.6 dBm. (*b*) Show that the maximum output power is 34.5 mW = 15.4 dBm.

In addition to pump power, the gain also depends on the fiber length. The maximum gain in a three-level laser medium of length L, such as an EDFA, is given by

$$G_{\text{max}} = \exp(\rho \sigma_e L) \tag{11.21}$$

where σ_e is the signal-emission cross section and ρ is the rare-earth element concentration. When determining the maximum gain, Eqs. (11.19) and (11.21) must be considered together. Consequently, the maximum possible EDFA gain is given by the lowest of the two gain expressions:

$$G \leq \min\left\{\exp(\rho\sigma_e L),\ 1 + \frac{\lambda_p}{\lambda_s}\frac{P_{p,in}}{P_{s,in}}\right\} \qquad (11.22)$$

Because $G = P_{s,out}/P_{s,in} = \exp(\rho\sigma_e L)$, it follows similarly that the maximum possible EDFA output power is given by the minimum of the two expressions:

$$P_{s,out} \leq \min\left\{P_{s,in}\exp(\rho\sigma_e L),\ P_{s,in} + \frac{\lambda_p}{\lambda_s}P_{p,in}\right\} \qquad (11.23)$$

Figure 11.7 illustrates the onset of gain saturation for various doped-fiber lengths as the pumping power increases from a low pump power (P1) to higher pump powers P2 and P3. As the fiber length increases for low pumping powers, the gain starts to decrease after a certain length because the pump does not have enough energy to create a complete population inversion in the downstream portion of the amplifier. In this case, the unpumped region of the fiber absorbs the signal, thus resulting in signal loss rather than gain in that section.

A detailed analysis of EDFA gain versus amplifier length has been presented in the literature to which the reader is referred for details [15]. Several points should be noted in relation to that research:

- The amplifier length that yields a maximum gain becomes longer with increasing signal wavelength, because photons at longer wavelengths have less energy and thus need less power to have the same gain as photons at shorter wavelengths.
- If a specific amplifier length is chosen, for example, 30 m, then the EDFA will amplify each wavelength differently, again because photon energy is wavelength

Fig. 11.7 The onset of gain saturation at various doped-fiber lengths as the pumping power increases

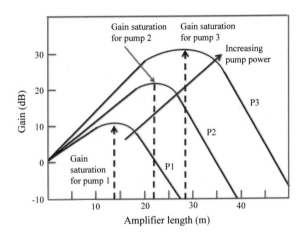

dependent. This leads to gain skew among different wavelengths as they emerge
from the amplifier.

- The 980-nm pumping yields a complete population inversion (maximum gain) at
  shorter amplifier lengths than 1480-nm pumping. This results in a lower amplifier
  noise figure when using 980-nm pumping.

11.4 Noises Generated in Optical Amplifiers

The dominant noise generated in an optical amplifier is called *amplified spontaneous
emission* (ASE) noise. The origin of this noise is the spontaneous recombination
of electrons and holes in the amplifier medium (transition 5 in Fig. 11.5). This
recombination occurs over a wide range of electron–hole energy differences and
thus gives rise to a broad spectral background of noise photons that get amplified
along with the optical signal as they travel through the EDFA. This is shown in
Fig. 11.8 for an EDFA amplifying a signal at 1540 nm. The spontaneous noise can
be modeled as a stream of random infinitely short pulses that are distributed all along
the amplifying medium. Such a random process is characterized by a noise power
spectrum that is flat with frequency. The power spectral density of the ASE noise is
[9]

$$S_{ASE}(f) = h\upsilon\, n_{Sp}[G(f) - 1] = P_{ASE}/B_O \tag{11.24}$$

where P_{ASE} is the ASE noise power in one polarization state in an optical bandwidth
B_o and n_{sp} is the *spontaneous-emission* or *population-inversion factor* defined as

$$n_{sp} = \frac{n_2}{n_2 - n_1} \tag{11.25}$$

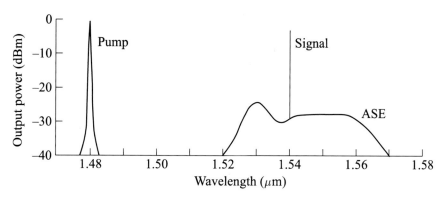

Fig. 11.8 Representative 1480-nm pump spectrum and a typical output signal at 1540 nm with the
associated amplified-spontaneous-emission (ASE) noise

Here n_1 and n_2 are the fractional densities or populations of electrons in a lower state 1 and an upper state 2, respectively. Thus, n_{sp} denotes how complete the population inversion is between two energy levels. From Eq. (11.25) it is seen that $n_{sp} \geq 1$. The equality holds for an ideal amplifier when the population inversion is complete. Typical values range from 1.4 to 4, depending on the wavelength and the pumping rate.

An important point is that the expression for P_{ASE} is for an individual spatial mode and an individual polarization state. For single-mode fibers the center term in Eq. (11.24) must be multiplied by a factor 2 to get the total ASE power, because a single-mode fiber has one spatial mode and two polarization modes. If an EDFA is made from a multimode fiber, P_{ASE} becomes much larger because such a fiber has many spatial modes. The ASE noise level depends on whether codirectional or counterdirectional pumping is used.

Because ASE originates ahead of the photodiode, it gives rise to three different noise components in an optical receiver in addition to the normal thermal noise of the photodetector. This occurs because the photocurrent consists of a number of beat signals between the signal and the optical noise fields, in addition to the squares of the signal field and the spontaneous-emission field. If the total optical field is the sum of the signal field E_s and the ASE noise field E_n, then the total photodetector current i_{tot} is proportional to the square of the electric field of the composite optical signal:

$$i_{tot} \propto (E_s + E_n)^2 = E_s^2 + E_n^2 + 2E_s E_n$$

Here the first two terms arise purely from the signal and noise, respectively. The third term is a mixing component (a *beat signal*) between the signal and noise, which can fall within the bandwidth of the receiver and degrade the signal-to-noise ratio. First, taking the ASE photons into account, the optical power incident on the photodetector becomes

$$P_{in} = GP_{s,in} + P_{ASE} = GP_{s,in} + S_{ASE}B_O \tag{11.26}$$

Note that B_o can be reduced significantly if an optical filter precedes the photodetector. Substituting P_{in} into Eq. (6.6) and inserting the resulting expression for the photocurrent i_p into Eq. (6.12) to find the shot noise then yields the total mean-square signal-plus-ASE shot-noise current

$$\langle i_{shot}^2 \rangle = \sigma_{shot}^2 = \sigma_{shot-s}^2 + \sigma_{shot-ASE}^2 = 2q\mathscr{R}GP_{s,in}B_e + 2q\mathscr{R}S_{ASE}B_0B_e \tag{11.27}$$

where B_e is the front-end receiver electrical bandwidth.

The other two noises arise from the mixing of the different optical frequencies contained in the light signal and the ASE, which generates two sets of beat frequencies. Because the signal and the ASE have different optical frequencies, the beat noise of the signal with the ASE noise that is in the same polarization state as the signal is

$$\sigma^2_{s-ASE} = 4(\mathscr{R}GP_{s,in})(\mathscr{R}S_{ASE}B_e) \tag{11.28}$$

In addition, because the ASE spans a wide optical frequency range, it can beat against itself giving rise to the noise current

$$\sigma^2_{ASE-ASE} = \mathscr{R}^2 S^2_{ASE}(2B_0 - B_e)B_e \tag{11.29}$$

The total mean-square receiver noise current then becomes

$$\langle i^2_{total}\rangle = \sigma^2_{total} = \sigma^2_{th} + \sigma^2_{shot-s} + \sigma^2_{shot-ASE} + \sigma^2_{s-ASE} + \sigma^2_{ASE-ASE} \tag{11.30}$$

where the thermal noise variance σ^2_{th} is given by Eq. (6.15).

The last four terms in Eq. (11.30) tend to be of similar magnitudes when the optical bandwidth B_0 is taken to be the optical bandwidth of the spontaneous emission noise, which covers a 30-nm spectrum. However, one generally uses a narrow optical filter at the receiver, so that B_0 is on the order of 125 GHz (a 1-nm spectral width at 1550 nm) or less. In that case, Eq. (11.30) can be simplified by examining the magnitudes of the various noise components. First, the thermal noise can generally be neglected when the amplifier gain is large enough. Furthermore, because the amplified signal power $GP_{s,in}$ is much larger than the ASE noise power $S_{ASE} B_o$, the ASE–ASE beat noise given by Eq. (11.29) is significantly smaller than the signal–ASE beat noise. This also means that the second term in Eq. (11.27) is small compared to the first term, so that

$$\sigma^2_{shot} \approx 2q\mathscr{R}GP_{s,in}B_e \tag{11.31}$$

Using these results together with the expression for S_{ASE} from Eq. (11.24) yields the following approximate signal-to-noise ratio (S/N) at the photodetector output:

$$\left(\frac{S}{N}\right)_{out} = \frac{\sigma^2_{ph}}{\sigma^2_{total}} = \mathscr{R}^2\frac{G^2 P^2_{s,in}}{\sigma^2_{total}} \approx \mathscr{R}\frac{P_{s,in}}{2q\,B_e}\frac{G}{1 + 2\eta n_{sp}(G - 1)} \tag{11.32}$$

where η is the quantum efficiency of the photodetector and, from Eq. (6.11), the mean-square input photocurrent is

$$\langle i^2_{ph}\rangle = \sigma^2_{ph} = \mathscr{R}^2 G^2 P^2_{s,in} \tag{11.33}$$

Note that the term

$$\left(\frac{S}{N}\right)_{in} = \mathscr{R}\frac{P_{s,in}}{2q\,B_e} \tag{11.34}$$

in Eq. (11.32) is the signal-to-noise ratio of an ideal photodetector at the input to the optical amplifier. From Eq. (11.32) one then can find the noise figure of the optical amplifier, which is a measure of the S/N degradation experienced by a signal after passing through the amplifier. Using the standard definition of *noise figure* as the ratio between the *S/N* at the input and the *S/N* at the amplifier output, then for an ideal photodetector with $\eta = 1$

$$\text{Noise figure} = F_{EDFA} = \frac{(S/N)_{in}}{(S/N)_{out}} = \frac{1 + 2n_{sp}(G-1)}{G} \qquad (11.35)$$

When G is large, this becomes $2n_{sp}$. A perfect amplifier would have $n_{sp} = 1$, yielding a noise figure of 2 (or 3 dB), assuming $\eta = 1$. That is, using an ideal receiver with a perfect amplifier would degrade the *S/N* by a factor of 2. In a real EDFA, for example, n_{sp} is around 2 so the input *S/N* gets reduced by a factor of about 4. This results in a noise figure of 4–5 dB for a practical EDFA [16].

11.5   Optical Signal-To-Noise Ratio (OSNR)

When analyzing a transmission link that has a series of optical amplifiers in it, an important point to remember is that the light signal entering any particular optical amplifier includes some ASE noise from the preceding amplifiers in the link. Thus the optical receiver may contain a significant level of ASE noise that has been added by the cascade of optical amplifiers. In this case one has to evaluate the *optical signal-to-noise ratio* (OSNR) [9, 10]. This parameter is defined as the ratio of the average EDFA optical signal output power P_{ave} to the unpolarized ASE optical noise power P_{ASE}. The OSNR is given by

$$\text{OSNR} = \frac{P_{ave}}{P_{ASE}} \qquad (11.36a)$$

or, in decibels

$$\text{OSNR(dB)} = 10 \log \frac{P_{ave}}{P_{ASE}} \qquad (11.36b)$$

In practice, the OSNR can be measured with an *optical spectrum analyzer* (OSA), which is described in Chap. 14. The OSNR does not depend on factors such as the data format, pulse shape, or optical filter bandwidth, but only on the average optical signal power P_{ave} and the average optical noise power. OSNR is a metric that can be used in the design and installation of networks, as well as to check the health and status of individual optical channels. Sometimes an optical filter is used to significantly reduce the total ASE noise seen by the receiver. Typically such a filter has an optical bandwidth that is large compared to the signal, so that it does not affect the signal, yet that is narrow compared to the bandwidth associated with the ASE background. The ASE noise filter does not change the OSNR. However, it reduces the total power in the ASE noise to avoid overloading the receiver front end.

In order to have a meaningful assessment of what the OSNR tells about system performance, a connection is needed between OSNR and bit-error rate (BER). A number of different relationships have been proposed for this in the literature. Furthermore, there are different approaches for interpreting the results of an OSA measurement, which can lead to a difference in results of several decibels. Using the expression for Q given by Eq. (7.14) and the definition of OSNR from Eq. (11.36b), the following relation between Q and OSNR can be derived [11]

$$Q = \frac{2\sqrt{2}\,\text{OSNR}}{1 + \sqrt{1 + 4\,\text{OSNR}}} \tag{11.37}$$

Solving Eq. (11.37) for OSNR yields

$$\text{OSNR} = \frac{1}{2}Q\left(Q + \sqrt{2}\right) \tag{11.38}$$

Example 11.4 In Chap. 7 it is shown that to achieve a BER $= 10^{-9}$ the factor Q must be 6. What is the OSNR for this BER?

Solution: Using Eq. (11.38) then yields

$$\text{OSNR}\left(\text{BER} = 10^{-9}\right) = 0.5(6)(6 + \sqrt{2}) = 22.24 \approx 13.5\,\text{dB}$$

Therefore if an OSA measures an OSNR $\leq$ 13.5 dB, then the corresponding error rates are equal to or higher than BER $= 10^{-9}$.

For an OSNR analysis, first consider the case for a single EDFA. Converting the ASE noise power given by Eq. (11.24) into a decibel format, then for unpolarized ASE noise and for large G

$$10 \log P_{\text{ASE}} = 10 \log[(h\nu)(B_O)] + 10 \log 2n_{\text{sp}} + 10 \log G \tag{11.39}$$

Here, $h\nu$ is the photon energy and B_0 is the optical frequency range in which the OSNR is measured, which typically is 12.5 GHz (a 0.1-nm spectral width at 1550 nm). At 1550 nm, one obtains $(h\nu)(B_o) = 1.58 \times 10^{-6}$ mW, so that $10 \log (h\nu)(B_o) = -58$ dBm. Assuming that $G \gg 1$ and taking $F_{\text{EDFA}}(\text{dB}) = 10 \log 2n_{\text{sp}}$ from Eq. (11.35) as the amplifier noise figure, then in decibels

$$P_{\text{ASE}}(\text{dBm}) = -58\text{dBm} + F_{\text{EDFA}}(\text{dB}) + G(\text{dB}) \tag{11.40}$$

Using Eq. (11.36) and taking the EDFA output power to be G times the optical input power, $P_{\text{out}} = GP_{\text{in}}$, one obtains the requirement that in order to have an acceptable BER, the OSNR must be at least

$$\text{OSNR}(\text{dB}) = P_{\text{in}}\,(\text{dBm}) + 58\text{dBm} - F_{\text{EDFA}}\,(\text{dB}) \tag{11.41}$$

Drill Problem 11.4 To achieve a BER $= 10^{-15}$ the factor Q must be 8. (*a*) Show that the OSNR required for this BER is 37.7 dB. (*b*) If the EDFA noise figure is $F_{EDFA} = 5$ dB, show that for this OSNR the maximum allowed input power is $P_{in} = -5.3$ dBm (29 μW).

11.6   Fiber Link Applications

When designing an optical fiber link that requires optical amplifiers, there are three possible locations where the amplifiers can be placed, as shown in Fig. 11.1. Although the physical amplification process is the same in all three configurations, the various uses require operation of the device over different input power ranges. This in turn implies use of different amplifier gains. This section will look at simple conceptual analyses and present generic operational values for the three possible locations of EDFAs in an optical link.

11.6.1   Power Amplifier Functions

For the power amplifier, the input power is high because the device immediately follows an optical transmitter. High pump powers are normally required for this application [17]. The amplifier inputs are generally -8 dBm or greater, and the power amplifier gain must be greater than 5 dB in order for it to be more advantageous than using a preamplifier at the receiver.

Example 11.5 Consider an EDFA used as a power amplifier with a 10-dB gain. Assume the amplifier input is a 0-dBm level from a 1540-nm laser diode transmitter. If the pump wavelength is 980 nm, what is the pump power?

Solution: From Eq. (11.16), for a 10-dBm output at 1540 nm, the pump power must be at least

$$P_{p,in} \geq \frac{\lambda_s}{\lambda_p}\left(P_{s,out} - P_{s,in}\right) = \frac{1540}{980}(10\,\text{mW} - 1\,\text{mW}) = 14\,\text{mW}$$

11.6.2 Use of In-Line Amplifiers

In a long transmission system, optical amplifiers are needed to periodically restore the power level after it has decreased due to attenuation in the fiber. Normally, the gain of each EDFA in this amplifier chain is chosen to compensate exactly for the signal loss incurred in the preceding fiber section of length L, that is, $G = \exp(+\alpha L)$. The accumulated ASE noise is the dominant degradation factor in such a cascaded chain of amplifiers.

Example 11.6 Consider Fig. 11.9, which shows the values of the per-channel signal power, the per-channel ASE noise, and the SNR along a chain of seven optical amplifiers in a WDM link. As shown by the solid line, the input signal level starts out at 6 dBm and decays due to fiber attenuation as it travels along the link. When its power level has dropped to –24 dBm, it gets boosted back to 6 dBm by an optical amplifier. For a given channel transmitted over the link, the SNR starts out at a high level and then decreases at each amplifier as the ASE noise accumulates through the length of the link. For example, following amplifier number 1, the SNR is 28 dB for a 6-dBm amplified signal level and a –22-dBm ASE noise level. After amplifier number 4, the SNR is 22 dB for a 6-dBm amplified signal level and a –16-dBm ASE noise level. The higher the gain in the amplifier, the faster the ASE noise builds up. However, although the SNR decreases quickly in the first several amplifications, the incremental effect of adding another EDFA diminishes rapidly with an increasing

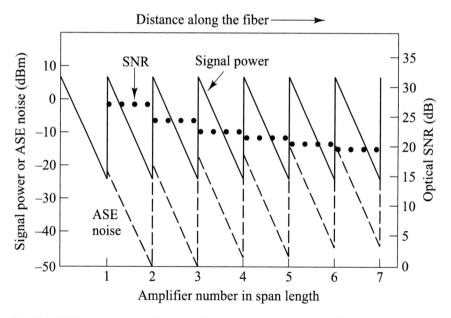

Fig. 11.9 SNR degradation as a function of link distance over which the ASE noise increases with the number of amplifiers; the curves show the signal level (solid lines), the ASE noise level (dashed lines), and the SNR (dotted lines) for a single channel in a WDM link

number of amplifiers. As a consequence, although the SNR drops by 3 dB when the number of EDFAs increases from one to two, it also drops by 3 dB when the number of amplifiers is increased from two to four, and by another 3 dB when the number of amplifiers is further increased to eight.

To compensate for the accumulated ASE noise, the signal power must increase at least linearly with the length of the link in order to keep a constant signal-to-noise ratio. If the total system length is $L_{tot} = NL$ and the system contains N optical amplifiers each having a gain $G = \exp(+\alpha L)$, then, using Eq. (11.24), the path-averaged ASE power along a chain of optical amplifiers is [10]

$$\langle P_{ASE}\rangle_{path} = \frac{N P_{ASE}}{L} \int_0^L \exp(-\alpha z)dz = \alpha L_{tot} h \nu\, n_{sp} F_{path}(G) B_0 \qquad (11.42)$$

where α is the fiber attenuation and the noise figure $F_{path}(G)$ is a power penalty defined as

$$F_{path}(G) = \frac{1}{G}\left(\frac{G-1}{\ln G}\right)^2 \qquad (11.43)$$

Basically, $F_{path}(G)$ gives the factor by which the path-average signal energy must be increased (as G increases) in a chain of N cascaded optical amplifiers to maintain a fixed SNR. For long-distance networks these optical amplifiers can be placed uniformly along the transmission path to yield the best combination of overall gain and final SNR. The input power levels for these in-line amplifiers nominally ranges from -26 dBm (2.5 µW) to -9 dBm (125 µW) with gains ranging from 8 to 20 dB. For metro networks typically only a single optical amplifier is needed to compensate for the path loss between two successive nodes [18].

Example 11.7 Consider an optical transmission path containing N cascaded optical amplifiers each having a 30-dB gain. (a) If the fiber has a loss of 0.2 dB/km, what is the span between optical amplifiers if there are no other system impairments? (b) How many amplifiers are needed for a 900-km link? (c) What is the noise penalty factor over the total path?

Solution: (a) For a fiber loss of 0.2 dB/km, the signal power attenuates by 30 dB every 150 km. (b) Thus, five amplifiers are needed within the transmission path. (c) From Eq. (11.43), the noise penalty factor over the total path is (in decibels)

$$10\log F_{path}(G) = 10\log\left[\frac{1}{1000}\left(\frac{1000-1}{\ln 1000}\right)^2\right] = 10\log 20.9 = 13.2\,\text{dB}$$

Drill Problem 11.5 Consider an optical transmission path containing N cascaded optical amplifiers each having a 20-dB gain. (*a*) If the fiber has a loss of 0.2 dB/km, what is the span between optical amplifiers if there are no other system impairments? (*b*) How many amplifiers are needed for a 900-km link? (*c*) What is the noise penalty factor over the total path?
[**Answers**: (*a*) 100 km; (*b*) 8 amplifiers; (*c*) 6.6 dB].

11.6.3 Optical Amplifier as a Preamplifier

An optical amplifier can be used as a preamplifier to improve the sensitivity of direct-detection receivers that are limited by thermal noise [19]. First, assume the receiver noise is represented by the electrical power level N. Let S_{min} be the minimum value of the electrical signal power S that is required for the receiver to perform with a specific acceptable bit-error rate. The acceptable signal-to-noise ratio then is S_{min}/N. If an optical preamplifier with gain G is used, the electrical received signal power is $G^2 S'$ and the signal-to-noise ratio is

$$\left(\frac{S}{N}\right)_{preamp} = \frac{G^2 S'}{N + N'} \tag{11.44}$$

where the noise term N' is the spontaneous emission from the optical preamplifier that gets converted by the photodiode in the receiver to an additional background noise. If S'_{min} is the new minimum detectable electrical signal level needed to maintain the same signal-to-noise ratio, then it is necessary to have

$$\frac{G^2 S'_{min}}{N + N'} = \frac{S_{min}}{N} \tag{11.45}$$

For an optical preamplifier to enhance the received signal level, one must have $S'_{min} < S_{min}$, so that

$$\frac{S_{min}}{S'_{min}} = \frac{G^2 N}{N + N'} > 1 \tag{11.46}$$

The ratio S_{min} to S'_{min} is the *improvement of minimum detectable signal* or *detector sensitivity.*

Example 11.8 Consider an EDFA used as an optical preamplifier. Assume that the receiver noise N is due to thermal noise and that the noise N' introduced by the preamplifier is dominated by signal–ASE beat noise. Under what conditions does Eq. (11.46) hold?

Solution: For sufficiently high gain G, Eq. (11.46) becomes

$$G^2 - 1 \approx G^2 > \frac{N'}{N} \approx \frac{\sigma^2_{s-ASE}}{\sigma^2_{th}}$$

Substituting Eq. (6.17) and Eq. (11.28) into this expression, using Eq. (11.24) for S_{ASE}, and solving for $P_{s,in}$ yields

$$P_{s,in} < \frac{k_B T \, h\nu}{R n_{sp} \eta^2 q^2}$$

If $T = 300$ K, $R = 50\,\Omega$, $\lambda = 1550$ nm, $n_{sp} = 2$, and $\eta = 0.65$, then $P_{s,in} < 490\,\mu$W. This level is much higher than any expected received signal, so the condition in Eq. (11.46) is always satisfied. However, note that this only specifies the upper bound on $P_{s,in}$. It does not mean that by making G sufficiently high, the improvement in sensitivity can be made arbitrarily large, because there is a minimum received optical power level that is needed to achieve a specific BER.

11.7 Raman Optical Amplifiers

11.7.1 Principle of Raman Gain

A *Raman optical amplifier* is based on a nonlinear effect called *stimulated Raman scattering* (SRS), which occurs in fibers at high optical powers [20–23]. Chapter 12 describes this characteristic in more detail. The SRS effect is due to an interaction between an optical energy field and the vibrational modes of the lattice structure in a material. Basically what happens is that an atom first absorbs a photon at a particular energy and then releases another photon at a lower energy, that is, at a longer wavelength than that of the absorbed photon. The energy difference between the absorbed and the released photons is transformed into a *phonon*, which is a vibrational mode of the material. The power transfer to higher wavelengths occurs over a broad spectral range of 80–100 nm. The shift to a particular longer wavelength is referred to as the *Stokes shift* for that wavelength. Figure 11.10 shows the Raman gain spectrum for a pump laser operating at 1445 nm and illustrates the SRS-induced power transfer to a signal at 1535 nm, which is 90 nm away from the pump wavelength. Depending on the link architecture, the SRS-generated signal can act as either an intentional amplification of a particular data wavelength or it could be an unwanted interference signal at that wavelength (see Chap. 12). The gain curve is given in terms of the *Raman gain coefficient* g_R units of 10^{-14} m/W.

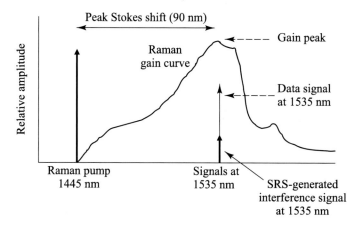

Fig. 11.10 Stokes shift and the resulting Raman gain spectrum from a pump laser operating at 1445 nm

Whereas an EDFA requires a specially constructed optical fiber for its operation, a Raman amplifier makes use of the standard transmission fiber itself as the amplification medium. The Raman gain mechanism can be achieved through either a lumped (or discrete) amplifier or a distributed amplifier. In the *lumped Raman amplifier* configuration, a spool of about 80 m of small-core fiber along with appropriate pump lasers is inserted into the transmission path as a distinct packaged unit.

For the *distributed Raman amplifier* application, optical power from one or more Raman pump lasers is inserted into the receiving end of the transmission fiber toward the transmitting end. This process converts the final 20–40 km of the transmission fiber into a preamplifier. Hence the word *distributed* is used because the gain is spread out over a wide distance. Figure 11.11 shows this effect on a single wavelength for several different pump levels. The figure shows the increase in signal power at the receiver amplifier output when the Raman pump lasers are turned on. As the optical

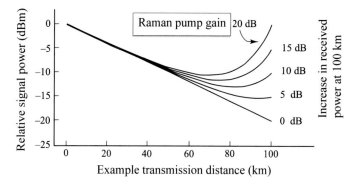

Fig. 11.11 Example of signal power evolution along a 100-km fiber link and the increase in signal power at the amplifier output for different values of Raman gain

power from the pumps travels upstream (from the receiver toward the sender), the SRS effect progressively transfers power from shorter pump wavelengths to longer signal wavelengths. This occurs over the characteristic Raman gain length $L_G = g_R P/A_{eff}$, where P is the pump laser power and A_{eff} is the effective area of the transmission fiber, which is approximately equal to the actual fiber cross-sectional area (see Chap. 12 for a detailed definition). In general, noise factors limit the practical gain of a distributed Raman amplifier to less than 20 dB.

11.7.2   Pump Lasers for Raman Amplifiers

Pump lasers with high output powers in the 1400-to-1500-nm region are required for Raman amplification of C-band and L-band signals. Lasers that provide fiber launch powers of up to 300 mW are available in standard 14-pin butterfly packages. Figure 11.12 shows the setup for a typical Raman amplification system. Here a pump combiner multiplexes the outputs from four pump lasers operating at different wavelengths (examples might be 1425, 1445, 1465, and 1485 nm) onto a single fiber. These pump-power couplers are referred to popularly as *14XX-nm pump-pump combiners*. Table 11.2 lists the performance parameters of a pump combiner based on fused-fiber coupler technology. This combined pump power then is coupled into the transmission fiber in a counterpropagating direction through a broadband WDM coupler, such as those listed in Table 11.3. The differences in the power levels measured between the two monitoring photodiodes shown in Fig. 11.12 give the amplification gain. The *gain-flattening filter* (GFF) is used to equalize the gains at different wavelengths (Table 11.3).

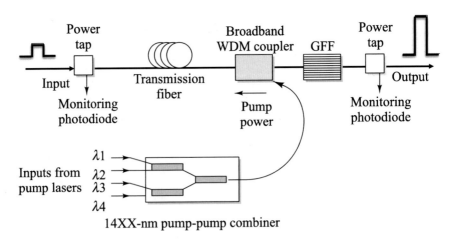

Fig. 11.12 Setup for a distributed Raman amplification system using four pump lasers

Table 11.2 Performance parameters of a 14XX-nm pump-pump combiner based on fused-fiber coupler technology

Parameter	Performance value
Device technology	Fused-fiber coupler
Wavelength range	1420–1500 nm
Channel spacing	Customized: 10–40 nm Standard: 10, 15, 20 nm
Insertion loss	<0.8 dB
Polarization dependent loss	<0.2 dB
Directivity	>55 dB
Optical power capability	3000 mW

Table 11.3 Performance parameters of broadband WDM couplers for combining 14XX-nm pumps and C-band or L-band signals

Parameter	Performance value	Performance value
Device technology	Micro-optics	Thin-film-filter
Reflection channelλ range (nm)	1420–1490	1440–1490
Pass channel λ range (nm)	1505–1630	1528–1610
Reflection channel insertion loss (dB)	0.30	0.6
Pass channel insertion loss (dB)	0.45	0.8
Polarization dependent loss (dB)	0.05	0.10
Polarization mode dispersion (ps)	0.05	0.05
Optical power capability (mW)	2000	500

11.8 Multiband Optical Amplifiers

The ever-growing demand for more bandwidth created an interest in developing wideband optical amplifiers that operate over several wavelength bands to handle a large number of WDM channels simultaneously. For example, a combination of two amplifier types can provide effective amplification in both the C- and L-bands or in the S- and C-bands. Extending this concept further, use of three amplifier types can provide signal gains in the S-, C-, and L-bands, in the C-, L-, and U-bands, or some other combination. The individual amplifiers could be based on thulium-doped silica fibers for the S-band, standard EDFAs for the C-band, gain-shifted EDFAs for the L-band, and different versions of Raman amplifiers [24, 25].

The amplifier combinations can be in parallel or in series, as shown in Figs. 11.13 and 11.14, respectively. In the parallel design a wideband demultiplexer splits the

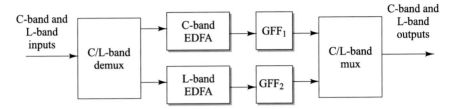

Fig. 11.13 Representation of C-band and L-band optical amplifiers arranged in parallel

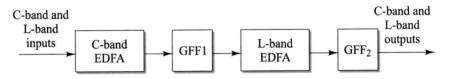

Fig. 11.14 Representation of C-band and L-band optical amplifiers arranged in series

incoming signal spectrum into two wavelength bands. The two bands then pass through corresponding optical amplifiers after which a wideband multiplexer recombines the two spectral bands. This setup requires the use of a guard band spanning several nanometers between the two spectral regions. This guard band prevents amplification overlap between the different paths and prevents noise power originating in one amplifier from interfering with signal amplification in an adjacent amplifier. In addition to having this unusable wavelength band, another disadvantage of the parallel configuration is that the two WDM devices needed before and after each amplifier add to the system insertion loss.

The series configuration is known as a *seamless wideband optical amplifier* because it does not require splitting the signals into separate paths. It also avoids the noise figure degradations of wavelength couplers and the additional costs of the couplers themselves. These amplifiers can be constructed either from a concatenation of two or more doped-fiber amplifiers or from a combination of a fiber amplifier and a Raman amplifier. However, the impact on the amplifier design due to nonlinear effects and amplification of Rayleigh scattering need to be considered for hybrid fiber amplifiers consisting of a combination of an EDFA and a Raman amplifier. These effects are not as strong for a hybrid optical amplifier consisting of a series of concatenated doped-fiber amplifiers, but in this case the gain characteristics of the different amplifier segments need to be matched carefully.

11.9 Overview of Optical Fiber Lasers

Optical fibers doped with rare-earth elements also can be used to create an optical fiber laser [3, 26, 27, 28]. Similar to the gain mechanisms in an optical fiber amplifier,

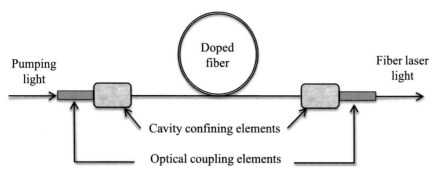

Fig. 11.15 Example of the lasing cavity architecture of a simple doped-fiber laser

elements such as erbium, ytterbium, neodymium, dysprosium, praseodymium, and thulium incorporated into an optical fiber can provide the gain and lasing medium needed for a fiber laser. In addition, fiber nonlinearities such as stimulated Raman scattering and four-wave mixing can be utilized as the gain mechanism in an optical fiber laser (see Chap. 12).

An example of a simple fiber laser is shown in Fig. 11.15. On the left side, light from a pump laser diode (e.g., with a 980-nm wavelength) is launched into the doped fiber, which might have a length of around 10 m. To create a lasing cavity, various types of reflecting cavity confining elements can be used at the fiber ends, such as dichroic mirrors, dielectric coatings that are deposited directly on the fiber ends, or a fiber loop mirror. One practical configuration is to add sections of fiber Bragg gratings at each end of the doped fiber, which act as reflecting mirror elements that provide optical feedback to build up the gain in the laser resonator cavity.

Optical fiber lasers have the following advantages compared to other types of lasers:

- Laser light is created within an optical fiber, so it is easy to couple the light output into a transmission fiber.
- Optical fiber lasers can have active regions that vary anywhere from nominally 10 m to several kilometers and hence they can produce high continuous output powers depending on the fiber length.
- The lasing fiber and its associated components (fiber Bragg gratings and optical couplers) can be coiled up in a compact and rugged housing.
- As a result of the high gain efficiency of doped fibers, a fiber laser can operate with very small pump powers.
- The doped fiber gain media have large gain bandwidths due to strongly broadened laser transitions in glasses, which can result in wide wavelength tuning ranges.

However, optical fiber lasers also have limitations compared to other lasers such as a more complex design process, limitations in achievable single-mode power as a result of nonlinear effects, and the possible risk of fiber damage due to high optical intensities.

11.10 Summary

The creation and development of optical amplifiers has provided significant increases in information capacity over longer transmission distances, which offer attractive economical advantages compared to electronic-based signal amplification. Optical amplifiers have found widespread use in applications ranging from ultra-long undersea links to short links in access networks.

The three main optical amplifier types are semiconductor optical amplifiers (SOAs), active fiber or doped-fiber amplifiers (DFAs), and Raman amplifiers. All optical amplifiers increase the power level of incident light through a stimulated emission or an optical power transfer process. In SOAs and DFAs the mechanism for creating a population inversion is the absorption of energy supplied from an external source called the pump. The pump supplies energy to electrons in an active medium, which raises them to higher energy levels to produce the population inversion. An incoming signal photon will trigger these excited electrons to drop to lower levels through a stimulated-emission process. Because one incoming trigger photon stimulates a cascade effect in which many excited electrons emit photons of equal energy as they drop to the ground state, the result is an amplified optical signal.

In contrast to the amplification mechanisms used in an SOA or DFA, in Raman amplification there is a transfer of optical power from a high-power pump wavelength (e.g., 500 mW at 1480 nm) to lightwave signals at longer wavelengths (e.g., a -25-dBm signal around 1550 nm). This Raman amplification mechanism is done without the need for a population-inversion process.

The dominant noise generated in an optical amplifier is called amplified spontaneous emission (ASE) noise. The origin of this noise is the spontaneous recombination of electrons and holes in the amplifier medium. This recombination occurs over a wide range of electron–hole energy differences and thus gives rise to a broad spectral background of noise photons that get amplified along with the optical signal as they travel through the EDFA. This is shown in Fig. 11.8 for an EDFA amplifying a signal at 1540 nm.

When analyzing a transmission link that has a series of optical amplifiers in it, an important point is that the light signal entering the optical receiver may contain a significant level of ASE noise that has been added by the preceding cascade of optical amplifiers. In this case one has to evaluate the optical signal-to-noise ratio (OSNR). This parameter is defined as the ratio of the average EDFA optical signal output power P_{ave} to the unpolarized ASE optical noise power P_{ASE}.

A Raman optical amplifier is based on a nonlinear effect called stimulated Raman scattering (SRS), which occurs in fibers at high optical powers. Whereas an EDFA requires a specially constructed optical fiber for its operation, a Raman amplifier makes use of the standard transmission fiber itself as the amplification medium. The Raman gain mechanism can be achieved through either a lumped (or discrete) amplifier or a distributed amplifier. In the lumped Raman amplifier configuration, a spool of about 80 m of small-core fiber along with appropriate pump lasers is inserted into the transmission path as a distinct packaged unit. For the distributed

Raman amplifier application, optical power from one or more Raman pump lasers is inserted into the end of the transmission fiber toward the transmitting end. This process converts the final 20–40 km of the transmission fiber into a preamplifier.

Problems

11.1 Consider an InGaAsP semiconductor optical amplifier that has the following parameter values:

Symbol	Parameter	Value
w	Active area width	5 μm
d	Active area thickness	0.5 μm
L	Amplifier length	200 μm
Γ	Confinement factor	0.3
τ_r	Time constant	1 ns
a	Gain coefficient	1×10^{-20} m^2
V_g	Group velocity	2.0×10^8 m/s
n_{th}	Threshold density	1×10^{24} m^{-3}

If a 100-mA bias current is applied, find

(a) The pumping rate R_p,
(b) The maximum (zero-signal) gain,
(c) The saturation photon density,
(d) The photon density if a 1-μW signal at 1310 nm enters the amplifier.

11.2 Verify that the gain expression in Eq. (11.13) follows from the integral relationship Eq. (11.12b).

11.3 The *output saturation power* $P_{out,sat}$ is defined as the amplifier output power for which the amplifier gain G is reduced by 3 dB (a factor of 2) from its unsaturated value G_0. Assuming $G_0 \gg 1$, show the amplifier output saturation power is

$$P_{out,sat} = \frac{G_0 \ln 2}{(G_0 - 2)} P_{amp,sat}$$

11.4 Assume the gain profile of an optical amplifier is $g(\lambda) = g_0 \exp\left[-(\lambda - \lambda_0)^2 / 2(\Delta\lambda)^2\right]$ where λ_0 is the peak-gain wavelength and $\Delta\lambda$ is the spectral width of the amplifier gain. If $\Delta\lambda = 25$ nm, find the FWHM (the 3-dB gain) of the amplifier gain if the peak gain at λ_0 is 30 dB.

11.5 Compare the maximum theoretical power conversion efficiency (PCE) for 980-nm and 1475-nm pumping in an EDFA for a 1545-nm signal. Contrast this with actual measured results of PCE = 50.0 and 75.6% for 980-nm and 1475-nm pumping, respectively.

11.6 Consider an EDFA power amplifier that produces $P_{s,out} = 27$ dBm for an input level of 2 dBm at 1542 nm.

(a) Find the amplifier gain.

(b) What is the minimum pump power that is required?

11.7 .

    (a) To see the relative contributions of the various noise mechanisms in an optical amplifier, calculate the values of the five noise terms in Eq. (11.30) for operational gains of $G = 20$ dB and 30 dB. Assume the optical bandwidth is equal to the spontaneous emission bandwidth (30-nm spectral width) and use the following parameter values:

(b) To
see the effect of using a narrowband optical filter at the receiver, let $B_o = 1.25 \times 10^{11}$ Hz (125 GHz at 1550 nm) and find the same five noise terms for $G = 20$ dB and 30 dB.

Symbol	Parameter	Value
η	Photodiode quantum efficiency	0.6
$\mathfrak{R}$	Responsivity	0.73 A/W
P_{in}	Input optical power	1 μW
λ	Wavelength	1550 nm
B_o	Optical bandwidth	3.77×10^{12} Hz
B_e	Receiver bandwidth	1×10^9 Hz
n_{sp}	Spontaneous emission factor	2
R_L	Receiver load resistor	1000 Ω

11.8 .

    (a) Consider a cascaded chain of k fiber-plus-EDFA segments. Each fiber segment has a length L and an attenuation α. The EDFA has a gain $G = 1/\alpha$. (a) Show that the path-averaged signal power is

$$\langle P \rangle_{path} = P_{s,in} \frac{G - 1}{G \ln G}$$

(b) Derive the path-averaged ASE power given by Eq. (11.42).

11.9 Consider a long-distance transmission system containing a cascaded chain of EDFAs. Assume each EDFA is operated in the saturation region and that the slope of the gain-versus-input power curve in this region is -0.5; that is, the gain changes by ± 3 dB for a ∓ 6-dB variation in input power. Let the link have the operational parameters shown in the table below. Suppose there is a sudden 6-dB drop in signal level at some point in the link. Find the power output levels after the degraded signal has passed through 1, 2, 3, and 4 succeeding amplifier stages.

Symbol	Parameter	Value
G	Nominal gain	7.1 dB
$P_{s,out}$	Nominal output optical power	3.0 dBm
$P_{s,in}$	Nominal input optical power	−4.1 dBm

11.10 Consider an EDFA with a gain of 26 dB and a maximum output power of 0 dBm.

    (a) Compare the output signal levels per channel for 1, 2, 4, and 8 wavelength channels, where the input power is 1 μW for each signal.

    (b) What are the output signal levels per channel in each case if the pump power is doubled?

Answers to Selected Problems

11.1   (a) $R_p = 1.26 \times 10^{27}$ (electrons/cm³)/s (b) $g_0 = 7.5$ cm⁻¹
      (c) $N_{ph,sat} = 1.67 \times 10^{15}$ photons/cm³ (d) $N = 1.32 \times 10^{10}$ photons/cm³.
11.3 $P_{out,sat} = 0.693\, P_{amp,sat}$
11.4   With the 3-dB gain $G = 27$ dB, then FWHM $= 0.50\ \Delta\lambda$
11.5   (a) PCE (980 nm) $\leq 63.4\%$; PCE (1475 nm) $\leq 95.5\%$
11.6   (a) $G = 25$ dB; (b) $P_{p,in} \geq 785$ mW
11.7 $\sigma_{th}^2 = 1.62 \times 10^{-14}$ A²; $\sigma_{shot-s}^2 = 2.34 \times 10^{-14}$ A²; $\sigma_{ASE-ASE}^2 = 2.26 \times 10^{-14}$ A²; $\sigma_{s-ASE}^2 = 5.47 \times 10^{-12}$ A²; $\sigma_{ASE-ASE}^2 = 7.01 \times 10^{-13}$ A²
11.9   Because the slope of the gain-versus -input power curve is –0.5, then for a 6-dB drop in the input signal, the gain increases by +3 dB.

    1. Thus at the first amplifier, a −10.1-dBm signal now arrives and experiences a +10.1-dB gain. This gives a 0-dBm output (versus a normal + 3-dBm output).

    2. At the second amplifier, the input is now −7.1 dBm (down 3 dB from the usual −4.1 dBm level). Hence the gain is now 8.6 dB (up 1.5 dB), yielding an output of

$$-7.1\,\text{dBm} + (7.1 + 1.5)\,\text{dB} = 1.5\,\text{dBm}$$

    3. At the third amplifier, the input is now −5.6 dBm (down 1.5 dB from the usual −4.1 dBm level). Hence the gain is up 0.75 dB, yielding an output of

$$-5.6\,\text{dBm} + (7.1 + 0.75)\,\text{dB} = 2.25\,\text{dBm}$$

    4. At the fourth amplifier, the input is now −4.85 dBm (down 0.75 dB from the usual −4.1 dBm level). Hence the gain is up 0.375 dB, yielding an output of

$$-4.85 \, \text{dBm} + (7.1 + 0.375) \, \text{dB} = 2.63 \, \text{dBm}$$

which is within 0.37 dB of the normal $+3$ dBm level.

11.10   (a) For N input signals, the output signal level is given by

$$P_{s,out} = G \sum_{i=1}^{N} P_{s,in}(i) \leq 1 \, \text{mW}.$$

The inputs are 1 μW (-30 dBm) each and the gain is 26 dB (a factor of 400). Thus for one input signal, the output is (400)(1 μW) = 400 μW or -4 dBm. For two input signals, the total output is 800 μW or -1 dBm. Thus the level of each individual output signal is 400 μW or -4 dBm.

For four input signals, the total input level is 4 μW or -24 Bm. The output then reaches its limit of 0 dBm, because the maximum gain is 26 dB. Thus the level of each individual output signal is 250 μW or -6 dBm.

Similarly, for eight input channels the maximum output level is 0 dBm, so the level of each individual output signal is 1/8(1 mW) = 125 μW or -9 dBm.

(b) When the pump power is doubled, the outputs for one and two inputs remain at the same level. However, for four inputs, the individual output level is 500 μW or -3 dBm, and for 8 inputs, the individual output level is 250 μW or -6 dBm.

References

1. D.R. Zimmerman, L.H. Spiekman, Amplifiers for the masses: EDFA, EDWA, and SOA amplets for metro and access applications. J. Lightw. Technol. **22**, 63–70 (2004)
2. R.E. Hunsperger, Optical amplifiers, chap. 13, in *Integrated Optics*, Springer, 6th edn. (2009), pp. 259–275
3. V. Ter-Mikirtychev, *Fundamentals of Fiber Lasers and Fiber Amplifiers* (Springer, 2014)
4. S. Singh, R.S. Kaler, Review on recent developments in hybrid optical amplifier for dense wavelength division multiplexed system. Opt. Eng. **54**, 100901(1–11) (2015)
5. M. Singh, A review on hybrid optical amplifiers. J. Opt. Commun. **39**(3), 267–272 (2018)
6. N.K. Dutta, Q. Wang, *Semiconductor Optical Amplifiers*, 2nd edn. (World Scientific, 2013)
7. B.L. Anderson, R.L. Anderson, *Fundamentals of Semiconductor Devices*, 2nd edn. (McGraw-Hill, 2018)
8. W. Miniscalco, Erbium-doped glasses for fiber amplifiers at 1500 nm. J. Lightw. Technol. **9**, 234–250 (1991)
9. P.C. Becker, N.A. Olsson, J.R. Simpson, *Erbium-Doped Fiber Amplifiers* (Academic Press, 1999)
10. E. Desurvire, *Erbium-Doped Fiber Amplifiers: Principles and Applications* (Wiley, New York, 2002)
11. E. Desurvire, D. Bayart, B. Desthieux, S. Bigo, *Erbium-Doped Fiber Amplifiers: Devices and System Developments* (Wiley, New York, 2002)
12. G.R. Khan, An analytical method for gain in erbium-doped fiber amplifier with pump excited state absorption. Opt. Fiber Technol. **18**(6), 421–424 (2012)

13. G. Sobon, P. Kaczmarek, K.M. Abramski, Erbium-ytterbium co-doped fiber amplifier operating at 1550 nm with stimulated lasing at 1064 nm. Opt. Commun. **285**(7), 1929–1933 (2012)
14. C. Harder, Pump diode lasers, chap. 5, in *Optical Fiber Telecommunications-V*, ed. by I.P. Kaminov, T. Li, A.E. Willner, vol. A (Academic Press, 2008), pp. 107–144
15. M.A. Ali, A.F. Elrefaie, R.E. Wagner, S.A. Ahmed, A detailed comparison of the overall performance of 980 and 1480 nm pumped EDFA cascades in WDM multiple-access light-wave networks. J. Lightw. Technol. **14**, 1436–1448 (1996)
16. D.M. Baney, P. Gallion, R.S. Tucker, Theory and measurement techniques for the noise figure of optical amplifiers. Opt. Fiber Tech. **6**(2), 122–154 (2000)
17. A. Hardy, R. Oron, Signal amplification in strongly pumped fiber amplifiers. IEEE J. Quantum Electron. **33**, 307–313 (1997)
18. A.V. Tran, R.S. Tucker, N.L. Boland, Amplifier placement methods for metropolitan WDM ring networks. J. Lightw. Technol. **22**, 2509–2522 (2004)
19. T.T. Ha, G.E. Keiser, R.L. Borchart, Bit error probabilities of OOK lightwave systems with optical amplifiers. J. Opt. Commun. **18**, 151–155 (1997)
20. L. Sirleto, M.A. Ferrara, Fiber amplifiers and fiber lasers based on stimulated Raman scattering: a review. Micromachines **11**(247), 26 (2020)
21. J. Bromage, Raman amplification for fiber communications systems. J. Lightw. Technol. **22**, 79–93 (2004)
22. J. Chen, C. Lu, Y. Wang, Z. Li, Design of multistage gain-flattened fiber Raman amplifiers. J. Lightw. Technol. **24**, 935–944 (2006)
23. L.A.M. Saito, P.D. Taveira, P.B. Gaarde, K. De Souza, E.A. De Souza, Multi-pump discrete Raman amplifier for CWDM system in the O-band. Opt. Fiber Technol. **14**(4), 294–298 (2008)
24. T. Sakamoto, S.-I. Aozasa, M. Yamada, M. Shimizu, Hybrid fiber amplifiers consisting of cascaded TDFA and EDFA for WDM signals. J. Lightw. Technol. **24**, 2287–2295 (2006)
25. H.H. Lee, P.P. Iannone, K.C. Reichmann, J.S. Lee, B. Pálsdóttir, A C/L-band gain-clamped SOA-Raman hybrid amplifier for CWDM access networks. Photon. Technol. Lett. **20**(3), 196–198 (2008)
26. Z.-R. Lin, C.-K. Liu, G. Keiser, Tunable dual wavelength erbium-doped fiber ring laser covering both C-band and L-band for high-speed communications. Optik **123**(1), 46–48 (2012)
27. Y. Feng, *Raman Fiber Lasers* (Springer, Berlin, 2017)
28. Z. Yan, X. Li, Y. Tang, P.P. Shum, X. Yu, Y. Zhang, Q.J. Wang, Tunable and switchable dual-wavelength Tm-doped mode-locked fiber laser by nonlinear polarization evolution. Opt. Express **23**(4), 4369–4376 (2015)

Chapter 12
Nonlinear Processes in Optical Fibers

Abstract This chapter discusses several different nonlinear effects that start to appear in an optical fiber link as the optical power increases beyond a certain level. This can occur in a fiber when the core area is made small or when several high-strength optical fields from different signal wavelengths are present in a fiber at the same time. Consequences of nonlinear effects for signal levels include power gain or loss at different wavelengths, wavelength conversions, and crosstalk between wavelength channels. In some cases the nonlinear effects can degrade WDM system performance, while in other situations they might provide a useful application.

The design of a lightwave transmission system requires careful planning and consideration of factors such as fiber selection, choice and tuning of optoelectronic components, optical amplifier placement, and path routing. The goal of these efforts is to create a network that meets the design criteria, is reliable, and is easy to operate and maintain. As the previous chapters describe, the design process must take into account all power penalties associated with optical signal-degradation processes.

Intuitively, it seems natural to let the input optical power be as large as possible to overcome the power penalty effects to achieve the link design goals. However, this works only if the fiber is a linear medium; that is, if the fiber properties are independent of the optical signal power level. In an actual fiber several different nonlinear effects start to appear as the optical power level increases beyond a certain level [1–5]. High optical power densities can occur in a fiber when the core area is made small or when several high-strength optical fields from different signal wavelengths are present in a fiber at the same time. Nonlinear effects arise when the electromagnetic fields from these high power densities interact with acoustic waves and molecular vibrations. To avoid nonlinear effects, the total power in the fiber must not exceed a specific level. This restriction limits the power per WDM channel. For example, if the nonlinear threshold for the total launched power into a fiber is 17 dBm (50 mW), then for a 64-channel DWDM link the power limit per wavelength is −1.0 dBm (0.78 mW). Consequences of nonlinear effects for signal levels of this magnitude or greater include power gain or loss at different wavelengths, wavelength conversions, and crosstalk between wavelength channels. In some cases the nonlinear effects can degrade WDM system performance, while in other situations they might provide a useful application.

© The Author(s), under exclusive license to Springer Nature Singapore Pte Ltd. 2021   477
G. Keiser, *Fiber Optic Communications*,
https://doi.org/10.1007/978-981-33-4665-9_12

This chapter first gives a general overview of nonlinear processes in optical fibers in Sect. 12.1. Because the nonlinearities arise above a certain optical power threshold, the effect becomes negligible once the signal has become sufficiently attenuated after traveling a certain distance along the fiber. This gives rise to the concept of an effective length and an associated parameter called effective area, as Sect. 12.2 describes. The next five sections discuss how the major nonlinear processes physically affect system performance. These nonlinearities are stimulated Raman scattering (Sect. 12.3), stimulated Brillouin scattering (Sect. 12.4), self-phase modulation (Sect. 12.5), cross-phase modulation (Sect. 12.6), and four-wave mixing (Sect. 12.7). Mitigation of four-wave mixing can be achieved by means of a special fiber design or by a chromatic dispersion-compensation method, which is the topic of Sect. 12.8. On the other hand, nonlinear effects also can have beneficial uses. Section 12.9 describes applications of cross-phase-modulation and four-wave-mixing techniques for performing wavelength conversion in WDM networks. Another application of nonlinear effects in a silica fiber is the use of solitons for optical communications, which depends on self-phase modulation effects. Section 12.10 addresses this topic.

12.1 Classifications of Nonlinearities

Optical nonlinearities can be classified into two general categories, which are summarized in Table 12.1. The first category encompasses the nonlinear inelastic scattering processes. These are *stimulated Brillouin scattering* (SBS) and *stimulated Raman scattering* (SRS). The second category of nonlinear effects arises from intensity-dependent variations in the refractive index in a silica fiber, which is known as the *Kerr effect*. These effects include *self-phase modulation* (SPM), *cross-phase modulation* (XPM), and *four-wave mixing* (FWM). In the literature, FWM sometimes is referred to as four-photon mixing (FPM), and XPM also is designated by CPM. Note that certain nonlinear effects are independent of the number of WDM channels.

SBS, SRS, and FWM result in gains or losses in a wavelength channel. The power variations depend on the optical signal intensity. These three nonlinear processes provide gains to some channels while depleting power from others, thereby producing crosstalk between the wavelength channels. In analog video systems, SBS significantly degrades the carrier-to-noise ratio when the scattered power is equivalent to

Table 12.1 Summary of nonlinear effects in optical fibers

Nonlinearity category	Single-channel	Multiple-channel
Scattering related	Stimulated Brillouin scattering	Stimulus Raman scattering
Index related	Self-phase modulation	Cross-phase modulation and four-wave mixing

Fig. 12.1 Nonlinear effects cause optical power reductions and lead to power penalties

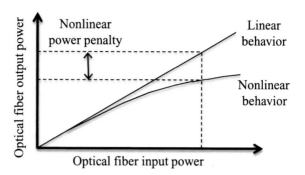

the signal power in the fiber. SPM and XPM affect only the phase of signals, which causes chirping in digital pulses. This can worsen pulse broadening due to dispersion, particularly in very high-rate systems, such as 40 Gb/s and higher. Techniques for mitigating FWM are described in Sect. 12.8.

When any of these nonlinear effects contribute to signal strength reduction, the amount of optical power reduction (in decibels) is known as the *power penalty* for that effect, as Fig. 12.1 illustrates. As the following sections show, factors that influence to what degree a particular nonlinearity affects optical fiber link performance include chromatic and polarization-mode dispersions, the effective core area of the fiber, the number and spacing of wavelength channels in a WDM system, the length of the transmission link, and the light source linewidth and emitted optical power level.

12.2 Effective Length and Effective Area

Modeling the nonlinear processes can be quite tedious because their effects depend on the transmission length, the cross-sectional area of the fiber, and the optical power level in the fiber. The difficulty arises from the fact that the impact of the nonlinearity on signal fidelity increases with distance. However, this is offset by the continuous exponential decrease in signal power along the fiber due to attenuation, as stated in Eq. (3.1a). Thus, the nonlinear effects tend to occur only at the beginning of a long fiber span. In practice, one can use a simple but sufficiently accurate model that assumes the power is constant over a certain fiber length, which is less than or equal to the actual fiber length. This *effective length* L_{eff} takes into account power absorption along the length of the fiber, that is, the fact that the optical power decays exponentially with length. For a link of length L with no optical amplifiers the effective length is given by

$$L_{eff} = \frac{\int_0^L P(z)dz}{P_{in}} = \frac{\int_0^L P_{in}e^{-\alpha z}dz}{P_{in}} = \int_0^L e^{-\alpha z}dz = \frac{1 - exp(-\alpha L)}{\alpha} \quad (12.1)$$

Fig. 12.2 The effective length is modeled as the transmission length at which the shaded area equals the area under the actual power-distribution curve

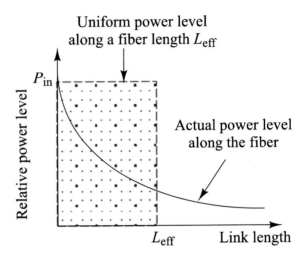

Here L is the length of the fiber span, $P(z)$ is the optical power at a distance z along the fiber, P_{in} is the power level launched into the fiber, and α is the fiber attenuation.

As Fig. 12.2 illustrates, this means that the area designated by the shaded pattern is equal to the area under the power-distribution curve. For an attenuation of 0.21 dB/km (or, equivalently, 4.80×10^{-2} km^{-1}) at 1550 nm, the effective length is about 21 km when $L \gg 1/\alpha$. When there are optical amplifiers in a link, the signal-impairments owing to the nonlinearities do not change as the signal passes through the amplifiers.

> **Drill Problem 12.1** For long fiber spans when $L >> 1/\alpha$ the condition exp$(-\alpha L) << 1$ holds. Show that when the fiber attenuation is 0.21 dB/km (or, equivalently, 4.80×10^{-2} km^{-1}) at 1550 nm, the effective length is 20.8 km.

The effects of nonlinearities increase with the light intensity in a fiber. For a given optical power, this intensity is inversely proportional to the cross-sectional area of the fiber core. As Fig. 12.3 shows, although the intensity is not distributed uniformly across the fiber-core area, for practical purposes one can use an *effective area* A_{eff}, which assumes a uniform intensity distribution across most of the cross-sectional area of the core. This area can be calculated from mode-overlap integrals and, in general, is close to the actual core cross-sectional area. The result is that the impact of most nonlinear effects in a fiber can be calculated based on the effective area of the fundamental mode in that fiber. For example, the effective intensity I_e of a light pulse becomes $I_e = P/A_{eff}$, where P is the optical power contained in the pulse. Table 12.2 lists the effective areas of some single-mode fibers.

Fig. 12.3 The effective area is modeled as a central area of the fiber core within which the intensity is assumed to be uniform

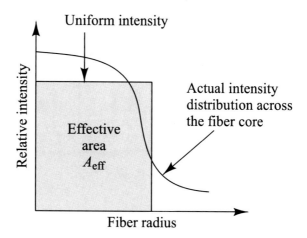

Table 12.2 Effective area and attenuation of some single-mode fibers

Fiber type	Attenuation (db/km)	Effective area (μm^2)
G.652 standard single-mode	0.35 at 1310 nm	72
G.652 C/D low-water-peak	0.20 at 1550 nm	72
Dispersion-compensating	0.40 at 1550 nm	21
G.655 single-mode	0.21 at 1550 nm	55

12.3 Stimulated Raman Scattering

Stimulated Raman scattering is an interaction between lightwaves and the vibrational modes of silica molecules [2–5]. If a photon with energy $h\nu_1$ is incident on a molecule that has a vibrational frequency v_m, the molecule can absorb some energy from the photon. In this interaction, the photon is scattered, thereby attaining a lower frequency v_2 and a corresponding lower energy $h\nu_2$. The modified photon is called a *Stokes photon*. Because the optical signal wave that is injected into a fiber is the source of the interacting photons, it is often called the *pump wave* because it supplies power for the generated wave.

The SRS process generates scattered light at a wavelength longer than that of the incident light (up to 125 nm). If another signal is present at this longer wavelength, the SRS light will amplify it and the pump-wavelength signal will decrease in power; Fig. 12.4 illustrates this effect. Consequently, SRS can severely limit the performance of a multichannel optical communication system by transferring energy from short-wavelength channels to neighboring higher-wavelength channels. This is a broadband effect that can occur in both directions in the fiber. Powers in WDM channels separated by up to 16 THz (125 nm) can be coupled through the SRS effect, as Fig. 12.5 illustrates in terms of the Raman gain coefficient g_R as a function of the channel separation Δv_s. This figure shows that, owing to SRS, the power transferred

Fig. 12.4 SRS transfers optical power from shorter wavelengths to longer wavelengths

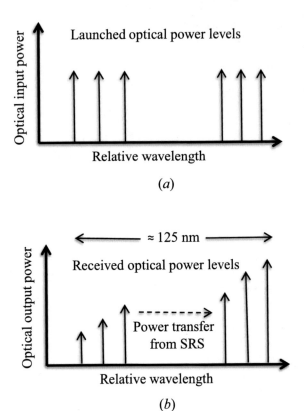

Fig. 12.5 Simplified linear approximation for the Raman gain coefficient as a function of channel spacing with a low-value decaying tail at channel separations greater than 16 THz

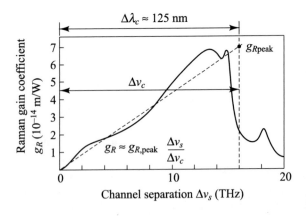

from a lower-wavelength channel to a higher-wavelength channel increases approximately linearly with channel spacing up to a maximum of about $\Delta v_c = 16$ THz (or $\Delta \lambda_c = 125$ nm in the 1550 nm window), and then drops off sharply for larger spacings.

To see the effects of SRS, consider a WDM system that has N channels equally spaced in a 30-nm band centered at 1545 nm. Channel 0, which is at the lowest (shortest) wavelength, is affected the worst because power gets transferred from this channel to all longer-wavelength channels. For simplicity, assume that the transmitted power P is the same on all channels, that the Raman gain increases linearly as shown by the dashed line in Fig. 12.5, and that there is no interaction between the other channels. If $F_{out}(j)$ is the fraction of power coupled from channel 0 to channel j, then the total fraction of power coupled out of channel 0 to all the other channels is [6].

$$F_{out} = \sum_{j=1}^{N-1} F_{out}(j) = \sum_{j=1}^{N-1} g_{R,peak} \frac{j \Delta v_s}{\Delta v_c} \frac{P L_{eff}}{2 A_{eff}}$$

(12.2)

$$= \frac{g_{R,peak} \Delta v_s P L_{eff}}{2 \Delta v_c A_{eff}} \frac{N(N-1)}{2}$$

The power penalty for this channel then is $-10 \log(1 - F_{out})$. To keep the penalty below 0.5 dB, it is necessary to have $F_{out} < 0.1$. Using Eq. 12.2, and with $A_{eff} = 55$ μm^2 and $g_{R,peak} = 7 \times 10^{-14}$ m/W from Fig. 12.5, gives the criterion

$$[NP][N-1][\Delta v_s] L_{eff} < 5 \times 10^3 \text{mW} \cdot \text{THz} \cdot \text{km}$$

(12.3)

Here, NP is the total power from N sources coupled into the fiber, $(N-1) \Delta v_s$ is the total occupied optical bandwidth, and L_{eff} is the effective length, which takes into account absorption along the length of the fiber.

Example 12.1 The power penalty in decibels per channel due to SRS can be calculated from Eq. (12.2) as $PP_{SRS} = -10 \log(1 - F_{out})$. Assume the following parameter values: $g_R = 4.7 \times 10^{-14}$ m/W, $L_{eff} = 21$ km, $\Delta v_c = 125$ GHz, and $A_{eff} = 55$ μm^2. Table 12.3 shows the maximum allowed launch power per channel for an SRS power penalty of 0.5 dB in links having channel counts of N = 8, 16, 40 and 80 and for DWDM channel spacing of $\Delta v_s = 50$ and 100 GHz. This table indicates that the SRS power penalty becomes less as the channel spacing decreases or as the number of DWDM channels decreases.

Table 12.3 Approximate allowed optical signal launch power to have a maximum SRS Power penalty of 0.5 dB

Maximum launch power (dBm)		
Channel count	50 GHz	100 GHz
8	21	18
16	14.7	11.6
40	6.6	3.6
80	0.5	−2.5

Drill Problem 12.2 Assume the following parameter values: $g_R = 4.7 \times 10^{-14}$ m/W, $L_{\text{eff}} = 21$ km, $\Delta v_c = 125$ GHz, and $A_{\text{eff}} = 55$ μm^2. Verify the examples given in Table 12.3 of the maximum launch power P for various channel counts and channel separations of $\Delta v_s = 50$ and 100 GHz for a SRS power penalty of 0.5 dB.

12.4 Stimulated Brillouin Scattering

Stimulated Brillouin scattering arises when a strong optical signal generates an acoustic wave that produces variations in the refractive index [2–5,7]. These index variations cause lightwaves to scatter in the backward direction along the fiber toward the transmitter. This backscattered light experiences gain from the forward-propagating signals, which leads to depletion of the signal power. The frequency of the scattered light at a wavelength λ experiences a Doppler shift given by

$$v_B = 2n V_s / \lambda \qquad (12.4)$$

where n is the index of refraction and V_s is the velocity of sound in the material. In silica, this interaction occurs over a very narrow *Brillouin linewidth* of $\Delta v_B = 20$ MHz at 1550 nm. For $V_s = 5760$ m/s in fused silica, the frequency of the backward-propagating light at 1550 nm is downshifted by 11 GHz (0.09 nm) from the original signal, as Fig. 12.6 illustrates. The SBS effect is confined within a single wavelength channel in a WDM system. Thus, the effects of SBS accumulate individually for each channel, and, consequently, occur at the same power level in each WDM channel, analogous to a single-channel system.

System impairment starts when the amplitude of the scattered wave is comparable to the signal power. For typical fibers, the threshold power for this process is around 10 mW for single-fiber spans. In a long fiber chain containing optical amplifiers,

Fig. 12.6 Stimulated Brillouin scattering produces a Doppler shift of about 11 GHz in the scattered light

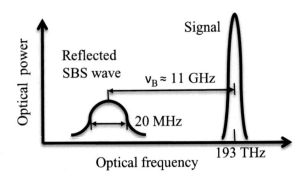

there are normally optical isolators to prevent backscattered signals from entering the amplifier. Consequently, the impairment due to SBS is limited to the degradation occurring in a single amplifier-to-amplifier span.

One criterion for determining at what point SBS becomes a problem is to consider the SBS threshold power P_{th}. This is defined to be the signal power at which the backscattered light equals the fiber-input power. The calculation of this expression is rather complicated, but an approximation is given by [8]

$$P_{th} \approx 21 \frac{A_{eff} b}{g_B L_{eff}} \left(1 + \frac{\Delta v_{source}}{\Delta v_B} \right) \qquad (12.5)$$

Here, A_{eff} is the effective cross-sectional area of the propagating wave, Δv_{source} is the source linewidth, and the polarization factor b lies between 1 and 2 depending on the relative polarizations of the pump and Stokes waves. The effective length L_{eff} is given in Eq. (12.1) and g_B is the *Brillouin gain coefficient*, which is approximately 4×10^{-11} m/W, independent of the wavelength. Equation (12.5) shows that the SBS threshold power increases as the source linewidth becomes larger.

Example 12.2 Consider an optical source with a 40-MHz linewidth. Using the values $\Delta v_B = 20$ MHz at 1550 nm, $A_{eff} = 55 \times 10^{-12}$ m^2 (for a typical dispersion-shifted single-mode fiber), $L_{eff} = 20$ km, and assuming a value of $b = 2$, then from Eq. (12.5) it follows that $P_{th} = 8.6$ mW $= 9.3$ dBm.

> **Drill Problem 12.3** Consider an optical source with a 60-MHz linewidth. Using the values $\Delta v_B = 20$ MHz at 1550 nm, $A_{eff} = 55 \times 10^{-12}$ m^2 (for a typical dispersion-shifted single-mode fiber), $L_{eff} = 20$ km, assuming a value of $b = 1$, and letting $g_B = 3.5 \times 10^{-11}$ m/W, show from Eq. (12.5) that $P_{th} = 5.0$ mW (or 7.0 dBm).

Figure 12.7 illustrates the effect of SBS on unmodulated signal power once the threshold is reached. The plots give the relative Brillouin scattered power and the signal power transmitted through a fiber as a function of the input power. Below a certain signal level called the *SBS threshold*, the transmitted power increases linearly with the input level. The effect of SBS is negligible for these low power levels but becomes greater as the optical power level increases.

At the SBS threshold, the SBS process becomes nonlinear and the launched signal loses an increasingly greater percentage of its power as the signal strength becomes larger. Beyond the SBS threshold, the percentage increase in signal depletion continues to grow with signal strength until the *SBS limit* is reached. Any additional optical power launched into the fiber after this point merely is scattered backward along the fiber due to the SBS effect. Thus above the SBS limit the transmitted power remains constant for higher inputs, because all the added power is extracted from the signal to feed the scattered wave.

Fig. 12.7 The effect of SBS on signal power in an optical fiber

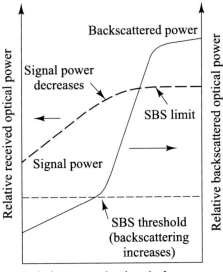

12.5   Self-Phase Modulation

The refractive index n of many optical materials has a weak dependence on *optical intensity I* (equal to the optical power per effective area in the fiber) given by

$$n = n_0 + n_2 I = n_0 + n_2 \frac{P}{A_{eff}} \qquad (12.6)$$

where n_0 is the *ordinary refractive index* of the material and n_2 is the *nonlinear index coefficient*. The factor n_2 is about 2.6×10^{-8} $\mu m^2/W$ in silica, between 1.2 and 5.1×10^{-6} $\mu m^2/W$ in tellurite glasses, and 2.4×10^{-5} $\mu m^2/W$ in $As_{40}Se_{60}$ chalcogenide glass. The nonlinearity in the refractive index is known as the *Kerr nonlinearity* [9]. This nonlinearity produces a carrier-induced phase modulation of the propagating signal, which is called the *Kerr effect*. In single-wavelength links, this gives rise to *self-phase modulation* (SPM), which converts optical power fluctuations in a propagating lightwave to spurious phase fluctuations in the same wave.

The main parameter is the *nonlinear coefficient* γ, which indicates the magnitude of the nonlinear effect for SPM. This parameter is given by

$$\gamma = \frac{2\pi}{\lambda} \frac{n_2}{A_{eff}} \qquad (12.7)$$

where λ is the free-space wavelength and A_{eff} is the effective core area. The value of γ ranges from 1 to 5 W^{-1} km^{-1} in silica depending on the fiber type and the wavelength. For example, $\gamma = 1.3\ \text{W}^{-1}\ \text{km}^{-1}$ at 1550 nm for a G.652 single-mode fiber that has an effective area equal to 72 μm^2. The frequency shift $\Delta\varphi$ arising from SPM is given by

$$\Delta\varphi = \frac{d\varphi}{dt} = \gamma L_{eff} \frac{dP}{dt} \tag{12.8}$$

Here L_{eff} is the effective length given by Eq. (12.1) and dP/dt is the derivative of the optical pulse power; that is, it shows that the frequency shift occurs when the optical pulse power is changing in time.

To see the effect of SPM, consider what happens to the optical pulse shown in Fig. 12.8 as it propagates in a fiber. Here the time axis is normalized to the parameter t_0, which is the pulse half-width at the $1/e$-intensity point. The edges of the pulse represent a time varying intensity, which rises rapidly from zero to a maximum value, and then returns to zero. In a medium having an intensity-dependent refractive index, a time varying signal intensity will produce a time varying refractive index. Thus the index at the peak of the pulse will be slightly different than the value in the wings of

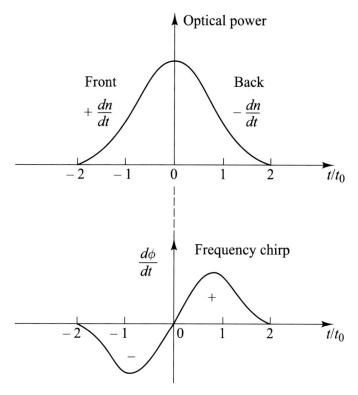

Fig. 12.8 Phenomenological description of spectral broadening of a pulse due to self-phase modulation

the pulse. The leading edge will see a positive dn/dt, whereas the trailing edge will see a negative dn/dt.

This temporally varying index change results in a temporally varying phase change, shown by $d\varphi/dt$ in Fig. 12.8. The consequence is that the instantaneous optical frequency differs from its initial value ν_0 across the pulse. That is, because the phase fluctuations are intensity-dependent, different parts of the pulse undergo different phase shifts. This leads to what is known as *frequency chirping*, in that the rising edge of the pulse experiences a red shift in frequency (toward lower frequencies or longer wavelengths), whereas the trailing edge of the pulse experiences a blue shift in frequency (toward higher frequencies or shorter wavelengths). Because the degree of chirping depends on the transmitted power, SPM effects are more pronounced for higher-intensity pulses.

For some types of fibers, the time-varying phase may result in a power penalty owing to a GVD-induced spectral broadening of the pulse as it travels along the fiber. In the *normal dispersion region* the chromatic dispersion is negative [that is, from Eq. (3.32) the GVD parameter $\beta_2 > 0$] and the group delay decreases with wavelength. This means that because red light has a longer wavelength than blue, the red light travels faster in silica because $n_{red} < n_{blue}$ (see Fig. 3.9). Therefore, in the normal dispersion region the red-shifted leading edge of the pulse travels faster and thus moves away from the center of the pulse. At the same time the blue-shifted trailing edge travels slower, and thus also moves away from the center of the pulse. In this case chirping worsens the effects of GVD-induced pulse broadening. On the other hand, in the *anomalous dispersion region* where chromatic dispersion is positive so that the group delay increases with wavelength, the red-shifted leading edge of the pulse travels slower and thus moves toward the center of the pulse. Similarly, the blue-shifted trailing edge travels faster, and also moves toward the center of the pulse. In this case, SPM causes the pulse to narrow, thereby partly compensating for chromatic dispersion.

12.6   Cross-Phase Modulation in WDM Systems

Cross-phase modulation (XPM) appears in WDM systems and has a similar origin as SPM. In this case the name derives from the fact that the refractive index nonlinearity converts optical intensity fluctuations in a particular wavelength channel to phase fluctuations in adjacent copropagating channels [2–5] In addition, because the refractive index seen by a particular wavelength is influenced by both the optical intensity of that wave itself and also by the optical power fluctuations of neighboring wavelengths, SPM is always present when XPM occurs. Analogous to SPM, for two interacting wavelengths the XPM-induced frequency shift $\Delta\varphi$ is given by

$$\Delta\varphi = \frac{d\varphi}{dt} = 2\gamma L_{eff}\frac{dP}{dt} \tag{12.9}$$

where the parameters are the same as for Eq. (12.8). When multiple wavelengths propagate in a fiber, the total phase shift for an optical signal with frequency ω_i is

$$\Delta\varphi_i = \gamma L_{eff}\left[\frac{dP_i}{dt} + 2\sum_{j \neq i}\frac{dP_j}{dt}\right] \tag{12.10}$$

The first term in the square brackets represents the SPM contribution and the second term arises from XPM. The factor 2 in the bracketed expression shows that the weight of XPM is twice that of SPM. However, XPM only appears when the two interacting light beams or pulses overlap in space and time. In general, pulses from two different wavelength channels will not remain superimposed because each has a different GVD. This greatly reduces the impact of XPM for direct-detection optical fiber transmission systems.

12.7 Four-Wave Mixing in WDM Channels

To transmit the high capacities of dense WDM channels over long distances requires operation in the 1550-nm window of dispersion-shifted fiber. In addition, to preserve an adequate signal-to-noise ratio, a high-speed system operating over long distances and having nominal optical repeater spacings of 100 km needs optical launch powers of around 1 mW per channel. For such WDM systems, the simultaneous requirements of high launch power and low dispersion give rise to the generation of new frequencies due to four-wave mixing [2–5].

Four-wave mixing (FWM) is a third-order nonlinearity in optical fibers that is analogous to intermodulation distortion in electrical systems. When wavelength channels are located near the zero-dispersion point, three optical frequencies (v_i, v_j, v_k) will mix to produce a fourth intermodulation product v_{ijk} given by

$$v_{ijk} = v_i + v_j - v_k \text{ with i, j} \neq \text{k} \tag{12.11}$$

When this new frequency falls in the transmission window of the original frequencies, it can cause severe crosstalk.

Figure 12.9 shows a simple example for two waves at frequencies v_1 and v_2. As these waves copropagate along a fiber, they mix and generate sidebands at $2v_1 - v_2$ and $2v_2 - v_1$. Similarly, three copropagating waves will create nine new optical sideband waves at frequencies given by Eq. (12.11). These sidebands will travel along with the original waves and will grow at the expense of signal-strength depletion. In general, for N wavelengths launched into a fiber, the number of generated mixing products M is

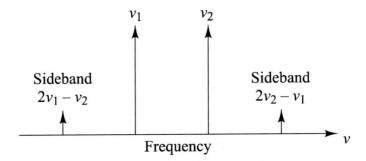

Fig. 12.9 Two optical waves at frequencies ν_1 and ν_2 mix to generate two third-order sidebands

$$M = \frac{N^2}{2}(N - 1) \tag{12.12}$$

If the channels are equally spaced, a number of the new waves will have the same frequencies as the injected signals. Thus the resultant crosstalk interference plus the depletion of the original signal waves can severely degrade multichannel system performance unless steps are taken to diminish it.

Example 12.3 Consider the case of three DWDM signals propagating in a fiber at optical frequencies ν_1, ν_2, and ν_3. What frequency components result from four-wave mixing?

Solution: From Eq. (12.12) for $N = 3$ channels there are $M = 9$ mixing products generated. These are as follows:

$$\nu_{123} = \nu_1 + \nu_2 - \nu_3 \text{ (identical to } \nu_{213} = \nu_2 + \nu_1 - \nu_3)$$

$$\nu_{321} = \nu_3 + \nu_2 - \nu_1 \text{ (identical to } \nu_{231} = \nu_2 + \nu_3 - \nu_1)$$

$$\nu_{312} = \nu_3 + \nu_1 - \nu_2 \text{ (identical to } \nu_{132} = \nu_1 + \nu_3 - \nu_2)$$

$$\nu_{112} = 2\nu_1 - \nu_2$$

$$\nu_{113} = 2\nu_1 - \nu_3$$

$$\nu_{221} = 2\nu_2 - \nu_1$$

$$\nu_{223} = 2\nu_2 - \nu_3$$

$$\nu_{331} = 2\nu_3 - \nu_1$$

$$v_{332} = 2v_3 - v_2$$

The efficiency of four-wave mixing depends on fiber dispersion and the channel spacings. Because the dispersion varies with wavelength, the signal waves and the generated waves have different group velocities. This destroys the phase matching of the interacting waves and lowers the efficiency at which power is transferred to newly generated frequencies. The higher the group velocity mismatches and the wider the channel spacings, the lower the four-wave mixing.

At the exit of a fiber of length L and attenuation α, the power P_{ijk} that is generated at the frequency v_{ijk} due to the interaction of signals at frequencies v_i, v_j, and v_k that have fiber-input powers P_i, P_j, and P_k, respectively, is

$$P_{ijk}(L) = \eta (D\kappa)^2 P_i(0) P_j(0) P_k(0) \exp(-\alpha L) \tag{12.13}$$

where the *nonlinear interaction constant* κ is

$$\kappa = \frac{32\pi^3 \chi_{1111}}{n_2 \lambda c} \left(\frac{L_{eff}}{A_{eff}} \right) \tag{12.14}$$

Here, χ_{1111} is the third-order nonlinear susceptibility; η is the efficiency of the four-wave mixing; n is the fiber refractive index; and D is the degeneracy factor, which counts the possible permutations of the three frequencies. The parameter D has the value of 3 or 6 for two waves mixing or three waves mixing, respectively. The effective length L_{eff} is given by Eq. (12.1) and A_{eff} is the effective cross-sectional area of the fiber. In G.652 single-mode fibers, only frequencies with separations less than 20 GHz will mix efficiently. In contrast, the FWM mixing efficiencies are greater than 20 percent for channel separations up to 50 GHz for G.653 dispersion-shifted fibers.

12.8 Mitigation Schemes for FWM

To reduce the effects of four-wave mixing in a DWDM link with close channel separations, for example, 100-GHz or less, it is important to have a high value of chromatic dispersion throughout the link. The reason is due to the fact that the efficiency of FWM depends on matching the phase relationship between the interacting DWDM signals. When there is chromatic dispersion in the fiber, the signals at different wavelengths travel with different group velocities. This means that the propagating waves move in and out of phase with each other. This condition greatly diminishes the FWM efficiency.

If the chromatic dispersion is low, or if there are regions of both positive and negative dispersion in the DWDM operating band, then a large number of FWM terms can be generated by the DWDM signals. This effect is particularly troublesome

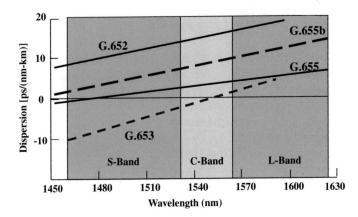

Fig. 12.10 Chromatic dispersion as a function of wavelength in various spectral bands for several different standard fiber types

if G.653 dispersion-shifted fibers are used for DWDM applications in the C-band. The main problem is that these fibers have both positive and negative dispersion regions around the zero-dispersion point at 1550 nm, as shown in Fig. 12.10. The consequence is that the DWDM channels on either side of this zero-dispersion point generate a large number of interfering in-band signals.

For standard G.652 single-mode fibers the high chromatic dispersion value of about 17 ps/(nm · km) in the C-band effectively suppresses four-wave mixing. However, for high data rates, such as 10 Gb/s and beyond, this high chromatic dispersion value quickly leads to large pulse-spreading effects along a G.652 fiber.

The limitations of the G.652 and G.653 fibers with respect to suppressing FWM led to the development of the G.655 fiber. As Fig. 12.10 shows, the G.655 fiber has a chromatic dispersion value ranging from about 3 to 9 ps/(nm · km) in the C-band. The ITU-T Recommendation G.655 specifies several different versions of this fiber. These include G.655B for use in both the S-band and C-band, G.655.D low-dispersion fibers with a chromatic dispersion ranging from 2.80 to 6.2 ps/(nm · km) at 1550 nm, and G.655.E medium-dispersion fibers with a dispersion value ranging from 6.06 to 9.31 ps/(nm · km) at 1550 nm. In either case, these chromatic dispersion values are sufficient to suppress FWM effects.

12.9 Basic Optical Wavelength Converters

One beneficial application of cross-phase modulation and four-wave mixing techniques is for performing wavelength conversion in WDM networks. An optical wavelength converter is a device that can translate information on an incoming wavelength directly to a new wavelength without entering the electrical domain. Such a device is an important component in all-optical networks, because the wavelength of the incoming signal may already be in use by another information channel residing on the destined outgoing path. Converting the incoming signal to a new wavelength will allow both information channels to traverse the same outbound fiber simultaneously. This section describes two classes of wavelength converters with one example from each class [10–15].

12.9.1 Wavelength Converters Using Optical Gatings

A wide variety of optical-gating techniques using devices such as semiconductor optical amplifiers, semiconductor lasers, or nonlinear optical-loop mirrors have been investigated to achieve wavelength conversion. The use of a semiconductor optical amplifier (SOA) in a cross-phase modulation (XPM) mode has been one of the most successful techniques for implementing single-wavelength conversion. The configurations for implementing this scheme include the Mach–Zehnder interferometer and the Michelson interferometer setups shown in Fig. 12.11.

The XPM scheme relies on the dependency of the refractive index on the carrier density in the active region of the SOA. As depicted in Fig. 12.11, the basic concept is that an incoming information carrying signal at wavelength λ_s and a continuous-wave (CW) signal at the desired new wavelength λ_c (called the *probe beam*) are simultaneously coupled into the device. The two waves can be either copropagating or counterpropagating. However, the noise in the copropagating case is lower. The

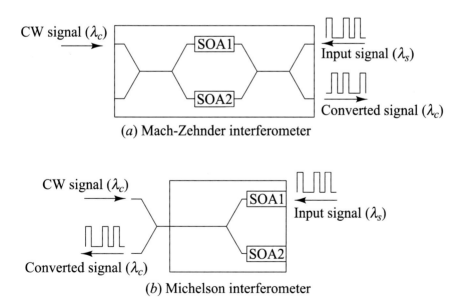

Fig. 12.11 **a** Mach-Zehnder interferometer and **b** Michelson interferometer setups using a pair of SOAs for implementing the cross-phase modulation wavelength-conversion scheme

signal beam modulates the gain of the SOA by depleting the carriers, which produces a modulation of the refractive index. When the CW beam encounters the modulated gain and refractive index, its amplitude and phase are changed, so that it now carries the same information as the input signal. As shown in Fig. 12.11, the SOAs are placed in an asymmetric configuration so that the phase change in the two amplifiers is different. Consequently, the CW light is modulated according to the phase difference. A typical splitting ratio is 69/31 percent. These types of converters readily handle data rates of at least 10 Gb/s.

A limitation of the XPM architecture is that it only converts one wavelength at a time. In addition, it has limited transparency in terms of the data format. Any information that is in the form of phase, frequency, or analog amplitude is lost during the wavelength conversion process. Consequently, this scheme is restricted to converting digital signal streams.

12.9.2 Wavelength Converters Based on Wave-Mixing

Wavelength conversion based on nonlinear optical wave mixing offers important advantages compared to other conversion methods. The advantages include a multi-wavelength conversion capability and transparency to the modulation format. The mixing arises from nonlinear interactions among optical waves traversing a nonlinear material. The outcome is the generation of another wave whose intensity is proportional to the product of the intensities of the interacting waves. The phase and frequency of the generated wave are a linear combination of these parameters of the interacting waves. Therefore, the wave mixing preserves both amplitude and phase information, and consequently is the only wavelength-conversion category that offers strict transparency to the modulation format.

Two wavelength-conversion schemes are four-wave mixing in either a passive waveguide or SOA, and difference-frequency generation in waveguides. For wavelength conversion, the FWM scheme employs the mixing of three distinct input waves to generate a fourth distinct output wave. In this method, an intensity pattern resulting from two input waves interacting in a nonlinear material forms a grating. For example, in SOAs there are three physical mechanisms that can form a grating. These are carrier-density modulation, dynamic carrier heating, and spectral hole burning. The third input wave in the material gets scattered by this grating, thereby generating an output wave. The frequency of the generated output wave is offset from that of the third wave by the frequency difference between the first two waves. If one of the three incident waves contains amplitude, phase, or frequency information and the other two waves are constant, then the generated wave will contain the same information.

Difference-frequency generation in waveguides is based on the mixing of two input waves. In this case, the nonlinear interaction of the material is with a pump and a signal wave.

12.10 Principles of Solitons

As Chap. 3 describes, group velocity dispersion (GVD) causes most pulses to broaden in time as they propagate through an optical fiber. However, a particular pulse shape known as a *soliton* takes advantage of nonlinear effects in silica, particularly self-phase modulation (SPM) resulting from the Kerr nonlinearity, to overcome the pulse-broadening effects of GVD [16–20].

The term "soliton" refers to special kinds of waves that can propagate undistorted over long distances and remain unaffected after collisions with each other. John Scott Russell made the first recorded observation of a soliton in 1838, when he saw a peculiar type of wave generated by boats in narrow Scottish canals [16]. The resulting water wave was of great height and traveled rapidly and undiminished over a long distance. After passing through slower waves of lesser height, the waves emerged from the interaction undistorted, with their identities unchanged.

In an optical communication system, solitons are very narrow, high-intensity optical pulses that retain their shape through the interaction of balancing pulse dispersion with the nonlinear properties of an optical fiber. If the relative effects of SPM and GVD are controlled just right, and the appropriate pulse shape is chosen, the pulse compression resulting from SPM can exactly offset the pulse broadening effect of GVD. Depending on the particular shape chosen, the pulse either does not change its shape as it propagates, or it undergoes periodically repeating changes in shape. The pulses that do not change in shape are called *fundamental solitons*, and those that undergo periodic shape changes are called *higher-order solitons*. In either case, attenuation in the fiber will eventually decrease the soliton energy. Because this weakens the nonlinear interaction needed to counteract GVD, periodically spaced optical amplifiers are required in a soliton link to restore the pulse energy.

12.10.1 Structures of Soliton Pulses

No optical pulse is completely monochromatic, because it excites a spectrum of frequencies. For example, as Eq. (10.1) shows, if an optical source emits power in a wavelength band $\Delta\lambda$, its spectral spread is $\Delta\nu$. This is important, because in an actual fiber a pulse is affected by both the GVD and the Kerr nonlinearity. This is particularly significant for high-intensity optical excitations. Because the medium is dispersive, the pulse width will spread in time with increasing distance along the fiber owing to GVD. In addition, when a high-intensity optical pulse is coupled to a fiber, the optical power modulates the refractive index seen by the optical excitation. This induces phase fluctuations in the propagating wave, thereby producing a chirping effect in the pulse, as shown in Fig. 12.8. The result is that the front of the pulse (at smaller times) has lower frequencies and the back of the pulse (at later times) has higher frequencies than the carrier frequency.

When such a pulse traverses a medium with a positive GVD parameter β_2 for the constituent frequencies, the leading part of the pulse is shifted toward a longer wavelength (lower frequencies), so that the speed in that portion increases. Conversely, in the trailing half, the frequency rises so the speed decreases. This causes the trailing edge to be further delayed. Consequently, in addition to a spectral change with distance, the energy in the center of the pulse is dispersed to either side, and the pulse eventually takes on a rectangular-wave shape. Figure 12.12 illustrates these intensity changes as the pulse travels along such a fiber. The plot is in terms of the normalized time. These effects will severely limit high-speed long-distance transmission if the system is operated in this condition.

On the other hand, when a narrow high-intensity pulse traverses a medium with a negative GVD parameter for the constituent frequencies, GVD counteracts the chirp produced by SPM. Now, GVD retards the low frequencies in the front end of the pulse and advances the high frequencies at the back. The result is that the high-intensity sharply peaked soliton pulse changes neither its shape nor its spectrum as it travels along the fiber. Figure 12.13 illustrates this for a fundamental soliton. Provided the pulse energy is sufficiently strong, this pulse shape is maintained as it travels along the fiber.

To derive the evolution of the pulse shape required for soliton transmission, one needs to consider the *nonlinear Schrödinger (NLS) equation*.

$$-j\frac{\partial u}{\partial z} = \frac{1}{2}\frac{\partial^2 u}{\partial t^2} + N^2|u|^2 u - j(\alpha/2)u \qquad (12.15)$$

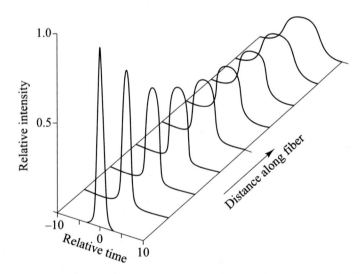

Fig. 12.12 Temporal changes in a narrow high-intensity pulse that is subjected to the Kerr effect as it travels through a nonlinear dispersive fiber that has a positive GVD parameter

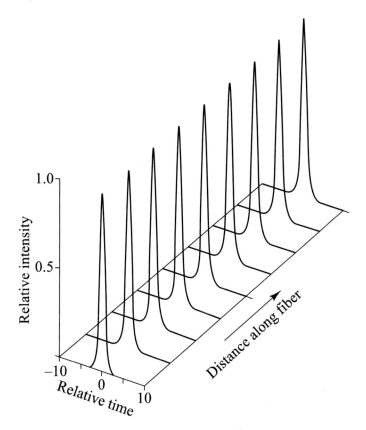

Fig. 12.13 Characteristics of a high-intensity sharply peaked soliton pulse that is subjected to the Kerr effect as it travels through a nonlinear dispersive fiber that has a negative GVD parameter

Here, $u(z, t)$ is the pulse envelope with power $|u|^2$, z is the propagation distance along the fiber, N is an integer designating the *order* of the soliton, and α is the coefficient of energy gain per unit length, with negative values of α representing energy loss. Following conventional notation, the parameters in Eq. (12.15) have been expressed in special soliton units to eliminate scaling constants in the equation.

These parameters (defined in Sect. 12.10.2) are the *normalized time* T_0, the *dispersion length* L_{disp}, and the *soliton peak power* P_{peak}.

For the three right-hand terms in Eq. (12.15):

(1) The first term represents GVD effects of the fiber. Acting by itself, dispersion tends to broaden pulses in time.

(2) The second nonlinear term denotes the fact that the refractive index of the fiber depends on the light intensity. Through the self-modulation process, this physical phenomenon broadens the frequency spectrum of a pulse.

(3) The third term represents the effects of energy loss or gain, for example, due to fiber attenuation or optical amplification, respectively.

Solving the NLS equation analytically yields a pulse envelop that is either independent of z (for the fundamental soliton with $N = 1$) or that is periodic in z (for higher-order solitons with $N \geq 2$). The general theory of solitons is mathematically complex and can be found in the literature [18–20]. Thus, this chapter presents only the basic concepts for fundamental solitons. The solution to Eq. (12.15) for the fundamental soliton is given by

$$u(z, t) = \text{sech}(t) \exp(jz/2) \tag{12.16}$$

where $\text{sech}(t)$ is the hyperbolic secant function. This is a bell-shaped pulse, as Fig. 12.14 illustrates. The time scale is given in units normalized to the $1/e$ width of the pulse. Because the phase term $\exp(jz/2)$ in Eq. (12.16) has no influence on the shape of the pulse, the soliton is independent of z and hence is nondispersive in the time domain.

When examining the NLS equation, one finds that the first-order effects of the dispersive and nonlinear terms are just complementary phase shifts. For a pulse given by Eq. (12.16), these phase shifts are

$$d\varphi_{\text{nonlin}} = |u(t)|^2 dz = \text{sech}^2(t)dz \tag{12.17}$$

for the nonlinear process, and

$$d\varphi_{disp} = \left(\frac{1}{2u} \frac{\partial^2 u}{\partial t^2} \right) dz = \left[\frac{1}{2} - \text{sech}^2(t) \right] dz \tag{12.18}$$

for the dispersion effect. Figure 12.15 shows plots of these terms and their sum, which is a constant. Upon integration, the sum simply yields a phase shift of $z/2$, which is common to the entire pulse. Because such a phase shift changes neither the temporal nor the spectral shape of a pulse, the soliton remains completely nondispersive in both the temporal and frequency domains.

Fig. 12.14 The hyperbolic secant function used for soliton pulses. The time scale is given in units normalized to the 1/e width of the pulse

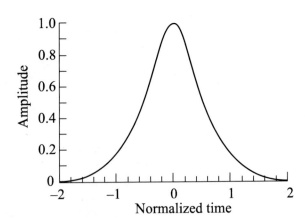

Fig. 12.15 Dispersive and
nonlinear phase shifts of a
soliton pulse, where their
sum is a constant that yields
a common phase shift for the
entire pulse

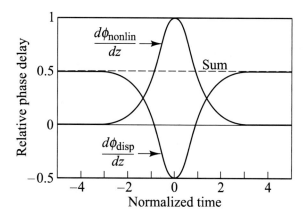

12.10.2 Fundamental Parameters for Solitons

Recall that the full-width half-maximum (FWHM) of a pulse is defined as the full
width of the pulse at its *half-maximum power level* (see Fig. 12.16). For the solution to
Eq. (12.15), the power is given by the square of the envelope function in Eq. (12.16).
Thus the FWHM T_s of the fundamental soliton pulse in normalized time is found from
the relationship $\text{sech}^2(\tau) = 1/2$ with $\tau = T_s/(2T_0)$, where T_0 is the basic normalized
time unit. This yields

$$T_0 = \frac{T_s}{2\cosh^{-1}\sqrt{2}} = \frac{T_s}{1.7627} \approx 0.567 T_s \qquad (12.19)$$

Example 12.4 Typical soliton FWHM pulse widths T_s range from 15 to 50 ps, so
that the normalized time T_0 is on the order of 9–30 ps.

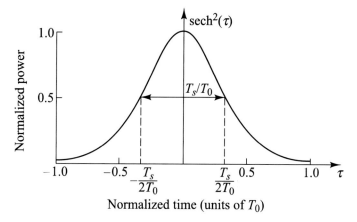

Fig. 12.16 Definition of the half-maximum soliton width in terms of normalized time units

The *normalized distance* parameter (also called *dispersion length*) L_{disp} is a characteristic length for the effects of the dispersion term. As described below, L_{disp} is a measure of the period of a soliton. This parameter is given by

$$L_{disp} = \frac{2\pi c}{\lambda^2}\frac{T_0^2}{D} = \frac{1}{\left[2cosh^{-1}\sqrt{2}\right]^2}\frac{2\pi c}{\lambda^2}\frac{T_s^2}{D} = 0.322\frac{2\pi c}{\lambda^2}\frac{T_s^2}{D} \qquad (12.20)$$

where c is the speed of light, λ is the wavelength in vacuum, and D is the dispersion of the fiber.

Example 12.5 Consider a dispersion-shifted fiber having $D = 0.5$ ps/(nm·km) at 1550 nm. If $T_s = 20$ ps, then.

$$L_{disp} = \frac{1}{(1.7627)^2}\frac{2\pi(3 \times 10^8 \text{ m/s})}{(1550 \text{ nm})^2}\frac{(20 \text{ ps})^2}{0.5 \text{ ps/(nm - km)}} = 202 \text{ km}$$

which shows that L_{disp} is on the order of hundreds of kilometers.
The parameter P_{peak} is the *soliton peak power* and is given by

$$P_{peak} = \frac{A_{eff}}{2\pi n_2}\frac{\lambda}{L_{disp}} = \left(\frac{1.7627}{2\pi}\right)^2\frac{A_{eff}\lambda^3}{n_2c}\frac{D}{T_s^2} \qquad (12.21)$$

where A_{eff} is the effective area of the fiber core cross section, n_2 is the nonlinear intensity-dependent refractive-index coefficient [see Eq. (12.6)], and L_{disp} is measured in km.

Example 12.6 For $\lambda = 1550$ nm, $A_{eff} = 50$ μm^2, $n_2 = 2.6 \times 10^{-16}$ cm^2/W, and with the value of $L_{disp} = 202$ km from Example 12.5, then using Eq. (12.21) the soliton peak power P_{peak} is

$$P_{peak} = \frac{(50 \text{ μm}^2)}{2\pi(2.6 \times 10^{-16} \text{ cm}^2/\text{W})}\frac{1550 \text{ nm}}{202 \text{ km}} = 2.35 \text{ mW}$$

This shows that when L_{disp} is on the order of hundreds of kilometers, P_{peak} is on the order of a few milliwatts.

For $N > 1$, the soliton pulse experiences periodic changes in its shape and its spectrum when propagating through the fiber. It resumes its initial shape at multiple distances of the *soliton period*, which is given by

$$L_{period} = \frac{\pi}{2}L_{disp} \qquad (12.22)$$

As an example, Fig. 12.17 shows the evolution of a second-order soliton, in which $N = 2$.

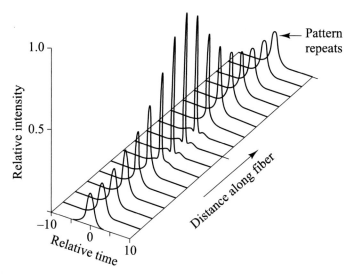

Fig. 12.17 Propagation characteristics of a second-order soliton (N = 2)

12.10.3 Width and Spacing of Soliton Pulses

The soliton solution to the NLS equation holds to a reasonable approximation only when individual pulses are well separated. To ensure this, the soliton width must be a small fraction of the bit slot. This eliminates use of the non-return-to-zero (NRZ) format that is commonly implemented in digital systems. Consequently, the return-to-zero (RZ) format is used. This condition thus constrains the achievable bit rate, because there is a limit on how narrow a soliton pulse can be generated.

If T_b is the width of the bit slot, then we can relate the bit rate B to the soliton half-maximum width T_s by

$$B = \frac{1}{T_b} = \frac{1}{2s_0 T_0} = \frac{1.7627}{2s_0 T_s} \tag{12.23}$$

where the factor $2s_0 = T_b/T_0$ is the *normalized separation* between neighboring solitons.

The physical explanation of the separation requirement is that the overlapping tails of closely spaced solitons create nonlinear interactive forces between them. These forces can be either attractive or repulsive, depending on the initial relative phase of the solitons. For solitons that are initially in phase and separated by $2s_0 > > 1$, the soliton separation is periodic with an *oscillation period*:

$$\Omega = \frac{\pi}{2} exp(s_0) \tag{12.24}$$

The mutual interactive force between in-phase solitons thus results in periodic attraction, collapse, and repulsion. The *interaction distance* L_I is

$$L_I = \Omega L_{disp} = L_{period} \exp(s_0) \tag{12.25}$$

This interaction distance, and particularly the ratio L_I/L_{disp}, determine the maximum bit rate allowable in soliton systems.

These types of interactions are not desirable in a soliton system, because they lead to jitter in the soliton arrival times. One method for avoiding this situation is to increase s_0, because the interaction between solitons depends on their spacing. Because Eq. (12.23) is accurate for $s_0 > 3$, this equation together with the criterion that $\Omega L_{disp} > > L_T$, where L_T is the total transmission distance, is suitable for system designs in which soliton interaction can be ignored.

Using Eq. (12.20) for L_{disp}, Eq. (12.23) for T_0, and Eq. (3.25) for D, the design condition $\Omega L_{disp} > > L_T$ becomes

$$B^2 L_T << \left(\frac{2\pi}{s_0 \lambda}\right)^2 \frac{c}{16D} \exp(s_0) = \frac{\pi}{8 s_0^2 |\beta_2|} \exp(s_0) \tag{12.26}$$

When written in this form, Eq. (12.26) shows the effects on the bit rate B or the total transmission distance L_T for selected values of s_0.

Example 12.7 Suppose a link needs to transmit information at a rate of 10 Gb/s over an 8600 km soliton link across the Pacific Ocean.

(a)   Because this is a high data rate over a long distance, one can start by selecting a value of $s_0 = 8$. Then, from Eq. (12.24) it follows that $\Omega = 4682$. Given that the dispersion length is at least 100 km, then $\Omega L_{disp} > 4.7 \times 10^5$ km, which for all practical purposes satisfies the condition $\Omega L_{disp} > > L_T = 8600$ km.

(b)   If $D = 0.5$ ps/(nm · km) at 1550 nm then for a 10-Gb/s data rate Eq. (12.26) yields

$$L_T << 2.87 \times 10^5 \text{ km}$$

This is satisfied, because the right-hand side is 33 times greater than the desired length.

(c)   Using Eq. (12.23), the FWHM soliton pulse width is found to be

$$T_s = \frac{0.881}{s_0 B} = \frac{0.881}{8(10 \times 10^9 \text{ b/s})} = 11 \text{ps}$$

(d)   The fraction of the bit slot occupied by a soliton when $s_0 = 8$ is

$$\frac{T_s}{T_B} = \frac{0.881}{s_0} = \frac{0.881}{8} = 11\%$$

Note that for a given value of s_0, this is independent of the bit rate. For example, if the data rate is 20 Gb/s, then the FWHM pulse width is 5.5 ps, which also occupies 11 percent of the bit slot.

12.11 Summary

For low optical signal power levels, an optical fiber behaves as a linear transmission medium. This means three things: (a) The properties of a fiber do not change as the optical signal power varies, (b) the wavelength does not change as the signal propagates along the fiber, and (c) a specific signal does not interact with other simultaneously propagating signals at different wavelengths. However, for power levels that are higher than about +3 dBm (2 mW) the fiber material starts to exhibit nonlinear power-dependent properties. These nonlinear effects are manifested as interactive changes and power loss in the propagating signal. The optical power loss gives rise to a nonlinear power penalty.

Optical nonlinearities can be classified into scattering-related and refractive index-related categories. The nonlinear inelastic scattering processes include stimulated Raman scattering (SRS) and stimulated Brillouin scattering (SBS). The second category of nonlinear effects arises from intensity-dependent variations in the refractive index in a silica fiber, which is known as the Kerr effect. These effects include self-phase modulation (SPM), cross-phase modulation (XPM), and four-wave mixing (FWM). Note that certain nonlinear effects are independent of the number of WDM channels.

SBS, SRS, and FWM result in gains or losses of signal power in a wavelength channel. The power variations depend on the optical signal intensity. These three nonlinear processes provide gains to some channels while depleting power from other channels, thereby producing crosstalk between the wavelength channels. SPM and XPM affect only the phase of signals, which causes chirping in digital pulses. This can worsen pulse broadening due to dispersion, particularly in very high-rate systems. FWM can be suppressed by avoiding the use of fibers with both positive and negative chromatic dispersion values in the operating wavelength band.

Problems

12.1 A 50-km single-mode fiber has an attenuation of 0.55 dB/km at 1310 nm and 0.28 dB/km at 1550 nm. Compare the effective lengths of this fiber at 1310 and 1550 nm.

12.2 Consider a 1550-nm optical source that has a 40-MHz linewidth. Suppose a given single-mode fiber has a 72-μm^2 effective area and a 0.2-dB/km attenuation at 1550 nm. Assuming that the polarization factor $b = 2$ and that the Brillouin gain coefficient is $g_B = 4 \times 10^{-11}$ m/W, what is the threshold power for stimulated Brillouin scattering at 1550 nm for a 40-km link? If the attenuation for this fiber is 0.4 dB/km at 1310 nm and all other parameters are the same, what is the threshold power for stimulated Brillouin scattering at 1310 nm?

12.3   Consider three copropagating optical signals at frequencies v_1, v_2, and v_3. If these frequencies are evenly spaced so that $v_1 = v_2 - \Delta v$ and $v_3 = v_2 + \Delta v$, where Δv is an incremental frequency change, list the third-order waves that are generated due to FWM and plot them in relation to the original three waves. Note that several of these FWM-generated waves coincide with the original frequencies.

12.4   A soliton transmission system operates at 1550 nm with fibers that have a dispersion of 1.5 ps/(nm · km) and an effective core area of 50 μm^2. Find the peak power required for fundamental solitons that have a 16-ps FWHM width. Use the value $n_2 = 2.6 \times 10^{-16}$ cm^2/W. What are the dispersion length and the soliton period? What is the required peak power for 30-ps pulses?

12.5   A telecommunications service provider wants a single-wavelength soliton transmission system that is to operate at 40 Gb/s over a 2000-km distance. How would you design such a system? You are free to choose whatever components and design parameters are needed.

12.6   Consider a WDM system that utilizes two soliton channels at wavelengths λ_1 and λ_2. Because different wavelengths travel at slightly different velocities in a fiber, the solitons of the faster channel will gradually overtake and pass through the slower-channel solitons. If the collision length L_{coll} is defined as the distance between the beginning and end of the pulse overlap at the half-power points, then

$$L_{coll} = \frac{2T_s}{D \Delta \lambda}$$

where $\Delta \lambda = \lambda_1 - \lambda_2$, T_s is the pulse FWHM, and D is the dispersion parameter.
(a) What is the collision length for $T_s = 16$ ps, $D = 0.5$ ps/(nm·km), and $\Delta \lambda = 0.8$ nm?
(b) Four-wave mixing effects arise between the soliton pulses during their collision, but then collapse to zero afterward. To avoid amplifying these effects, the condition $L_{coll} \geq 2L_{amp}$ should be satisfied, where L_{amp} is the amplifier spacing. What is the upper bound for L_{amp} for the above case?

12.7   Based on the conditions described in Prob. 12.6, what is the maximum number of allowed wavelength channels spaced 0.4 nm apart in a WDM soliton system when $L_{amp} = 25$ km, $T_s = 20$ ps, and $D = 0.4$ ps/(nm · km)?

Answers to Selected Problems

12.1 First using Eq. (3.3) change 0.55 dB/km to 0.127 km^{-1} and 0.28 dB/km to 0.0645 km^{-1}. Then from Eq. (12.1) $L_{eff}(1310) = 7.9$ km and $L_{eff}(1550) = 14.9$ km.

12.2 Using Eq. (3.3) change 0.20 dB/km to 0.046 km^{-1} and 0.40 dB/km to 0.092 km^{-1}. Then using Eq. (12.1) it follows that $L_{eff}(1550) = 18.3$ km and $L_{eff}(1310) = 10.6$ km. Then using Eq. (12.5), $P_{th}(1550) = 12.4$ mW and $P_{th}(1310) = 21.4$ mW.

12.3 The following nine 3rd-order waves are generated due to FWM:

$v113 = 2(v2 - \Delta v) - (v2 + \Delta v) = v2 - 3\Delta v.$

$v112 = 2(v2 - \Delta v) - v2 = v2 - 2\Delta v.$

$v123 = (v2 - \Delta v) + v2 - (v2 + \Delta v) = v2 - 2\Delta v.$

$v223 = 2v2 - (v2 + \Delta v) = v2 - \Delta v = v1.$

$v132 = (v2 - \Delta v) + (v2 + \Delta v) - v2 = v2.$

$v221 = 2v2 - (v2 - \Delta v) = v2 + \Delta v = v3.$

$v231 = v2 + (v2 + \Delta v) - (v2 - \Delta v) = v2 + 2\Delta v.$

$v331 = 2(v2 + \Delta v) - (v2 - \Delta v) = v2 + 3\Delta v.$

$v332 = 2(v2 + \Delta v) - v2 = v2 + 2\Delta v.$

12.4 $P_{peak} = 11.0$ mW; $L_{disp} = 43$ km, $L_{period} = 67.5$ km; $P_{peak} = 3.1$ mW.

12.6 (a) From the equation $L_{coll} = 80$ km;

(b) From the given condition $L_{amp} = 0.5\ L_{coll} \leq 40$ km.

12.7 From the equation and conditions in Problem 12.6, $\Delta\lambda_{max} = \frac{T_s}{DL_{amp}} = 2$ nm. Therefore, the maximum number of channels = 2.0 nm/0.4 nm = 5.

References

1. A.D. Ellis, M.E. McCarthy, M.A.Z. Khateeb, I.M. Sorokina, N.J. Doran, Performance limits in optical communications due to fiber nonlinearity: Tutorial. Adv. Optics Photonics **9**(3), 429–502 (2017)
2. J. Toulouse, Optical nonlinearities in fibers: Review, recent examples, and systems applications. J. Lightw. Technol. **23**, 3625–3641 (2005)
3. R.H. Stolen, The early years of fiber nonlinear optics. J. Lightw. Technol. **26**(9), 1021–1031 (2008)

4. M. F. Fereira, *Nonlinear Effects in Optical Fibers*, Wiley (2011)
5. R. Boyd, *Nonlinear Optics*, Academic Press, 4th ed. (2020)
6. J. A. Buck, *Fundamentals of Optical Fibers*, Wiley, 2nd ed. (2004)
7. A. Kobyakov, M. Sauer, D. Chowdhury, Stimulated Brillouin scattering in optical fibers. Adv. Opt. Photonics **2**(1), 1–59 (2010)
8. F. Forghieri, R. W. Tkach, A. R. Chraplyvy, Fiber nonlinearities and their impact on transmission systems. in I. P. Kaminow, T. L. Koch, eds. *Optical Fiber Telecommunications–III*, Vol. A, Academic Press (1997)
9. S. O. Kasap, *Principles of Electronic Materials and Devices*, McGraw-Hill, 4th ed. (2018)
10. M. J. Connelly, *Semiconductor Optical Amplifiers*, Springer (2002)
11. J. T. Hsieh, P. M. Gong, S. L. Lee, J. Wu, Improved dynamic characteristics on SOA-based FWM wavelength conversion in light-holding SOAs," IEEE J. Selected Topics Quantum Electron. **10**, 1187–1196 (2004)
12. X. Yi, R. Yu, J. Kurumida, S.J.B. Yoo, A theoretical and experimental study on modulation-format-independent wavelength conversion. J. Lightw. Technol. **28**(4), 587–595 (2010)
13. R. Ramaswami, K.N. Sivarajan, *G*, 3rd edn. (Susaki, Optical Networks, Morgan Kaufmann, 2009)
14. W. Wang, L.G. Rau, D.J. Blumenthal, 160 Gb/s variable length packet/10 Gb/s-label all-optical label switching with wavelength conversion and unicast/multicast operation. J. Lightw. Technol. **23**(1), 211–218 (2005)
15. N.Y. Kim, X. Tang, J.C. Cartledge, A.K. Atieh, Design and performance of an all-optical wavelength converter based on a semiconductor optical amplifier and delay interferometer. J. Lightw. Technol. **25**(12), 3730–3738 (2007)
16. J. S. Russell, *Reports of the Meetings of the British Assoc. for the Advancement of Science*, p. 1844
17. H. Haus, W.S. Wong, Solitons in optical communications. Rev. Mod. Physics **68**, 432–444 (1996)
18. A. Hasegawa, "Theory of information transfer in optical fibers: A tutorial review" (emphasis is on solitons). Opt. Fiber Technol. **10**(2), 150–170 (2004)
19. M. F. Ferreira, M. V. Facão, S. V. Latas, M. H. Sousa, Optical solitons in fibers for communication systems. Fiber Integr. Optics **24**(3–4), 287–313 (2005)
20. L. F. Mollenauer, J. P. Gordon, *Solitons in Optical Fibers: Fundamentals and Applications*, Academic Press (2006)

Chapter 13
Fiber Optic Communication Networks

Abstract Various types of optical fiber networks have been conceived, designed, and built to satisfy a wide range of transmission capacities and speeds. The link lengths between users can vary from short localized connections within a building or a campus environment to networks that span continents and run across oceans. This chapter defines basic terminology and general network concepts, illustrates different fiber optic network architectures, discusses the concept of network layering, defines data packet switching elements, describes how these elements route signals along wavelength channels, and shows how network configuration flexibility can offer connection protection in case there are link or node failures.

This chapter covers performance and implementation issues related to optical fiber link configurations that can be utilized in various types of networks to connect users having a wide range of transmission capacities and speeds. The links between these users can range in length from short localized connections within a building, a data center, or a campus environment to networks that span continents and go across oceans. A major motivation for developing sophisticated communication networks has been the rapid proliferation of information exchange desired by institutions involved in fields such as commerce, finance, education, scientific and medical research, health care, national and international security, social media, and entertainment. The potential for this information exchange arose from the ever-increasing power of computers and data storage devices, which need to be interconnected by high-speed, high-capacity networks.

Section 13.1 defines basic terminology and general network concepts, discusses the concept of network layering, and describes fiber optic network topologies. Section 13.2 illustrates the commonly used optical network configurations, which include star, tree, and mesh network layouts.

For the terrestrial and undersea long-haul category, the closely coupled SONET and SDH physical layer standards provide a mechanism for multiplexing and

---

The original version of this chapter was revised: All the incorrect sentences have been corrected. The correction to this chapter can be found at https://doi.org/10.1007/978-981-33-4665-9_15

G. Keiser, *Fiber Optic Communications*,
https://doi.org/10.1007/978-981-33-4665-9_13

507

transmitting optical signals so they can be shared between networks within the global telecom infrastructure. Section 13.3 discusses the physical layer aspects of SONET/SDH ring network. To increase the capacity of optical fiber links, research and development engineers and scientists are devising increasingly higher-rate transmission methods and sophisticated data encoding schemes. Section 13.4 gives some examples for high-speed optical fiber transceivers and optical interconnects operating from 10 to 400 Gb/s.

Two critical elements for implementing high-capacity WDM networks include fixed and reconfigurable optical add/drop multiplexers (OADMs) and optical cross-connects (OXCs). Sections 13.5 and 13.6, respectively, define these elements and describe how they route signals along wavelength channels or lightwave paths. Section 13.6 also introduces the concepts of wavelength routing, optical packet switching, and optical burst switching. Section 13.7 illustrates applications of network elements, such as OADMs and OXCs, to various categories of WDM networks. In addition this section also addresses the concept of elastic networks. The architecture and operations of passive optical networks (PON) are examined in Sect. 13.8.

13.1   Concepts of Optical Networks

This section provides some background material concerning the concepts of optical networks [1–3]. The discussion illustrates different network architectures, notes what organizations own and operate various network segments, and defines some network terminology.

13.1.1   Terminology Used for Networks

To start, it is helpful to define some terminology using Fig. 13.1 as a guideline.

Data Packet (or Simply Packet) A data packet is a group of information bits plus overhead bits for transportation management of the data.

Stations Devices that network subscribers use to communicate are called *stations*. These may be computers, monitoring equipment, telephones, fax machines, or other telecom equipment.

Networks To establish connections between these stations, transmission paths run between them to form a collection of interconnected stations called a *network*.

Node Within this network, a *node* is a point where one or more communication lines terminate and/or where stations are connected. Stations also can be connected directly to a transmission line.

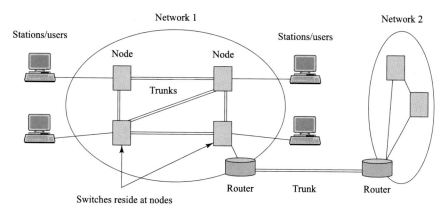

Fig. 13.1 Definitions of various elements of a network

Server A network server is a powerful computer that provides users with access to shared software or hardware resources. The shared resources can include disk space, hardware access, and email services. A server typically has a large amount of random access memory (RAM) and runs continuously on a robust operating system.

Trunk The term *trunk* normally refers to a transmission line that runs between nodes or networks and supports large traffic loads.

Topology The *topology* is the logical manner in which nodes are linked together by transmitting channels to form a network.

Switching and Routing The transfer of information from source to destination is achieved through a series of *switches* residing at intermediate nodes and the process is called *switching*. The selection of a suitable path through a network is accomplished by means of *routers* and the process is referred to as *routing*.

Thus a *switched communication network* consists of an interconnected collection of nodes, in which information streams that enter the network from a station are routed to the destination by being switched from one transmission path to another at a node.

13.1.2 Generic Network Categories

Networks can be divided into several broad generic categories, as Fig. 13.2 illustrates. Note that there is nothing unique about this illustration, because it merely shows general interconnection hierarchies. The following definitions are useful to know:

Local Area Network (LAN) A *local area network* (LAN) interconnects users in a localized area such as a large room or work area, a department, a home, a building, an office or factory complex, or a small group of buildings. A LAN employs relatively inexpensive hardware that allows users to share common expensive resources such

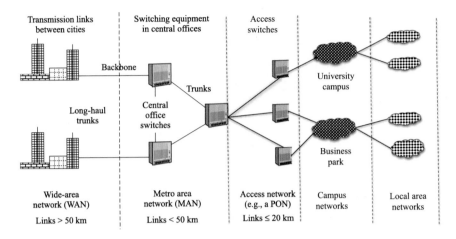

Fig. 13.2 Definitions of some terms used in describing different segments of a public network

as servers, high-performance printers, specialized instrumentation, or other equipment. Ethernet is the most popular networking technology used in LANs. Local area networks usually are owned, used, and operated privately by a single organization.

Campus Network A *campus network* is an extension of a LAN and can be considered as an interconnection of a collection of LANs in a localized area. Similar to a LAN, it is an autonomous network owned and managed by a single organization that exists within a local geographical area. In networking terminology the word *campus* refers to any group of buildings that are within reasonable walking distance of each other. Thus a campus network could be deployed in a university campus, a business park, a government center, a research center, or a medical center. Typically a campus network uses routers to provide an access path into a larger network, such as a metropolitan area network or the Internet.

Metropolitan Area Network (MAN) A MAN is commonly referred to as a *metro network* and spans a larger area than a LAN or a campus network. The interconnected facilities could range from buildings located in several city blocks to an entire city and the metropolitan area surrounding it. Thus the distances between central switching offices for a metro network range from a few to several tens of kilometers. Metro network resources are owned and operated by many telecommunication organizations.

Access Network An access network lies between a metro network and a LAN or campus network. This network category encompasses connections that extend from a centralized switching facility to individual businesses, organizations, and homes. One function of an access network is to concentrate the information flows that originate in a local network and send this aggregated traffic to the switching facility. This path is called the *upstream direction*. In the other transmission direction (*downstream* or toward the user), the network provides voice, data, video, and other services to subscribers. The transmission distances tend to be up to 20 km. Typically a single telecommunication service provider owns a particular access network.

Wide Area Network (WAN) A WAN spans a large geographical area. The transmission distances can range from links between switching facilities in neighboring cities to cross-country terrestrial or intercontinental undersea lines. The large collection of WAN resources are owned and operated by either private organizations or telecommunication service providers.

Enterprise and Public Networks When a private organization (for example, a company, a government entity, a medical facility, a university, or a commercial enterprise) owns and operates a network, it is called an *enterprise network*. These networks only provide services to the members of the organization. On the other hand, networks that are owned by the telecom carriers provide services such as leased lines or real-time telephone connections to the general public. These networks are referred to as *public networks*, because the services are available to any users or organizations that wish to subscribe to them.

Central Office A centralized switching facility in a public network is called a *central office* (CO) or a *point of presence* (POP). The CO houses a series of large telecom switches that establish temporary connections for the duration of a requested service time between subscriber lines or between users and network resources.

Backbone The term *backbone* means a link that connects multiple network segments. For example, a backbone handles internetwork traffic; that is, traffic that originates in one network segment and is transmitted to another segment. Backbones can be short or long links.

Long-Haul Network A *long-haul network* interconnects distant cities or geographical regions and spans hundreds to thousands of kilometers between central offices. For example, this could be a high-capacity link carrying terabits of information between New York and San Francisco, between countries in Africa, or between Australia, China, and Singapore.

Data Center Network (DCN) The exponential rise of Internet traffic resulting from applications such as streaming video, social networking, search engines, cloud storage, and cloud computing has created the establishment of powerful data centers. Such a center can be a building, dedicated space within a building, or a group of buildings used to house telecom associated components such as storage systems. These applications are data-intensive and require high interaction between the servers in the data center. A specialized optical network called a *data center network* (DCN) is used to transmit large amounts of highly dynamic data traffic between data center servers. Section 13.7 gives more details on configurations and operations of a DCN.

Passive Optical Network Many different transmission media can be used in an access network, including twisted-pair copper wires, coaxial cable, optical fibers, and radio links. Optical distribution networks that do not require any active optoelectronic components in the access region offer a number of operating advantages over other media. This implementation is called a *passive optical network* (PON) and is the basis for the fiber-to-the-premises (FTTP) networks described in Sect. 13.8.

13.1.3 Layered Structure Approach to Network Architectures

When discussing the design and implementation of a telecom system, the term *network architecture* is used to describe the general physical arrangement and operational characteristics of communicating equipment together with a common set of communication protocols. A *protocol* is a set of rules and conventions that governs the generation, formatting, control, exchange, and interpretation of information that is transmitted through a telecom network or that is stored in a database.

A traditional approach to setting up a protocol is to subdivide it into a number of individual pieces or layers of manageable and comprehensible size. The result is a layered structure of services, which is called a *protocol stack*. In this scheme each layer is responsible for providing a set of functions or capabilities to the layer above it by using the functions or capabilities of the layer below. A user at the highest layer is offered all the capabilities of the lower levels for interacting with other users and peripheral equipment distributed on the network.

As an example of a structured approach for simplifying the complexity of modern networks, the International Standards Organization (ISO) developed an *open system interconnect (OSI) reference model* for dividing the functions of a network into seven operational layers [4–6]. As Fig. 13.3 illustrates, by convention these layers are viewed as a vertical sequence with the numbering starting at the bottom layer.

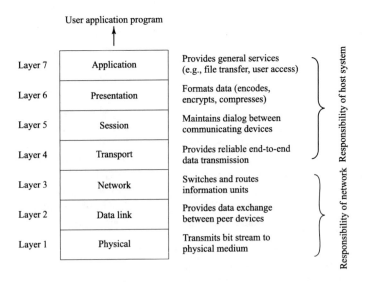

Fig. 13.3 General structure and functions of the seven-layer OSI reference model

Each layer performs specific functions using a standard set of protocols. A given layer is responsible for providing a service to the layer above it by using the services of the layer below it. Thereby an increasingly larger number of functions are provided when moving up the protocol stack, and the level of functional abstraction increases. The lower layers govern the *communication facilities*. These deal with the physical connections, data link control, and routing and relaying functions that support the actual transmission of data. The upper layers support *user applications* by structuring and organizing data for the needs of the user.

In the classical OSI model, the various layers carry out the following functions:

Physical Layer The *physical layer* refers to a physical transmission medium, such as a wire or an optical fiber, which can handle a certain amount of bandwidth. It provides different types of physical interfaces to equipment, and its functions are responsible for actual transmission of bits across an optical fiber or metallic wire.

Data Link Layer The purpose of the *data link layer* is to establish, maintain, and release links that directly connect two nodes. Its functions include framing (defining how data is structured for transport), multiplexing, and demultiplexing of data. Examples of data link protocols include the *point-to-point protocol* (PPP), Ethernet, and the *high-level data link control* (HDLC) protocol.

Network Layer The function of the *network layer* is to deliver data packets from source to destination across multiple network links. Typically, the network layer must find a path through a series of connected nodes, and the nodes along this path must forward the packets to the appropriate destination. The dominant network layer protocol is the *Internet Protocol* (IP).

Transport Layer The *transport layer* is responsible for reliably delivering the complete message from the source to the destination to satisfy a *quality of service* (QoS) requested by the upper layer. The QoS parameters include throughput, transit delay, bit-error rate, delay time to establish a connection, cost, information security, and message priority. The *transmission control protocol* (TCP) used in the Internet is an example of a transport layer protocol.

Higher Layers The higher layers (session, presentation, application) support user applications, which are not covered here.

Note that there is nothing inherently unique about using seven layers or about the specific functionality in each layer. In actual applications some of the layers may be omitted and other layers can be subdivided into further sublayers. Thus the layering mechanism should be viewed as a framework for discussions of implementation and not as an absolute requirement.

Fig. 13.4 The optical layer describes the wavelength connections that lie on top of the physical layer

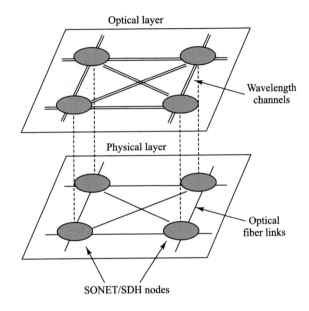

13.1.4 Optical Layer Functions

When dealing with optical network concepts, the term *optical layer* is used to describe various network functions or services. The optical layer is a wavelength-based concept and lies just above the physical layer, as shown in Fig. 13.4. This means that whereas the physical layer provides a physical connection between two nodes, the optical layer provides *lightpath services* over that link. A *lightpath* is an end-to-end optical connection that may pass through one or more intermediate nodes. For example, in an eight-channel WDM link there are eight lightpaths, which may go over a single physical line. Note that for a specific multiple-segment lightpath the wavelengths between various node pairs in the overall link may be different.

The optical layer may carry out processes such as wavelength multiplexing, adding and dropping wavelengths, and support of optical crossconnects or wavelength switching. Networks that have these optical-layer functions are referred to as *wavelength-routed networks*. Section 13.6 has further details on these types of networks.

13.2 Common Network Topologies

Figure 13.5 shows the four common topologies used for fiber optic networks [1, 5, 6]. These are the *linear bus, ring, star,* and *mesh configurations*. Each has its own particular advantages and limitations in terms of reliability, expandability, and performance characteristics.

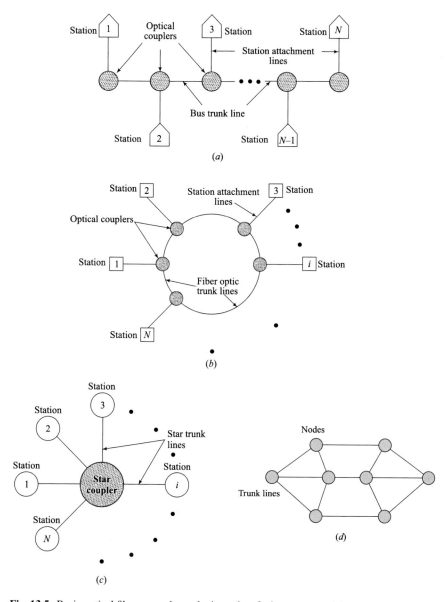

Fig. 13.5 Basic optical fiber network topologies: **a** bus, **b** ring, **c** star, and **d** mesh configurations

The primary advantages of a non-optical bus network (such as coaxial cable based Ethernet) are the passive nature of the coaxial cable transmission medium and the ability to easily install low-loss taps on the coaxial line. In contrast, a fiber-optic-based bus network is more difficult to implement, because there are no low-perturbation optical-tap equivalents to coax taps for efficiently coupling optical signals into and out of the main optical fiber trunk line. Access to the bus is achieved by means of an optical coupling element, which can be either active or passive. An *active coupler* converts the optical signal on the data bus to its electric baseband counterpart before any data processing (such as injecting additional data into the signal stream or merely passing on the received data) is carried out. A *passive coupler* employs no electronic elements. It is used passively to tap off a portion of the optical power from the bus. Examples of these taps are the 2×2 couplers described in Chap. 10.

In a ring topology, consecutive nodes are connected by point-to-point links that are arranged to form a single closed path. Information in the form of data packets is transmitted from node to node around the ring. The interface at each node is an active device that has the ability to recognize its own address in a data packet in order to accept messages. The active node forwards those messages that do not contain its own address on to its next neighbor.

In a star architecture, all nodes are joined at a single point called the *central node* or *hub*. The central node can be an active or a passive device. Using an active hub, one can control all routing of messages in the network from the central node. This is useful when most of the communications are between the central and the outlying nodes, as opposed to information exchange between the attached stations. If there is a great deal of message traffic between the outlying nodes, then a heavy switching burden is placed on an active central node. In a star network with a passive central node, a power splitter is used at the hub to divide the incoming optical signals among all the outgoing lines to the attached stations.

In a mesh network, point-to-point links connect the nodes in an arbitrary fashion that can vary greatly from one application to another. This topology allows significant network configuration flexibility and offers connection protection in case there are multiple link or node failures. Link protection in mesh networks is accomplished by means of a mechanism that first determines where a failure has occurred and then restores the interrupted services by redirecting the traffic from a failed link or node to travel over another link in the mesh (see Sect. 3.3 for examples) [7, 8].

13.2.1   Performance of Passive Linear Buses

As shown in Fig. 13.5a, a linear bus uses a single cable, which connects all the included nodes of the entire network. The main losses are the cable attenuations and the optical power loss arising at each tap coupler. If a fraction F_c of optical power is extracted at each port of the coupler, then the *connecting loss* L_c is

$$L_c = -10\log(1 - F_c) \tag{13.1}$$

For example, if this fraction F_c is 20%, then L_c is about 1 dB. That is, the optical power is reduced by 1 dB at any coupling junction. The total loss of a linear bus increases linearly with the number of network nodes. As a result of the high signal reduction at each node, bus topologies typically are used only when a network installation is small, simple, or temporary.

13.2.2 Performance of Star Networks

To see how a star coupler can be applied to a given network, consider the various optical power losses associated with the coupler. Section 10.2.4 gives the details of how an individual star coupler works. As a quick review, the excess loss is defined as the ratio of the input power to the total output power. That is, it is the fraction of power lost in the process of coupling light from the input port to all the output ports. From Eq. (10.25), for a single input power P_{in} and N output powers, the excess loss in decibels is given by

$$\text{Fiber star excess loss} = L_{excess} = 10\log\left(\frac{P_{in}}{\sum_{i=1}^{N} P_{out,i}}\right) \tag{13.2}$$

Here $P_{out,i}$ is the optical power output from branch i. In an ideal star coupler the optical power from any input is evenly divided among the output ports so that all values of $P_{out,i}$ are equal. The total loss of the device consists of its splitting loss plus the excess loss in each path through the star. The *splitting loss* is given in decibels by

$$\text{Splitting loss} = L_{split} = -10\log\left(\frac{1}{N}\right) = 10\log N \tag{13.3}$$

Drill Problem 13.1 Assuming that there is no excess loss, show that the splitting losses of ideal 8×8, 16×16, and 32×32 star couplers are 9.03, 12.04, and 15.05 dB, respectively.

To find the power-balance equation, the following parameters are used:

- P_S is the fiber-coupled output power from a source in dBm

- P_R is the minimum optical power in dBm required at the receiver to achieve a specific bit-error rate
- α is the fiber attenuation
- All stations are located at the same distance
  L from the star coupler
- L_c is the connector loss in decibels.

Then, the power-balance equation for a particular link between two stations in a star network is

Fig. 13.6 Total optical power loss as a function of the number of attached stations for linear-bus and star architectures

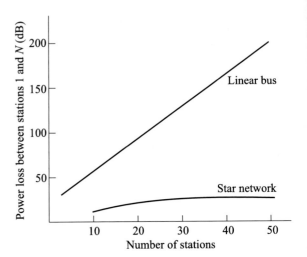

$$P_S - P_R = \text{Link Loss} + \text{Margin}$$
$$= L_{\text{excess}} + \alpha(2L) + 2L_c + L_{\text{split}} + \text{Margin}$$
$$= L_{\text{excess}} + \alpha(2L) + 2L_c + 10\log N + \text{Margin} \qquad (13.4)$$

Here, it is assumed that there are connector losses at the transmitter and the receiver. The term "Margin" designates additional available optical power for unforeseen link losses. This equation shows that, in contrast to a passive linear bus where the loss increases in direct proportion to N, for a star network the loss increases much slower as log N. Figure 13.6 compares the performance of the two architectures.

Example 13.1 Consider two star networks that have 10 and 50 stations, respectively. Let the optical source be an LED with a -10-dBm fiber-coupled power and assume a receiver sensitivity of -48 dBm. Assume each station is located 500 m from the star coupler and that the fiber attenuation is 0.4 dB/km. Assume the excess loss is 0.75 dB for the 10-station network and 1.25 dB for the 50-station network. Let the connector loss be 1.0 dB. What is the power margin for 10 and for 50 stations?

Solution: For $N = 10$, from Eq. (13.4) the power margin between the transmitter and the receiver is

$$P_S - P_R = 38\,\text{dB} = [0.75 + 0.4(1.0) + 2(1.0) + 10\log 10]\,\text{dB} + \text{Margin}$$
$$= 13.2\,\text{dB} + \text{Margin}$$

Therefore the power margin is 24.8 dB. For N = 50:

$$P_S - P_R = 38\,\text{dB} = [1.25 + 0.4(1.0) + 2(1.0) + 10\log 50]\,\text{dB} + \text{Margin}$$
$$= 20.6\,\text{dB} + \text{Margin}$$

Therefore here the power margin is 17.4 dB.

Drill Problem 13.2 Consider a star network that operates at 1 Gb/s and uses a 16 × 16 star coupler with an excess loss of 1.35 dB. Assume each station is located 5 km from the star coupler and that the fiber attenuation is 0.25 dB/km at 1550 nm. Let the connector loss at each end of the link be 1.0 dB. Suppose the optical source emits 0 dBm into the fiber and that the *pin* photodetector has a −27-dBm sensitivity at 1 Gb/s. Show that the system margin is 9.11 dB.

13.3 Basic SONET/SDH Concepts

With the advent of fiber optic transmission lines, the next step in the evolution of the digital time-division-multiplexing (TDM) scheme was a standard signal format called *synchronous optical network* (SONET) in North America and *synchronous digital hierarchy* (SDH) in other parts of the world. This section addresses the basic concepts of SONET/SDH, its optical interfaces, and fundamental network implementations. The aim here is to discuss only the physical-layer aspects of SONET/SDH as they relate to optical transmission lines and optical networks. Topics such as the detailed data format structure, SONET/SDH operating specifications, and the relationships of switching methodologies such as carrier Ethernet service aggregation with SONET/SDH are beyond the scope of this text. These can be found in numerous sources [9–12].

13.3.1 SONET/SDH Frame Formats and Speeds

Figure 13.7 shows the basic structure of a SONET frame. This is a two-dimensional structure consisting of 90 columns by 9 rows of bytes, where one byte is eight bits. Here, in standard SONET terminology, a *section* connects adjacent pieces of equipment, a *line* is a longer link that connects two SONET devices, and a *path* is a complete end-to-end connection. The fundamental SONET frame has a 125-μs duration. Thus, the transmission bit rate of the basic SONET signal is

$$\text{STS-1} = (90\,\text{bytes/row})(9\,\text{rows/frame})$$
$$(8\,\text{bits/byte})/(125\,\mu\text{s/frame}) = 51.84\,\text{Mb/s}$$

This is called an STS-1 signal, where STS stands for *synchronous transport signal*. All other SONET signals are integer multiples of this rate, so that an STS-N signal has a bit rate equal to N times 51.84 Mb/s. When an STS-N signal is used to modulate an optical source, the *logical STS-N signal* is first scrambled to avoid long strings of ones and zeros and to allow easier clock recovery at the receiver. After undergoing

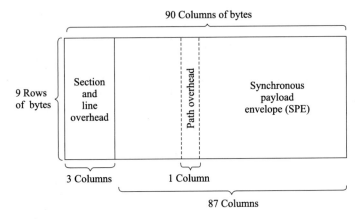

Fig. 13.7 Basic structure of an STS-1 SONET frame

electrical-to-optical conversion, the resultant *physical-layer optical signal* is called OC-*N*, where OC stands for *optical carrier*. In practice, it has become common to refer to SONET links as OC-*N* links.

In SDH the basic rate is equivalent to STS-3, or 155.52 Mb/s. This is called the *synchronous transport module—level 1* (STM-1). Higher rates are designated by STM-*M*. (Note: Although the SDH standard uses the notation "STM-*N*," here the designation "STM-*M*" is used to avoid confusion when comparing SDH and SONET rates.) Values of *M* supported by the ITU-T recommendations are *M* = 1, 4, 16, and 64. These are equivalent to SONET OC-*N* signals, where *N* = 3*M* (i.e., *N* = 3, 12, 48, and 192). This shows that, in practice, to maintain compatibility between SONET and SDH, *N* is a multiple of three. Analogous to SONET, SDH first scrambles the logical signal. In contrast to SONET, SDH does not distinguish between a logical electrical signal (e.g., STS-*N* in SONET) and an optical signal (e.g., OC-*N*), so that both signal types are designated by STM-*M*. Table 13.1 lists commonly used values of OC-*N* and STM-*M*.

Referring to Fig. 13.7, the first three columns comprise transport overhead bytes that carry network management information. The remaining field of 87 columns is

Table 13.1 Commonly used SONET and SDH transmission rates

SONET level	Electrical level	SDH level	Line rate (Mb/s)	Common rate name
OC-N	STS-N	–	N × 51.84	–
OC-1	STS-1	–	51.84	–
OC-3	STS-3	STM-1	155.52	155 Mb/s
OC-12	STS-12	STM-4	622.08	622 Mb/s
OC-48	STS-48	STM-16	2488.32	2.5 Gb/s
OC-192	STS-192	STM-64	9953.28	10 Gb/s
OC-768	STS-768	STM-256	39813.12	40 Gb/s

called the *synchronous payload envelope* (SPE) and carries user data plus nine bytes of *path overhead* (POH). The POH supports performance monitoring by the end equipment, status, signal labeling, a tracing function, and a user channel. The nine path-overhead bytes are always in a column and can be located anywhere in the SPE. An important point to note is that the synchronous byte-interleaved multiplexing in SONET/SDH (unlike the asynchronous bit interleaving used in earlier TDM standards) facilitates add/drop multiplexing of individual information channels in optical networks.

For values of N greater than 1, the columns of the frame become N times wider, with the number of rows remaining at nine, as shown in Fig. 13.8a. Thus, an STS-3 (or STM-1) frame is 270 columns wide with the first nine columns containing overhead information and the next 261 columns being payload data. The basic frame

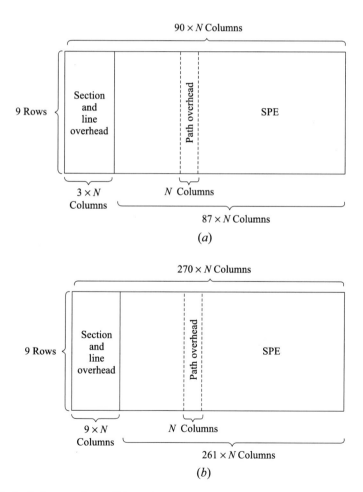

Fig. 13.8 Basic formats of **a** an STS-N SONET frame, **b** an STM-N SDH frame

structure for SDH is shown in Fig. 13.8b. An STM-N frame has a 125-μs duration and consists of nine rows, each of which has a length of $270 \times N$ bytes. Because the line and section overhead bytes differ somewhat between SONET and SDH, a translation mechanism is needed to interconnect SONET and SDH equipment.

Drill Problem 13.3 Using Fig. 13.8 and Table 13.1, show that 37152 voice channels (64-kb/s per voice channel) can be carried by an STM-16 SDH system.

13.3.2 Optical Interfaces in SONET/SDH

To ensure interconnection compatibility between equipment from different manufacturers, the SONET and SDH specifications provide details for the optical source characteristics, the receiver sensitivity, and transmission distances for single-mode fibers. Seven transmission ranges are defined with different terminology for SONET and SDH, as Table 13.2 indicates. The ITU-T G.957 Recommendation also designates the SDH categories by codes such as I-1, S-1.1, L-.1, and so on as indicated in the table. Table 13.3 shows the wavelength and attenuation ranges specified for transmission distances up to 80 km.

Depending on the attenuation and dispersion characteristics for each hierarchical level shown in Table 13.2, feasible optical sources include LEDs, multimode lasers, and various single-mode lasers. The system objectives are to achieve a bit-error rate (BER) of less than 10^{-10} for rates less than 1 Gb/s and 10^{-12} for higher rates and/or higher-performance systems.

Table 13.2 Transmission distances and their SONET and SDH designations, where x denotes the STM-x level

Transmission distance	SONET terminology	SDH terminology
≤2 km	Short-reach (SR)	Intraoffice (I-1)
15 km at 1310 nm	Intermediate-reach (IR-1)	Short-haul (S-x.1)
15 km at 1550 nm	Intermediate-reach (IR-2)	Short-haul (S-x.2)
40 km at 1310 nm	Long-reach (LR-1)	Long-haul (L-x.1)
80 km at 1550 nm	Long-reach (LR-2)	Long-haul (L-x.3)
120 km at 1550 nm	Very long-reach (VR-1)	Very long (V-x.3)
160 km at 1550 nm	Very long-reach (VR-2)	Ultra long (U-x.3)

Table 13.3 Wavelength ranges and attenuation for transmission distances up to 80 km

Distance (km)	Wavelength range at 1310-nm (nm)	Wavelength range at 1550-nm (nm)	Attenuation at 1310 nm (dB/km)	Attenuation at 1550 nm (dB/km)
≤15	1260–1360	1430–1580	3.5	Not specified
≤40	1260–1360	1430–1580	0.8	0.5
≤80	1280–1335	1480–1580	0.5	0.3

The receiver sensitivities designated in G.957 are the worst-case, end-of-life values. They are defined as the minimum-acceptable, average, received power needed to achieve a 10^{-10} BER. The values take into account extinction ratio, pulse rise and fall times, optical return loss at the source, receiver connector degradations, and measurement tolerances. The receiver sensitivity does not include power penalties associated with dispersion, jitter, or reflections from the optical path; these are included in the maximum optical path penalty. Table 13.4 lists the receiver sensitivities for various link configurations up through long-haul distances (80 km). Note that the specifications and recommendations are updated periodically, so the reader should refer to the latest version of the documents for specific details.

Longer transmission distances are possible using higher-power lasers. To comply with eye-safety standards, an upper limit is imposed on fiber-coupled powers. If the maximum total output power (including ASE) is set at the Class-3A laser limit of

Table 13.4 Source output, attenuation, and receiver sensitivity ranges for various rates and distances up to 80 km (see ITU-T G.957)

Parameter	Intraoffice	Short-haul (1)	Short-haul (2)	Long-haul (1)	Long-haul (3)
Wavelength (nm)	1310	1310	1550	1310	1550
Fiber		SM	SM	SM	SM
Distance (km)	≤2	15	15	40	80
Designation	I-1	S-1.1	S-1.2	L-1.1	L-1.3
Source range (dBm)					
155 Mb/s	−15 to −8	−15 to −8	−15 to −8	0 to 5	0 to 5
622 Mb/s	−15 to −8	−15 to −8	−15 to −8	−3 to +2	−3 to +2
2.5 Gb/s	−10 to −3	−5 to 0	−5 to 0	−2 to +3	−2 to +3
Attenuation (dB)					
155 Mb/s	0 to 7	0 to 12	0 to 12	10 to 28	10 to 28
622 Mb/s	0 to 7	0 to 12	0 to 12	10 to 24	10 to 24
2.5 Gb/s	0 to 7	0 to 12	0 to 12	10 to 24	10 to 24
Receiver sensitivity (dBm)					
155 Mb/s	−23	−28	−28	−34	−34
622 Mb/s	−23	−28	−28	−28	−28
2.5 Gb/s	−18	−18	−18	−27	−27

	Number of wavelengths (channels)	Nominal power per channel (dBm)
Table 13.5 Maximum nominal optical power per wavelength channel based on the total optical power being +17 dBm (see ITU-T Recommendation G.692)	1	17.0
	2	14.0
	3	12.2
	4	11.0
	5	10.0
	6	9.2
	7	8.5
	8	8.0

$P3A = +17$ dBm, then for ITU-T G.655 fiber this allows transmission distances of 160 km for a single-channel link. Using this condition, then for M operational WDM channels the maximum nominal channel power should be limited to Pchmax $= P3A - 10 \log M$. Table 13.5 lists the maximum nominal optical powers per channel up to $M = 8$.

13.3.3 SONET/SDH Rings

A key characteristic of SONET and SDH is that they can be configured as either a ring or a mesh architecture. This is done to create *loop diversity* for uninterrupted service protection purposes in case of link or equipment failures. The SONET/SDH rings are commonly called *self-healing rings* because the traffic flowing along a certain path can automatically be switched to an alternate or standby path following failure or degradation of the link segment.

Three main features, each with two alternatives, classify all SONET/SDH rings, thus yielding eight possible combinations of ring types. First, there can be either two or four fibers running between the nodes on a ring. Second, the operating signals can travel either clockwise only (which is termed a *unidirectional ring*) or in both directions around the ring (which is called a *bidirectional ring*). Third, protection switching can be performed either via a line-switching scheme or a path-switching scheme [13–15]. Upon link failure or degradation, *line switching* moves all signal channels of an entire OC-N channel to a protection fiber. Conversely, *path switching* can move individual payload channels within an OC-N channel (e.g., an STS-1 subchannel in an OC-12 channel) to another path.

Of the eight possible combinations of ring types, the following two architectures have become popular for SONET and SDH networks:

- Two-fiber, unidirectional, path-switched ring (two-fiber UPSR).
- Two-fiber or four-fiber, bidirectional, line-switched ring (two-fiber or four-fiber BLSR).

The common abbreviations of these configurations are given in parentheses. They are also referred to as unidirectional or bidirectional self-healing rings (USHRs or BSHRs), respectively.

Figure 13.9 shows a two-fiber unidirectional path-switched ring network. By convention, in a unidirectional ring the normal working traffic travels clockwise around the ring, on the *primary path*. For example, the connection from node 1 to node 3 uses links 1 and 2, whereas the traffic from node 3 to node 1 traverses links 3 and 4. Thus, two communicating nodes use a specific bandwidth capacity around the entire perimeter of the ring. If nodes 1 and 3 exchange information at an OC-3 rate in an OC-12 ring, then they use one-quarter of the capacity around the ring on all the primary links. In a unidirectional ring the counterclockwise path is used as an alternate route for protection against link or node failures. This *protection path* utilizes links 5 through 8 as indicated by the dashed lines. To achieve protection, the signal from a transmitting node is dual-fed into both the primary and protection fibers. This establishes a designated protection path on which traffic flows counterclockwise: namely, from node 1 to node 3 via links 5 and 6, as shown in Fig. 13.9a.

Consequently, two identical signals from a particular node arrive at their destination from opposite directions, usually with different delays, as denoted in Fig. 13.9b. The receiver normally selects the signal from the primary path. However, it continuously compares the fidelity of each signal and chooses the alternate signal in case of severe degradation or loss of the primary signal. Thus each path is individually switched based on the quality of the received signal. For example, if path 2 breaks or equipment in node 2 fails, then node 3 will switch to the protection channel to receive signals from node 1.

Figure 13.10 illustrates the architecture of a four-fiber bidirectional line-switched ring. Here, two primary fiber loops (with fiber segments labeled 1*p* through 8*p*) are used for normal bidirectional communication, and the other two secondary fiber loops

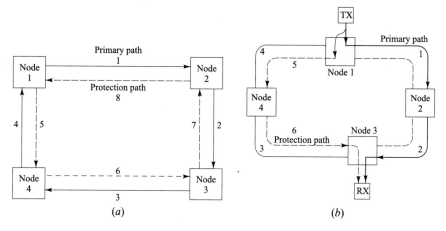

Fig. 13.9 **a** Generic two-fiber unidirectional path-switched ring (UPSR) with a counter-rotating protection path; **b** Flow of primary and protection traffic from node 1 to node 3

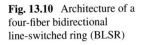
Fig. 13.10 Architecture of a
four-fiber bidirectional
line-switched ring (BLSR)

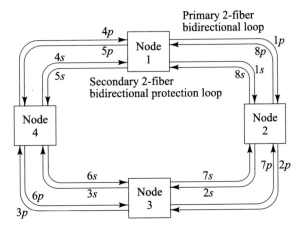

are standby links for protection purposes (with fiber segments labeled 1s through
8s). In contrast to the two-fiber UPSR, the four-fiber BLSR has a capacity advantage
because it uses twice as much fiber cabling and because traffic between two nodes
is sent only partially around the ring. To see this, consider the connection between
nodes 1 and 3. The traffic from node 1 to node 3 flows in a clockwise direction along
links 1p and 2p. Now, however, in the return path the traffic flows counterclockwise
from node 3 to node 1 along links 7p and 8p. Thus the information exchange between
nodes 1 and 3 does not tie up any of the primary channel bandwidth in the other half
of the ring.

To see the function and versatility of the standby links in the four-fiber BLSR,
consider first the case where a transmitter or receiver circuit card used on the primary
ring fails in either node 3 or 4. In this situation, the affected nodes detect a loss-
of-signal condition and switch both primary fibers connecting these nodes to the
secondary protection pair, as shown in Fig. 13.11. The protection segment between
nodes 3 and 4 now becomes part of the primary bidirectional loop. The exact same

Fig. 13.11 Reconfiguration
of a four-fiber BLSR under
transceiver or line failure

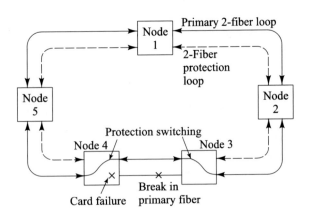

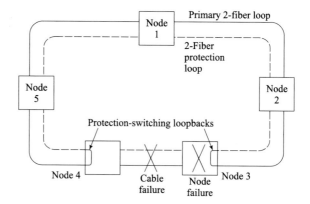

Fig. 13.12 Reconfiguration of a four-fiber BLSR under node or fiber-cable failure

reconfiguration scenario will occur when the primary fiber connecting nodes 3 and 4 breaks. Note that in either case the other links remain unaffected.

Now suppose an entire node fails, or both the primary and the protection fibers in a given span are severed, which could happen if they are in the same cable duct between two nodes. In this case, the nodes on either side of the failed internodal span internally switch the primary-path connections from their receivers and transmitters to the protection fibers, in order to loop traffic back to the previous node. This process again forms a closed ring, but now with all of the primary and protection fibers in use around the entire ring, as shown in Fig. 13.12.

13.3.4 SONET/SDH Network Architectures

Commercially available SONET/SDH equipment allows the configuration of a variety of network architectures, as shown in Fig. 13.13. For example, one can build point-to-point links, linear chains, unidirectional path-switched rings (UPSR), bidirectional line-switched rings (BLSR), and interconnected rings. The OC-192 four-fiber BLSR could be a large national backbone network with a number of OC-48 rings attached in different cities. The OC-48 rings can have lower-capacity localized OC-12 or OC-3 rings or chains attached to them, thereby providing the possibility of attaching equipment that has an extremely wide range of rates and sizes. Each of the individual rings has an independent failure-recovery mechanism and SONET/SDH network management procedures.

A fundamental SONET/SDH network element is the *add/drop multiplexer* (ADM). This piece of equipment is a fully synchronous, byte-oriented multiplexer that is used to add and drop subchannels within an OC-*N* signal. Figure 13.14 shows the functional concept of an ADM. Here, various OC-12s and OC-3s are multiplexed into an OC-48 stream. Upon entering an ADM, these subchannels can be individually dropped by the ADM and others can be added. For example, in Fig. 13.14, one OC-12 and two OC-3 channels enter the left-most ADM as part of an OC-48 channel. The

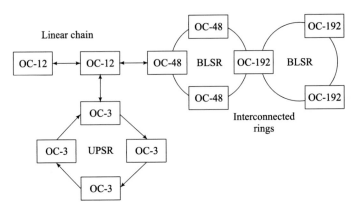

Fig. 13.13 Generic configuration of a large SONET or SDH network consisting of linear chains and various types of interconnected rings

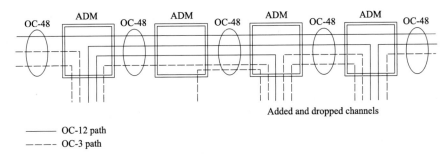

Fig. 13.14 Functional concept of an electronic add/drop multiplexer for SONET/SDH applications

OC-12 is passed through and the two OC-3s are dropped by the first ADM. Then, two more OC-12s and one OC-3 are multiplexed together with the OC-12 channel that is passing through and the aggregate (partially filled) OC-48 is sent to another ADM node downstream.

The SONET/SDH architectures can also be implemented with multiple wavelengths. For example, Fig. 13.15 shows a dense WDM deployment on an OC-192 trunk ring for n wavelengths (e.g., one could have $n = 16$). The different wavelength outputs from each OC-192 transmitter are passed first through a variable attenuator to equalize the output powers. These are then fed into a wavelength multiplexer, possibly amplified by a post-transmitter optical amplifier, and sent out over the transmission fiber. Additional optical amplifiers might be located at intermediate points and/or at the receiving end.

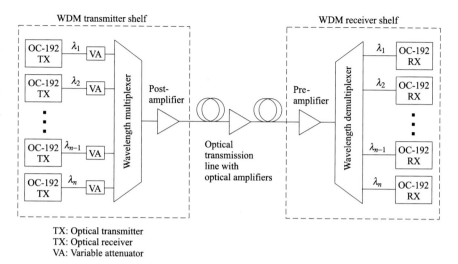

TX: Optical transmitter
TX: Optical receiver
VA: Variable attenuator

Fig. 13.15 DWDM deployment of n wavelengths in an OC-192/STM-64 trunk ring

13.4 High-Speed Lightwave Transceivers

A challenge to creating efficient and reliable optical networks that satisfy an ever-growing demand for bandwidth is the development of high-speed optical fiber transceivers. A variety of transceivers exist that incorporate both an optical transmitter and receiver in the same miniaturized package. For example, the *small form factor pluggable* (SFP) transceiver shown in Fig. 13.16 can be used for a wide range of applications. Of interest for such devices are a *hot-pluggable* capability (meaning they can be inserted and removed from transmission equipment line cards without turning off the electric power) and the incorporation of sophisticated wavelength control into the package. Depending on the particular internal electronics and the type of optical connector used, these standard-sized SFP units can be designed for systems ranging from 2.5 Gb/s over a 10-km distance using a pair of single-mode fibers to 100 Gb/s over a 2-km distance using a multiple fiber connector.

The transmission rates and their related standards activities are changing rapidly to accommodate more and more higher-bandwidth traffic. For example, the IEEE 802.3 Ethernet Working Group ratified the 200–400 Gb/s Ethernet specifications in 2018, and large data centers started deploying 400 Gb/s links beginning in 2020. Further Ethernet specifications for up to 800 Gb/s and 1.6 Tb/s connectivity will be ratified by 2023.

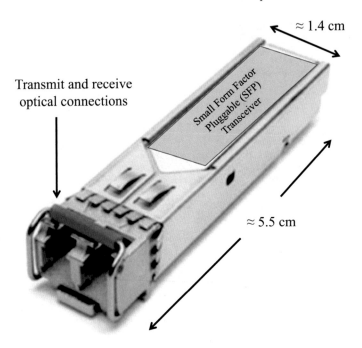

Transmit and receive
optical connections

≈ 1.4 cm

Small Form Factor
Pluggable (SFP)
Transceiver

≈ 5.5 cm

Fig. 13.16 A standard SFP transceiver package with transmit and receive optical connections

13.4.1 *Links Operating at 10 Gb/s*

Owing to product improvement efforts, several multimode fibers with different bandwidth grades exist for short-distance 10-Gb/s and beyond use. To identify these different fiber categories, the ISO/IEC 11801 Structured Cabling Standard gives four classifications of multimode fiber in terms of bandwidth. As Table 13.6 shows, these fibers are referred to as grades OM1 through OM5. Note that the values of the bandwidth depend on the measurement standard used, and the maximum transmission distances may vary depending on the exact fiber design from different manufacturers. To summarize this table, note the following characteristics:

Table 13.6 Multimode fiber classifications and their use with 1- and 10-Gb/s Ethernet

Fiber class and size	BW @ 850 nm (MHz-km)	BW @ 1300 nm (MHz-km)	Max distance for 1 Gb/s @ 850 nm (m)	Max distance for 1 Gb/s @ 1300 nm (m)	Max distance for 10 Gb/s @ 850 nm (m)
OM1 62.5/125	200	500	300	550	33
OM2 50/125	500	800	750	200	82
OM3 50/125	2000	500	950	600	300
OM4 50/125	4700	500	1040	600	550

- **OM1 grade fiber** is the original (often called *legacy*) multimode fiber that was designed for use with LEDs. Most of these legacy fibers have a 62.5-μm-diameter core, although in some cases the early installed fibers had a 50-μm-diameter core. The bandwidth of these fibers is 200 MHz-km at 850 nm and 500 MHz-km at 1310 nm. The data rates with LEDs are limited to about 100 Mb/s.

- **OM2 grade fiber** has an improved bandwidth and can be used to expand networks that contain existing legacy 50-μm core diameter fibers. If only this fiber is used, then at 850 nm it is possible to send 1-Gb/s signals over about 750 m distances and 10 Gb/s over 82 m.

- **OM3 grade fiber** has a higher bandwidth and can support 10-Gb/s data rates over distances up to 300 m.

- **OM4 grade fiber** has a bandwidth of 4700 MHz-km and increases the transmission distance to 550 m when using commercial 850-nm VCSELs for existing 1 and 10-Gb/s applications as well as for 40- and 100-Gb/s Ethernet systems.

- **OM5 grade fiber** has performance values similar to OM4 for insertion loss and transmission distances. However, OM5 fiber is designed for use at wavelengths of 850, 880, 910, and 940 nm, so that that it can support four simultaneous WDM wavelengths. An attenuation value of 2.3 dB/km is specified for 953-nm operation, but testing of installed links only needs to be done at 850 and 1300 nm wavelengths.

For a short-reach 10-Gb/s network to be compliant with installation standards, all segments of the network should use the same grade of multimode fiber. However, in some situations it may be impractical or too expensive to replace existing low-grade multimode fiber with a higher-grade fiber. In this case most likely a link will contain a mixture of, for example, OM2 and OM3 fibers that are spliced together. If the link is to transport 10-Gb/s data, then the bandwidths of the fibers will determine the resulting *effective maximum link length*. If all geometric parameters of the interconnected OM2 and OM3 fibers are the same, then a general expression for determining the effective maximum length L_{max} is given by

$$L_{max} = L_{OM2}\frac{BW_{OM3}}{BW_{OM2}} + L_{OM3} \qquad (13.5)$$

where L_{OMx} and BW_{OMx} are the length and bandwidth, respectively, of the OM*x* grade fiber. To be compliant with the maximum allowed dispersion, the effective maximum link length calculated by Eq. (13.5) must be less than the achievable link length if only OM3 fiber is used. For standard OM3 fiber this length is 300 m. With OM4 fibers, the maximum length is 550 m at an 850-nm wavelength for operation at 10 Gb/s.

Example 13.2 An engineer wants to create a link consisting of 40 m of OM2 fiber that has a 500-MHz bandwidth and 100 m of OM3 fiber that has a 2000-MHz bandwidth. What is the effective maximum link length?

Solution: From Eq. (13.5), the effective maximum link length is

$$L_{max} = (40\,\text{m})(2000/500) + 100\,\text{m} = 260\,\text{m}$$

Because the calculated length is less than 300 m, this link would comply with the installation standard.

Drill Problem 13.4 An engineer wants to create a link consisting of 50 m of OM2 fiber that has a 500-MHz bandwidth and 90 m of OM3 fiber that has a 1500-MHz bandwidth. Show that the effective maximum link length is 240 m.

If transmission at 10 Gb/s is attempted over legacy OM1 multimode fibers, the signal is restricted to distances of only about 33 m. To achieve 300-m transmission distances at 10-Gb/s over legacy multimode fibers, an alternative scheme is to send four 3.125-Gb/s data streams over four different wavelengths. A number of manufacturers offer 850-nm VCSEL sources that can be modulated directly at 10 Gb/s. Both the transmitter and receiver can be housed in the same package.

For access network applications ranging from 7 to 20 km, which the 10-GbE specification calls *long-reach* (LR), the links need to use InGaAsP-based distributed feedback (DFB) lasers operating at 1310 nm over single-mode fibers. These links operate near the 1310-nm dispersion minimum of G.652 single-mode fibers and the light sources can be modulated directly.

For metro network applications ranging from 40 to 80 km, which the 10-GbE specification calls *extended-reach* (ER), the links need to use externally modulated DFB lasers operating at 1550 nm over single-mode fibers. A variety of transceiver packages exist for both LR and ER applications. Three of the several configurations include the 300-pin, XFP (similar to SFP modules), and SFP modules. The 300-pin device, shown in Fig. 13.17, is about $114 \times 90 \times 12$ mm ($L \times W \times H$) in size and contains other internal electronics besides the normal circuitry used for the light source and

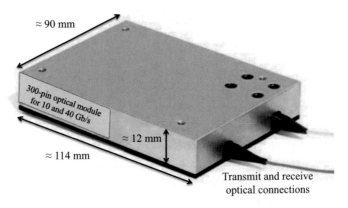

Fig. 13.17 Example of an industry standard 300-pin optical transmit/receive module for 10 and 40 Gb/s rates

photodetection operations. These electronics include clocking functions for transmit and receive signals, a line loopback capability for diagnostic signals, and circuitry for multiplexing sixteen 622-Mb/s SONET/SDH inputs into a 10-Gb/s stream, which becomes the electric signal input to the laser. In the receive direction, a photodetector changes the incoming 10-Gb/s optical signal to an electrical format, which then is separated back into sixteen 622-Mb/s electrical signals. External electronics need to provide these functions for the smaller transceiver modules.

13.4.2 Transceivers for 40 Gb/s Links

New challenges in terms of transceiver response characteristics, chromatic dispersion control, and polarization-mode dispersion compensation arise when transitioning to higher capacity links, such as 40 Gb/s data rates [16–18]. For example, compared to a 10-Gb/s system, when using a conventional on-off keying (OOK) modulation format a link operating at 40 Gb/s is sixteen times more sensitive to chromatic dispersion, is four times more sensitive to polarization-mode dispersion, and needs an optical signal-to-noise ratio (OSNR) that is at least 6 dB higher to reach an equivalent BER.

Therefore, alternate modulation schemes besides OOK have been considered. One method is *differential binary phase-shift keying*, which is known as DBPSK or simply DPSK (see Sect. 8.5). One advantage of DPSK is that when using a balanced receiver the OSNR needed to reach a specific BER is about 3 dB lower than that required for OOK. (A *balanced receiver* uses a pair of matched photodiodes to achieve a higher sensitivity.) Because a factor of 3 dB means only half the amount of optical power is needed for an equivalent BER, the lower OSNR that is needed for DPSK can be allocated to reducing the optical power level at the receiver, to relaxing the loss specifications for various link components, or to extending the transmission distance. For example, the transmission distance could be doubled if there are no other signal impairments, such as additional nonlinear effects. In addition, DPSK is quite resilient to nonlinear effects, which are a problem when using OOK for data rates greater than 10 Gb/s. This resilience is due to the fact that compared to OOK the optical power is more evenly distributed in DPSK (that is, there is optical power in each bit slot, as the next paragraph describes) and the optical peak power is 3 dB lower for the same average optical power. These factors reduce nonlinear effects that are dependent on the bit pattern and on optical power levels.

In contrast to OOK where the information is conveyed through amplitude changes, DPSK carries the information in the optical phase. Optical power appears in each DPSK bit slot and can occupy the entire slot for NRZ-DPSK, or it can occupy part of the slot in the form of a pulse in an RZ-DPSK format. Binary data is encoded as either a 0 or a π optical phase shift between adjacent slots. For example, the information bit 1 may be transmitted by a 180° carrier phase shift relative to the carrier phase in the previous slot, while the information bit 0 is transmitted by no phase shift relative to the carrier phase in the previous signaling interval.

13.4.3 Transceivers for 100 Gb/s Links

A number of different coherent, self-coherent, and direct-detection-based optical transmission techniques have been proposed for links operating at 100 Gb/s. These techniques typically involve a combination of polarization division multiplexing (PDM) with differential quadrature phase-shift keying (DQPSK) or orthogonal frequency division multiplexing (OFDM) [19]. Of these proposals, the data communications and telecom industries have selected polarization multiplexed quadrature phase-shift keying (PM-QPSK) as the preferred format for deployment within 50-GHz-spaced 100-Gb/s applications. This agreement arose because of the capability of PM-QPSK-formatted signals to pass through multiple optical add/drop multiplexers (see Sect. 13.5) and the tolerance of such signals to polarization-mode dispersion effects.

The transceiver module for these applications is the C form-factor pluggable (CFP) configuration, which Fig. 13.18 illustrates. The device size is about 145 × 77 × 13.6 mm (L × W×H) and operates off a single 3.3-V power supply. This transceiver is the result of a multiple-source agreement (MSA) between competing manufacturers. The use of the letter "C" is derived from the Latin symbol C that expresses the number 100, because the standard for this module form factor was developed primarily for 100-Gb/s systems. The CFP is a hot-pluggable module that supports a wide range of 40 and 100-Gb/s applications, such as 40 and 100G Ethernet, OC-768/ STM-256, OTU3, and OTU4. Here the acronyms OTU3 (optical transport unit 3) and OTU4 (optical transport unit 4) refer to line rates of 43 and 112 Gb/s, respectively, that have been standardized by the ITU-T Recommendation G.709, which is commonly called the Optical Transport Network (OTN) or digital wrapper technology. Different CFP module versions (for example, smaller CFP2 and CFP4 formats) are available to support various link distances over either multimode or single-mode fibers. The CFP module includes features such as advanced thermal

Fig. 13.18 Example of a CFP optical transmit/receive module for 100 Gb/s rates

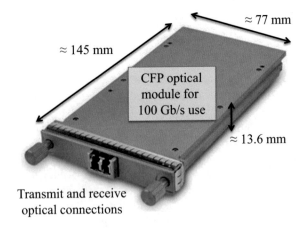

≈ 77 mm

≈ 145 mm

CFP optical module for 100 Gb/s use

≈ 13.6 mm

Transmit and receive optical connections

management, electromagnetic interference (EMI) management, and a management data input/output (MDIO) interface for Ethernet data management.

13.4.4 Links Operating at 400 Gb/s and Higher

Building on the success of single-channel 10-, 40-, and 100-Gb/s links, the next development step was the creation of ultrafast gigabit and terabit links running at 400 and 800 Gb/s and higher [20–22]. The implementation of 400-Gb/s systems achieves a spectral efficiency up to 8 b/s/Hz and provides a fourfold increase in the transport capacity of a 100-Gb/s system. Similar to the links operating up to 100 Gb/s, the development of 400-Gb/s links is based on the standard 50-GHz DWDM grid, thereby making them compatible with existing reconfigurable optical add/drop multiplexing (ROADM) optical networks (see Sect. 13.5). The modulation techniques for achieving such high-speed networks include various high-order quadrature amplitude modulation (QAM) formats and the concept of optical orthogonal frequency-division multiplexing (OFDM).

13.5 Schemes for Optical Add/Drop Multiplexing

An *optical add/drop multiplexer* (OADM) is a device that allows the insertion or extraction of one or more wavelengths from a fiber at a network node. For example, an OADM might have the capability to drop and insert three wavelengths from a set of N being transported over a single fiber. The remaining $(N - 3)$ wavelengths pass unaffected through (also called *express through*) the OADM to the next node. Without an OADM, if only three wavelengths out of $N \gg 3$ are needed at the local node, then all the other wavelengths also would need to be processed there by means of optoelectronic transceivers. This will greatly increase the equipment costs at that node. Thus the advantage of the add/drop function is that no signal processing is required for the set of wavelengths that passes straight through the OADM.

The OADM can reside at an optical amplifier site in a long-haul network or it can be located at a node in a metro network. Depending on how it is designed, an OADM can operate either statically in a fixed add/drop configuration or it can be reconfigured dynamically from a remote network management site. A fixed optical add/drop multiplexer is simply referred to as an OADM. A dynamic device is called a *reconfigurable OADM* (ROADM) [23–25]. A fixed OADM obviously is not as flexible as a ROADM and may require a change of hardware components if a different set of wavelengths needs to be dropped or added.

Depending on whether an engineer is designing a metro or a long-haul network, different performance specifications need to be addressed when implementing an optical add/drop capability in the network. In general, because of the nature of the services provided, changes in the add/drop configuration for a long-haul network occur less frequently than in a metro network where there tends to be a high turnover rate in service requests and in wavelengths being transported. In addition, compared to a metro network the wavelength spacing usually is much narrower in a long-haul network and the optical amplifiers used must cover a wider spectral band. Section 13.7 has some examples of this topic.

13.5.1   Configurations of OADM Equipment

A number of different OADM configurations are possible. Most OADMs are constructed using the WDM elements described in Chap. 10. These can be a series of dielectric thin-film filters, an arrayed waveguide grating, a set of liquid crystal devices, or a series of fiber Bragg gratings used in conjunction with optical circulators. The architecture selected for a particular application depends on implementation factors such as the number of wavelengths to be dropped and added at a node, the desired modularity of the OADM (e.g., how easy is it to upgrade this device), and the constraint of whether random individual wavelengths or a neighboring group of wavelengths should be processed. Here it will be assumed that N wavelengths traveling on a single fiber enter the OADM and a subset M of them is processed at the node. After processing, these M wavelengths can be reinserted onto the fiber line to rejoin the through-going $N - M$ channels.

In the configuration illustrated in Fig. 13.19, all N incoming wavelengths are separated into individual channels at the OADM input by means of a wavelength demultiplexer. This gives a very versatile architecture, because any of the N wavelengths can be dropped, processed at the node, and then reinserted onto the outgoing fiber by means of a second wavelength multiplexer. In the illustration given here, M wavelengths are dropped and the remaining $N - M$ channels individually pass through the OADM. The M dropped wavelengths are labeled λi through λk to indicate that any combination of M wavelengths selected from the N incoming light

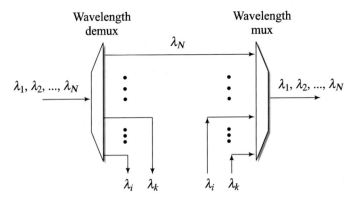

Fig. 13.19 Concept of a simple passive optical add/drop multiplexer

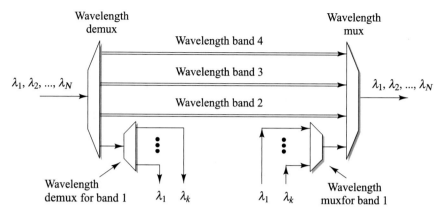

Fig. 13.20 Modular expandable passive OADM architecture

channels can be dropped. Such a configuration is useful if a large fraction of the N wavelengths needs to be dropped and added. However, it is not cost-effective for dropping and adding a small fraction of the incoming wavelengths.

Figure 13.20 shows a more modular OADM architecture. Here the N incoming light channels are split into several *bands of wavelengths*. This function can be achieved using either a set of thin-film filters or an AWG. The band to be dropped can be sent through a second demultiplexer followed by processing of the individual wavelengths. As an example, if $N = 12$ and three wavelengths need to be dropped, the incoming wavelengths can be divided into four bands of three wavelengths each. Added lightwave channels go through two stages of multiplexing to rejoin the pass-through bands. An advantage of this approach is that for future OADM upgrades to drop another wavelength band at this node, network engineers can incorporate another second-stage demultiplexer to handle the next desired wavelength band.

13.5.2 Reconfiguring OADM Equipment

The task of reconfiguring a fixed OADM manually may require several days of planning and implementation of hardware changes. In contrast, the use of ROADMs gives service providers the ability to reconfigure a network within minutes from a remote network management console. The dynamic flexibility to drop and add selected wavelengths quickly at a particular node upon customer demand for new or expanded services is known popularly as *service provisioning on the fly*. This is particularly important in metro networks where changes in service requests tend to be significantly more dynamic than in long-haul networks and clients expect very rapid responses to their requests. Such service requests can have fairly diverse origins such as time-varying business applications, on-demand entertainment, and emergency or disaster-response communications.

A variety of ROADM architectures are possible. This section shows three example architectures based on *wavelength blockers, arrays of small switches,* and *wavelength-selective switches.* Section 13.6 describes the concept of an optical cross-connect, which uses a more complex architecture of multiple wavelength-selective switches. Each ROADM type can have a number of alternative configurations and different operational characteristics. The design of a particular ROADM depends on factors such as cost, reliability, technology maturity, desired network operating flexibility, and upgrade capability of the equipment. ROADM characteristics include:

- **Wavelength dependence**. If the architectures are wavelength-dependent they are said to be *colored* or have *colored ports.* When the ROADM operation is independent of wavelength, it is referred to as being *colorless* or having *colorless ports.*
- **ROADM degree**. The *degree* of a ROADM refers to the number of bidirectional multiple-wavelength interfaces the device supports. Thus a degree-2 ROADM has two bidirectional WDM interfaces and a degree-4 ROADM can support up to four bidirectional WDM interfaces, for example, in north, south, east and west directions.
- **Remote reconfiguration**. The ability to change the ROADM configuration from a remotely located network management workstation is an important feature, because it greatly reduces operations costs by eliminating the need for a service person to visit the ROADM site in order to manually upgrade the device.
- **Express channels**. The ability to have *express channels* that allow a selected set of wavelengths to pass through the node without the need for optical-to-electrical-to-optical conversion saves the expense of having optical transceivers at the ROADM for those wavelengths.
- **Modular expansion**. To avoid an initial high setup cost to connect transmitters and receivers to each add/drop port, service providers usually first activate the minimum number of ports needed to support current traffic and then later add more channels as the service demand increases. This is known popularly as the *pay-as-you-grow approach.*
- **Minimum optical impairment**. Having an express wavelength feature requires a careful engineering design to avoid the accumulation of optical signal impairments as a particular wavelength set passes through several sequential ROADMs. These include effects such as crosstalk between channels, wavelength-dependent attenuation, ASE noise, and polarization-dependent loss.

Wavelength Blocker Configuration

Figure 13.21 shows the simplest ROADM configuration, which uses a *broadcast-and-select approach.* In this degree-2 architecture a passive optical coupler splits the incoming lightwave signal power into two paths. One branch is an express path and the other branch is diverted to a drop site. Located in the express path is a device called a *wavelength blocker* that can be configured to block those wavelengths that are to be received at the node. The drop segment contains a $1 \times N$ optical power splitter, which divides and directs the light signal into N tunable filters that allow selection of

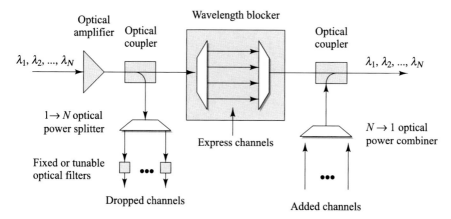

Fig. 13.21 Example of a ROADM concept based on the use of a wavelength blocker

any desired wavelength. The add segment contains N tunable laser sources that allow the inserted wavelengths to join the express wavelengths on the main fiber line by means of an $N \times 1$ optical power combiner and another passive optical coupler. An attractive feature of this ROADM is that initially only the two passive couplers and the wavelength blocker are needed. Later different sizes of optical power splitters and combiners can be selected to drop and add other desired wavelengths, which now need to be prevented from traversing the express path by proper settings of the wavelength blocker. One drawback of this ROADM is that the losses of the optical power splitters and combiners increase as their size grows, that is, as the number of wavelengths to be dropped and added increases.

Switch Array Configuration

Figure 13.22 shows a colored *switching-based ROADM* configuration, which is also known as a *demux-switch-mux approach*. Here the N incoming wavelengths first pass through a demultiplexer. As Fig. 13.23 shows, individual 2×2 or 1×2 switches then allow each wavelength to either bypass the node or be dropped. The dropped wavelength can be added back onto the outgoing fiber using either the same 2×2 switch as used by the incoming wavelength or a 2×1 switch analogous to the drop device. After the N lightwaves emerge from the 2×2 or 2×1 switches, a wavelength multiplexer combines them onto the outgoing fiber.

In a less versatile variation of this colored architecture, the ROADM may be designed to express through a certain set of K wavelengths, say λ_1 through λ_K, without using switches for them. One implementation alternative for this is to replace the set of 2×2 switches with an $N \times M$ switch that has N ports for N incoming wavelengths and M add and drop ports. This could be accomplished with an $N \times M$ array of MEMS (micro-electromechanical system) mirrors. Figure 13.24 gives an example for a 4×4 configuration. To drop and add a wavelength, a miniature mirror in the switch path of that channel is set at an angle to divert the incoming light to a

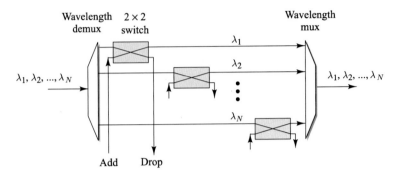

Fig. 13.22 A colored (wavelength-dependent) switching-based ROADM

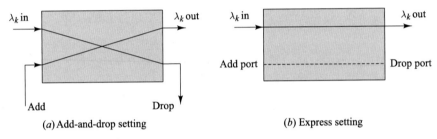

(a) Add-and-drop setting   (b) Express setting

Fig. 13.23 Internal connections of a 2 × 2 switch for **a** add/drop and **b** express functions

drop port and to insert the same wavelength via an add port. Figure 13.24 shows this for port 3.

In either case (full or partial add/drop capability), a limitation of the switching-based ROADM architecture is that it can be costly and complex to implement. If the ROADM is designed to have a full add/drop capability of all wavelengths, then the transceivers of all express wavelengths will not be used. In case of a partial add/drop capability, a smaller set of transceivers is needed, but careful planning is required to determine what the $K = N - M$ express wavelengths should be and the ROADM implementation becomes constrained.

Figure 13.25 shows a variation of the switching-based ROADM architecture, which is colorless and has more design and implementation flexibility. In contrast to the parallel architecture shown in Fig. 13.22, this is a serial configuration. This configuration has a series of $M \leq N$ wavelength-tunable devices that allow any single wavelength to be added or dropped by any device. Here M is the maximum number of wavelengths that can be used at the node. Note that in this *tuning-based ROADM* architecture, there is no constraint on which particular wavelengths are express or add/drop channels, because the tunable elements are colorless; that is, they can be set to any wavelength value. However, this configuration usually is limited to a small

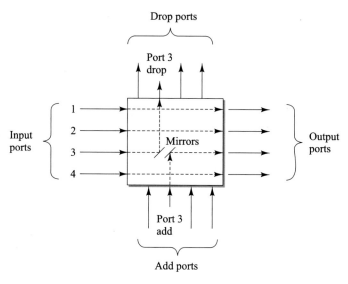

Fig. 13.24 Example of ROADM based on a 4 × 4 array of MEMS mirrors

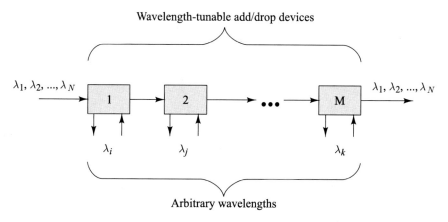

Fig. 13.25 Wavelength-tunable ROADM for adding and dropping small channel counts

number of drop channels; otherwise the accumulated optical loss through the series of switching elements may be fairly large.

Wavelength-Selective Switch Configuration

In the evolution of add/drop multiplexer technology to enable more sophisticated mesh network applications, *wavelength-selective switch* (WSS) technology was introduced [26, 27]. A key feature of a WSS is that it can direct each wavelength entering a common input port to any one of a multiple number of output ports, as indicated generically in Fig. 13.26. A basic ROADM is formed using one WSS module

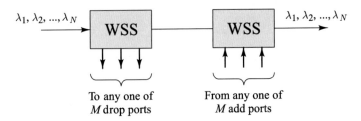

$\lambda_1, \lambda_2, ..., \lambda_N$

$\lambda_1, \lambda_2, ..., \lambda_N$

To any one of
M drop ports

From any one of
M add ports

Fig. 13.26 Flexible ROADM architecture based on a pair of wavelength-selective switches

for dropping wavelengths and another for adding them. Each module contains a set of wavelength selective switches. To designate the size of such a switch, it is labeled as a $1 \times M$ switch, where M indicates the number of output ports to which a wavelength can be directed. If there are N incoming wavelengths that are to be switched to any of M output ports, then one WSS module contains N wavelength selective switches of size $1 \times M$. Available WSS configurations can send wavelengths to between four and ten ports.

Figure 13.27 illustrates the switching concept using 16 input wavelengths (λ_1 through λ_{16}), which can be directed to any of four output lines with sixteen 1×4 wavelength selective switches. These four output lines include one express port for through-going wavelengths and three ports for dropping wavelengths. In Fig. 13.27 the incoming wavelengths are divided into individual channels by a demultiplexer and then pass through a variable optical attenuator (VOA). The function of the VOAs is to ensure that all wavelengths leaving the ROADM have the same optical power level. The 1×4 WSS following a VOA either sends a wavelength to the express port or switches it to any of the three drop ports. A versatile feature is that any collection of wavelengths can be switched to a given drop port simultaneously, thereby creating an *any-to-any switching capability*. Because each ROADM port has a multiple-wavelength capability, this increases the degree of a WSS-based ROADM to more than 2.

13.6   Optical Switching Architectures

The appearance of enhanced multimedia services requiring huge bandwidths, such as broadcast high-definition television (HDTV) and video-on-demand, created a need for transitioning from basic interconnected rings to extremely high-capacity rings adjoined to mesh networks that can support clusters of up to 50 nodes in a metropolitan area. Therefore, another network element with more sophisticated switching capabilities than a ROADM is needed. This element, which is called an *optical crossconnect* (OXC), provides switched pass-through paths for express traffic that does not terminate at the node and an interface for dropping and adding optical signals at the node. The express traffic can be switched from any input fiber to any output fiber. Internally such a device can have either an electrical or optical switching

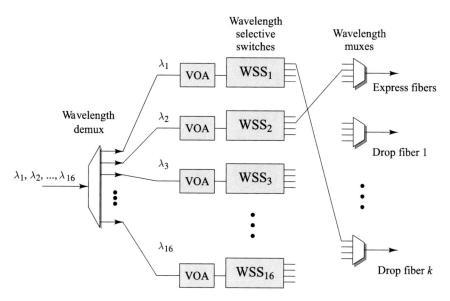

Fig. 13.27 WSS-based ROADM that can switch any combination of 16 input wavelengths simultaneously to any output port

fabric. To understand basic optical switching technology, Sect. 13.6.1 looks at general OXC configurations, Sect. 13.6.2 considers the performance impact when wavelength conversion is used, and Sect. 13.6.3 describes the standard implementations of wavelength routing or optical circuit switching.

As network traffic volume rises, the number of wavelengths per fiber will increase. This means that changes are needed for the earlier-generation switching methods in which optical signals are converted to electrical signals, switched electronically, and then converted back to an optical format. The main reason is that the performance of the electronic equipment used in this *optical-to-electrical-to-optical* (O/E/O) conversion process is strongly dependent on the data rate and protocol. To overcome these limitations, the concept of all-optical switching was introduced. Two approaches to this concept are optical burst switching (OBS) and optical packet switching (OPS). Sections 13.6.4 and 13.6.5, respectively, discuss these concepts. As a final topic, Sect. 13.6.6 describes the elastic optical networking scheme.

13.6.1 Concept of an Optical Crossconnect

A high level of path-configuration modularity, capacity scaling, and flexibility in adding or dropping channels at a client site can be achieved by introducing the concept of an optical crossconnect switch in the physical path structure of an optical network. These *optical crossconnect* (OXC) switches sit at junctions in ring and

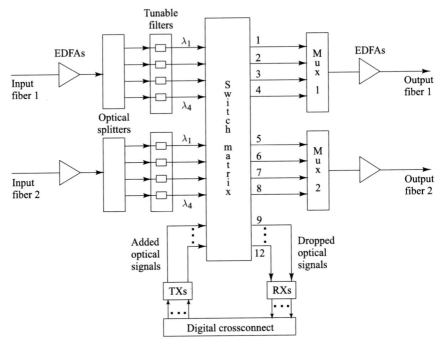

Fig. 13.28 Optical cross-connect architecture using optical space switches and no wavelength converters

mesh networks that are interconnected through many hundreds of optical fibers, each carrying dozens of wavelength channels. At such a junction an OXC can dynamically route, set up, and take down very high-capacity lightpaths [28, 29].

To visualize the operations of an OXC, consider first the OXC architecture shown in Fig. 13.28 that uses a switching matrix for directing the incoming wavelengths that arrive on a series of M input fibers. The switch matrix can operate electrically or optically; that is, it can switch the incoming signals in either the electric or the optical domain. Each of the M input fibers carries N wavelengths, any or all of which could be added or dropped at a node. For illustration simplicity, here assume $M = 2$ and $N = 4$. At each fiber input, the arriving aggregate of N signal wavelengths is demultiplexed and then enters the switch matrix. The switch matrix directs the channels either to one of the eight output lines if it is a through-traveling signal, or to a particular receiver attached to the OXC at output ports 9 through 12 if it has to be dropped to a user at that node. Signals that are generated locally by a user get connected electrically via the *digital crossconnect matrix* (DXC) to an optical transmitter. From here the wavelengths enter the switch matrix, which directs them to the appropriate output line. The M output lines, each carrying separate wavelengths, are fed into a wavelength multiplexer to form a single aggregate output stream. An optical amplifier to boost the signal level for transmission over the trunk fiber normally follows the multiplexer.

13.6.2 Considerations for Wavelength Conversion

Contentions for lightpath use could arise in the architecture shown in Fig. 13.28 when channels having the same wavelength but traveling on different input fibers enter the OXC and need to be switched simultaneously to the same output fiber. This could be resolved by assigning a fixed wavelength to each optical path throughout the network, or by dropping one of the incoming channels at the node and retransmitting it at another wavelength. However, in the first case, wavelength reuse and network scalability (expandability) are reduced, and in the second case the add/drop flexibility of the OXC is lost. Example 13.3 illustrates how wavelength conversion at any output of the OXC can eliminate these blocking characteristics [30–32].

Example 13.3 Consider the 4×4 OXC shown in Fig. 13.29. Here two input fibers are each carrying two wavelengths. Either wavelength can be switched to any of the four output ports. The OXC consists of three 2×2 switch elements. Here λ_1 on input fiber 1 passes through to output fiber 1 and λ_2 on input fiber 2 passes through to output fiber 2. For the other two wavelengths, suppose that λ_2 on input fiber 1 needs to be switched to output fiber 2 and that λ_1 on input fiber 2 needs to be switched to output fiber 1. This is achieved by having the first two switch elements set in the bar state (the straight-through configuration) and the third element set in the cross state, as indicated in Fig. 13.29. Without wavelength conversion there would be wavelength contention at both output ports. By using wavelength converters, the cross-connected lightwaves can be prevented from contending for the same output fiber.

To illustrate the effect of wavelength conversion, consider a simple model that is based on standard series independent-link assumptions commonly used in circuit-switched networks [30]. In this simplified example, during a request for establishing a lightpath connection between two stations, the usage of a wavelength on a fiber is statistically independent of other fiber links and other wavelengths. Here a *lightpath*

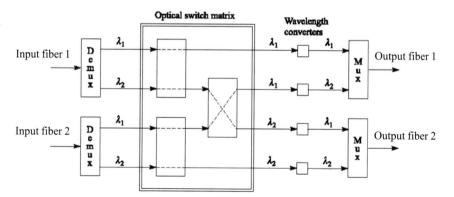

Fig. 13.29 Example of a simple 4×4 optical cross-connect architecture using optical space switches and wavelength converters

is defined as a temporary point-to-point optical connection. Although this model tends to overestimate the probability that a wavelength is blocked along a path, it provides insight into the network performance improvement when using wavelength conversion.

First consider the network configuration shown in Fig. 13.30. Assume there are $H + 1$ OXC nodes in the network, so that there are H links (or hops) between node 1 and node $H + 1$. Suppose the number of available wavelengths per fiber link is F, and let ρ be the probability that a wavelength is used on any fiber link. Then, because ρF is the expected number of busy wavelengths on any link, ρ is a measure of the *wavelength utilization* along the path, and $1 - \rho$ is the probability that a particular wavelength is not occupied on a given hop. Thus $(1 - \rho)^H$ is the probability that a wavelength slot is not blocked for all H hops from the signal source to the desired destination at node $H + 1$.

Without wavelength conversion, a connection request between node 1 and node $H + 1$ can be honored only if there is a free wavelength along the entire lightpath, that is, if there is a wavelength that is unused on each of the H intervening fibers. Thus, the probability P_b that the connection request from node 1 and node $H + 1$ is blocked is the probability that each wavelength is used on at least one of the H links, so that

$$P_b = \left[1 - (1 - \rho)^H\right]^F \tag{13.6}$$

Now consider a network with wavelength conversion. In this case, a connection request between node 1 and node $H + 1$ is blocked if one of the H intervening fibers is full; that is, the fiber is already supporting F independent sessions on different wavelengths. Thus, the probability $P_{b,conv}$ that the connection request from node 1 to node $H + 1$ is blocked is the probability that there is a fiber link in this path with all F wavelengths in use, so that

$$P_{b,conv} = 1 - \left(1 - \rho^F\right)^H \tag{13.7}$$

As a simple example, Fig. 13.31 shows the blocking probabilities as a function of the wavelength utilization ρ for a lightpath of $H = 10$ hops ($H + 1$ nodes) without and with wavelength conversion. In this example, the number of wavelengths is taken to be $F = 20$. It is clear that wavelength conversion significantly reduces the blocking probability. Despite its dramatic possible performance improvement, the

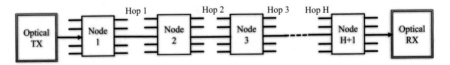

Fig. 13.30 Concept of a lightpath traversing $H + 1$ switching nodes in H hops

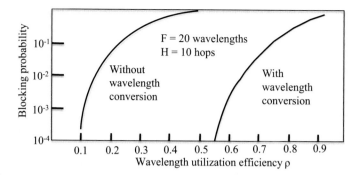

Fig. 13.31 Packet blocking probabilities with and without wavelength conversion for the case of 20 wavelengths and 10 hops

use of wavelength blocking can be expensive in large networks if it is applied to all WDM channels and at all nodes.

13.6.3 Methodologies for Wavelength Routing

To send information quickly and reliably across a network, service providers use various techniques to establish a *circuit-switched lightpath* between communicating end equipment. An OXC is a key element to set up express paths through intermediate nodes for this process. Because an OXC is a large complex switch, it is used in extended mesh backbone networks, where there is a heavy volume of traffic between nodes, to connect equipment such as SONET/SDH terminals, IP routers, and ROADMs. In such a network the lightpath normally is set up for long periods of time. Depending on the desired service running between distant nodes, this time connection can range from minutes to months and even longer.

Lightpaths running from a source node to a destination node may traverse many fiber link segments along the route. At intermediate points along the connection route, the lightpaths may be switched between different links, and sometimes the lightpath wavelength may need to be changed when entering another link segment. As noted in Sect. 13.6.2, this wavelength conversion is necessary if two lightpaths entering some segment happen to have the same wavelength.

The process of establishing a lightpath is called by various names, such as wavelength routing, optical circuit switching, or lightpath switching. The more popular terms are *wavelength routing* and *wavelength routed network* (WRN). Many different static and dynamic approaches have been proposed and implemented for establishing a lightpath. Because a method for setting up a lightpath requires deciding which path to traverse and what wavelength to use, it involves a *routing and wavelength assignment* (RWA) procedure. In general the RWA problem is fairly complex and special software algorithms have been developed for solving it [33–35].

13.6.4   Optical Packet Switching

The success of electronic packet-switched networks lies in their ability to achieve reliable high packet throughputs and to adapt easily to traffic congestion and transmission link or node failures. Various studies have been undertaken to extend this capability to all-optical networks in which no *O/E/O* conversion takes place along a lightpath. In the concept of an *optical packet-switched* (OPS) network, user traffic is routed and transmitted through the network in the form of optical packets along with in-band control information that is contained in a specially formatted header or label. For OPS systems the header processing and routing functions are carried out electronically, and the switching of the optical payloads is done in the optical domain for each individual packet. This decoupling between header or label processing and payload switching allows the packets to be routed independent of payload bit rate, coding format, and packet length.

Optical label swapping (OLS) is a technique for realizing a practical OPS implementation [36–38]. In this procedure, optically formatted packets (which contain a standard IP header and an information payload as shown in Fig. 13.32) first have an optical label or control packet attached to them before they enter the OPS network. Note that in some OPS schemes the wavelength used to transmit the label may be different from that used by the packet. When the payload-plus-label packet travels through an OPS network, the optical packet switches at intermediate nodes process only the optical label electronically. The offset time shown in Fig. 13.32 is needed to allow time for the optical packet switch to be set after the optical label is processed. This is done to extract routing information for the packet and to determine other factors such as the wavelength on which the packet is being transmitted and the bit rate of the encapsulated payload. Because the payload remains in an optical format as it moves through the network, it may use any modulation scheme and may be encoded at a very high bit rate.

The limitation of an OPS network is related to the technology for creating practical optical buffers. Similar to other switching methodologies, these buffers are needed to store the optical packets temporarily during the time it takes to set up an output path through an intermediate optical packet switch and to resolve any port contentions that may arise between two or more incoming packets destined for the same output port. This technology limitation can be circumvented through the optical burst-switching concept described in the next section.

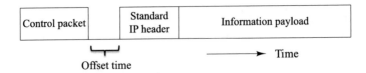

Fig. 13.32 Optically formatted packet used for optical label swapping

13.6.5   Optical Burst Switching

Optical burst switching (OBS) was conceived to provide an efficient solution for sending high-speed bursty traffic over WDM networks [39–41]. Traffic is considered as being *bursty* if there are long idle times between the busy periods in which a large number of packets arrive from users. This format is typical of data traffic in contrast to voice traffic, which is characterized by a more continuous bit stream nature. There are two advantages to OBS. First, it offers the high bandwidth and packet-sized granularity of optical packet-switched networks without the need for complex optical buffering. Second, it provides the low packet-processing overhead that is characteristic of wavelength-routed networks. Thus, the performance characteristics of OBS lie between those of a wavelength-routed network and an optical packet-switched network.

In the OBS network concept, a collection of optical burst switches are interconnected with WDM links to form the central core of the network. Devices called *edge routers* collect traffic flows from various sources at the periphery (edge) of a WDM network, as illustrated in Fig. 13.33. The flows then are sorted into different classes according to their destination address and grouped into variable-sized elementary switching units called *bursts*. The characteristics of an edge router play a critical role in an OBS system, because the overall network performance depends on how a burst is assembled based on particular types of traffic statistics.

Before a burst is transmitted, the edge router generates a *control packet* and sends it to the destination to set up a lightpath for this burst. As the control packet travels toward the destination, each OBS along the lightpath reads the burst size and arrival time from the control packet. Then, in advance of the burst arrival, the burst switch schedules an appropriate time period on a wavelength that the next lightpath

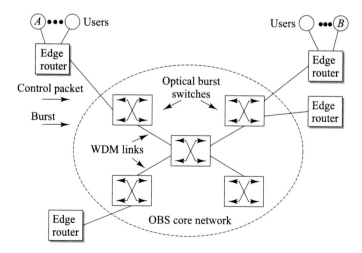

Fig. 13.33 Generic structure of an optical burst switching (OBS) network

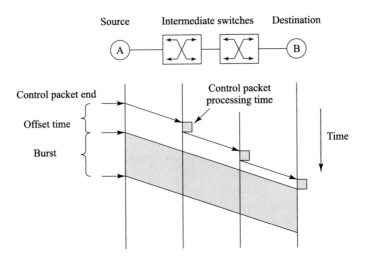

Fig. 13.34 Timing diagram for the progression of a control packet and a burst through an OBS network

segment will use to carry the burst. This reservation for the arriving burst is called *burst scheduling*.

Figure 13.34 shows an example of a *timing diagram* for the connection setup through two intermediate burst switches from source node A to destination node B. This figure shows that following the transmission of the control packet, the burst itself is sent after a specified delay called an *offset time*. The time it takes to set up a connection depends on the end-to-end propagation time of the control packet, the sum of all the processing delays t_{proc} of the control packet at intermediate nodes, and the time t_{conf} it takes to configure the link. Once the link is configured, the travel time for the burst to reach the destination node is equal to the propagation time, because no further processing is required for the burst at any intermediate node. Therefore, the start of the burst has to be delayed by an offset time following transmission of the control packet in order to give the link enough time to be configured. If there are N intermediate nodes in the link, then the offset time must be at least $Nt_{proc} + t_{conf}$. Note that the edge router sends the burst immediately after expiration of the offset time without waiting for an acknowledgment from the destination to indicate that the link is complete. Furthermore, the burst transmission may start before the control packet reaches its destination.

Solutions for burst-assembly algorithms include formatting a burst based on a fixed assembly time, a fixed burst length, or a hybrid time-length/burst-length method. The main parameters are a maximum assembly time threshold T, a maximum burst length B, and a minimum burst length B_{min}. When the incoming packet flow is low, the threshold T time ensures that a packet is not delayed too long in the assembly queue. When the incoming packet flow is high, the upper limit B on the burst size also limits packet delay times by restricting the time needed to assemble a burst. Otherwise, if the burst size is not limited, long assembly times could result during heavy incoming traffic loads. If the assembled burst length is smaller than B_{min} at the scheduled burst transmission time, then padding bits are added to bring the length to B_{min}.

13.6.6   Elastic Optical Networks

The historical development of sophisticated dense WDM networks saw the creation and specification of fixed spectral grids to accommodate a wide variety of data rates and modulation formats. In the development process, the ITU-T standardization of these grids specified increasingly narrower spectral widths moving from 200 to 100 to 50 and then to 25 GHz. These spectral widths correspond to 1.6, 0.8, 0.4, and 0.2 nm, respectively, in the 1550-nm wavelength window. Typical data rates that can be supported by a 50-GHz grid range from 10 Gb/s to 0.8 Tb/s. As Fig. 13.35a shows, wavelength channels carrying data rates of 10, 40, and 100 Gb/s fit into 50-GHz fixed grid slots. Higher rates such as 400 Gb/s and 1 Tb/s have to occupy two and three 50-GHz slots, respectively. Thus, the fixed grid scheme causes wastage of bandwidth, especially at low bit rates.

This limitation of the fixed grid scheme can be alleviated by means of an *elastic optical network* (EON) grid, which has a flexible bandwidth and an adaptive channel spacing that changes dynamically according to the bandwidth requirement of the transmitted client signals [42–45]. In an EON the spectrum is divided into narrow slots with much finer granularity than the fixed ITU-T grids and optical connections are allocated a different number of slots. Figure 13.35b shows the spectrum saving when using the flexible grid scheme.

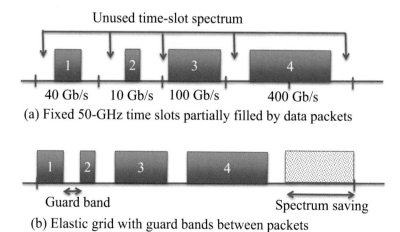

Fig. 13.35 Illustration of spectrum saving when using an elastic optical network scheme instead of having fixed transmission time slots

13.7 WDM Network Implementations

This section presents some WDM network implementation examples.

13.7.1 Long-Distance WDM Networks

Long-distance or long-haul DWDM networks consist of a collection of ROADMs, switches, and optical crossconnects interconnected with point-to-point high-capacity trunk lines. This collection of equipment can be configured through any combination of ring or mesh networks, as Fig. 13.36 shows.

Each long-haul trunk cable contains a large number of single-mode fibers, as Fig. 13.37 illustrates. The cables often are based on fiber-ribbon configurations, which are available with up to 864 optical fibers housed in a cable structure that is less than one inch (24.4 mm) in diameter (see Fig. 2.30 for a cable example). The individual fibers each can transport many closely spaced wavelengths simultaneously, and each of these independent lightwave channels can support multigigabit data rates. For example, depending on the service needs, a single fiber in a standard long-haul link can carry up to 160 channels of 2.5, 10, or 40-Gb/s of traffic using 160 individual wavelengths. In addition, transmission systems capable of sending data rates of 400 Gb/s and 1 Tb/s per wavelength are being installed worldwide.

A long-haul trunk line also can have various ROADMs of degree-N for inserting and extracting traffic at remote intermediate points. Standard transmission distances

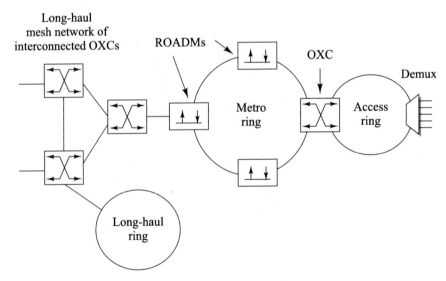

Fig. 13.36 Wideband DWDM networks can be configured through any combination of ring or mesh networks

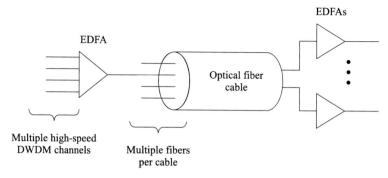

Fig. 13.37 Long-haul trunk cables contain a large number of high-capacity single-mode fibers

in long-haul terrestrial DWDM links are 600 km with an 80-km nominal spacing between optical amplifiers.

If the separations in 160 DWDM channels are 50 GHz, then the required frequency span is 8 THz (8000 GHz), which is equivalent to a wavelength spectral band of about 65 nm. This requires operation over both the C-band and either the S-band or L-band simultaneously. Consequently the active and passive components intended for long-haul applications must meet high performance requirements, such as the following:

- The optical amplifiers must operate over a wide spectral band;
- The optical amplifiers need high-power pump lasers to amplify a large number of channels;
- Each wavelength has to exit an optical amplifier with the same power level to prevent an increasing skew in power levels from one wavelength to another as the signals pass through successive amplifiers;
- Strict temperature stabilization and optical frequency controls are required of laser transmitters to prevent crosstalk between lightwave channels;
- High-rate transmission over long distances requires optical signal-conditioning techniques such as chromatic-dispersion compensation, polarization-mode dispersion compensation, and Rayleigh backscattered noise elimination [46–49].

A typical erbium-doped fiber amplifier (EDFA) is limited to the 1530-to-1560-nm C-band in which a standard erbium-doped fiber has a high gain response. By adding a Raman amplification mechanism, the gain response can be extended into both the S-band and the L-band. Figure 13.38 illustrates one amplification concept for operation in both the C-band and S-band. Here a multiple-wavelength distributed Raman amplifier pump unit is added ahead of a band-splitting device, which separates the S-band and
C-band. The Raman amplification boosts the power levels in both the S-band and the C-band. After passing through the band splitter, the gains of the S-band wavelengths traveling in the bottom path can be enhanced further with an S-band amplifier (e.g., by the TDFA described in Chap. 11). An EDFA boosts the power levels of the C-band

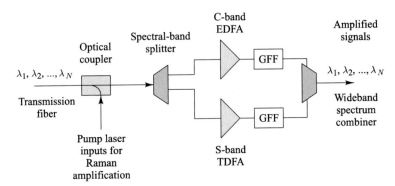

Fig. 13.38 An optical amplification concept for combined operation in the C-band and S-band

wavelengths in the top path in Fig. 13.38. Following the amplification processes, all the wavelengths are recombined with a wideband multiplexing unit. The gain-flattening filters (GFFs) that are used for equalizing the final output powers of all the wavelengths can be either passive or active devices.

A network operator can use remotely controllable degree-4 ROADMs to reconfigure the number of added or dropped wavelengths in different parts of the network when necessary. This function can supply another wavelength to a node for additional capacity or can shut down specific wavelength connections that no longer need service. The degree-4 OXCs can set up new routes for specific wavelengths, can close discontinued connections, or can reroute high-capacity lightpaths in case of link congestion or faults.

13.7.2 Metro WDM Networks

Nominally a *metro network* is configured as a mesh network or a SONET/SDH bidirectional, line-switched ring (BLSR), as Fig. 13.10 illustrates. A ring consists of 2.5-Gb/s or 10-Gb/s (OC-48/STM-16 or OC-192/STM-64) point-to-point connections between central offices that are spaced 10–20 km apart. A metro network may comprise from three to eight nodes and the ring circumference nominally is less than 80 km. At switching nodes within the metro ring, the degree-2 ROADMs allow multiple selected wavelengths to be extracted and inserted at the metro central offices. A degree-2 ROADM or a degree-4 OXC provides interconnections to a long-haul network. Within the metro central offices, the SONET/SDH equipment has STS-1 grooming capabilities. The term *grooming* means that low-speed SONET/SDH rates, such as 51.84-Mb/s STS-1 or 155.52-Mb/s STM-1 rates, are packaged into higher-rate circuits, such as 2.5-Gb/s or 10-Gb/s.

Attached to a metro network is the *access network*, which consists of links between end users and a central office. The access network could be either a ring or a star-based passive optical network or PON (see Sect. 13.8). An *access ring* configuration ranges from 10 to 40 km in circumference and typically contains three or four nodes, whereas a maximum of 32 users located up to 20 km away can be attached to a single PON. Within the access network ring, either a fixed OADM or a ROADM provides the capability to add or drop multiple wavelengths to local users or to other regional networks. A degree-4 OXC provides the connection from the access ring to a metro network.

In contrast to the stringent performance specifications imposed on wideband long-haul DWDM systems, the shorter spans in metro and access networks relax some of the requirements. In particular, if CWDM technology is employed, the 20-nm spectral band tolerances allow the use of optoelectronic devices that are not temperature-controlled. However, other requirements unique to metro applications arise, such as the following:

- A *high degree of connectivity* is required to support meshed traffic in which various wavelengths are inserted and extracted at different points along the path;
- A *modular* and *flexible switching device* such as a ROADM is needed because the wavelength add/drop patterns and link capacities vary dynamically with various levels of new and completed service demands from different nodes;
- Because the add/drop functions can change dynamically from node to node, *special components*, such as a *variable optical attenuator* (VOA), are needed to equalize the power levels of newly added wavelengths with those that already are on the fiber;
- *Optical amplifiers* optimized for metro network use are needed, because interconnection losses can be fairly high and the power level of express wavelengths can change as the light signal passes through successive nodes.

13.7.3 Data Center Networks

The tremendous worldwide growth of dynamic data communications and cloud-based services created a need to move the information products of computation intensive functions from localized devices such as personal computers, cell phones, security cameras, and environmental monitors to large centralized storage and computer clusters. These facilities are known as *data centers* (also spelled "*datacenters*") and can contain thousands of servers, storage units, and network equipment that need to be interconnected inside the data centers and also must be linked to metro and wide area networks. The interconnections form a *data center network* (DCN) [50–55].

Figure 13.39 presents a high level snapshot of a traditional DCN architecture, which consists of a three-layer tree switching topology. A typical data center contains a large number of equipment racks that each generally hold up to 48 vertically stacked servers. These servers are all interconnected through an edge switch that is mounted at the top of the rack and hence is referred to as Top-of-Rack (ToR) switch. The ToR switch is called an edge switch because it collects traffic flows from various sources at the periphery of the DCN. As such, the ToR switches function as the access point to other racks in the data center. At the next higher level numerous

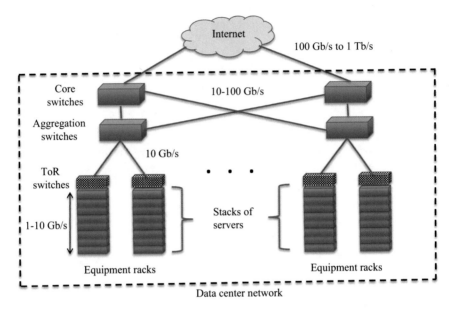

Fig. 13.39 Configuration of a traditional data center network

ToR switches interface to aggregation switches, which in turn are interconnected to outside networks through core switches.

As new transmission and switching technologies and methodologies are emerging, the simple three-layer model shown in Fig. 13.39 continues to evolve. This entails the replacement of legacy electronic switches in the data center with optical switching technologies to provide larger transmission capacities and faster data rates between the servers and the external networks. Typical data rates vary from 1 to 10 Gb/s for interconnections between servers within a rack, from 10 to 100 Gb/s in the aggregation layer, and from 100 Gb/s to 1 Tb/s in the core layer.

13.8 Passive Optical Networks

A *passive optical network* (PON) is based on using multiple wavelengths and bidirectional transmission on a single optical fiber [56–61]. In a PON there are no active components between the central office and the customer premises. Instead, only passive optical components are placed in the network transmission path to guide the traffic signals contained within specific optical wavelengths to the user endpoints and back to the central office.

13.8.1 Basic Architectures for PONs

Figure 13.40 illustrates the architecture of a basic traditional PON in which a fiber optic network connects switching equipment in a central office with a number of service subscribers. In the central office, data and digitized voice are combined and sent downstream to customers over an optical link by using a 1490-nm wavelength. The upstream (customer to central office) return path for the data and voice uses a 1310-nm wavelength on the same fiber. Video services are sent downstream using a 1550-nm wavelength. There is no video service in the upstream direction. All users attached to a particular PON share the same 1310-nm upstream wavelength. This requires carefully timed assignments of transmission permits to the users, so that upstream traffic streams from different users do not interfere with each other. Note that whereas the original PON implementation used fixed wavelengths, later versions specified spectral bands around the three key wavelengths.

Starting at the central office, one single-mode optical fiber strand runs to a passive *optical power splitter* near a housing complex, an office park, or some other campus environment. At this point the passive splitting device simply divides the optical power into N separate paths to the subscribers. If the splitter is designed to divide the incident optical power evenly and if P is the optical power entering the splitter, then the power level going to each subscriber is P/N. Designs of power dividers with nonuniform splitting ratios are also possible depending on the application. The number of splitting paths can vary from 2 to 64, but in a basic PON they typically are 8, 16, or 32. From the optical splitter, individual single-mode fibers run to each building or serving equipment. The optical fiber transmission span from the central

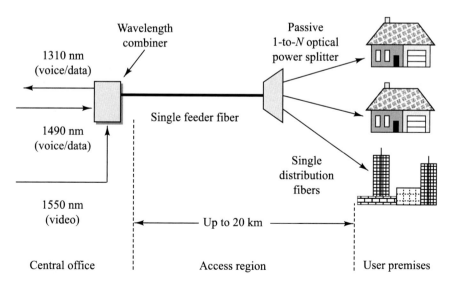

Fig. 13.40 Architecture of a typical passive optical network

office to the user can be up to 20 km. Thus active devices exist only in the central office and at the user end.

Several alternative PON implementation schemes have been used since the passive network concept inception [61–65]. The main ones are as follows:

- *Broadband* PON (BPON) is based on the G.983.1 to G.983.5 series of ITU-T Recommendations that specified the asynchronous transport mode (ATM) technique as the transport and signaling protocol. This legacy PON has been phased out due to its relatively high cost compared to Ethernet and its incompatibility with current IP over DWDM technology.
- *Gigabit* PON (GPON) was the next evolution of the BPON standard. GPON provides downstream speeds of 2.5 Gb/s and upstream speeds of 1.25 Gb/s. It is based on the G.984.1 to G.984.6 series of ITU-T Recommendations. Typically 32 users share the bandwidth. A drawback of GPON is that it uses an inefficient and complicated data encapsulation method that is time-consuming and adds cost to the terminal equipment in both the central office and at the user locations.
- *Ethernet* PON (EPON) or *gigabit Ethernet* PON (GE-PON) uses 1-Gb/s Ethernet as the underlying protocol. The EPON operation is described in the IEEE 802.3 EPON standard (several versions of 802.3 have been ratified by IEEE during its evolution depending on the latest data rate capabilities). Because Ethernet devices are used universally from the home or business all the way through worldwide backbone networks, EPON provides efficient connectivity for any type of IP-based traffic. Thus EPON can be much more cost-effective compared to GPON.
- Moving beyond the capabilities of EPON are the 10-Gb/s PON technologies. The two main versions are called XG-PON (or 10G-PON) and XGS-PON. The XG-PON version specifies an upstream rate of 2.5 Gb/s and a downstream rate of 10 Gb/s. XGS-PON is a symmetric version that transmits at 10 Gb/s in both the upstream and downstream directions. These versions are based on the IEEE 802.3av EPON standard and on the ITU-T Recommendation G.987 series for sending both upstream and downstream traffic at 10 Gb/s. The 10 Gigabit PON wavelengths are 1577 nm for downstream and 1270 nm for upstream transmissions. These differ from the 1490-nm downstream/1310-nm upstream wavelengths of GPON and EPON, thereby allowing 10G-PON systems to coexist on the same fiber with either of the earlier PON versions.
- A further extension of PON capabilities is the NG-PON2 (Next-Generation PON2), which operates at 40 Gb/s in both the upstream and downstream directions (IEEE 802.3ca). To enable simultaneous implementation on the same fiber as all earlier generation PONs, the NG-PON2 wavelengths are 1524 to 1540 nm for upstream transmissions and 1596–1603 nm for downstream traffic.
- *WDM* PON uses a different wavelength for each user to greatly enhance network capacity. In this architecture a wavelength multiplexer (usually an AWG) is used in place of the power splitter shown in Fig. 13.40. Its flexible service offering is a major advantage of a WDM PON compared to other PON types. Because each user has a dedicated wavelength that is not shared with others, a customer with very high bandwidth demands can easily be

Table 13.7 A selection of acronyms for FTTx

Acronym	Meaning
FTTB	*Fiber-to-the-building* or *Fiber-to-the-business* The optical fiber line reaches the boundary of a building
FTTC	*Fiber-to-the-curb/kerb* or *Fiber-to-the-cabinet* The street cabinet is within 100 m of the user premises
FTTH	*Fiber-to-the-home* The optical fiber reaches the outside boundary of the living space
FTTN	*Fiber-to-the-node* or *Fiber-to-the-neighborhood* The fiber ends in a street cabinet that can be several km from the user
FTTP	*Fiber-to-the-premises* This is a generic term covering both FTTB and FTTH

accommodated without affecting other lower-usage customers. The dedicated wavelength also provides for a higher level of information security compared to other PONs where users share wavelengths.

The application of PON technology for providing broadband connectivity in the access network to homes, multiple-occupancy units, and small businesses commonly is called *fiber-to-the-x* (FTTx) [56, 66, 67]. Here *x* is a letter indicating how close the fiber endpoint comes to the actual user. Table 13.7 lists the acronyms for various FTTx concepts. FTTP for *fiber-to-the-premises* has become the prevailing term that encompasses the various FTTx concepts. FTTP networks can use any of the various PON technologies.

13.8.2 Active PON Modules

This section gives a snapshot of the basic functions and compositions of the optoelectronic equipment located in the central office and at or near the endpoint users.

Optical Line Termination (OLT)

The *optical line termination* (OLT) equipment is located in a central office and controls the bidirectional flow of information across the network. An OLT must be able to support transmission distances of up to 20 km. In the downstream direction the function of an OLT is to take in voice, data, and video traffic from a long-distance or a metro network and broadcast it to all the users on the PON. In the reverse direction (upstream), an OLT accepts and distributes multiple types of voice and data traffic from the network users.

A typical OLT is designed to control more than one PON. Figure 13.41 gives an example of an OLT that is capable of serving four independent passive optical networks. In this case, if there are 32 connections to each PON, then the OLT can distribute information to 128 users.

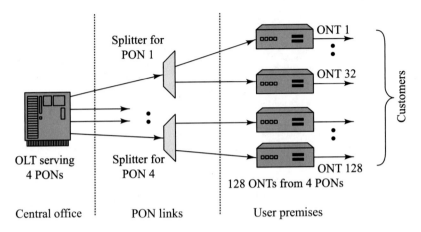

Fig. 13.41 An OLT that is capable of serving four independent passive optical networks

Simultaneous transmission of separate service types on the same PON fiber is possible by using different wavelengths for each direction. As described in Sect. 13.8.1, various generations of PON technology use different spectral bands for bidirectional transmissions, so that the various PON technologies can be used simultaneously on the same fiber. Depending on the particular PON standard being used, the downstream and upstream transmission equipment operates at rates ranging from 1 to 40 Gb/s. In some cases the transmission rates are the same in either direction (a *symmetric* network). In other PON standards the downstream rate may be higher than the upstream rate, which is called an *asymmetric* implementation. A number of different transmission formats can be used for the downstream video transmission at 1550 nm.

Optical Network Termination (ONT)

The *optical network termination* (ONT) equipment is located directly at the customer premises. The purpose of an ONT is to provide an optical connection to the PON on the upstream side and to interface electrically to the local customer equipment. Depending on the communication requirements of the customer or block of users, the ONT typically supports a mix of telecommunication services, including various Ethernet rates, SONET/SDH connections, and video formats.

A wide variety of ONT functional designs and chassis configurations are available to accommodate the needs of different levels of demand. The size of an ONT can range from a simple box that may be attached to the outside of a house to a fairly sophisticated unit mounted in a standard indoor electronics rack for use in large apartment complexes or office buildings. At the high-performance end, an ONT can aggregate, groom, and transport various types of information traffic coming from the user site and send it upstream over a single-fiber PON infrastructure. The term *grooming* means that the switching equipment looks inside of a time-division multiplexed data stream, identifies the destinations of the individual multiplexed channels, and then reorganizes the channels so that they can be delivered efficiently to their destinations. In conjunction with the OLT, an ONT allows dynamic bandwidth allocation based on varying demands of users to enable smooth delivery of data traffic that typically arrives in bursts from the users.

Optical Network Unit (ONU)

An *optical network unit* (ONU) is similar to an ONT but normally is housed in an outdoor equipment shelter near the user premises. These installations would include shelters located at a curb or in a centralized place within an office park. Thus the ONU equipment must be environmentally rugged to withstand wide temperature variations. The shelter for the outdoor ONU must be water-resistant, vandal-proof, and able to withstand high winds. In addition, there has to be a local power source to run the equipment, together with emergency battery power backup. The link from the ONU to the customer premises can be a twisted-pair copper wire, a coaxial cable, an independent optical fiber link, or a wireless connection.

13.8.3   Controlling PON Traffic Flows

Two key network functions of an OLT are to control user traffic and to assign bandwidth dynamically to the ONT modules. Up to 32 ONTs use the same wavelength and share a common optical fiber transmission line, so some type of transmission synchronization must be used to avoid collisions between traffic coming from different ONTs. The simplest method is to use *time-division multiple access* (TDMA) wherein each user transmits information within a specific assigned time slot at a prearranged data rate. However, this does not make efficient use of the available bandwidth because many time slots will be empty if a number of network users do not have information to be sent back to the central office.

A more efficient process is *dynamic-bandwidth allocation* (DBA), wherein time slots of an idle or low-utilization user are assigned to a more active customer. The exact DBA scheme implemented through an OLT in a particular network depends on factors such as user priorities, the quality of service guaranteed to specific customers, the desired response time for bandwidth allocation, and the amount of bandwidth requested (and paid for) by a customer.

As shown in Fig. 13.42, the OLT uses *time-division multiplexing* (TDM) to combine incoming voice and data streams that are destined for users on the PON. As a simple example of this, if there are N independent information streams coming into the OLT, each of which is running at a data rate of R b/s, then the TDM scheme interleaves them electrically into a single information stream operating at a higher rate of $N \times R$ b/s. The resulting multiplexed downstream signal is broadcast to all the ONTs. Each ONT discards or accepts the incoming information packets depending on the packet header address. Encryption may be necessary to maintain privacy because the downstream signal is broadcast, and every ONT receives all the information destined for each end terminal.

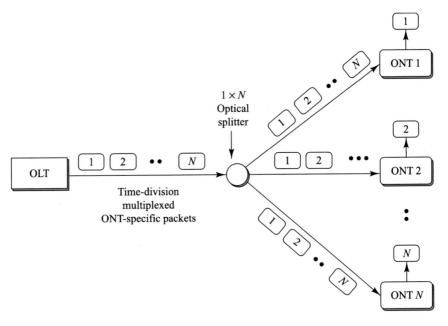

Fig. 13.42 Operation of a downstream time-division multiplexing process

Sending traffic in the upstream direction is more complicated because all users have to time-share the same wavelength. To avoid collisions between the transmissions of different users, the system uses a *time-division multiple access* (TDMA) protocol. Figure 13.43 gives a simple example of this. The OLT controls and coordinates the traffic from each ONT by sending permissions to them to transmit during a specific time slot. The time slots must be synchronized because transit times vary between ONTs (see Example 13.5).

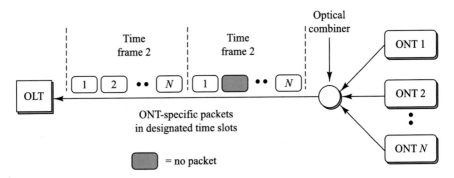

Fig. 13.43 Operation of an upstream time-division multiple access protocol

Example 13.5 Consider two ONTs located 3 and 20 km from the OLT, respectively. For these two links, what are the roundtrip propagation times for messages coming from and returning to the OLT? Let the speed of light in the fiber be 2×10^8 m/s.

Solution: The roundtrip propagation times are

$$t(3\,\text{km}) = 2(3\,\text{km})/(2 \times 10^8\,\text{m/s}) = 30\,\mu\text{s}$$
$$t(20\,\text{km}) = 2(20\,\text{km})/(2 \times 10^8\,\text{m/s}) = 200\,\mu\text{s}$$

Drill Problem 13.5 Because the ranging procedure in a PON has a limited accuracy, a guard time is placed between consecutive bursts to avoid collisions of the independent packets, as Fig. 7.18 illustrates. Verify that a 25.6-ns guard time consumes 64 bits at a 2.488-Gb/s data rate.

13.8.4 Protection Switching for PON Configurations

The ITU-T Recommendation G.984.1 describes the use of a protection switching mechanism. This allows several different types of PON configurations that include redundancy of links and equipment for network protection. Among these are a fully redundant $1 + 1$ protection and a partially redundant $1{:}N$ protection. Figure 13.44 shows that in $1 + 1$ protection the traffic is transmitted simultaneously over two separate fiber lines from the source to the destination. Typically these two paths do not overlap at any point, so that a cable cut would affect only one fiber transmission path. In the $1 + 1$ protection scheme, the receiving equipment selects one of the links as the *working fiber* for reception of information. If a fiber in that link is cut or the transmission equipment on that link fails, then the receiver switches over to the *protection fiber* and continues to receive information. This protection method provides rapid switchover during failures and does not require a protection signaling protocol between the source and destination. However, it requires duplicate fibers and redundant transmission equipment for each link.

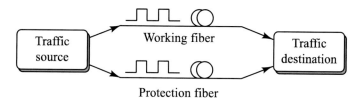

Fig. 13.44 Fully redundant $1 + 1$ protection of links

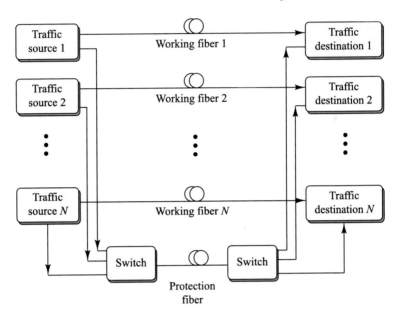

Fig. 13.45 The 1:N protection procedure

The 1:N protection procedure offers a more economical use of fibers and equipment. As shown in Fig. 13.45, here one protection fiber is shared among N working fibers. This arrangement offers protection in the event that one of the working fibers fails. For most operational networks this level of protection is adequate as failures of multiple fibers are rare (unless all the fibers are in the same cable). In contrast to the 1 + 1 protection method, in the 1:N protection scheme traffic is transmitted only over the working fiber in normal operation. When there is a failure on a particular link, the source and destination both switch over to the protection fiber. This requires an automatic switching protocol between the endpoints to enable use of the protection link.

13.8.5   WDM PON Architectures

The ever-increasing demand for higher-capacity services can quickly lead to demands well in excess of 100 Mb/s per subscriber. Because a standard three-wavelength FTTP network will not be able to satisfy such demands, an enhancement is to use more wavelengths to create a WDM PON. This method uses a separate wavelength for each transmitting ONT, so that an ONT can send its information continuously over the shared upstream fiber without having to wait for a specific assigned transmit time slot [58, 65–70].

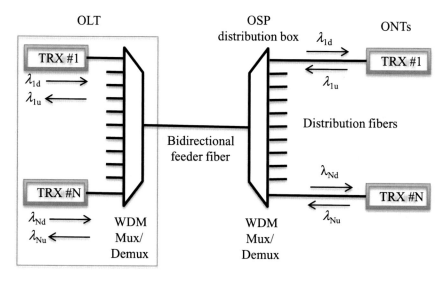

Fig. 13.46 The basic WDM PON architecture

The basic WDM PON architecture uses a WDM mux/demux device at both the central office and at the outside plant (OSP) distribution box, as shown in Fig. 13.46. In the central office N laser transmitters send independent wavelengths to a WDM multiplexer, such as an AWG. The downstream wavelengths are designated by λ_{1d} through λ_{Nd}. The WDM mux combines these wavelengths into a single traffic stream to be sent over a single fiber in the downstream direction. The OSP distribution cabinet, which normally houses the optical power splitter in a standard PON (see Fig. 13.40), now contains a WDM demultiplexer, again such as an AWG. When the downstream wavelengths arrive at this OSP cabinet, the demultiplexer separates these N wavelengths into individual fibers with a unique wavelength going to each ONT.

In the upstream direction, each ONT sends out a unique wavelength in order not to interfere with the traffic from other ONTs. The N upstream wavelengths, designated by λ_{1u} through λ_{Nu}, are combined onto the feeder fiber with the WDM multiplexer located in the OSP distribution cabinet. Upon arriving at the OLT in the central office, a WDM demultiplexer separates these N wavelengths into individual OLT optical receivers.

Its service-offering flexibility is a major advantage of a WDM PON compared to other PON types. Because each user has a dedicated wavelength that is not shared with others, a customer with very high bandwidth demands can easily be accommodated without affecting other lower-usage customers. The dedicated wavelength also provides for a higher level of information security compared to other PON types.

In such a WDM PON it is desirable to have a *colorless* ONT, which means that no ONT should have a fixed transmission wavelength assigned to it. An obvious but extremely expensive solution is to use a tunable laser at each ONT, but low-cost end equipment is a driving factor in PON implementations, so this is not a viable solution. Thus a major challenge for a WDM PON implementation is to have a low-cost, high-output optical source at each ONT.

One method is to use spectral splicing of a single broadband relatively inexpensive light source. Various techniques are being explored to achieve this. One idea is to have each ONT contain a source with a broad optical output spectrum within the transmitter, such as a *superluminescent light emitting diode* (SLED). The broad spectral output of the source at the ONT is connected to one port on a local WDM device, such as a thin-film filter or an AWG. Only those optical spectral components from the SLED that can pass through the WDM channel are transmitted through to the central office. Although all the ONTs have identical SLEDs, each user is connected to a different port on the local WDM device, so it is possible to slice a different part of the available optical spectrum for each ONT. This scheme thereby provides each ONT with a different transmitting wavelength.

Another concept is to use a broadband source (for example, a superluminescent laser diode or a broadband EDFA source) at the central office and send the source output downstream through an AWG. Unique spectrally sliced wavelengths travel to individual ONTs and seed a relatively inexpensive source at each ONT, such as a Fabry-Perot (FP) laser diode. The seeding action forces the FP laser to operate in a quasi single mode, which is one of the oscillating modes shown in Fig. 4.23. Because a different seed wavelength arrives at each ONT, the FP laser located at that ONT can transmit data upstream at its uniquely locked wavelength.

13.9 Summary

Various types of optical networks have been conceived, designed, and built to satisfy a wide range of transmission capacities and speeds. The link lengths between users can vary from short localized connections within a building or a campus environment to networks that span continents and run across oceans. A major motivation for developing sophisticated communication networks has been the rapid proliferation of information exchange desired by institutions involved in fields such as commerce, finance, education, scientific and medical research, health care, national and international security, and entertainment.

The main network topologies for optical networks are the linear bus, ring, star, and point-to-point mesh configurations. Each configuration has its own particular advantages and limitations in terms of reliability, expandability, and performance characteristics. Some of these configurations use basic point-to-point links, which are simple connections between individual users or between a pair of multiplexing equipment.

Linear bus networks using passive couplers typically are limited to connections between several nodes because of the rapid accumulation of optical coupler losses from node to node along the bus.

In a ring topology, consecutive nodes are connected by point-to-point links that are arranged to form a single closed path. Information in the form of data packets is transmitted from node to node around the ring. The interface at each node is an active device that has the ability to recognize its own address in a data packet in order to accept messages. The node forwards those messages that do not contain their own address on to their next neighbor.

In a star architecture, all nodes are joined at a single point called the central node or hub. The hub can be an active or a passive device. Using an active hub, all routing of messages in the network is controlled from the central node. This is useful when most of the communications are between the central and the outlying nodes, as opposed to information exchange between the attached stations. In a star network with a passive central node, a power splitter is used at the hub to divide the incoming optical signals among all the outgoing lines to the attached stations. Star topologies are widely used in passive optical networks (PONs) and in active Ethernet networks.

In a mesh network, point-to-point links connect the nodes in an arbitrary fashion that can vary greatly from one application to another. This topology allows significant network configuration flexibility and offers connection protection in case there are multiple link or node failures. Link protection is accomplished by means of a mechanism that first determines where a failure has occurred and then restores the interrupted services by redirecting the traffic from a failed link or node to travel over another link in the mesh.

Problems

13.1   Consider an N-node star network in which 0 dBm of optical power is coupled from any given transmitter into the star. Let the fiber loss be 0.3 dB/km. Assume the stations are located 2 km from the star, the receiver sensitivity is -38 dBm, each connector has a 1-dB loss, the excess loss in the star coupler is 3 dB, and the
link margin is 3 dB.

   (a)   Show that the maximum number of stations
          that can be incorporated on this network is $N = 380$.
   (b)   Show that a maximum number of $N = 95$ stations can be attached if
          the
          receiver sensitivity is -32 dBm.

13.2   A two-story office building has two 10-feet-wide hallways per floor that connect four rows of offices with eight offices per row as is shown in Fig. 13.47. Each office is a 15 feet $\times$ 15 feet square. The office ceiling height is 9 feet with a false ceiling hung 1 foot below the actual ceiling. Also, as shown in Fig. 13.47, there is a wiring room for LAN interconnection and control equipment in one corner of each floor. Every office has a local-area network socket on each of the two walls that are perpendicular to the hallway wall. If the cables can be run only in the walls and in the ceilings, estimate the length of cable (in feet) that is required for the following configurations:

Fig. 13.47 Figure 13.47 for
Problem 13.2

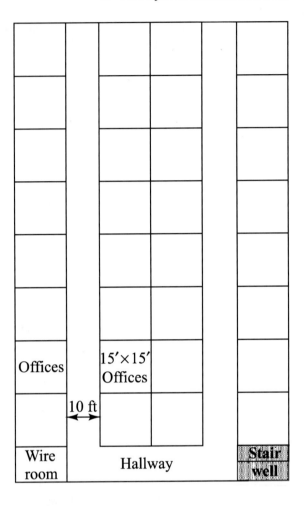

(a)   A coaxial cable bus with a twisted-pair wire drop from the ceiling to
each outlet.

(b)   A fiber optic star that connects each outlet to the wiring room on the
corresponding floor and a vertical fiber optic riser that connects the
stars in each wiring room.
[Answer: Cable/floor = 2916 ft. Total cable in the building = 5850
ft.]

13.3   Consider the $M \times N$ grid of stations shown in Fig. 13.48 that are to be
connected by a local-area network. Let the stations be spaced a distance
d apart and assume that interconnection cables will be run in ducts that
connect nearest-neighbor stations (i.e., ducts are not run diagonally in
Fig. 13.48). Show that for the following configurations, the cable length
for interconnecting the stations is as stated:

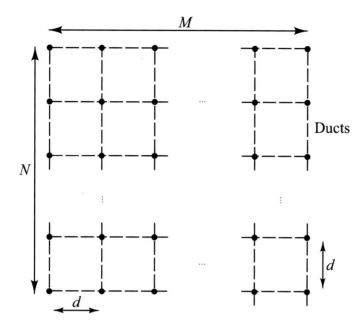

Fig. 13.48   Figure 13.48 for Problems 13.3 and 13.4

(a) $(MN - 1)\, d$ for a bus configuration.

(b) MNd for a ring topology.

(c) $MN(M + N - 2)d/2$ for a star topology where each subscriber is connected *individually* to the network hub located *in one corner* of the grid.

13.4   Consider the $M \times N$ rectangular grid of computer stations shown in Fig. 13.48, where the spacing between stations is d. Assume these stations are to be connected by a star-configured LAN using the duct network shown in the figure. Furthermore, assume that each station is connected to the central star by means of its own dedicated cable.

(a)   If m and n denote the relative position of the star, show that the total cable length L needed to connect the stations is given by

$$L = [MN(M + N + 2)/2 - Nm \times (M - m + 1) - Mn(N - n + 1)]d$$

(b)   Show that if the star is located in one corner of the grid, then this expression becomes

$$L = M N(M + N - 2)\, d/2$$

(c)   Show that the shortest cable length is obtained when the star is at the center of the grid.

13.5   In this problem do not resort to wavelength conversion but assume that wavelengths can be reused in different parts of a network. Show that the minimum number of wavelengths required to connect N nodes in a WDM network are as follows:

(a) $N - 1$ for a star network
(b) $N(N - 1)/2$ for a ring network

13.6   (a) Show that the number of 64-kb/s voice channels that can be carried by an STM-1, an STM-16, and an STM-64 system are 2340 channels, 37,440 channels, and 149,760 channels, respectively.
(b) Show that the number of 20-Mb/s digitized video channels that can be transported over these systems are STM-1: 7 channels, STM-16: 119 channels, and STM-64: 479 channels (see Drill Problem 13.3).

13.7   Compare the system margins for 40 and 80-km long-haul OC-48 (STM-16) links at 1550 nm for the minimum and maximum source output ranges. Assume there is a 1.5-dB coupling loss at each end of the link. Use Tables 13.3 and 13.4.
[Answers: Minimum power at 40 km: Margin = +2 dB; maximum power at 40 km: Margin = +4 dB. Minimum power at 80 km: Margin = −2 dB; maximum power at 80 km: Margin = +3 dB]

13.8   Verify that the maximum optical powers per wavelength channel given in Table 13.5 yield a total power level of +17 dBm in an optical fiber.

13.9   Consider the four-node network shown in Fig. 13.49. Each node uses a different combination of three wavelengths to communicate with the other nodes, so that there are six different wavelengths in the network. Given that node 1 uses λ_2, λ_4, and λ_6 for information exchange with the other nodes (i.e., these wavelengths are added and dropped at node 1, and the remaining wavelengths from other nodes pass through), establish wavelength assignments for the other nodes.
[Answer: Add/drop wavelengths at node 2: 3, 5, 6; node 3: 1, 2, 3; node 4: 1, 4, 5]

13.10 Suppose an engineer wants to attach a 50-m long standard 50-μm fiber that has a bandwidth of 500 MHz-km to a 100-m higher-grade 50-μm fiber that has a bandwidth of 2000 MHz-km. A 10-m length of the lower grade fiber is used at each end in order to connect to transmission equipment. Show that this link will not work at 10 Gb/s.
[Answer: From Eq. (13.5), $L_{max} = 380$ m > OM3 allowed 300-m limit. Thus the link will not work at 10 Gb/s.]

13.11 Consider an OBS system in which the processing and configuration times are both 1 ms. Similar to Fig. 13.34, suppose the burst goes between edge routers A and B through four intermediate OXCs and that all nodes are 10 km apart. Show that the offset time is 5 ms.

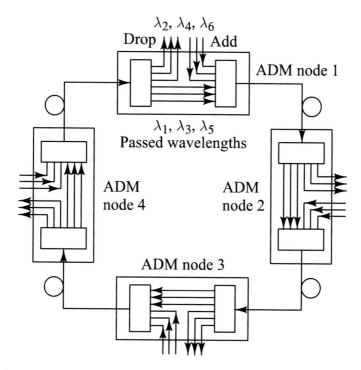

Fig. 13.49 Figure 13.49 for Problem 13.9

13.12 Consider the passive optical network shown in Fig. 13.40. Assume the following:

(a) Laser sources at 1310 and 1490 nm can launch optical powers of 2.0 and 3.0 dBm, respectively, into a fiber.

(b) The passive optical splitter is 10 km from the central office..

(c) The users are 5 km from the power splitter..

(d) The insertion losses are 13.5 and 16.6 dB for 1×16 and 1×32 splitters, respectively.

(e) The fiber attenuation is 0.6 dB/km at 1310 nm and 0.3 dB/km at 1550 nm.

(f) For simplicity assume there are no connector, splice, or other losses in the link.

Find the total link loss for the following four situations:

(i) A 1310-nm laser transmits upstream (user to central office) through a 1×16 splitter.

(ii) A 1310-nm laser transmits upstream through a 1×32 splitter.

(iii) A 1550-nm laser transmits downstream through a 1×16 splitter.

(iv) A 1550-nm laser transmits downstream through a 1×32 splitter
[Answers: (i) 22.5 dB; (ii) 25.6 dB; (iii) 18.0 dB; (iv) 21.1 dB].

13.13 Suppose the optical power emerging from a feeder fiber is to be distributed among eight individual houses. Assume that they are separated by 100 m and lie along a straight line going out from the end of the feeder fiber. One way to distribute the power is to use a 1 × 8 star coupler and run individual fibers to each house. Why is this preferable to the configuration where a single fiber runs along the line to the homes and individual tap couplers in the fiber line each extract 10% of the power from the line for each house that the line passes?
[Answer: The star coupler will distribute an equal amount of optical power to each house. The other configuration results in dramatically different power levels between the first and last houses on the line.]

13.14 Because the ranging procedure in a PON has limited accuracy, a guard time is placed between consecutive bursts from the ONTs to avoid collisions of the independent packets, as Fig. 7.18 illustrates. Verify that a 25.6-ns guard time consumes 16 bits at 622 Mb/s, 32 bits at 1244 Mb/s, and 64 bits at 2.488 Gb/s.
[Example, at 622 Mb/s one bit occupies a $1/(622 \times 106)$ s $= 1.6$ ns time slot. Thus 16 bits occupy 25.6 ns.]

13.15 An ONT intended for indoor installation requires a 12-V dc power supply and consumes 15 W of electric power during normal operation. If there is an electric power outage, the ONT disables nonessential services and now consumes 7 W of power supplied by backup batteries. Using vendor data sheets, select an indoor uninterruptible power supply (UPS) that satisfies these requirements.

13.16 An ONT intended for outdoor installation requires a 12-V dc power supply and consumes 15 W of electric power during normal operation. If there is an electric power outage, the ONT disables nonessential services and now consumes 7 W of power supplied by backup batteries. Using vendor data sheets, select an outdoor uninterruptible power supply (UPS) that satisfies these requirements. Assume the UPS must operate over temperature extremes of -25 to $+35$ °C.

References

1. B.A. Forouzan, *Data Communications and Networking*, 5th edn. (McGraw-Hill, New York, 2013)
2. The Fiber Optic Association, The FOA Reference for Fiber Optics, www.thefoa.org/tech/ref/OSP/jargon.html. Accessed June 2020
3. T. Anttalainen, V. Jaaskelainen, *Introduction to Communication Networks* (Artech House, 2015)
4. ITU-T Rec. X.210, *Open Systems Interconnection Layer Service Definition Conventions* (Nov. 1993)
5. R. Ramaswami, K.N. Sivarajan, G. Susaki, *Optical Networks*, 3rd edn. (Morgan Kaufmann, Burlington, 2009)

6. L. Peterson, B. Davie, *Computer Networks*, 5th edn. (Morgan Kauffman, Burlington, 2012)
7. M. Nurujjaman, S. Sebbah, C.M. Assi, M. Maier, Optimal capacity provisioning for survivable next generation Ethernet transport networks. J. Opt. Commun. Netw. **4**(12), 967–977 (2012)
8. C.S. Ou, B. Mukherjee, *Survivable Optical WDM Networks* (Springer, Berlin, 2005)
9. R.K. Jain, *Principles of Synchronous Digital Hierarchy* (CRC Press, Boca Raton, 2013)
10. H. van Helvoort, *The ComSoc Guide to Next Generation Optical Transport: SDH/SONET/OTN* (Wiley-IEEE Press, Hoboken, 2010)
11. Alliance for Telecommunications Industry Solutions (ATIS), *Synchronous Optical Network (SONET)—Basic Description Including Multiplex Structure, Rates, and Formats*, 2015 edn. (April 2015)
12. ITU-T sample SDH recommendations: (a) G.692, *Optical Interfaces for Multichannel Systems with Optical Amplifiers* (Jan. 2007); (b) G.841, *Types and Characteristics of SDH Network Protection Architectures* (Oct. 1998); (c) G.957, *Optical Interfaces for Equipments and Systems Relating to the Synchronous Digital Hierarchy* (March 2006)
13. H.T. Mouftah, P.H. Ho, *Optical Networks: Architecture and Survivability* (Springer, Berlin, 2003)
14. H.G. Perros, *Connection-Oriented Networks: SONET/SDH, ATM, MPLS and Optical Networks* (Wiley, Hoboken, 2005)
15. M.J. Li, M.J. Soulliere, D.J. Tebben, L. Nederlof, M.D. Vaughn, R.E. Wagner, Transparent optical protection ring architectures and applications. J. Lightw. Technol. **23**, 3388–3403 (2005)
16. A. Scavennec, O. Leclerc, Toward high-speed 40-Gb/s transponders. Proc. IEEE **94**(5), 986–996 (2006)
17. O. Bertran-Pardo, J. Renaudier, G. Charlet, H. Mardoyan, P. Tran, M. Salsi, S. Bigo, Overlaying 10 Gb/s legacy optical networks with 40 and 100 Gb/s coherent terminals. J. Lightw. Technol. **30**(14), 2367–2375 (2012)
18. R.J. Essiambre, R.W. Tkach, Capacity trends and limits of optical communication networks. Proc. IEEE **100**, 1035–1055 (2012)
19. P.J. Winzer, G. Raybon, H. Song, A. Adamiecki, S. Corteselli, A.H. Gnauck, D.A. Fishman, C.R. Doerr, S. Chandrasekhar, L.L. Buhl, T.J. Xia, G. Wellbrock, W. Lee, B. Basch, T. Kawanishi, K. Higuma, Y. Painchaud, 100-Gb/s DQPSK transmission: from laboratory experiments to field trials. J. Lightw. Technol. **26**(20), 3388–3402 (2008)
20. X. Zhou, L. Nelson, DSP for 400 Gb/s and beyond optical networks. J. Lightw. Technol. **32**(16), 2716–2725 (2014)
21. L. Mehedy, M. Bakaul, A. Nirmalathas, Single channel directly detected optical-OFDM towards higher spectral efficiency and simplicity in 100 Gb/s Ethernet and beyond. J. Opt. Commun. Netw. **3**(5), 426–434 (2011)
22. Y.-K. Huang et al., High-capacity fiber field trial using terabit/s all-optical OFDM super-channels with DP-QPSK and DP-8QAM/DP-QPSK modulation. J. Lightw. Technol. **31**(4), 546–553 (2013)
23. Y. Wang, X. Cao, Multi-granular optical switching: A classified overview for the past and future. *IEEE Commun. Surveys & Tutor.* **14**(3), 698–713 (2012) (Third Quarter)
24. S.L. Woodward, M. Feuer, P. Palacharia, ROADM-node architectures for reconfigurable photonic networks. Chap. 15 in *Optical Fiber Telecommunications Volume VIB*, 6th edn. (Academic Press, Cambridge, 2013)
25. K.G. Vlachos, F.M. Ferreira, S.S. Sygletos, A reconfigurable OADM architecture for high-order regular and offset QAM based OFDM super-channels. J. Lightw. Technol. **37**(16), 4008–4016 (2019)
26. T.A. Strasser, J.L. Wagener, Wavelength-selective switches for ROADM applications. IEEE J. Sel. Topics Quantum Electron. **16**(5), 1150–1157 (2010)
27. D.J.F. Barros, J.M. Kahn, J.P. Wilde, T.A. Zeid, Bandwidth-scalable long-haul transmission using synchronized colorless transceivers and efficient wavelength-selective switches. J. Lightw. Technol. **30**(16), 2646–2660 (2012)
28. F. Naruse, Y. Yamada, H. Hasegawa, K.-I. Sato, Evaluations of OXC hardware scale and network resource requirements of different optical path add/drop ratio restriction schemes. IEEE/OSA J. Opt. Commun. Netw. **4**(11), B26–B34 (2012)

29. J.L. Strand, Integrated route selection, transponder placement, wavelength assignment, and restoration in an advanced ROADM architecture. IEEE/OSA J. Opt. Commun. Netw. **4**(3), 282–288 (2012)

30. R.A. Barry, P. Humblet, Models of blocking probability in all-optical networks with and without wavelength conversion. IEEE J. Select Areas Commun. **14**(5), 858–867 (1996)

31. Z. Xu, Q. Jin, Z. Tu, & S. Gao, All-optical wavelength conversion for telecommunication mode-division multiplexing signals in integrated silicon waveguides. Appl. Opt. **57**(18), 5036–5040 (2018)

32. E. Stassen, C. Kim, D. Kong, H. Hu, M. Galili, L.K. Oxenløwe, K. Yvind, M. Pu, Ultra-low power all-optical wavelength conversion of high-speed data signals in high-confinement AlGaAs-on-insulator microresonators. APL Photon. **4**, 100804 (2019)

33. B.C. Chatterjee, N. Sarma, P.P. Sahu, Priority based routing and wavelength assignment with traffic grooming for optical networks. IEEE/OSA J. Opt. Commun. Netw. **4**(6), 480–489 (2012)

34. N. Charbonneau, V.M. Vokkarane, A survey of advance reservation routing and wavelength assignment in wavelength-routed WDM networks. IEEE Commun. Surv. Tutor. **14**(4), 1037–1064 (Fourth Quarte 2012)

35. A.G. Rahbar, Review of dynamic impairment-aware routing and wavelength assignment techniques in all-optical wavelength-routed networks. *IEEE Commun. Surv. Tutor.* **14**(4), 1065–1089 (Fourth Quarter 2012)

36. D.J. Blumenthal, B.E. Olsson, G. Rossi, T.E. Dimmick, L. Rau, M. Masanovic, O. Lavrova, R. Doshi, O. Jerphagnon, J.E. Bowers, V. Kaman, L.A. Coldren, J. Barton, All-optical label swapping networks and technologies. J. Lightw. Technol. **18**(12), 2058–2075 (2000)

37. A. Pattavina, Architectures and performance of optical packet switching nodes for IP networks. J. Lightw. Technol. **23**(3), 1023–1032 (2005)

38. T. Ismail, Optical packet switching architecture using wavelength optical crossbars. IEEE/OSA J. Opt. Commun. Netw. **7**, 461–469 (2015)

39. M. Wang, S. Li, E.W.M. Wong, M. Zukerman, Evaluating OBS by effective utilization. IEEE Commun. Lett. **17**(3), 576–579 (2013)

40. T. Venkatesh, C. Siva Ram Murthy, *An Analytical Approach to Optical Burst Switched Networks* (Springer, Berlin, 2010)

41. C. F. Li, *Principles of All-Optical Switching* (Wiley, Hoboken, 2013)

42. M. Jinno, Elastic optical networking: roles and benefits in beyond 100-Gb/s era. J. Lightw. Technol. **35**(5), 1116–1124 (2017)

43. Ujjwal, J. Thangaraj, Review and analysis of elastic optical network and sliceable bandwidth variable transponder architecture. Opt. Eng. **57**, 110802 (2018)

44. V. López, L. Velasco, *Elastic Optical Networks* (Springer, Berlin, 2016)

45. D.M. Marom, P.D. Colbourne, A. D'Errico, N.K. Fontaine, Y. Ikuma, R. Proietti, L. Zong, J.M. Rivas-Moscoso, I. Tomkos, Survey of photonic switching architectures and technologies in support of spatially and spectrally flexible optical networking. J. Opt. Commun. Netw. **9**, 1–26 (2017)

46. L. Yan, X.S. Yao, M.C. Hauer, A.E. Willner, Practical solutions to polarization-mode-dispersion emulation and compensation. J. Lightw. Technol. **24**(11), 3992–4005 (2006)

47. H. Bülow, F. Buchali, A. Klekamp, Electronic dispersion compensation. J. Lightw. Technol. **26**(1), 158–167 (2008)

48. A.B. Dar, R.K. Jha, Chromatic dispersion compensation techniques and characterization of fiber Bragg grating for dispersion compensation. Opt. Quantum Electron. **49**, article 108 (2017)

49. C.H. Yeh, J.R. Chen, W.Y. You, W.P. Lin, C.W. Chow, Rayleigh backscattering noise alleviation in long-reach ring-based WDM access communication. IEEE Access **8**, 105065–105070 (2020)

50. C. Kachris, I. Tomkos, A survey on optical interconnects for data centers. IEEE Commun. Surv. Tutor. **14**(4), 1021–1036 (Fourth Quarter 2012)

51. C. Kachris, K. Bergman, I. Tomkos, *Optical Interconnects for Future Data Center Networks* (Springer, Berlin, 2013)

52. T. Segawa, Y. Muranaka, R. Takahashi, High-speed optical packet switching for photonic datacenter networks. NTT Tech. Review **14**, 1–7 (2016)

53. D.J. Blumenthal, H. Ballani, R.O. Behunin, J.E. Bowers, P. Costa, D. Lenoski, P. Morton, S.B. Papp, P. T. Rakich, Frequency-stabilized links for coherent WDM fiber interconnects in the datacenter. J. Lightw. Technol. **38** (Apr. 2020)

54. G. Kanakis et al., High-speed VCSEL-based transceiver for 200 GbE short-reach intra-datacenter optical interconnects. Appl. Sci. **9** (2019)

55. C. Xie, L. Wang, L. Dou, M. Xia, S. Chen, H. Zhang, Z. Sun, J. Cheng, Open and disaggregated optical transport networks for data center interconnects. J. Opt. Commun. Netw. **12**, C12–C22 (2020)

56. G. Keiser, *FTTX Concepts and Applications* (Wiley, Hoboken, 2006)

57. C.F. Lam (ed.), *Passive Optical Networks: Principles and Practice* (Academic Press, Cambridge, 2007)

58. Y.C. Chung, Y. Takushima, Wavelength-division-multiplexed passive optical networks (WDM PONs), in Chap. 23 in *Optical Fiber Tele-communications Volume VIB*, 6th edn. (Academic Press, Cambridge, 2013)

59. H.S. Abbas, M.A. Gregory, The next generation of passive optical networks: a review. J. Netw. Comput. Appl. **67**, 53–74 (2016)

60. D.A. Khotimsky, NG-PON2 transmission convergence layer: a tutorial. J. Lightw. Technol. **34**(5), 1424–1432 (1 March 2016)

61. D. Zhang, D. Liu, X. Wu, D. Nesset, Progress of ITU-T higher speed passive optical network (50G-PON) standardization: Review. J. Opt. Commun. Network. **12**(10), D99–D108 (2020)

62. ITU-T Recommendation G.983.1, *Broadband Optical Access Systems Based on Passive Optical Network (PON)* (Jan 2005)

63. ITU-T Recommendation G.984.1, *Gigabit-capable passive optical network (GPON): General characteristics* (Mar. 2008)

64. ITU-T Recommendation G.987, *10-Gigabit-Capable Passive Optical Network (XG-PON): Definitions, Abbreviations, and Acronyms* (June 2012)

65. IEEE 802.3ca 25G/50G-EPON standard (June 2020)

66. C.-L. Tseng, C.-K. Liu, J.-J. Jou, W.-Y. Lin, C.-W. Shih, S.-C. Lin, S.-L. Lee, G. Keiser, Bidirectional transmission using tunable fiber lasers and injection-locked Fabry-Pérot laser diodes for WDM access networks. Photonics Technol. Lett. **20**(10), 794–796 (2008)

67. S.-C. Lin, S.-L. Lee, H.-H. Lin, G. Keiser, R.J. Ram, Cross-seeding schemes for WDM-based next-generation optical access networks. J. Lightw. Technol. **29**(24), 3727–3736 (2011)

68. E. Wong, Next-generation broadband access networks and technologies. J. Lightw. Technol. **30**(4), 597–608 (2012)

69. F. Xiong, W.-D. Zhong, H. Kim, A broadcast-capable WDM passive optical network using offset polarization multiplexing. IEEE J. Lightw. Technol. **30**(14), 2329–2336 (2012)

70. L.B. Du, X. Zhao, S. Yin, T. Zhang, A.E.T. Barratt, J. Jiang, D. Wang, J. Geng, C. DeSanti, C.F. Lam, Long-reach wavelength-routed TWDM PON: Technology and deployment. J. Lightw. Technol. **37**(3), 688–697 (1 Feb. 2019)

Chapter 14
Basic Measurement and Monitoring Techniques

Abstract This chapter discusses measurement techniques that have been developed for characterizing the operational behavior of devices and fibers, for ensuring that the correct components have been selected for a particular application, and for verifying that the network is configured properly. In addition, various operational methods for performance monitoring are needed to verify that all the design and operating specifications of a link are met when it is running. A wide selection of sophisticated test equipment exists for these measurement requirements and procedures.

Engineers from many diverse disciplines need to make performance measurements at all stages of the design, installation, and operation of an optical fiber communication network. Numerous levels of measurement techniques have been developed for characterizing the operational behavior of devices and fibers, for ensuring that the correct components have been selected for a particular application, and for verifying that the network is configured properly. In addition, various methods for monitoring operational performance are needed to verify that all the design and operating specifications of a link are met when it is running. A wide selection of sophisticated test equipment exists for each of these measurement categories.

During the link design phase an engineer can find the operational parameters of many passive and active components on vendor data sheets. These include fixed parameters for fibers, passive optical devices, and optoelectronic components such as light sources, photodetectors, and optical amplifiers. For example, fixed fiber parameters include core and cladding diameters, refractive-index profile, mode-field diameter, and cut-off wavelength. Once such fixed parameters are known, generally there is no need for a link design engineer to measure them again.

However, variable parameters of communication system elements, such as optoelectronic components, may change with operational conditions and need to be measured before, during, and after a link is fielded. Of particular importance are accurate and comprehensive measurements of the optical fiber, because this component cannot be replaced readily once it has been installed. Although many physical properties of fiber remain constant, the attenuation and dispersions of a fiber can change during fiber cabling and cable installation. In single-mode fibers, chromatic

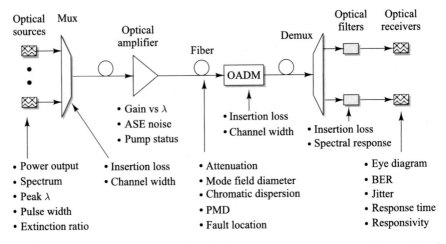

Fig. 14.1 Components of a typical WDM link and some performance-measurement parameters of user interest

and polarization-mode dispersions are important factors that can limit the transmission distance or data rate. Chromatic dispersion effects are of particular importance in high-speed DWDM links, because their behavior depends on the link configuration. Measurement and monitoring of polarization-mode dispersion is important for data rates at and above 10 Gb/s, because its statistical behavior ultimately can limit the highest achievable data rate.

When a link is being installed and tested, the operational parameters of interest include bit-error rate, timing jitter, and signal-to-noise ratio as indicated by the eye pattern. During actual operation, measurements are needed for maintenance and monitoring functions to determine factors such as fault locations in fibers and the status of remotely located optical amplifiers and other active devices.

This chapter discusses measurements and performance monitoring tests of interest to designers, installers, and operators of fiber optic links and networks. Of particular interest here are measurements for WDM links. Figure 14.1 shows some of the relevant test parameters and at what points in a WDM link they are of importance. The operational impact or impairment for many of these factors can be accounted for and controlled through careful network design. Other parameters may need to be monitored and possibly compensated for dynamically during network operation. In either case, all of these parameters must be measured at some point during the time period ranging from network design concept to service provisioning.

First, Sect. 14.1 addresses internationally recognized measurement standards for component and system evaluations. Next, Sect. 14.2 lists basic test instruments for optical fiber communication link characterizations. A fundamental unit in lightwave communications is optical power and its measurement with optical power meters, which is the topic of Sect. 14.3. Turning to measurement techniques, Sect. 14.4 gives an overview of methods and specialized equipment for characterizing optical fiber parameters. In addition to determining geometric parameters, this equipment also can measure attenuation and chromatic dispersion. During and after link installation, several basic parameters need to be checked. For example, the error performance can be estimated through eye pattern measurements, which is the topic of Sect. 14.5. The physical integrity of the fielded link typically is checked with an optical time-domain reflectometer (OTDR), as Sect. 14.6 explains. Section 14.7 discusses optical performance monitoring, which is essential for managing high-capacity lightwave transmission networks. Network functions that need such monitoring include amplifier control, channel identification, and assessment of the integrity of optical signals. Section 14.8 describes some basic performance measurements procedures.

14.1   Overview of Measurement Standards

Before examining measurement techniques, it is helpful to look at what standards exist for fiber optics. As summarized in Table 14.1, the three basic classes are primary, component testing, and system standards.

Primary standards refer to measuring and characterizing fundamental physical parameters such as attenuation, bandwidth, mode-field diameter for single-mode fibers, and optical power. In the United States the main group involved in primary standards is the *National Institute of Standards and Technology* (NIST) [1]. This organization carries out fiber optic and laser standardization work, and addresses high-speed telecom measurement standards. Another goal is to support and accelerate the development of emerging technologies. Other national organizations include the *National Physical Laboratory* (NPL) in the United Kingdom [2] and the *Physikalisch-Technische Bundesanstalt* (PTB) in Germany [3].

As Table 14.2 summarizes, several international organizations are involved in formulating component and system testing standards. The major organizations that deal with measurement methods for links and networks are the *Institute for Electrical and Electronic Engineers* (IEEE), the *International Electrotechnical Commission* (IEC), and the *Telecommunication Standardization Sector of the International Telecommunication Union* (ITU-T).

Table 14.1   Three standards classes, the involved organizations, and their functions

Standards class	Involved organizations	Functions of the organizations
Primary	• NIST (USA) • NPL (UK) • PTB (Germany)	• Characterize physical parameters • Support and accelerate development of emerging technologies (NIST)
Component testing	• TIA • ITU-T • IEC	• Define component evaluation tests • Establish equipment-calibration procedures
System testing	• ANSI • IEEE • ITU-T	• Define physical-layer test methods • Establish measurement procedures for links and networks

Table 14.2 Major standards organizations and their functions related to testing

Organization	Internet address	Testing-related activities
IEEE	www.ieee.org	Establish and publish measurement procedures for links and networks • Define physical-layer test methods • IEEE 802.3ah Ethernet in First Mile (EFM)
ITU-T	www.itu.int/ITU-T	Create and publish standards in all areas of telecommunications • Series G for telecommunication transmission systems and media, digital systems, and networks • Series L for construction, installation, and protection of cables and outside plant elements • Series O for specifications of measuring equipment
TIA	www.tiaonline.org	Created over 200 test specifications under the designation Fiber Optic Test Procedures (FOTP) • Define physical-layer test methods • TIA -455-XX or FOTP-XX documents

Component testing standards define relevant tests for fiber optic component performance, and they establish equipment calibration procedures. A key organization for component testing is the *Telecommunication Industry Association* (TIA). The TIA has a list of over 200 fiber optic test standards and specifications under the general designation TIA-455-XX, where XX refers to a specific measurement technique. These standards are also called *Fiber Optic Test Procedures* (FOTP), so that TIA-455-XX becomes FOTP-XX. These include a wide selection of recommended methods for testing the response of fibers, cables, passive devices, and electro-optic components to environmental factors and operational conditions. A full catalog of TIA specifications can be found at https://www.tiaonline.org.

System standards refer to measurement methods for links and networks. Some of the major organizations involved here are the *American National Standards Institute* (ANSI), the *Institute for Electrical and Electronic Engineers* (IEEE), the European Telecommunications Standards Institute (ETSI), and the ITU-T. Of special interest for fiber optics systems are test standards and recommendations from the ITU-T that are aimed at all aspects of optical networking.

14.2   Survey of Test Equipment

As optical signals pass through the various parts of an optical link, they need to be measured and characterized in terms of fundamental parameters such as optical power, polarization, and spectral content. The basic instruments for carrying out such measurements on optical fiber components and systems include optical power meters, attenuators, tunable laser sources, spectrum analyzers, and optical time-domain reflectometers. These come in a variety of capabilities, with sizes ranging from portable, handheld units for field use to sophisticated bench-top or rack-mountable instruments for laboratory and manufacturing applications. In general, the field units do not need to have the extremely high precision of laboratory instruments, but they need to be more rugged to maintain reliable and accurate measurements under extreme environmental conditions of temperature, humidity, dust, and mechanical stress. However, even the handheld instruments for field use have reached a high degree of sophistication with automated microprocessor-controlled test features, computer interfaces, and Internet access capabilities.

More sophisticated instruments, such as polarization analyzers and optical communication analyzers, are available for measuring and analyzing polarization-mode dispersion (PMD), eye diagrams, and pulse waveforms. These instruments enable a variety of statistical measurements to be made at the push of a button, after the user has keyed in the parameters to be tested and the desired measurement range.

Table 14.3 lists some essential test equipment and their functions for installation and operation of optical communication systems. This section defines a selection of the first six instruments in this table. Later sections give more details concerning optical power meters, bit-error rate testers, and optical time-domain reflectometers.

Table 14.3 Some widely used optical-system test instruments and their functions

Test instrument	Function
Test-support lasers (multiple-wavelength or broadband)	Assist in tests that measure the wavelength-dependent response of an optical component or link
Optical spectrum analyzer	Measures optical power as a function of wavelength
Multifunction optical test system	Factory or field instruments with exchangeable modules for performing a variety of measurements
Optical power attenuator	Reduces power level to prevent instrument damage or to avoid overload distortion in the measurements
Conformance analyzer	Measures optical receiver performance in accordance with standards-based specifications
Visual fault indicator	Uses visible light to give a quick indication of a break in an optical fiber
Optical power meter	Measures optical power over a selected wavelength band
BER test equipment	Uses standard eye-pattern masks to evaluate the data-handling ability of an optical link
OTDR (field instrument)	Measures attenuation, length, connector/splice losses, and reflectance levels; helps locate fiber breaks
Optical return loss tester	Measures total reverse power in relation to total forward power at a particular point

Table 14.4 Characteristics of laser-source instruments used for test support

Parameter	Tunable source	Broadband source
Spectral output range	Selectable: e.g., 1370–1495 nm or 1460–1640 nm	Peak wavelength ±25 nm
Total optical output power	Up to 8 dBm	>3.5 mW (5.5 dBm) over a 60-nm range
Power stability	<±0.02 dB	<±0.05 dB
Wavelength accuracy	<±10 pm	(Not applicable)

14.2.1 Lasers Used for Test Support

Specialized light sources are desirable for testing optical components. Table 14.4 lists the characteristics of two such laser source instruments used for test support.

Tunable laser sources are important instruments for measurements of the wavelength-dependent response of an optical component or link. A number of vendors offer such light sources that generate a true single-mode laser line for every selected wavelength point. Typically the source is an external-cavity semiconductor laser. A movable diffraction grating may be used as a tunable filter for wavelength selection. Depending on the source and grating combination, an instrument may be tunable over (for example) the 1280-to-1330-nm, the 1370-to-1495-nm, or the 1460-to-1640-nm band. Wavelength scans can be done automatically with an output power that is flat across the scanned spectral band. The minimum output power of such an instrument usually is –10 dBm, and the absolute wavelength accuracy is typically ±0.01 nm (±10 pm).

A *broadband incoherent light source* with a high output power coupled into a single-mode fiber is desirable to evaluate passive DWDM components. Such an instrument can be realized by using the amplified spontaneous emission (ASE) of an erbium-doped fiber amplifier. The power spectral density of the output is up to one hundred times (20 dB) greater than that of edge-emitting LEDs and up to 100,000 times (50 dB) greater than white-light tungsten lamp sources. The instrument can be specified to have a total output power of greater than 3.5 mW (5.5 dBm) over a 50 nm range with a spectral density of −13 dBm/nm (50 μW/nm). The relatively high-power spectral density allows test personnel to characterize devices with medium or high insertion loss. Peak wavelengths might be 1280, 1310, 1430, 1550, or 1650 nm.

14.2.2 Optical Spectrum Analyzer

The widespread implementation of WDM systems calls for making optical spectrum analyses to characterize the behavior of various telecom network elements as a function of wavelength. One widely used instrument for doing this is an *optical spectrum analyzer* (OSA), which measures optical power as a function of wavelength. The

Fig. 14.2 Operation of a grating-based optical spectrum analyzer

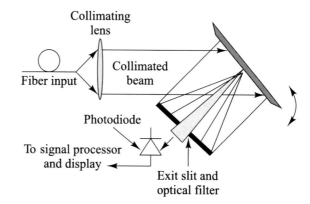

most common implementation uses an optical filter based on a diffraction grating, which yields wavelength resolutions to less than 0.1 nm. Higher wavelength accuracy (±0.001 nm) is achieved with wavelength meters based on Michelson interferometry.

Figure 14.2 illustrates the operation of a grating-based optical spectrum analyzer. Light emerging from a fiber is collimated by a lens and is directed onto a diffraction grating that can be rotated. The exit slit selects or filters the spectrum of the light from the grating. Thus it determines the *spectral resolution* of the OSA. The term *resolution bandwidth* describes the width of this optical filter. Typical OSAs have selectable filters ranging from 10 nm down to 0.1 nm. The optical filter characteristics determine the *dynamic range,* which is the ability of the OSA to simultaneously view large and small signals in the same sweep. The bandwidth of the amplifier is a major factor affecting the sensitivity and sweep time of the OSA. In the O-band through L-band the photodiode is usually an InGaAs device.

The OSA normally sweeps (scans in a certain time interval) across a spectral band making measurements at discretely spaced wavelength points. This spacing depends on the bandwidth-resolution of the instrument and is known as the *trace-point spacing.*

14.2.3 Multipurpose Test Equipment

To reduce the number of individual pieces of test gear that a field support engineer needs to carry around, manufacturers are producing compact test equipment with units for multiple functions. Of particular interest are multipurpose handheld test units for the installation, turn-up, and maintenance of optical fiber links in enterprise networks, metro networks, and PONs. A typical single handheld instrument might incorporate functions such as a power meter, a bidirectional dual-wavelength loss tester, an optical return-loss tester, a visual fault indicator for locating breaks and failures in a fiber cable, and a talk set for full-duplex communications between field personnel. Test results can be displayed on a typical 5-in. high-resolution color touchscreen or they can be sent via Bluetooth links or wireless connections to storage devices or more sophisticated data analysis instruments.

14.2.4　Optical Attenuators

In many laboratory or production tests, the characteristics of a high optical signal level may need to be measured. For example, if the level is a strong output from an optical amplifier, the signal may need to be attenuated precisely before being measured. This is done to prevent instrument damage or to avoid overload distortion in the measurements. An *optical attenuator* allows a test engineer to reduce an optical signal level up to, for example, 60 dB (a factor of 10^6) in precise steps at a specified wavelength, which is usually 1310 or 1550 nm. At the lower performance end, these attenuators are small devices (approximately $2 \times 5 \times 10$ cm) intended for quick field measurements that may only need to be accurate to 0.5 dB. Laboratory instruments may have an attenuation precision of 0.001 dB.

14.2.5　OTN Tester for Performance Verification

A major concern in the implementation of an optical transport network (OTN) is how to verify that the various network elements are functioning properly. In relation to this, the ITU-T has published the following guidelines:

- G.709, Interfaces for the OTN (June 2020)
- G.798, Characteristics of OTN Hierarchy
  Functional Blocks (Dec. 2017; Amend. 3, Jan. 2021)

Of particular importance is verification that the OTN elements that may have been designed, produced, and installed by different vendors conform to these ITU-T recommendations. Such tests, which are known as *conformance tests*, include the following:

- Verifying the correctness of the interface specifications of the elements
- Checking the correct responses of the device under test (DUT)
- Verifying the correct behavior of forward-error-correcting (FEC) modules
- Examining that mapping of client signals are performed correctly

Multifunction, multiport, multiuser portable laptop-sized instruments for field conformance testing can characterize parameters such as physical-layer jitter, wander, and bit-error rate testing from 1.5 Mb/s to 100 Gb/s with the results displayed in various formats on a 9-in. or larger screen.

14.2.6 Visual Fault Indicator

A *visual fault locator* (VFL) is a handheld pen-sized instrument that uses a visible laser light source to locate events such as fiber breaks, overly tight bends in a fiber (for example, in an equipment rack), or poorly mated connectors. The source emits a bright beam of red light (e.g., 650 nm) into a fiber, thereby allowing the user to see a fiber fault or a high-loss point as a glowing or blinking red light. In using such a device, the events must occur where the fiber or connector is in the open so that visual observation of the emitted red light is possible.

The nominal light output is 1 mW, so the light will be visible through a fiber jacket at a fault point. This power level allows a user to detect a fiber fault visually. The device generally is powered by one 1.5-V size AA battery and operates in either a continuously on or a blinking mode.

14.3 Optical Power Measurement Methods

Optical power measurement is the most basic function in fiber optic metrology. However, this parameter is not a fixed quantity and can vary as a function of other parameters such as time, distance along a link, wavelength, phase, and polarization.

14.3.1 Physical Basis of Optical Power

To get an understanding of optical power, it is instructive to look at its physical basis and how it relates to other optical quantities such as energy, intensity, and radiance.

- Light particles, called *photons*, have a certain energy associated with them, which changes with wavelength. The relationship between the energy E of a photon and its wavelength λ is given by the equation $E = hc/\lambda$, which is known as *Planck's Law*. In terms of wavelength (measured in units of μm), the energy in electron volts is given by the expression $E(\text{eV}) = 1.2406/\lambda$ (μm). Note that $1\,\text{eV} = 1.60218 \times 10^{-19}$ J.
- *Optical power P* measures the rate at which photons arrive at a detector. Thus it is a measure of energy transfer per time. Because the rate of energy transfer varies with time, the optical power is a function of time. It is measured in *watts* or joules per second (J/s).
- As noted in Chap. 4, *radiance* is a measure, in watts, of how much optical power radiates into a designated solid angle per unit of emitting surface.

Because optical power can vary with time, its measurement also changes with time. Figure 14.3 shows a plot of the power level in a signal pulse stream as a function of time. Different instantaneous power level readings are obtained depending on the

Fig. 14.3 Peak and average powers in a series of general, NRZ, and RZ optical pulses

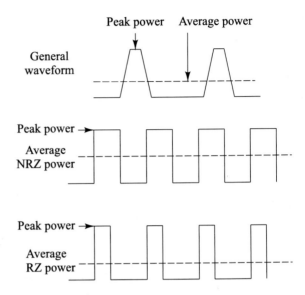

exact instant when the measurement is made. Therefore, two standard classes of power measurements can be specified in an optical system. These are the peak power and the average power. The *peak power* is the maximum power level in a pulse, which might be sustained for only a very short time.

The *average power* is a measure of the power level averaged over a relatively long time period compared to the duration of an individual pulse. For example, the measurement time period could be one second, which contains many signal pulses. As a simple example, in a nonreturn-to-zero (NRZ) data stream there will be an equal probability of 1 and 0 pulses over a long time period. In this case, as shown in Fig. 14.3, the average power is half the peak power. If a return-to-zero (RZ) modulation format is used, the average power over a long sequence of pulses will be one-fourth the peak power because there is no pulse in a 0 time slot and a 1 time slot is only half filled.

The sensitivity of a photodetector normally is expressed in terms of the average power level impinging on it, because the measurements in an actual fiber optic system are done over many pulses. However, the output level for an optical transmitter normally is specified as the peak power. This means that the average power coupled into a fiber, which is the power level that a photodetector measures, is at least 3 dB lower than if the link designer uses the peak source output in power-budget calculations as the light level entering the fiber.

14.3.2   Optical Power Meters

The function of an *optical power meter* is to measure total power over a selected wavelength band. Some form of optical power detection is in almost every piece of lightwave test equipment. Handheld instruments come in a wide variety of types with different levels of capability. Multiwavelength optical power meters using several photodetectors are the most common instrument for measuring optical signal power levels. Usually the meter outputs are given in dBm (where 0 dBm = 1 mW) or dBμ (where 0 dBμ = 1 μW).

For example, using a Ge photodetector typically allows a measuring range of +18 to −60 dBm in the 780-to-1600-nm wavelength band, whereas an InGaAs photodetector allows a measuring range of +3 to −73 dBm in the 840-to-1650-nm wavelength band. In each case, the power measurements can be made at a number of calibrated wavelengths (for example, 11 calibrated wavelengths). User-selectable threshold settings can let the instrument show a pass/fail on a built-in display. Connections can be made to a smart app via a USB port or a Bluetooth interface for data reporting from the field, cloud storage, and workflow management. Interfaces for email or messaging are other connection options.

14.4   Characterization of Optical Fibers

Many millions of kilometers of optical fibers have been fabricated and installed worldwide. Various types of equipment for factory use have been developed to characterize the physical and performance parameters of these fibers. Whereas early equipment tended to specialize on measuring only one or two parameters, modern sophisticated instruments require only one simple fiber preparation to characterize optical fibers accurately during the manufacturing process. These parameters include mode-field diameter, attenuation, cut-off wavelength, chromatic dispersion versus wavelength, refractive-index profile, effective area, and geometric properties such as core and cladding diameters, core-to-cladding concentricity error, and fiber circularity. The two basic measurement methods used by this specialized equipment are the refracted near-field method and the transmitted near-field techniques. This section first describes these two techniques and then looks at some standard ways to measure attenuation.

14.4.1   Refracted Near-Field Method

The *refracted near-field* measurement method is recommended by the ITU-T and TIA for determining the refractive-index profile (RIP) [4]. This method determines the index profile by moving a focused laser across the fiber end face and examining the distribution of the light that is refracted sideways out of the core as a function of the radial position of the laser spot. The variation in the detected optical signal level is proportional to the index change at the fiber end face. The RIP parameter can be used to calculate the geometrical parameters of a fiber and to estimate all transmission properties (e.g., chromatic dispersion and the cut-off wavelength) except attenuation and polarization-mode dispersion.

14.4.2 Transmitted Near-Field Technique

The *transmitted near-field* measurement method is recommended by the ITU-T and TIA for measuring mode-field characteristics [5, 6]. Knowledge of the mode-field diameter (MFD) is important because it describes the radial optical field distribution across the fiber core. Detailed information of the MFD enables one to calculate characteristics such as source-to-fiber coupling efficiency, splice and joint losses, microbending loss, and dispersion. A transmitted near-field scan directly provides the intensity distribution $E^2(r)$ at the fiber exit. From this distribution one then can calculate the MFD using the Petermann II equation. The Petermann II expression is given by Eq. (2.73) in terms of the field intensity distribution as [7, 8]

$$\text{MFD} = 2\left[\frac{2\int_0^\infty E^2(r)r^3 dr}{\int_0^\infty E^2(r)r dr}\right]^{1/2} \tag{14.1}$$

Because it is easy to program this equation, the measurement equipment software can calculate the MFP directly from the near-field data.

14.4.3 Optical Fiber Attenuation Measurements

Attenuation of optical power in a fiber waveguide is a result of absorption processes, scattering mechanisms, and waveguide effects. The manufacturer is generally interested in the magnitude of the individual contributions to attenuation, whereas the system engineer who uses the fiber is more concerned with the total transmission loss of the fiber. Here only measurement techniques for total transmission loss are treated.

Three basic methods are available for determining attenuation in fibers. The earliest devised and most common approach involves measuring the optical power transmitted through a long and a short length of the same fiber using identical input couplings. This method is known as the cutback technique. A less accurate but nondestructive method is the insertion-loss method, which is useful for cables with

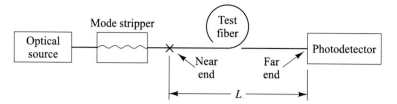

Fig. 14.4 Schematic experimental setup for determining fiber attenuation by the cutback technique

connectors on them. These two methods are described in this section. Section 14.6 describes the third technique, which involves the use of an OTDR.

The Cutback Technique
The *cutback technique* [9], which is a destructive method requiring access to both ends of the fiber, is illustrated in Fig. 14.4. Measurements may be made at one or more specific wavelengths, or, alternatively, a spectral response may be required over a range of wavelengths. To find the transmission loss, the optical power is first measured at the output (or far end) of the fiber. Then, without disturbing the input condition, the fiber is cut off a few meters from the source, and the output power at this near end is measured. If P_F and P_N represent the output powers of the far and near ends of the fiber, respectively, the average loss α in decibels per kilometer is given by

$$\alpha = \frac{10}{L} \log \frac{P_N}{P_F} \qquad (14.2)$$

where L (in kilometers) is the separation of the two measurement points. The reason for these steps is that it is extremely difficult to calculate the exact amount of optical power launched into a fiber. By using the cutback method, the optical power emerging from the short fiber length is the same as the input power to the fiber of length L. The function of the mode stripper in Fig. 14.4 is to remove cladding modes that can interfere with the measurement of attenuation in the core.

Example 14.1 An engineer wants to find the attenuation at 1310 nm of a 4.95-km long fiber. The only available instrument is a photodetector, which gives an output reading in volts. Using this device in a cutback-attenuation setup, the engineer measures an output of 6.58 V from the photodiode at the far end of the fiber. After cutting the fiber 2 m from the source, the output voltage from the photodetector now reads 2.21 V. What is the attenuation of the fiber in dB/km?

Solution Because the output voltage from the photodetector is proportional to the optical power, Eq. (14.2) can be written as

$$\alpha = \frac{10}{L_1 - L_2} \log \frac{V_2}{V_1}$$

where L_1 is the length of the original fiber, L_2 is the length after cut off, and V_1 and V_2 are the voltage output readings from the long and short lengths, respectively. Then the attenuation in decibels is

$$\alpha = \frac{10}{4.950 - 0.002} \log \frac{6.58}{2.21} = 0.95 \text{ dB/km}$$

Drill Problem 14.1 A technician plans to use the cutback method to measure the attenuation at 1550 nm of a 9.60-km long single-mode fiber. The measured optical power at the far end is 410 μW. After cutting the fiber 2 m from the source, the optical output measured is 680 μW. Show that the attenuation of the fiber at this wavelength is 0.229 dB/km.

In carrying out this measurement technique, special attention must be paid to how optical power is launched into the fiber because in a multimode fiber different launch conditions can yield different loss values. The effects of modal distributions in the multimode fiber that result from different numerical apertures and spot sizes on the launch end of the fiber are shown in Fig. 14.5. If the spot size is small and its NA is less than that of the fiber core, the optical power is concentrated in the

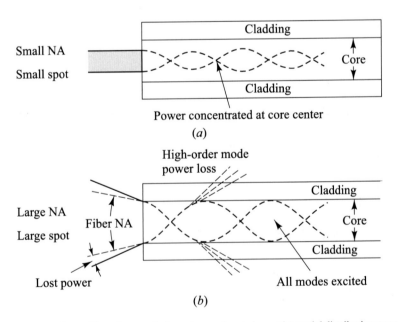

Fig. 14.5 The effects of launch numerical aperture and spot size on the modal distribution: **a** under-filling the fiber excites only lower-order modes; **b** an overfilled fiber has excess attenuation from higher-order mode loss

center of the core, as Fig. 14.5a shows. In this case, the attenuation contribution arising from higher-order mode power loss is negligible. In Fig. 14.5b the spot size is larger than the fiber core and the spot NA is larger than that of the fiber. For this overfilled condition, those parts of the incident light beam that fall outside the fiber core and outside the fiber NA are lost. In addition, there is a large contribution to the attenuation arising from higher-mode power loss (see Sects. 5.1 and 5.3).

Steady-state equilibrium-mode distributions are typically achieved by the *mandrel-wrap* method. In this procedure, excess higher-order cladding modes that are launched by initially overexciting the fiber are filtered out by wrapping several turns of fiber around a mandrel, which is about 1.0–1.5 cm in diameter. In single-mode fibers, this type of mode filter or mode stripper is used to eliminate cladding modes from the fiber.

Insertion-Loss Method

For cables with connectors, one cannot use the cutback method. In this case, one commonly uses an *insertion-loss technique* [9]. This is less accurate than the cutback method but is intended for field measurements to give the total attenuation of a cable assembly in decibels.

The basic setup is shown in Fig. 14.6, where the launch and detector couplings are made through connectors. The wavelength-tunable light source is coupled to a short length of fiber that has the same basic characteristics as the fiber to be tested. For multimode fibers, a mode scrambler is used to ensure that the fiber core contains an equilibrium-mode distribution. In single-mode fibers, a cladding mode stripper is employed so that only the fundamental mode is allowed to propagate along the fiber. A wavelength-selective device, such as an optical filter, is generally included to find the attenuation as a function of wavelength.

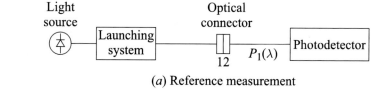

(*a*) Reference measurement

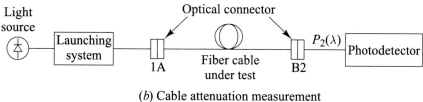

(*b*) Cable attenuation measurement

Fig. 14.6 Test setup for using the insertion-loss technique to measure attenuation of cables that have attached connectors

To carry out the attenuation tests, the connector of the short-length launching fiber is attached to the connector of the receiving system and the launch power level $P_1(\lambda)$ is recorded. Next, the cable assembly to be tested is connected between the launching and receiving systems, and the received power level $P_2(\lambda)$ is recorded. The attenuation of the cable in decibels is then

$$A = 10 \log \frac{P_1(\lambda)}{P_2(\lambda)} \qquad (14.3)$$

This attenuation is the sum of the loss of the cabled fiber and the connector between the launch connector and the cable.

Example 14.2 The insertion-loss technique also can be used for measuring the loss through an optical device that has fiber flyleads attached. Suppose an optical filter with attached flyleads is inserted into the link in Fig. 14.6a. Consider the case when the power at the photodetector prior to inserting the filter is $P_1 = 0.51$ mW and the power level with the optical filter in the link is $P_2 = 0.43$ mW. What is the insertion loss of the device?

Solution From Eq. (14.3)
Insertion loss $= 10 \log P_1/P_2 = 10 \log 0.51/0.43 = 0.74$ dB.

14.5 Concept of Eye Diagram Tests

The use of an *eye diagram* is a traditional technique for quickly and intuitively assessing the quality of a received signal. Modern bit-error rate (also called *bit-error ratio*) measurement instruments construct such eye diagrams by generating a pseudorandom pattern of ones and zeros at a uniform rate but in a random manner. When the pulses in this pattern are superimposed simultaneously, an eye pattern as shown in Fig. 14.7 is formed [10, 11]. The word *pseudorandom* means that the generated combination or sequence of ones and zeros will eventually repeat but that it is sufficiently random for test purposes. A *pseudorandom binary sequence* (PRBS) comprises sequences of 2^N different N-bit long combinations. For example, these can be four different 2-bit long combinations, eight different 3-bit long combinations, sixteen different 4-bit long combinations, etc. up to a limit set by the instrument. These combinations are randomly selected. The PRBS pattern length is of the form $2^N - 1$, where N is an integer. This choice assures that the pattern-repetition rate is not harmonically related to the data rate. Typical values of N are 7, 10, 15, 20, 23, and 31. After this limit has been reached, the data sequence will repeat.

Ideally, if the signal impairments are small, the received pattern should look like that shown in Fig. 14.7. However, time-varying signal impairments in the transmission path can lead to amplitude variations within the signal and timing skews between the data signal and the associated clock signal. Note that a clock signal,

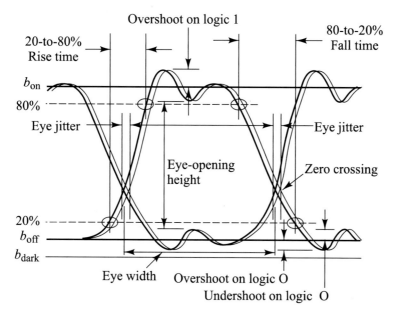

Fig. 14.7   General configuration of a fairly clean eye diagram showing definitions of fundamental measurement parameters

which typically is encoded within a data signal, is used to help the receiver interpret the incoming data correctly. Thus, in an actual link the received pattern will become wider or distorted on the sides and on the top and bottom, as shown in Fig. 14.8.

Fig. 14.8 Signal-distorting effects cause the eye opening to get smaller

14.5.1 Standard Mask Testing

Interpretation of the characteristics of a distorted eye diagram is done by means of *mask testing*. Depending on which protocol standard is used, industry-defined masks can take the shape of a polygon or a square, which must fit within the eye diagram opening as shown in Fig. 14.9. In some cases (e.g., 622-Mb/s SONET) the mask is a six-sided polygon sitting in the middle of the eye, whereas for other protocols the mask shape is a rectangle (e.g., OC-48 and OC-192) or a diamond (e.g., Gigabit Ethernet). The *mask height* is sized in proportion to the power level of the signal. This height indicates the minimum separation that is needed between the logic 1 and 0 levels in order to achieve a specific bit-error rate, as can be derived from the Q factor described in Chap. 7. The *slopes* of the polygon edges indicate the allowed 10-to-90% rise and fall times. The *mask width* is proportional to the bit rate; that is, the width is narrower for higher bit rates. This is related to the *jitter parameter* shown in Fig. 14.9, which is half the peak-to-peak jitter tolerance associated with the signal. The *overshoot* and *undershoot* parameters bound the amplitudes in terms of the logic 1 and 0 levels, respectively. In Fig. 14.9 the eye measurement parameters are defined as follows:

- P_1 is the mean optical power level associated with a long string of 1 bits
- P_0 is the mean optical power level associated with a long string of 0 bits
- A is the lowest inner upper eye level
- B is the highest inner lower eye level.

The operating software of most modern bit-error rate test instruments has a wide selection of built-in masks for different protocols. In addition, the instrument user can key in custom masks for any application or to check the test results differently. Table 14.5 lists the five mask-parameter values of several protocols. The parameter values are given in terms of *unit intervals* (UI), where the pattern height $(P_1 - P_0)$

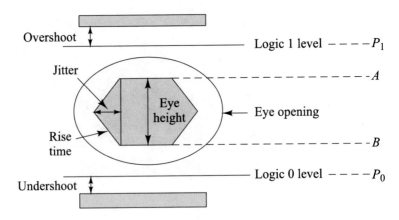

Fig. 14.9 The two bars and the six-sided polygon define the eye-pattern mask

Table 14.5 Standard NRZ eye-mask parameters for several protocols given in terms of unit intervals

Protocol	Jitter	Rise time	Eye height	Overshoot	Undershoot
OC-3	0.15	0.200	0.60	0.20	0.20
OC-12	0.25	0.150	0.60	0.20	0.20
OC-48/192	0.40	0.000	0.50	0.25	0.25
Gigabit Ethernet	0.22	0.155	0.60	0.30	0.20
Fibre Channel	0.15	0.200	0.60	0.30	0.20

has UI $= 1.0$. Note that the rise-time parameter for OC-48 is zero, because the mask is a rectangle.

14.5.2 Stressed Eye Opening

The standards for many high-speed transmission protocols specify a test that uses what is called a *stressed eye*. Among these standards are Gigabit Ethernet, 10-Gigabit Ethernet (l0GigE), Fibre Channel, and SONET OC-48 and OC-192. The concept of this test is to assume that all different possible jitter and intersymbol-interference impairments that might occur to a signal in a fielded link will close the eye down to a diamond shape, as shown in Fig. 14.10. If the eye opening of the optical receiver under test is greater than this diamond-shaped area of assured error-free operation, then it is expected to operate properly in an actual fielded system. The stressed-eye template height typically is between 0.10 and 0.25 UI. See Sect. 14.8.4 for a discussion of stressed eye tests.

Diamond-shaped
eye opening

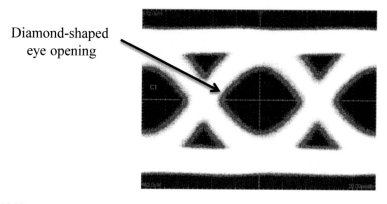

Fig. 14.10 The inclusion of all possible signal distortion effects results in a stressed eye with only a small diamond-shaped opening, which defines the mask area

14.5.3   BER Contours

Associated with the stressed eye is a parameter called the *BER contour*. Basically the BER contours are analogous to geographical contours that indicate the height and steepness profile of a hill. As indicated in Fig. 14.11, the BER contours show different levels of error probability within an eye diagram. From such a plot it can be seen that the different BER contours get closer to each other as the slope becomes steeper. This means that errors are more likely to occur if the receiver operates close to the edge of such a steep contour plot. Thus the farther a receiver decision point is within a contour boundary, the better its performance will be. This is referred to as having a *healthier eye*.

14.6   Optical Time-Domain Reflectometer

An *optical time-domain reflectometer* (OTDR) is a versatile portable instrument that is used widely to evaluate the characteristics of an installed optical fiber link. In addition to identifying and locating faults or anomalies within a link, this instrument measures parameters such as fiber attenuation, length, optical connector and splice losses, and light reflectance levels (see Ref. [12] and also vendor websites for descriptions and operations).

An OTDR is fundamentally an optical radar. As shown in Fig. 14.12, the OTDR operates by periodically launching narrow laser pulses into one end of a fiber under test by using either a directional coupler or a circulator. The properties of the optical fiber link then are determined by analyzing the amplitude and temporal characteristics of the waveform of the reflected and back-scattered light. A typical OTDR consists of

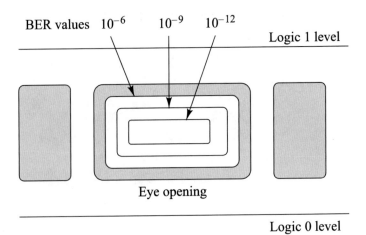

Fig. 14.11 An eye-contour diagram gives a three-dimensional view of the BER

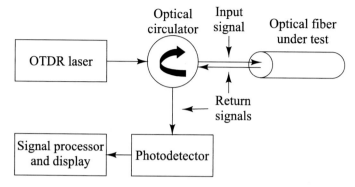

Fig. 14.12   Operational principle of an OTDR using an optical circulator

a light source and receiver, data acquisition and processing modules, an information storage unit for retaining data either in the internal memory or on an external disk, and a display.

14.6.1   OTDR Trace Characterization

Figure 14.13 shows a typical trace as seen on the display screen of an OTDR. The scale of the vertical axis is logarithmic and measures the returning (back-reflected) signal in decibels. The horizontal axis denotes the distance between the instrument

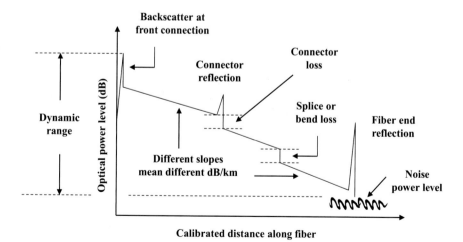

Fig. 14.13   Representative trace of backscattered and reflected optical power as displayed on an OTDR screen and the meanings of various trace features; *Note* Points where there are connectors, splices, fiber breaks, or other physical changes in the link are called "events"

and the measurement point in the fiber. In addition to the trace, an OTDR can place a number next to an event on the display and give a list of these numbers and their corresponding measurement information in a table below the trace.

The backscattered waveform has four distinct features:

- A large initial pulse resulting from Fresnel reflection at the input end of the fiber
- A long decaying tail resulting from Rayleigh scattering in the reverse direction as the input pulse travels along the fiber
- Abrupt shifts in the curve caused by optical loss at joints or connectors in the fiber line
- Positive spikes arising from Fresnel reflection at the far end of the fiber, at fiber joints, and at fiber imperfections.

Fresnel reflection and Rayleigh scattering principally produce the backscattered light. *Fresnel reflection* occurs when light enters a medium having a different index of refraction. For a glass–air interface, when light of power P_0 is incident perpendicular to the interface, the reflected power P_{ref} is

$$P_{ref} = P_0 \left(\frac{n_{fiber} - n_{air}}{n_{fiber} + n_{air}} \right)^2 \tag{14.4}$$

where n_{fiber} and n_{air} are the refractive indices of the fiber core and air, respectively. A perfect fiber end reflects about 4% of the power incident on it. However, because fiber ends generally are not polished perfectly flat and perpendicularly to the fiber axis, the reflected power tends to be much lower than the maximum possible value. In particular, this is the case if an *angle-polished connector* (APC) is used.

The detection and measurement accuracy of an event depend on the SNR that an OTDR can achieve at that point. This is defined as the ratio between the back-reflected signal and the noise level. The SNR depends on factors such as the OTDR pulse width, how often the OTDR samples the signal, and the distance to the measurement point.

Two important performance parameters of an OTDR are dynamic range and measurement range. *Dynamic range* is defined as the difference between the initial backscatter power level at the front connector and the noise level peak at the far end of the fiber. It is expressed in decibels of one-way fiber loss. Dynamic range provides information on the maximum fiber loss that can be measured and denotes the time required to measure a given fiber loss. A basic limitation of an OTDR is the tradeoff between dynamic range and *event location resolution*. For high spatial resolution, the pulse width has to be as small as possible. However, this reduces the signal-to-noise ratio and thus lowers the dynamic range. Typical distance resolution values range from 8 cm for a 10-ns pulse to 5 m for a 50 μs pulse.

Measurement range deals with how far away an OTDR can identify events in the link, such as splice points, connection points, or fiber breaks. The maximum range R_{max} depends on the fiber attenuation α and on the pulse width, that is, on the dynamic range D_{OTDR}. If the attenuation is given in dB/km, then the maximum range in km is

$$R_{max} = D_{OTDR}/\alpha \qquad (14.5)$$

Example 14.3 Consider an OTDR that has a dynamic range of 36 dB. If a cable installation engineer wants to use this instrument to characterize a fiber with an 0.5-dB/km attenuation, what is the maximum fiber range R_{max} that can be tested?

Solution From Eq. (14.5) the maximum range is $R_{max} = D_{OTDR}/\alpha = 72$ km.

14.6.2 Attenuation Measurements with an OTDR

Rayleigh scattering reflects light in all directions throughout the length of the fiber. This factor is the dominant loss mechanism in most high-quality fibers. The optical power that is Rayleigh-scattered in the reverse direction inside the fiber can be used to determine attenuation.

The optical power at a distance x from the input coupler can be written as

$$P(x) = P(0)exp\left[-\int_0^x \beta(y)dy\right] \qquad (14.6)$$

Here, $P(0)$ is the fiber input power and $\beta(y)$ is the fiber loss coefficient in km^{-1}, which may be position-dependent; that is, the loss may not be uniform along the fiber. The parameter 2β can be measured in natural units called *nepers*, which are related to the loss $\alpha(y)$ in decibels per kilometer through the relationship (see Appendix B)

$$\beta\left(km^{-1}\right) = 2\beta(nepers) = \frac{\alpha(dB)}{10\log e} = \frac{\alpha(dB)}{4.343} \qquad (14.7)$$

Under the assumption that the scattering is the same at all points along the optical waveguide and is independent of the modal distribution, the power $P_R(x)$ scattered in the reverse direction at the point x is

$$P_R(x) = SP(x) \qquad (14.8)$$

Here, S is the fraction of the total power that is scattered in the backward direction and trapped in the fiber. Thus the backscattered power from point x that is seen by the OTDR photodetector is

$$P_D(x) = P_R(x)exp\left[-\int_0^x \beta_R(y)dy\right] \qquad (14.9)$$

where $\beta_R(y)$ is the loss coefficient for the reverse-scattered light. Because the modes in the fiber excited by the backscattered light can be different from those launched in the forward direction, the parameter $\beta_R(y)$ may be different from $\beta(y)$.

Substituting Eqs. (14.7) and (14.8) into Eq. (14.9) yields

$$P_D(x) = SP(0)exp\left[-\frac{2\overline{\alpha}(x)x}{10\log e}\right] \tag{14.10}$$

where the average attenuation coefficient $\overline{\alpha}(x)$ of the forward and reverse losses is defined as

$$\overline{\alpha}(x) = \frac{1}{2x}\int_0^x [\alpha(y) + \alpha_R(y)]dy \tag{14.11}$$

Using this equation, the average attenuation coefficient can be found from a data trace such as the one shown in Fig. 14.13. For example, the average attenuation between two points x_1 and x_2, where $x_1 > x_2$, is

$$\overline{\alpha} = -\frac{10\left[\log P_D(x_2) - \log P_D(x_1)\right]}{2(x_2 - x_1)} \tag{14.12}$$

Example 14.4 An OTDR is used to measure the attenuation of a long length of fiber. If the optical power level measured by the OTDR at the 8-km point is 0.5 of the measured value at the 3-km point, what is the fiber attenuation?

Solution Eq. (14.12) can be expressed as

$$\alpha = \frac{10\log\left[\frac{P_D(x_2)}{P_D(x_1)}\right]}{2(x_2 - x_1)} = \frac{10\log 0.5}{2(8-3)} = 0.3\,\text{dB/km}$$

Drill Problem 14.3 An OTDR is used to measure the attenuation of a 300-m long plastic optical fiber. If the optical power level measured at the 300-m point is 0.025 of the power measured at the 100-m point, show that the fiber attenuation is 0.040 dB/m or 40 dB/km.

14.6.3   OTDR Dead Zone

The concept of a dead zone is another important OTDR specification. *Dead zone* is the distance over which the photodetector in an OTDR is saturated momentarily after

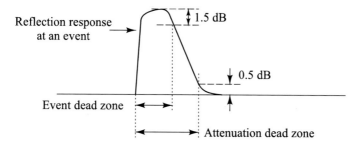

Fig. 14.14 Two specifications for dead zone are the *event dead zone* and the *attenuation dead zone*

it measures a strong reflection. As Fig. 14.14 shows, there are two specifications for dead zone. An *event dead zone* specifies the minimum distance over which an OTDR can detect a reflective event that follows another reflective event. Typically vendors specify this as the distance between the start of a reflection and the −1.5-dB point on the falling edge of the reflection. A short pulse width is used when measuring the event dead zone. For example, a 30-ns pulse width would give a 3-m event dead zone.

The *attenuation dead zone* indicates over what distance the photodetector in an OTDR needs to recover following a reflective event before it is again able to detect a splice. This means that the receiver has to recover to within 0.5 dB of the backscatter value. Nominal attenuation dead zones range from 10 to 25 m.

Typically an OTDR dead zone is the same length as the distance that the optical pulse covers in a fiber plus a few meters. Thus OTDR vendors have started employing a special length of fiber called an *optical pulse suppressor* (OPS), which is inserted between the OTDR and the fiber. The OPS moves the dead zone from the beginning of the fiber under test to this special fiber. This can reduce the event dead zone to about 1 m, so that anomalies occurring within a short distance, for example, within the cabling system of a central office, may be detected and measured.

14.6.4 Locating Fiber Faults

To locate breaks and imperfections in an optical fiber, the fiber length L (and, hence, the position of the break or fault) can be calculated from the time difference between the pulses reflected from the front of the fiber and the event location. If this time difference is t, then the length L is given by

$$L = \frac{ct}{2n_1} \tag{14.13}$$

where n_1 is the core refractive index of the fiber. The number "2" in the denominator accounts for the fact that light travels a length L from the source to the break point and then another length L on the return trip back to the source.

Example 14.5 Consider a long optical fiber with a core refractive index $n_1 = 1.460$. Suppose that an engineer uses OTDR to locate a break in the fiber. If the break is located 15 km away, what is the return time of an OTDR test pulse?

Solution Using Eq. (14.13) the return time is.

$$t = \frac{2n_1 L}{c} = \frac{2(1.460)(15\,\text{km})}{3 \times 10^5\,\text{km/s}} = 0.146\,\text{ms}$$

The fault-location accuracy dL of an OTDR can be found by differentiating Eq. (14.13) to get

$$dL = \frac{c}{2n_1}dt \qquad (14.14)$$

Here dt is the accuracy to which the time difference between the original and reflected pulses must be measured. For $dL \le 0.5$ m and with $n_1 = 1.480$, it is necessary to have.

$$dt = \frac{2n_1}{c}dL \le \frac{2(1.480)}{3 \times 10^8\,\text{m/s}}(0.5) = 4.9\,\text{ns}$$

To measure dt to this accuracy, the pulse width must be $\le 0.5dt$ (because the time difference is measured between the original and reflected pulse widths). Thus a pulse width of 2.5 ns or less is needed to locate a fiber fault within 0.5 m of its true position.

14.6.5   Measuring Optical Return Loss

Reflections of the light in a backward direction occur at various points in optical links that use laser transmitters. This can occur at connectors, fiber ends, optical coupler interfaces, and within the fiber itself due to Rayleigh scattering. The percent of power reflected back from a particular point in a light path is called *back reflection*. If it is not controlled, the back reflections can cause optical resonance in the laser source and result in erratic operation and increased laser noise. In addition, the back reflections can undergo multiple reflections in the transmission line and increase the bit-error rate when they enter the receiver.

Therefore it is desirable to measure the *optical return loss* (ORL), which is the percent of total reverse power in relation to total forward power at a particular point. The ORL is expressed as a ratio of reflected power P_{ref} to incident power P_{inc}

$$\text{ORL} = 10\log(P_{\text{ref}}/P_{\text{inc}}) \qquad (14.15)$$

One can use either an OTDR or an ORL meter to measure this parameter. Although an OTDR can give precise reflectance values at individual events along a fiber transmission path, it has a limitation in measuring the back reflections near and within the OTDR dead zone. Because such an event can be a major contributor to ORL, it is better to use a return loss meter. Such meters are available commercially as compact handheld devices.

14.7   Optical Performance Monitoring

Modern communication networks have become an essential part of society with applications ranging anywhere from simple web browsing to high-profile business transactions. Due to the importance of these networks to everyday life, users have come to expect the network to always be available and to function properly. To offer services with an extremely high degree of reliability, operators need to have a means to monitor the health and status of all parts of their network continuously. In a modern network this monitoring function is the performance management subset of a larger set of network management functions. Basically the network health is assessed by means of a continuous inline BER measurement. The information obtained from this test is used to assure that the quality-of-service (QoS) requirements are met. In addition, another standard network management function is fault monitoring, which checks to see where and why a network failure has occurred or is about to take place.

Optical performance monitoring (OPM) adds to these standard network management concepts by checking the status of elements in the physical layer to examine the temporal behavior of the basic performance factors that affect the signal quality. Depending on the desired network control complexity and the system cost constraints, optical performance monitoring can range anywhere from simply checking the optical power level of each WDM channel to a highly sophisticated system that identifies the origins of a wide range of signal impairments and assesses their impact on network performance [13–15].

First Sect. 14.7.1 gives an overview of generic network management functions to show their relationship to OPM. Then Sect. 14.7.2 discusses management functions defined by the ITU-T for the multiple wavelengths in the optical layer. These functions are an extension of the standard SONET/ SDH procedures used for managing a single wavelength. Next Sect. 14.7.3 describes three levels of monitoring functions that can be carried through different categories of OPM. Sections 14.7.4 through 14.7.5 then give some general examples of OPM procedures. These include network maintenance, fault management, and OSNR monitoring. Section 14.8 describes some specific measurement methods.

14.7.1   Network Management Systems and Functions

Once the hardware and software elements of an optical network have been installed properly and integrated successfully, they need to be managed to ensure that the required level of network performance is being met. In addition, the network devices must be monitored to verify that they are configured properly and to ensure that corporate policies regarding network use and security procedures are being followed. This is carried out through *network management*, which is a service that uses a variety of hardware and software tools, applications, and devices to assist human network managers in monitoring and maintaining networks.

Figure 14.15 shows the components of a typical network management system and their relationships. The *network management console* is a specialized workstation that serves as the interface for the human network manager. There can be several of these workstations that perform different functions in a network. From such a console a network manager can view the health and status of the network to verify that all devices are functioning properly, that they are configured correctly, and that their application software is up to date. A network manager also can see how the network is performing, for example, in terms of traffic loads and fault conditions. In addition, the console allows control of the network resources.

The *managed elements* are network components, such as optical transmitters and receivers, optical amplifiers, optical add/drop multiplexers (OADMs), and optical crossconnects (OXCs). Each such device is monitored and controlled by its *element management system* (EMS). Management software modules, called *agents*, residing in a microprocessor within the elements continuously gather and compile information on the status and performance of the managed devices. The agents store this

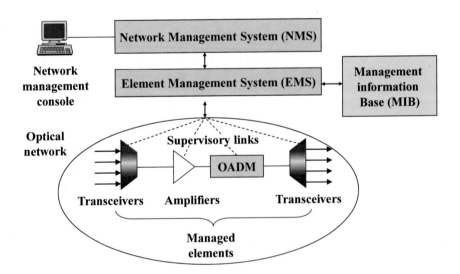

Fig. 14.15 Components of a typical network management system and their relationships

information in a *management information base* (MIB), and then provide the data to *management entities* within a *network management system* (NMS) that resides in the management workstation. A MIB is a logical base of information that defines data elements and their appropriate syntax and identifier, such as the fields in a database. This information may be stored in tables, counters, or switch settings. The MIB does not define how to collect or use data elements. It only specifies what the agent should collect and how to organize these data elements so that other systems can use them. The information transfer from the MIB to the NMS is done via a *network management protocol* such as the widely used *simple network management protocol* (SNMP).

When agents notice problems in the element they are monitoring (for example, link or component faults, wavelength drifts, reduction in optical power levels, or excessive bit-error rates), they send alerts to the management entities. Upon receiving an alert, the management entities can initiate one or more actions such as operator notification, event logging, system shutdown, or automatic attempts at fault isolation or repair. The EMS also can query or poll the agents in the elements to check the status of certain conditions or variables. This polling can be automatic or operator-initiated. In addition, there are *management proxies* that provide management information on behalf of devices that are not able to host an agent.

Network management functions can be classified into the five generic categories listed in Table 14.6. These are performance, configuration, accounting, fault, and security management.

Table 14.6 The purposes of five basic network management functions

Management function	Purpose
Performance management	Monitor and control parameters that are essential to the proper operation of a network in order to guarantee a specific quality of service to network users
Configuration management	Monitor network setup information and network device configurations to track and manage the effects on network operation of the various hardware and software elements
Accounting management	Measure network-utilization parameters so that individuals or groups of users on the network can be regulated and billed for services appropriately
Fault management	Detect fault or degradation symptoms, determine the origin and possible cause of faults, issues instructions on how to resolve the fault
Security management	Develop security policies, set up a network security architecture, implement firewall and virus-protection software, establish access-authentication procedures

14.7.2 Optical Layer Management

To deal with standardized management functions in the optical layer, the ITU-T defined a three-layer *optical transport network* (OTN) model in ITU-T Rec. G.709, which also is referred to as the *Digital Wrapper* standard. Just as the SONET/SDH standard enabled the management of single-wavelength optical networks using equipment from many different vendors, the G.709 standard enables the broad adoption of technology for managing multiwavelength optical networks. The structure and layers of the OTN closely parallel the path, line, and section sublayers of SONET.

The model is based on a client/server concept. The exchange of information between processes running in two different devices connected through a network may be characterized by a *client/server interaction*. The terms *client* and *server* describe the functional roles of the elements in the network, as Fig. 14.16 illustrates. The process or element that requests or receives information is called the *client*, and the process or element that supplies the information is called the *server*.

Figure 14.17 illustrates the three-layer model for a simple link. Client signals such as IP, Ethernet, or OC-N/STM-M are mapped from an electrical digital format into an optical format in an optical channel (OCh) layer. The OCh deals with single wavelength channels as end-to-end paths or as subnetwork connections between routing nodes. The *optical multiplex section* (OMS) layer represents a link carrying groups of wavelengths between multiplexing equipment or OADMs. The *optical transport section* (OTS) layer relates to a link between two optical amplifiers. Figure 14.18 shows where these sections fit into a link.

The OCh is divided further into the three sublayers shown in Fig. 14.17: the *optical channel transport unit* (OTU), the *optical channel data unit* (ODU), and the *optical channel payload unit* (OPU). Each of these sublayers has specific functions and associated overhead, which are as follows.

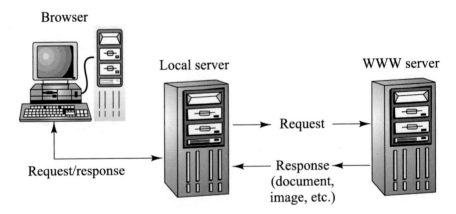

Fig. 14.16 The terms *client* and *server* describe the functional roles of communicating elements in the network; here the browser is the client

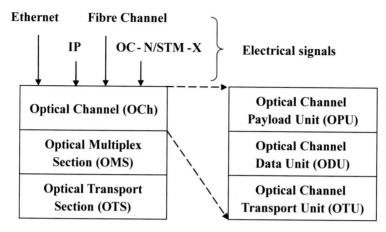

Fig. 14.17   Three-layer model for a simple link in an OTN and the three OCh sublayers

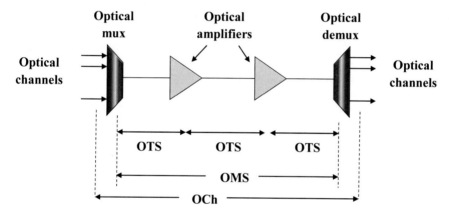

Fig. 14.18   The OMS layer represents a link carrying wavelengths between multiplexers or OADMs; the OTS layer relates to a link between two optical amplifiers

Optical channel payload unit The OPU frame structure contains the client signal payload and the overhead necessary for mapping any client signal into the OPU. Mapping of client signals may include rate adaptation of the client signal to a constant bit-rate signal. Examples of common signals are IP, various forms of Ethernet, Fibre Channel, and SONET/SDH. The three payload rates associated with the OPU sublayer are 2.5, 10, and 40 Gb/s. These correspond to standard SONET/SDH data rates (OC-48/STM-16, OC-192/STM-64, and OC-768/STM-256, respectively) but may be used for any client signal.

Optical channel data unit The ODU is the structure used to transport the OPU. The ODU consists of the OPU and the associated ODU overhead and provides path-layer connection monitoring functions. The ODU overhead contains information that enables maintenance and operation of optical channels. This information includes maintenance signals, path monitoring, tandem connection monitoring, automatic protection switching, and designation of fault type and location.

Optical channel transport unit The optical channel transport unit (OTU) contains the ODU frame structure, the OTU overhead, and appended forward error correction (FEC). The OTU changes the digital format of the ODU into a light signal for transport over an optical channel. It also provides error detection and correction and section layer connection monitoring functions.

14.7.3   Fundamental OPM Function

The fundamental function of optical performance monitoring is to examine the temporal behavior of performance factors that may affect the health of an optical signal. This process involves checking the operational status of elements in the physical layer and assessing the quality of the optical signals in each WDM channel. OPM can be viewed by means of the following three layers.

Transport-layer monitoring deals with optical-domain characteristics that relate to WDM channel management. This involves real-time examinations of factors such as the presence of a channel, whether the wavelength has been registered by the system, and the optical power level, spectral content, and OSNR of each WDM channel.

Optical signal monitoring examines the quality of each WDM channel. This measurement function examines the signal quality features of an individual channel. These features include the Q factor, the electronic SNR, and various eye-diagram statistics such as openness and distortions resulting from dispersion or nonlinear effects.

Protocol performance monitoring deals with digital measurements such as the BER.
    The main factors that OPM is watching for are component faults and signal impairments. *Component faults* can result from malfunctions or degradations of elements, improperly installed or configured equipment, or damage to a network (such as a backhoe digging up a cable or a storm destroying a fiber line). *Signal impairments* can arise from many diverse factors. Among these are noises and transients from optical amplifiers, chromatic and polarization-mode dispersions, nonlinear effects, and timing jitter.
    All these factors taken together present a big challenge to devising a comprehensive OPM system. However, a single OPM system does not need to check all possible degradation mechanisms. In fact, cost constraints prevent the deployment of such a super-sophisticated performance monitoring procedure. In a practical network, the simplest OPM system might monitor only the optical power levels of each channel at a particular point in a WDM network. Advanced OPM systems will include using a miniaturized spectrometer to control the outputs of devices such as optical amplifiers

and variable optical attenuators. More complex OPM systems are needed in reconfigurable networks to track the amount of accumulated dispersion on a per channel basis, because the effect on system performance of this impairment will vary as the network configuration changes.

14.7.4 OPM Architecture for Network Maintenance

An OPM taps off a small portion of the light signals in a fiber and separates the wavelengths or scans them onto a detector or detector array. This enables the measurement of individual channel powers, wavelength, and OSNR. These devices have an important role in controlling DWDM networks. For example, as shown in Fig. 14.19, most long-haul DWDM networks incorporate automated end-to-end power-balancing algorithms that use a high-performance OPM to measure the optical power level of each wavelength at optical amplifiers and at the receiver and to adjust the individual laser outputs at the transmitter. This information is exchanged by means of a separate supervisory channel, which uses a wavelength that lies outside of the signal spectrum but within the response band of the amplifier. In addition, manufacturers may embed an OPM function into dynamic elements such as an EDFA, an OADM, or an OXC to provide feedback for active control of total output power and to balance the power levels between channels. Other functions of an OPM include determining if a particular channel is active, verifying whether wavelengths match the specified channel plan, and checking whether optical power and OSNR levels are sufficient to meet the QoS requirements.

An OPM may have the following operational characteristics:

- Measures absolute channel power to within ± 0.5 dBm
- Identifies channels without prior knowledge of the wavelength plan
- Makes full S-, C-, or L-band measurements in less than 0.5 s

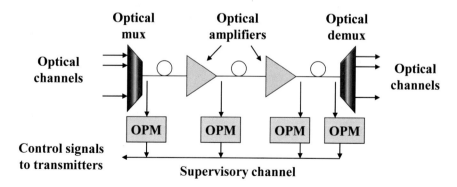

Fig. 14.19 DWDM networks might use an automated OPM to measure the light level of each wavelength at various network points and to adjust the individual laser outputs at the transmitter

- Measures center wavelength accuracy to better than ± 50 pm
- Determines OSNR with a 35-dB dynamic range to a ± 0.1-dB accuracy.

14.7.5 Detecting Network Faults

Faults in a network, such as physical cuts in a fiber transmission line or failure of a circuit card or optical amplifier, can cause portions of a network to be inoperable. Because network faults can result in system downtime or unacceptable network degradation, *fault management* is one of the most widely implemented and essential network management functions. As Fig. 14.20 illustrates, fault management involves the following processes:

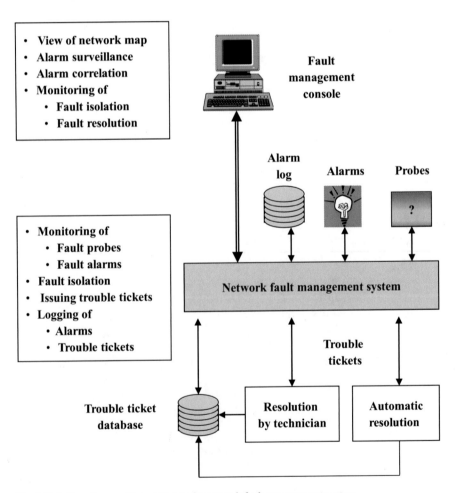

Fig. 14.20 Functions and interactions of a network fault management system

- Detecting faults or degradation symptoms. This can be done with *alarm surveillance*, which involves reporting alarms that may have different levels of severity and indicating possible causes of these alarms. Fault management also provides a summary of unresolved alarms and allows the network manager to retrieve and view the alarm information from an alarm log.
- Determining the origin and possible cause of faults either automatically or through the intervention of a network manager. To determine the location or origin of faults, the management system might use *fault-isolation* techniques such as alarm correlation from different parts of the network and diagnostic testing.
- After the faults are isolated, the system issues *trouble tickets* that indicate what the problem is and possible means of how to resolve the fault. These tickets go to either a technician for manual intervention or an automatic fault-correction mechanism. When the fault or degradation is corrected, this fact and the resolution method are indicated on the trouble ticket, which then is stored in a database.
- After the problem has been fixed, the repair is operationally tested on all major subsystems of the network. *Operational testing* involves requesting performance tests, tracking the progress of these tests, and recording the results. The classes of tests that might be performed include echo tests and connectivity examinations.

A basic factor in troubleshooting faults is to have a comprehensive physical and logical map of the network. Ideally this map should be part of a software-based management system that can show the network connectivity and the operational status of the constituent elements of the network on a display screen. With such a map, failed or degraded devices can be viewed easily and corrective action can be taken immediately.

14.8 Optical Fiber Network Performance Testing

The evolution of optical fiber communication technology has resulted in highly reliable telecom transmission systems for applications that include high-capacity long-distance links and optical metro, access, and in-building networks. A wide variety of communication protocols, data modulation formats, performance monitoring techniques, and performance testing methods have been devised to keep these systems running smoothly and reliably. In terms of performance testing, the major measurement methods include the bit-error rate (BER), the optical signal-to-noise ratio (OSNR), the Q factor, timing jitter, and the optical modulation amplitude. This section provides an overview of these techniques. Further extensive details can be found in the literature [13–20].

14.8.1   BER Measurements

The BER is an important performance quality indicator of a digital communication link. Because BER is a statistical parameter, its value depends on the measurement time and on the factors that cause the errors, such as signal dispersion, accumulated excess noise, and timing jitter. When BER measurements are made, both the number of misinterpreted bits and the total number of received bits are counted in a specific time window ΔT, which is called the *gating time*. If the errors are due to Gaussian noise in a relatively stable transmission link, then the BER does not fluctuate significantly over time, as is illustrated in Fig. 14.21a. In this case, a gating time window in which about 100 errors occur is needed to ensure a statistically valid BER. When bursts of errors occur, as shown in Fig. 14.21b, longer measurement times may be needed to accumulate 100 errors in order for the test to be statistically accurate.

From Eq. (7.5) it follows that when N_e errors occur in a time window ΔT at a bit rate B, then the BER is given by

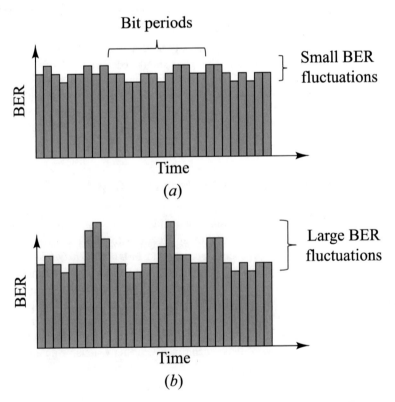

Fig. 14.21 Sequence of bit periods with **a** a relatively stable BER and **b** a bursty BER

$$BER = \frac{N_e}{B \, \Delta T} \qquad (14.16)$$

Thus the gating time window needed to measure $N_e = 100$ errors is $\Delta T = 100/(BER \times B)$. For high- speed communications greater than 1 Gb/s, the required bit-error rate typically needs to be 10^{-12} or lower.

Example 14.6 A given 10-Gb/s link is designed to operate with a 10^{-12} BER. What is the gating time window for accumulating 100 bit errors?

Solution From Eq. (14.16) the gating time is $\Delta T = 100/(BER \times B) = 100/(10^{-12} \times 10^{10}) = 10^4$ s ≈ 2.7 h.

However, a 10^{-12} BER level may be unacceptable for a 10-Gb/s data rate, so even lower bit-error rates, such as 10^{-15}, may be required to assure customers of a high grade of service. To accumulate 100 errors for such a BER would require over 100 days of measurement time. Because this is not practical, modern BER measuring instruments add an extra precisely calibrated amount of noise into the system, thereby accelerating the occurrence of errors. The extra noise decreases the receiver threshold, which increases the probability of errors and thus greatly reduces the gating time window. Although some accuracy is lost in this method, it reduces the test times to minutes instead of hours or days.

Example 14.7 In a real communication system, network operators also are interested in the frame error rate P_{frame}. If the total number of bits in a frame is k and P_e is the BER, then the probability that no bit in the frame has an error is.

$$1 - P_{frame} = (1 - P_e)^k \approx 1 - P_e k \qquad (14.17)$$

where the approximation results from the condition that $P_e \ll 1$. If $P_e = 10^{-12}$, what is the frame-error rate for an Ethernet frame length of 1518 bytes?

Solution Because there are 8 bits per byte, this frame has 12,144 bits. From Eq. (14.17) it follows that $P_{frame} = P_e \, k = (10^{-12}) \times 12{,}144 = 1.2144 \times 10^{-8}$.

14.8.2 OSNR Measurements

Measuring the SNR and its associated BER is straightforward for single-wavelength links with no optical amplifiers. However, in multispan optically amplified DWDM networks, the system performance is limited by the OSNR rather than by the optical signal power that arrives at the receiver. Although one could demultiplex the incoming DWDM traffic and then do BER estimates on each individual wavelength channel, an optical spectrum measurement can be performed with an *optical spectrum analyzer* (OSA) to derive the OSNR for each individual channel. The OSNR derived from the optical spectrum is an average-power, low-speed measurement, so it does not give information about the effects of temporal impairments on channel performance. However, because it can be correlated to the BER, the OSNR provides indirect BER information for preliminary performance diagnosis of a multichannel system or for giving advance warning of a possible BER degradation on a given DWDM channel.

As described in Sect. 11.5, the OSNR is given by

$$\text{OSNR} = \frac{P_{ave}}{P_{ASE}} \tag{14.18}$$

or in decibels,

$$\text{OSNR(dB)} = 10 \log \frac{P_{ave}}{P_{ASE}} \tag{14.19}$$

Example 14.8 Consider an optical signal level of -15 dBm ($32\ \mu\text{W}$) arriving at a *pin* optical receiver in a 10-Gb/s link. If the noise power is -34.5 dBm ($0.35\ \mu\text{W}$), what is the OSNR?

Solution From Eq. (14.18) it follows that OSNR $= 32/0.35 = 91$ or, in decibels, OSNR(dB) $= 10 \log 91 = 19.6$ dB.

The OSNR does not depend on factors such as the data format, pulse shape, or optical filter bandwidth, but only on the average optical signal power P_{ave} measured by the OSA and on the average ASE noise power P_{ASE}. OSNR is a metric that can be used for performance verification in the design and installation of networks, as well as to check the health and status of individual optical channels. Sometimes an optical filter is used to significantly reduce the total ASE noise seen by the receiver. Typically such a filter has an optical bandwidth that is large compared to the signal, so that it does not affect the signal, yet that is narrow compared to the bandwidth associated with the ASE background. The ASE noise filter does not change the OSNR. However, it reduces the total power in the ASE noise to avoid overloading the receiver front end.

The IEC Standard 61280-2-9 defines the OSNR as the ratio of the signal power at the peak of a channel to the noise power interpolated at the position of the peak. This document defines OSNR by the following equation

$$\text{OSNR} = 10 \log \frac{P_i}{N_i} + 10 \log \frac{B_m}{B_0} \tag{14.20}$$

where

- P_i is the optical signal power in watts in the ith channel,
- N_i is the interpolated value of the noise power in watts measured in the resolution bandwidth B_m at the midchannel spacing point,
- B_m is the resolution bandwidth of the measurement,
- B_o is the reference optical bandwidth, which typically is chosen to be 0.1 nm.

The second term in Eq. (14.20) is used to give an OSNR value that is independent of the resolution bandwidth B_m of the instrument. This allows a comparison of OSNR results that may have been obtained with different OSA instruments. The IEC 61280-2-9 standard also notes that in order to achieve an adequate OSNR measurement, the wavelength measurement range of the OSA must be wide enough to include all DWDM channels plus one-half of an ITU-T grid spacing at each end of the spectral range. In addition, the resolution bandwidth must be wide enough to include the entire signal power spectrum of each modulated channel because this has a direct impact on the accuracy of the noise measurement.

14.8.3   Q Factor Estimation

As noted by Eq. (7.13), the probability of error P_e in a digital communication link is related to the Q factor through the expression

$$P_e = BER = \frac{1}{2}erfc\left(\frac{Q}{\sqrt{2}}\right) = \frac{1}{2}\left[1 - erf\left(\frac{Q}{\sqrt{2}}\right)\right]$$

$$\approx \frac{1}{\sqrt{2\pi}}\frac{1}{Q}exp\left(-Q^2/2\right) \tag{14.21}$$

Recall from Eq. (7.14) that Q is proportional to the power difference between the logic 1 and 0 levels. Thus a simple way to examine the error probability versus the Q factor is to vary Q by changing the optical power level at the receiver. For a clear eye diagram, the decision threshold is midway between the 0 and 1 levels, and the noise variance contributed by the receiver remains constant as the input power is varied. These conditions generally hold for *pin* optical receivers where thermal noise in a transimpedance amplifier is the dominant noise.

A different approach is needed when the eye pattern is distorted. In this case, the eye-pattern mask technique described in Sect. 14.5.1 can be used to estimate system performance. This is particularly useful in multispan optically amplified DWDM networks in which the system performance is limited by the OSNR. When optical amplifiers are used in a transmission link, the optical power level at the receiver usually is high enough so that thermal noise and dark current noise can be neglected compared to the signal-ASE noise and the ASE-ASE beat noise. For a distorted eye pattern, the Q factor can be expressed as [20]

$$Q = \frac{2\mathscr{R}(A - B)P_{ave}}{\sqrt{(G_1 A + G_2)B_e} + \sqrt{(G_1 B + G_2)B_e}} \tag{14.22}$$

where $G_1 = 4\mathscr{R}(q + \mathscr{R}S_{ASE})P_{ave}$ and $G_2 = (S_{ASE}\mathscr{R})^2(2B_o - B_e)$. The dimensionless parameters A and B are the upper and lower eye mask boundaries, respectively, as illustrated in Fig. 14.9. Furthermore, $P_{ave} = (P_1 + P_2)/2$, $\mathscr{R}$ is the responsively, S_{ASE}

is the power spectral density of the ASE noise (see Sect. 11.4), B_e is the receiver electrical bandwidth, and B_o is the optical bandwidth, which usually is taken to be 0.1 nm.

For OSNR values greater than about 15 dB, the signal-ASE noise is the dominating noise factor, so that the contributions from ASE-ASE beat noise and shot noise can be neglected. In this case, the Q factor for a distorted eye pattern can be expressed by the simplified relationship [20]

$$Q = \frac{\left(\sqrt{A} - \sqrt{B}\right)\sqrt{P_{ave}}}{\sqrt{S_{ASE} B_e}} = \frac{\left(\sqrt{A} - \sqrt{B}\right)}{\sqrt{B_e}}\sqrt{OSNR} \qquad (14.23)$$

where the relationship between P_{ave} and OSNR is given by Eqs. (11.24) and (11.36).

Example 14.9 Consider an amplified transmission link for which the receiver Q factor expression for a distorted eye pattern is given by Eq. (14.23) when OSNR > 15 dB. For an OSNR value of 16, compare the values of the Q factor when (a) there is no eye closure penalty, that is, $A = 1$ and $B = 0$ and (b) with an eye closure penalty that has $A = 0.81$ and $B = 0.25$.

Solution From Eq. (14.23) it follows that

(a) $Q = 1 \times \dfrac{4}{\sqrt{B_e}}$

(b) $Q = (0.9 - 0.5) \times \dfrac{4}{\sqrt{B_e}} = 0.4 \times \dfrac{4}{\sqrt{B_e}}.$

14.8.4 OMA Measurement Method

Compared to long-haul lightwave networks, a different approach must be taken when testing optical Ethernet links that are based on the 10-Gb/s IEEE 802.3ae standard. One reason is that the lasers used for long-haul links usually are high-quality devices that operate with high extinction ratios. Therefore, in the long-haul case, to characterize the sensitivity of an optical receiver to input signals, engineers simply can use a slow responding power meter to measure the average optical signal power and hence determine the BER.

In contrast, the need to reduce cost in optical Ethernet links means that often one needs to use less expensive lasers that have lower extinction ratios but provide adequate performance at 10-Gb/s in metro, access, and campus networks. In this case, the low extinction ratio will result in the stressed (partially closed) eye shown in Fig. 14.22, and an average optical power measurement does not give a good indication of receiver performance. Consequently, the traditional tests used to characterize long-haul receivers need to be modified for 10-Gb/s optical Ethernet links. This

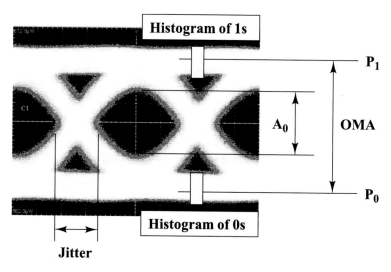

Fig. 14.22 Definitions of measurement parameters for analyzing a stressed eye

requirement led to the concept of using an *optical modulation amplitude* (OMA) method to measure the characteristics of a stressed eye.

The parameters for measuring the stressed eye are shown in Fig. 14.22. To measure the OMA, a transmitter puts out a repetitive square-wave pattern of typically five 1 bits and five 0 bits (…1111100000111110000…). The three key parameters derived from this pattern at the receiver are:

- A logic 1 amplitude P_1, which is taken from a histogram mean across the middle bit of the run of 1 bits. Only the middle bit of each sequence of 1 bits is selected so that the measured data points are far from any bit edge.
- A logic 0 amplitude P_0, which is taken from a histogram mean across the middle bit of the run of 0 bits when using the above repetitive square-wave pattern.
- A_0 is the height of the eye opening.

The *optical modulation amplitude* is defined as the difference between the high and low power levels:

$$OMA = P_1 - P_0 \qquad (14.24)$$

The metric spelled out in the IEEE 802.3ae standard for characterizing optical receivers is the *vertical eye closure penalty* (VECP). The VECP measures the vertical opening at the center 1% of the eye, which is given by the parameter A_0, and compares that value to the measured OMA value. Thus in decibels,

$$VECP = 10 \log \left(\frac{OMA}{A_0} \right) \qquad (14.25)$$

Table 14.7 Some IEEE
802.3 receiver test
requirements

10G Ethernet type	10G-Base-S	10G-Base-L	10G-Base-E
Extinction ratio (dB)	3.0	3.5	3.0
VECP (dB)	3.5	2.2	2.7

Table 14.7 lists the stressed receiver extinction ratio and VECP requirements for short-reach, long-reach, and extended-reach 10-Gb/s Ethernet receivers, which are designated by 10G-Base-S, 10G-Base-L, and 10G-Base-E, respectively. More extensive details on additional test parameters and methods are given in the IEEE802.3ae standard.

14.8.5 Measurement of Timing Jitter

In digital communication systems, *time jitter*, also called *timing jitter* or simply *jitter*, is defined as an instantaneous unintentional deviation in the ideal timing between binary symbols. Basically jitter occurs when the transition from one symbol state to the next state occurs earlier or later than the exact end of the bit time interval. Many different factors can contribute to jitter, including random amplitude and noise variations in a signal, periodic noise from switching power supplies, and charge storage mechanisms in circuits and photonic components.

Timing jitter is an especially important issue for high-speed optical fiber transmission systems because pulses are spaced very close together. In this case, incorrect interpretation of the edges of bit periods can lead to high BER values. In digital transmission systems timing jitter can be random or deterministic. Random jitter is caused by noises such as thermal and shot noises in the receiver, and from ASE noise accumulated throughout the transmission link. Deterministic jitter arises from pattern distorting effects due to factors such as chromatic dispersion, self-phase modulation, and interchannel crosstalk.

For a bit sequence with a data rate B, a jittered waveform can be expressed as

$$P_{jitter}(t) = P\left[t + \frac{\Delta\varphi(t)}{2\pi B}\right] = P[t + \Delta t(t)] \tag{14.26}$$

Here $\Delta\varphi(t)$ is the phase variation introduced by time jitter, which can be designated in degrees or radians, and $P(t)$ is the waveform in the absence of time jitter. The time deviation Δt, which is given by,

$$\Delta t(t) = \frac{\Delta\varphi(T)}{2\pi B} \tag{14.27}$$

can be expressed in a convenient measure called the *unit interval* (UI). This parameter is the ratio between the time jitter and the bit period $T = 1/B$, and is given by

$$\Delta t_{UI} = \frac{\Delta \varphi(T)}{2\pi} \qquad (14.28)$$

A number of different techniques can be used to measure jitter, including the use of BER testers, sampling oscilloscopes, and jitter detectors. Instruments such as a network performance analyzer have built-in highly accurate jitter measurement capabilities that satisfy the jitter test conditions specified in the ITU-T O.172 Recommendation.

14.9 Summary

Numerous levels of measurement techniques have been developed for characterizing the operational behavior of devices and fibers, for ensuring that the correct components have been selected for a particular application, and for verifying that the network is configured properly. In addition, various operational methods for performance monitoring are needed to verify that all the design and operating specifications of an optical link are met when it is running.

Optical power measurement is the most basic function in fiber optic metrology. However, this parameter is not a fixed quantity and can vary as a function of other parameters such as time, distance along a link, wavelength, phase, and polarization. Because optical power can vary with time, different instantaneous power level readings are obtained depending on the exact instant when the measurement is made. The two standard classes of power measurements are the peak power and the average power. The peak power is the maximum power level in a pulse, which might be sustained for only a very short time. The average power is a measure of the power level averaged over a relatively long time period compared to the duration of an individual pulse.

The use of an eye diagram is a traditional technique for quickly and intuitively assessing the quality of a received signal. Modern BER measurement instruments construct such eye diagrams by generating a pseudorandom pattern of ones and zeros at a uniform rate but in a random manner. When the pulses in this pattern are superimposed simultaneously, an eye pattern is formed on a display screen. Interpretation of the characteristics of a distorted eye diagram is done by means of mask testing. The operating software of most modern BER instruments has a wide selection of built-in masks for different protocols. In addition, the instrument user can key in custom masks for any application or to check the test results differently.

An optical time-domain reflectometer (OTDR) is a versatile portable instrument that is used to evaluate the characteristics of an installed optical fiber link. In addition to identifying and locating faults or anomalies within a link, this instrument measures parameters such as fiber attenuation, length, optical connector, and splice losses, and light reflectance levels.

To offer services with an extremely high degree of reliability, communication network operators need to have a means to continuously monitor the health and status of all parts of their network. Basically the network health is assessed by means of a continuous in-line BER measurement. The information obtained from this test is used to assure that the quality-of-service (QoS) requirements are met. Another standard network management function is fault monitoring, which checks to see where and why a network failure has occurred or is about to take place.

Optical performance monitoring (OPM) adds to these standard network management concepts by checking the status of elements in the physical layer to examine the temporal behavior of the basic performance factors that affect signal quality. Depending on the desired network control complexity and the system cost constraints, optical performance monitoring can range anywhere from simply checking the optical power level of each WDM channel to a highly sophisticated system that identifies the origins of a wide range of signal impairments and assesses their impact on network performance.

Problems

14.1   Consider the NRZ and RZ waveforms shown in Fig. 14.3. If the peak power in each waveform is 0.5 mW, show that the average powers are 0.25 and 0.125 mW for the NRZ and RZ patterns, respectively.

14.2   An engineer wants to find the attenuation at 1310 nm of an 1895-m long fiber. The only available instrument is a photodetector, which gives an output reading in volts. Using this device in a cutback-attenuation setup, the engineer measures an output of 3.31 V from the photodiode at the far end of the fiber. After cutting the fiber 2 m from the source, the output voltage from the photodetector now reads 3.78 V. Show that the attenuation of the fiber is 0.31 dB/km.

14.3   A field engineer has a 2400-m long optical cable that has connectors on both ends. Using an insertion loss technique and an optical power meter, the emerging optical power at the output end of the fiber is measured as 0.150 mW. If the power launched into the fiber is 0.65 mW, show that the cable attenuation (including connectors) is 6.37 dB.

14.4   Suppose an optical network element with attached flyleads is inserted into the link in Fig. 14.6a. Assume the flyleads are terminated with optical connectors. Consider the case when the power at the photodetector prior to inserting the component is $P_1 = 0.42$ mW and the power level with the optical element in the link is $P_2 = 0.35$ mW. Show that the insertion loss of the network element is 0.79 dB.

14.5   The optical power in a fiber at a distance x from the input end is given by Eq. (14.6). By assuming that the loss coefficient is uniform along the fiber, use this equation to derive Eq. (14.2).

14.6   Assuming that Rayleigh scattering is approximately isotropic (uniform in all directions), show that the fraction S of scattered light trapped in a multimode fiber in the backward direction is given by

$$S = \frac{\pi (NA)^2}{4\pi n^2} = \frac{1}{4}\left(\frac{NA}{n}\right)^2$$

where NA is the fiber numerical aperture, n is the core refractive index, and NA/n represents the half-angle of the cone of captured rays. If $NA = 0.20$ and $n = 1.50$, show that fraction of the scattered light that is recaptured by the fiber in the reverse direction is $S = 0.004$.

14.7 Suppose a single-mode OTDR has a usable dynamic range of 30 dB. Assuming typical fiber attenuation of 0.20 dB/km at 1550 nm and splices every 2 km (loss of 0.1 dB per splice), show that such a unit will be able to accurately certify distances of up to 120 km.

14.8 Three 5-km-long fibers have been spliced together in series and an OTDR is used to measure the attenuation of the resultant fiber. The reduced data of the OTDR display is shown in Fig. 14.23. (a) What are the attenuations in decibels per kilometer of the three individual fibers? (b) What are the splice losses in decibels? (c) What are some possible reasons for the large splice loss occurring between the second and third fibers? [Answers: (a) Attenuations: 0.40 dB/km for fiber 1, 0.36 dB/km for fiber 2, and 0.59 dB/km for fiber 3; (b) Splice losses: –0.5 dB for splice 1 and –2.0 dB for splice 2; (c) Large splice losses could occur because of mismatched fiber geometries or poor fiber end-face preparation.]

14.9 Let α be the attenuation of the forward-propagating light, α_s the attenuation of the backscattered light, and S the fraction of the total output power scattered in the backward direction, as described in Eq. (14.8). Show that the backscatter response of a rectangular pulse of width W from a point a distance L down the fiber is

$$P_S(L) = S\frac{\alpha_s}{\alpha} P_0 e^{-2\alpha L}\left(1 - e^{-\alpha W}\right)$$

Fig. 14.23 OTDR trace for Problem 14.8

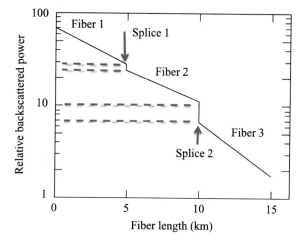

when $L \geq W/2$, and

$$P_S(L) = S\frac{\alpha_S}{\alpha} P_0 e^{-\alpha W}\left(1 - e^{-2\alpha L}\right)$$

for $0 \leq L \leq W/2$.

14.10   Using the expression given in Prob. 14.9 for the backscattered power $P_S(L)$ from a rectangular pulse of width W, show that for very short pulse widths the backscattered power is proportional to the pulse duration. Note: This is the basis of operation of an OTDR.

References

1. National Institute of Standards and Technology (NIST), Boulder, CO, USA. https://www.nist.gov
2. National Physical Laboratory (NPL), Teddington, UK. https://www.npl.co.uk
3. Physikalisch-Technische Bundesanstalt (PTB), Braunschweig, Germany. https://www.ptb.de
4. ITU-T Rec. G.650.2, Definitions and test methods for statistical and nonlinear related attributes of single-mode fibre and cable, Aug 2015
5. TIA-455-191 (FOTP- 191), IEC-60793-1-45 optical fibres-part 1–45: measurement methods and test procedures-mode field diameter, 14 April 2020
6. ITU-T Rec. G.652, Characteristics of a single-mode optical fibre and cable, Nov 2016
7. K. Petermann, Constraints for fundamental mode spot size for broadband dispersion-compensated single-mode fibers. Electron. Lett. **19**, 712–714 (1983)
8. R. Hui, M. O'Sullivan, *Fiber Optic Measurement Techniques* (Academic Press, 2009)
9. ITU-T Rec. G.650.1, Definitions and test methods for linear, deterministic attributes of single-mode fibre and cable, Mar 2018
10. ITU-T Rec. O.201, Q-factor test equipment to estimate the transmission performance of optical channels, July 2003
11. E. Ciaramella, A. Peracchi, L. Banchi, R. Corsini, G. Prati, BER estimation for performance monitoring in high-speed digital optical signals. J. Lightw. Technol. **30**(13), 2117–2124 (2012)
12. The Fiber Optic Association (FOA), Optical time domain reflectometer (OTDR) (2013). www.thefoa.org/tech/ref/testing/OTDR/OTDR.html
13. Z. Pan, C. Yu, A.E. Willner, Optical performance monitoring for the next generation optical communication networks. Opt. Fiber Technol. **16**, 20–45 (2010)
14. C.C.K. Chan, *Optical Performance Monitoring: Advanced Techniques for Next-Generation Photonic Networks* (Academic Press, 2010)
15. ITU-T Rec. G.697, Optical monitoring for DWDM systems, Nov 2016
16. ITU-T Rec. O.172, Jitter and Wander measuring equipment for digital systems which are based on the synchronous digital hierarchy (SDH), Apr 2005; Amendment 2 July 2010
17. C.-L. Yang, S.-L. Lee, OSNR monitoring using double-pass filtering and dithered tunable reflector. IEEE Photonics Technol. Lett. **16**, 1570–1572 (2004)
18. L. Li, J. Li, J. Qiu, Y. Li, W. Li, J. Wu, J. Lin, Investigation of in-band OSNR monitoring technique using power ratio. J. Lightw. Technol. **31**(1), 118–124 (2013)
19. C. Wang, S. Fu, H. Wu, M. Luo, X. Li, M. Tang, D. Liu, Joint OSNR and CD monitoring in digital coherent receiver using long short-term memory neural network. Opt. Express **27**(5), 6936–6945 (2019)
20. B. Szafraniec, T.S. Marshall, B. Nebendahl, Performance monitoring and measurement techniques for coherent optical systems. J. Lightw. Technol. **31**(4), 648–663 (2013)

Correction to: Fiber Optic Communications

Correction to:
G. Keiser, Fiber Optic Communications,
https://doi.org/10.1007/978-981-33-4665-9

The original version of the book was inadvertently published with incorrect sentences in Chapters 1, 3, 5, 8, 10 and 13. The sentences have now been corrected. The book has also been updated with these changes.

The updated versions of these chapters can be found at
　　https://doi.org/10.1007/978-981-33-4665-9_1
　　https://doi.org/10.1007/978-981-33-4665-9_3
　　https://doi.org/10.1007/978-981-33-4665-9_5
　　https://doi.org/10.1007/978-981-33-4665-9_8
　　https://doi.org/10.1007/978-981-33-4665-9_10
　　https://doi.org/10.1007/978-981-33-4665-9_13

Appendix A
International Units and Physical Constants

International Units

Quantity	Unit	Symbol	Dimensions
Length	Meter	m	
Mass	Kilogram	kg	
Time	Second	s	
Temperature	Kelvin	K	
Current	Ampere	A	
Frequency	Hertz	Hz	$1/s$
Force	Newton	N	$(kg\text{-}m)/s^2$
Pressure	Pascal	Pa	N/m^2
Energy	Joule	J	$N\cdot m$
Power	Watt	W	J/s
Electric charge	Coulomb	C	$A\cdot s$
Potential	Volt	V	J/C
Conductance	Siemens	S	A/V
Resistance	Ohm	Ω	V/A
Capacitance	Farad	F	C/V
Magnetic flux	Weber	Wb	$V\cdot s$
Magnetic induction	Tesla	T	Wb/m^2
Inductance	Henry	H	Wb/A

Physical Constants

Constant	Symbol	Value (mks units)
Speed of light in vacuum	c	2.99793×10^8 m/s
Electron charge	q	1.60218×10^{-19} C

(continued)

(continued)

Constant	Symbol	Value (mks units)
Planck's constant	h	6.6256×10^{-34} J·s
Boltzmann's constant	k_B	1.38054×10^{-23} J/K
$k_B T/q$ at T = 300 K	–	0.02586 eV
Permittivity of free space	ε_0	8.8542×10^{-12} F/m
Permeability of free space	μ_0	$4\pi \times 10^{-7}$ N/A^2
Electron volt	eV	$1\ \text{eV} = 1.60218 \times 10^{-19}$ J
Angstrom unit	Å	$1\ \text{Å} = 10^{-4}\ \mu\text{m} = 10^{-8}$ cm
Base of natural logarithm	e	2.71828
Pi	π	3.14159

Appendix B
Decibels

B.1 Definition

In designing and implementing an optical fiber link, it is of interest to establish, measure, and/or interrelate the signal levels at the transmitter, at the receiver, at the cable connection and splice points, at the input and output of a link component, and in the cable. A convenient method for this is to reference the signal level either to some absolute value or to a noise level. This is normally done in terms of a power ratio measured in *decibels* (dB) defined as

$$\text{Power ratio in dB} = 10 \log \frac{P_2}{P_1} \tag{B.1}$$

where P_1 and P_2 are electric or optical powers.

The logarithmic nature of the decibel allows a large ratio to be expressed in a fairly simple manner. Power levels differing by many orders of magnitude can be compared easily when they are in decibel form. Some very helpful figures to remember are given in Table B.1. For example, doubling the power means a 3-dB gain (the power level increases by 3 dB), halving the power means a 3-dB loss (the power level decreases by 3 dB), and power levels differing by factors of 10^N or 10^{-N} have decibel differences of $+10N$ dB and $-10N$ dB, respectively.

Table B.1 Examples of decibel measures of power ratios

Power ratio	10^N	10	2	1	0.5	0.1	10^{-N}
dB	$+10N$	$+10$	$+3$	0	-3	-10	$-10N$

© The Editor(s) (if applicable) and The Author(s), under exclusive license to Springer Nature Singapore Pte Ltd. 2021
G. Keiser, *Fiber Optic Communications*,
https://doi.org/10.1007/978-981-33-4665-9

Table B.2 Examples of dBm units (decibel measure of power relative to 1 mW)

Power (mW)	100	10	2	1	0.5	0.1	0.01	0.001
Value (dBm)	+20	+10	+3	0	−3	−10	−20	−30

B.2 The dBm

The decibel is used to refer to ratios or relative units. For example, one can say that a certain optical fiber has a 6-dB loss (the power level gets reduced by 75% in going through the fiber) or that a particular connector has a 1-dB loss (the power level gets reduced by 20% at the connector). However, the decibel gives no indication of the absolute power level. One of the most common derived units for doing this in optical fiber communications is the *dBm*. This is the decibel power level referred to 1 mW. In this case, the power in dBm is an absolute value defined by

$$\text{Power level} = 10 \log \frac{P}{1 \text{ mW}} \tag{B.2}$$

A useful relationship to remember is that 0 dBm = 1 mW. Negative dBm numbers designate power levels less than 1 mW, whereas positive dBm values indicate power levels greater than 1 mW. Some examples are shown in Table B.2.

B.3 The Neper

The *neper* (N) is an alternative unit that is sometimes used instead of the decibel. If P_1 and P_2 are two power levels, with $P_2 > P_1$, then the power ratio in nepers is given as the natural (or naperian) logarithm of the power ratio:

$$\text{Power ratio in nepers} = \frac{1}{2} ln \frac{P_2}{P_1} \tag{B.3}$$

where ln e = ln 2.71828 = 8.686.

To convert nepers to decibels, multiply the number of nepers by $20 \log e = 8.686$.

Appendix C
Acronyms

AGC	Automatic gain control
AM	Amplitude modulation
ANSI	American National Standards Institute
APD	Avalanche photodiode
ARQ	Automatic repeat request
ASE	Amplified spontaneous emission
ASK	Amplitude shift keying
ATM	Asynchronous transfer mode
AWG	Arrayed waveguide grating
BER	Bit error rate
BH	Buried heterostructure
BLSR	Bidirectional line-switched ring
BPON	Broadband PON
BS	Base station
CAD	Computer-aided design
CATV	Cable TV
CNR	Carrier-to-noise ratio
CO	Central office
CRC	Cyclic redundancy check
CRZ	Chirped return-to-zero
CS	Control station
CSO	Composite second order
CTB	Composite triple beat

(continued)

(continued)

CW	Continuous wave
CWDM	Course wavelength division multiplexing
DBA	Dynamic bandwidth assignment
DBR	Distributed Bragg reflector
DCE	Dynamic channel equalizer
DCF	Dispersion compensating fiber
DCM	Dispersion compensating module
DFA	Doped-fiber amplifier
DFB	Distributed feedback (laser)
DGD	Differential group delay
DGE	Dynamic gain equalizer
DPSK	Differential phase-shift keying
DQPSK	Differential quadrature phase-shift keying
DR	Dynamic range
DS	Digital system
DSF	Dispersion-shifted fiber
DUT	Device under test
DWDM	Dense wavelength division multiplexing
DXC	Digital cross-connect matrix
EAM	Electro-absorption modulator
EDFA	Erbium-doped fiber amplifier
EDWA	Erbium-doped wave guide amplifier
EH	Hybrid electric-magnetic mode
EHF	Extremely high frequency (30-to-300 GHz)
EIA	Electronics Industries Alliance
EM	Electromagnetic
EMS	Element management system
EO	Electro-optical
EPON	Ethernet PON
ER	Extended reach
FBG	Fiber Bragg grating
FDM	Frequency division multiplexing
FEC	Forward error correction
FM	Frequency modulation
FOTP	Fiber Optic Test Procedure
FP	Fabry-Perot
FSK	Frequency shift keying

(continued)

(continued)

FSR	Free spectral range
FTTH	Fiber to the home
FTTP	Fiber to the premises
FTTx	Fiber to the x
FWHM	Full-width half-maximum
FWM	Four-wave mixing
GE-PON	Gigabit Ethernet PON
GFF	Gain-flattening filter
GPON	Gigabit PON
GR	Generic Requirement
GUI	Graphical user interface
GVD	Group velocity dispersion
HDLC	High-Level Data Link Control
HE	Hybrid magnetic-electric mode
HFC	Hybrid fiber/coax
IEC	International Electrotechnical Commission
IEEE	Institute for Electrical and Electronic Engineers
ILD	Injection laser diode
IM	Intermodulation
IMD	Intermodulation distortion
IM-DD	Intensity-modulated direct-detection
IP	Internet Protocol
ISI	Intersymbol interference
ISO	International Standards Organization
ITU	International Telecommunications Union
ITU-T	Telecommunication Sector of the ITU
LAN	Local area network
LEA	Large effective area
LED	Light-emitting diode
LO	Local oscillator
LP	Linearly polarized
MAN	Metro area network
MCVD	Modified chemical vapor deposition
MEMS	Micro electro-mechanical system
MFD	Mode-field diameter
MIB	Management information base
MQW	Multiple quantum well

(continued)

(continued)

MZI	Mach-Zehnder interferometer
MZM	Mach-Zehnder modulator
NA	Numerical aperture
NF	Noise figure
NIST	National Institute of Standards and Technology
NMS	Network management system
NPL	National Physical Laboratory
NRZ	Nonreturn-to-zero
NZDSF	Non-zero dispersion-shifted fiber
O/E/O	Optical-to-electrical-to-optical
OADM	Optical add/drop multiplexer
OBS	Optical burst switching
OC	Optical carrier
ODU	Optical channel data unit
OLS	Optical label swapping
OLT	Optical line terminal
OMA	Optical modulation amplitude
OMI	Optical modulation index
OMS	Optical multiplex section
ONT	Optical network terminal
ONU	Optical network unit
OOK	On-off keying
OPM	Optical performance monitor
OPS	Optical pulse suppressor
OPS	Optical packet switching
OPU	Optical channel payload unit
ORL	Optical return loss
OSA	Optical spectrum analyzer
OSI	Open system interconnect
OSNR	Optical signal-to-noise ratio
OST	Optical standards tester
OTDM	Optical time-division multiplexing
OTDR	Optical time domain reflectometer
OTN	Optical transport network
OTS	Optical transport section
OTU	Optical channel transport unit
OVPO	Outside vapor-phase oxidation

(continued)

(continued)

OXC	Optical crossconnect
P2P	Point-to-point
PBG	Photonic bandgap fiber
PC	Personal computer
PCE	Power conversion efficiency
PCF	Photonic crystal fiber
PCVD	Plasma-activated chemical vapor deposition
PDF	Probability density function
PDH	Plesiochronous digital hierarchy
PDL	Polarization-dependent loss
PHY	Physical layer
pin	(p-type)-intrinsic-(n-type)
PLL	Phase-locked loop
PM	Phase-modulation
PMD	Polarization mode dispersion
PMMA	Polymethylmethacrylate
POF	Polymer (plastic) optical fiber
POH	Path overhead
PON	Passive optical network
POP	Point of presence
PPP	Point-to-point protocol
PRBS	Pseudorandom binary sequence
PSK	Phase shift keying
PTB	Physikalisch-Technische Bundesanstalt
PVC	Polyvinyl chloride
QCE	Quantum conversion efficiency
QoS	Quality of service
RAPD	Reach-through avalanche photodiode
RC	Resistance-capacitance
RF	Radio-frequency
RFA	Raman fiber amplifier
RIN	Relative intensity noise
RIP	Refractive index profile
rms	Root mean square
ROADM	Reconfigurable OADM
ROF	Radio-over-fiber
RS	Reed-Solomon

(continued)

(continued)

RWA	Routing and wavelength assignment
RZ	Return-to-zero
SAM	Separate-absorption-and-multiplication (APD)
SBS	Stimulated Brillouin scattering
SCM	Subcarrier modulation
SDH	Synchronous digital hierarchy
SFDR	Spur-free dynamic range
SFF	Small-form-factor
SFP	Small-form-factor (SFF) pluggable
SHF	Super-high frequency (3-to-30 GHz)
SLED	Superluminescent light emitting diode
SLM	Single longitudinal mode
SNMP	Simple network management protocol
SNR	Signal-to-noise ratio
SOA	Semiconductor optical amplifier
SONET	Synchronous optical network
SOP	State of polarization
SPE	Synchronous payload envelope
SPM	Self-phase modulation
SRS	Stimulated Raman scattering
SSMF	Standard single mode fiber
STM	Synchronous transport module
STS	Synchronous transport signal
SWP	Spatial walk-off polarizer
TCP	Transmission control protocol
TDFA	Thulium-doped fiber amplifier
TDM	Time-division multiplexing
TDMA	Time-division multiple access
TE	Transverse electric
TEC	Thermoelectric cooler
TFF	Thin-film filter
TIA	Telecommunications Industry Association
TM	Transverse magnetic
UHF	Ultra-high frequency (0.3-to-3 GHz)
UI	Unit interval
UPSR	Unidirectional path-switched ring
VAD	Vapor-phase axial deposition

(continued)

(continued)

VCSEL	Vertical-cavity surface-emitting laser
VECP	Vertical eye-closure penalty
VFL	Visual fault locator
VOA	Variable optical attenuator
VSB	Vestigial-sideband
WAN	Wide area network
WDM	Wavelength-division multiplexing
WRN	Wavelength routed network
WSS	Wavelength-selective switch
XPM	Cross-phase modulation
YIG	Yttrium iron garnet

Appendix D
List of Important Roman Symbols

Symbol	Definition
a	Fiber radius
A_{eff}	Effective area
B	Bandwidth
B_e	Receiver electrical bandwidth
B_o	Optical bandwidth
c	Speed of light $= 2.99793 \times 10^8$ m/s
C_j	Detector junction capacitance
d	Hole diameter in a PCF
D	Dispersion
D_{mat}	Material dispersion
D_n	Electron diffusion coefficient
D_p	Hole diffusion coefficient
D_{wg}	Waveguide dispersion
DR	Dynamic range
E	Energy ($E = h\nu$)
E	Electric field
E_g	Bandgap energy
E_{LO}	Local oscillator field
f	Frequency of a wave
F	Finesse of a filter
F(M)	Noise figure for APD with gain M
f(s)	Probability density function
F_{EDFA}	EDFA noise figure
g	Gain coefficient (Fabry-Perot cavity)

(continued)

(continued)

Symbol	Definition
G	Amplifier gain
g_B	Brillouin gain coefficient
h	Planck's constant $= 6.6256 \times 10^{-34}$ J-s $= 4.14$ eV-s
H	Magnetic field
I	Optical field intensity
i_B	Bias current
i_D	Photodetector bulk dark current
I_{DD}	Directly detected optical intensity
i_M	Multiplied photocurrent
i_p	Primary photocurrent
$i_p(t)$	Signal photocurrent
i_{th}	Threshold current
$\langle i_s^2 \rangle$	Mean-square signal current
$\langle i_{shot}^2 \rangle$	Mean-square shot-noise current
$\langle i_{dark}^2 \rangle$	Mean-square detector dark noise current
$\langle i_{th}^2 \rangle$	Mean-square thermal noise current
J	Current density
J_{th}	Threshold current density
k	Wave propagation constant $(k = 2\pi/\lambda)$
K	Stress intensity factor
k_B	Boltzmann's constant $= 1.38054 \times 10^{-23}$ J/K
L	Fiber length
L_c	Connection loss
L_{disp}	Dispersion length
L_{eff}	Effective length
L_F	Fiber coupling loss
L_i	Intrinsic loss
L_n	Electron diffusion length
L_p	Hole diffusion length
L_{period}	Soliton period
L_{split}	Splitting loss
L_{tap}	Tap loss
m	Modulation index or Modulation depth
m	Order of a grating
M	Avalanche photodiode gain
M	Number of modes
m_e	Effective electron mass

(continued)

(continued)

Symbol	Definition
m_h	Effective hole mass
n	Index of refraction
$\overline{N}$	Average number of electron-hole pairs
NA	Numerical aperture
n_i	Intrinsic n-type carrier concentration
N_{ph}	Photon density
n_{sp}	Population inversion factor
P	Optical power
$P_0(x)$	Probability distribution for a 0 pulse
$P_1(x)$	Probability distribution for a 1 pulse
$P_{amp,sat}$	Amplifier saturation power
P_{ASE}	ASE noise power
P_e	Probability of error
p_i	Intrinsic p-type carrier concentration
P_{in}	Incident optical power
P_{LO}	Local oscillator optical power
P_{peak}	Soliton peak power
PP_x	Power penalty for impairment x
P_{ref}	Reflected power
$P_{sensitivity}$	Receiver sensitivity
P_{th}	SBS threshold power
q	Electron charge $= 1.60218 \times 10^{-19}$ C
Q	BER parameter
Q	Q factor of a grating
R	Bit rate or Data rate
R	Reflectivity or Fresnel reflection
r	Reflection coefficient
$\mathcal{R}$	Responsivity
$\mathcal{R}_{APD}$	APD responsivity
R_{nr}	Nonradiative recombination rate
R_r	Radiative recombination rate
R_{sp}	Spontaneous emission rate
$S(\lambda)$	Dispersion slope
T	Period of a wave
T	Absolute temperature
T_{10-90}	10-to-90% rise time
T_b	Bit interval, Bit period, or Bit time

(continued)

(continued)

Symbol	Definition
t_{GVD}	Rise time from GVD
t_{rx}	Receiver rise time
t_{sys}	System rise time
V	Mode V number

Appendix E
List of Important Greek Symbols

Symbol	Definition
α	Refractive index profile shape
α	Optical fiber attenuation
α	Laser linewidth enhancement factor
$\alpha_s(\lambda)$	Photon absorption coefficient at a wavelength λ
β	Mode propagation factor
β_3	Third-order dispersion
Γ	Optical field confinement factor
Δ	Core-cladding index difference
ΔL	Array waveguide path difference
$\Delta \nu_B$	Brillouin linewidth
$\Delta \nu_{opt}$	Optical bandwidth
η	Light coupling efficiency
η	Quantum efficiency
η_{ext}	External quantum efficiency
η_{int}	Internal quantum efficiency
θ_A	Acceptance angle
λ	Wavelength
Λ	Period of a grating
Λ	Hole spacing or pitch of a PCF
λ_B	Bragg wavelength
λ_c	Cutoff wavelength
ν	Frequency
σ_{dark}	Detector dark noise current variance
σ_s	Signal current variance

(continued)

© The Editor(s) (if applicable) and The Author(s), under exclusive license
to Springer Nature Singapore Pte Ltd. 2021
G. Keiser, *Fiber Optic Communications*,
https://doi.org/10.1007/978-981-33-4665-9

(continued)

Symbol	Definition
σ_{shot}	Shot noise current variance
σ_T	Thermal noise current variance
σ_{wg}	Waveguide-induced pulse spreading
τ	Carrier lifetime
τ_{ph}	Photon lifetime
φ	Phase of a wave
φ_c	Critical angle
Φ	Photon flux

Printed in the United States
by Baker & Taylor Publisher Services